ANCESTRY AND SEX IN HUMAN CRANIA

ANCESTRY AND SEX IN HUMAN CRANIA

A Comparative Photographic Atlas

By

ROBERT W. MANN, PhD, D-ABFA, F-CPP

CHARLES C THOMAS • PUBLISHER • LTD.
Springfield • Illinois • U.S.A.

Published and Distributed Throughout the World by

CHARLES C THOMAS • PUBLISHER, LTD.
2600 South First Street
Springfield, Illinois 62704

ISBN 978-0-398-09364-8 (hard)
ISBN 978-0-398-09365-5 (ebook)

Library of Congress Catalog Card Number: 2021005660 (print)
2021005661 (ebook)

With THOMAS BOOKS *careful attention is given to all details of manufacturing and design. It is the Publisher's desire to present books that are satisfactory as to their physical qualities and artistic possibilities and appropriate for their particular use.* THOMAS BOOKS *will be true to those laws of quality that assure a good name and good will.*

Printed in the United States of America
MM-S-2

Library of Congress Cataloging-in-Publication Data

Names: Mann, Robert W., 1949– author.
Title: Ancestry and sex in human crania : a comparative photographic atlas / by Robert W. Mann.
Description: Springfield, Illinois : Charles C Thomas, Publisher, Ltd., [2021] | Includes bibliographical references and index.
Identifiers: LCCN 2021005660 (print) | LCCN 2021005661 (ebook) | ISBN 9780398093648 (hardback) | ISBN 9780398093655 (ebook)
Subjects: LCSH: Skull–Anatomy–Atlases. | Skull–Examination. | Forensic anthropology–Atlases. | Forensic osteology–Atlases.
Classification: LCC QM105 .M36 2021 (print) | LCC QM105 (ebook) | DDC 611/.715–dc23
LC record available at https://lccn.loc.gov/2021005660
LC ebook record available at https://lccn.loc.gov/2021005661

INTRODUCTION

This atlas consists of 29 male and 21 female crania and is intended to serve as a reference source of documented-identity crania in the Mann-Labrash Osteology Collection at the John A. Burns School of Medicine (JABSOM) of the University of Hawaii, Manoa. This collection reflects the unique population diversity of contemporary Hawaii and was begun in 2014 as part of the JABSOM Department of Anatomy, Biochemistry, and Physiology under the direction of former Chair Dr. Scott Lozanoff. It is among the newest documented-identity human skeletal collection in the U.S. and has grown to more than 255 adult crania and 20 complete skeletons at the time of this writing. The primary purpose of the collection, as part of the Willed Body Program, is to train and instruct JABSOM medical students in gross anatomy. It also serves as a diverse reference collection of skeletal and dental disease, trauma, surgical intervention, bone healing, human skeletal variation, musculoskeletal anatomy, congenital malformations, DNA, and genetics.

The Mann-Labrash collection includes males and females of various ancestries and ethnicities including European (Caucasian, White), African American (African/American Black, Black, Ethiopian), Asian, Pacific Islander (e.g., Hawaiian, Micronesian, Samoan), and Hispanic. Donors range in age from 21 to 107 years old and are from Hawaii where they were either born, resided, or were visiting at the time of death. All donors died between 1974 and 2019. Most donors are accompanied by medical records that include cause of death and, occasionally, history of illnesses. As in most populations, the ancestry (race) of these individuals was assigned at birth and self-reported on their donor forms. The following table shows the population composition (1,420,491 people) of Hawaii as of July 2018:

TABLE 1.
Population of Hawaii as of July 2018 (United States Census Bureau, www.census.gov/facts/HI)

Ancestry	Percentage
White	25.6
Black or African American alone	2.2
Asia alone	37.6
Native Hawaiian or Other Pacific Islander	10.2

continued

TABLE 1—*Continued.*

Two or more races (mixed ancestry)	24.0
Hispanic or Latino	10.7
White alone (not Hispanic or Latino)	21.8

This atlas is intended for anyone who does not have access to a comparative reference collection of documented or known-identity crania in their laboratory. Researchers seeking to establish ancestry based on visual or metric analysis may also want to refer to or utilize databases such as Fordisc 3 (Jantz and Ousley 2005), OSTEOWARE (2018), the Macromorphoscopic Databank (Hefner 2018), and 3D-ID (Slice and Ross 2010).[1] Many other resources are available that utilize non-metric and macromorphoscopic (a term coined by Ousley and Hefner [Plemons and Hefner] 2016), cranial and postcranial traits to estimate ancestry. The author emphasizes that this atlas is not intended to present, replace, or cover the full spectrum of variation of cranial ancestry and sex features, a virtually impossible task. It is, however, intended to add to our knowledge of the complexity and range of cranial variation and to provide examples based on contemporary known-identity individuals (see Jantz and Meadows Jantz 2000, Jantz 2001, and Jantz and Jantz 2016 for information on secular trends in craniofacial morphology).

At times, a researcher must rely on multiple sources to find a sufficient number of examples of crania for comparison with unknown-identity crania. To aid in such analyses, this atlas provides large, color photographs for researchers to reference when estimating ancestry, sex, and, to a limited extent, age. As many readers are aware, the largest contemporary documented osteological collections in the United States include the Hamann-Todd Osteological Collection in Cleveland, Ohio, the Robert J. Terry Osteology Collection and George S. Huntington Anatomical Collection in Washington, DC, and the William M. Bass Donated Collection in Knoxville, Tennessee. These collections consist primarily of individuals of European and African ancestry and few Asians. The Atlas of Ancestry and Sex, therefore, increases our comparative reference samples and diversity to include individuals of Asian and Pacific Islander ancestry.

Most crania in this atlas are presented using six anatomical views: anterior, right lateral, left lateral, inferior (basilar), superior, and posterior (occipital) in the Frankfort horizontal plane. A superior view of each mandible is included to provide the reader with size and shape features of the teeth and bone. Additional photos are included to highlight other features such as dental morphology and anatomical variants. Many photographs are labeled to identify specific features, while others are not, leaving interpretation to the reader. Figure captions reflect the author's opinion but are intended to allow readers to interpret features for themselves and draw their own conclusions based on the photographs of each cranium.

1. For more information see the References and Recommended Resources sections of this book, specifically: Bass (2005); Brooks et al. (1990); Buikstra and Ubelaker (1994); Hauser and DeStefano (1989); Hefner (2003, 2009, 2015); Hefner and Linde (2018); Hefner et al. (2015); Hrdlicka (1928); Krogman (1939); Langley et al. (2016 and 2017); Moore-Jansen et al. (1994); Plemons and Hefner (2016); Stewart (1979).

This atlas contains full-page color photographs and as many as 24 standard landmark measurements[2] of some crania to aid readers in their assessment of ancestry and sex. The photographs are intended to provide readers with the most holistic and integrated perspective of each feature. In other words, each feature is viewed as part of the whole without requiring the reader to place them along a fixed continuum or sequence based on size or shape. Photographs also allow readers to examine each feature as an integrated piece of the "puzzle."

Cranial measurements were obtained using a Mitutoyo SC-6"C digital sliding caliper and a Paleo-Tech spreading caliper. All measurements are recorded in millimeters and rounded to the nearest millimeter. Fordisc (Versions 3.0 and 3.1) was used for the analysis. Nasal bone contour was obtained using a General contour gauge placed approximately 1cm from nasion. Photographs of the contour gauge afford readers an opportunity to compare nasal bone morphology with their own estimates of nasal bone contour and shape.[3]

Crania were selected not as exemplars, but as being representative of the complex diversity present in European (White, Caucasian), African American (African, American Black, Black, Ethiopian), Asian, Pacific Islander, and Hispanic adult crania. Each individual's ancestry is reported in the order it is reported on their Willed Body Donor forms: their primary, or first, ancestry such as Caucasian (i.e., "European" or "White" in this book), followed by all secondary ancestries (e.g., Portuguese, Spanish). Individuals of mixed ancestry, sometimes referred to as multiracial, mixed races, or racial admixture such as Chinese European are not hyphenated, in accordance with *The Chicago Manual of Style* (2010). See Hefner and Spradley (2018) for a discussion of ancestral diversity that goes beyond the traditional three-ancestry groups (Hooton 1930) and why some Hispanic crania may misclassify as Japanese (Dudzik and Jantz 2016), and recommendations and best practices as presented by the Scientific Working Group in Anthropology (SWGANTH 2013). A few crania exhibiting disease and trauma were included in this atlas to show how these conditions can alter cranial features and, as a result, confuse or obscure estimates of ancestry and sex.

2. See Buikstra and Ubelaker (1994); Moore-Jansen et al. (1994).
3. For more on nasal bone structure with a contour gauge see Hefner (2003, 2009).

ACKNOWLEDGMENTS

First, and most importantly, I offer my sincerest thanks to our "silent teachers." This book would not be possible without the generosity and vision of our donors. They will be forever remembered for their kindness and their gift.

Special thanks to Steven Labrash for initiating the JABSOM osteology collection in 2004 and for serving as Director of the Willed Body Program. Thanks also to Stacy Lenze, Lee Labrash, Kiana Miller, Alex Wong, Wilson "Sully" Sullivan, Jesse Thompson, Michael Andrews-Newman, Jennifer Lai-Hipp, Jose ("Joey") Garcia III, Catherine Link, and Jessica Burden for their tireless efforts and contributions in helping to build and maintain this osteological collection. Chair of Anatomy, Dr. Takashi Matsui, former Chair Dr. Scott Lozanoff, and Dean Dr. Jerris Hedges were instrumental in resourcing and curating this collection for generations to come. Thanks to Beth Lozanoff for her talented medical illustrations. I thank Michele Tom and Mari Kuroyama-Ton for their years of dedicated service to the Department of Anatomy and Willed Body Program.

Thanks to Professor John R. DeFreytas, LtCol, USMC (ret.) for his in-depth editing, comments, and careful attention to detail during preparation of this book manuscript. Dr. Nicholas Passalacqua for suggesting the title of this atlas and Dr. Marin Pilloud and Dr. J DeMeo DMD for confirming the presence of palatoradicular grooves in one of the individuals. Dr. Michael W. Kenyhercz, one of the newest Diplomates (#141) of the American Board of Forensic Anthropology (ABFA) and, I am proud to say, one of my former Forensic Science Academy Fellows, for providing suggestions on relevant ancestry references. My thanks to Michael P. Thomas of Charles C. Thomas Publisher for his superb attention to detail and guidance. Thank you all for your contributions to this book.

All photos in this atlas were taken by the author. Most cranial measurements were performed by JABSOM osteology intern and research assistant Kiana Miller of Chaminade University of Honolulu and verified by the author.

My most profound thanks and love to Vara J. Mann, my wife, strength, and foundation.

CONTENTS

ANCESTRY AND SEX IN HUMAN CRANIA

I. CEPHALOMETRIC ANALYSIS OF THE DENTOALVEOLAR COMPLEX: A CLINICAL PERSPECTIVE ON ALVEOLAR PROGNATHISM

J. S. A. DeMeo, DMD, MS

Diplomate, American Board of Orthodontics

Adjunct Professior of Anatomy, Biochemistry and Physiology, JABSOM, University of Hawaii-Manoa, Honolulu, HI

Adjunct Professor of Orthodontics, Air Force Postgraduate Dental School, USUHS, San Antonio, TX

The study of the proportions of the human skull can be traced back to early history with the ancient cultures of the Egyptians, Greeks, Chinese and Indians all developing complex systems (Moorrees 2006; Ghafari 2006). An accurate understanding of the ratios and relationships of the components of the skull and the face can enhance the realism of art, which was the impetus for this area of study. During the Renaissance, artists such as Leonardo da Vinci and Albrecht Dürer studied facial proportions to create systems that would allow for realistic representation of their subjects and these systems are the basis of many of our current methodologies. In particular, Dürer was able to utilize landmarks and facial features to create facial angles that differentiated between profile types or classes (Figure 1). Later work by Petrus Camper in the 1700s, elaborated on this concept and created a standard for craniology utilizing facial angles that led to the facial types of prognathic and orthognathic (Figure 2). These initial angular measurements have been fundamental to the categorization of craniofacial deformities and how they are treated in orthodontics and maxillofacial surgery (Duterloo and Planche 2011; Moorrees 2006; Ghafari 2006).

Orthodontic cephalometry was derived from long-established anthropologic craniometric studies and converted for use by the advent of the radiograph by B. Holly Broadbent and Herbert Hofrath in 1931 (Duterloo and Planche 2011; Moorrees 2006). This cephalometer revolutionized the diagnosis and treatment planning of orthodontic patients and allowed for longitudinal growth studies that are fundamental to the understanding of human cranial growth and development. Subsequently, this led to a multitude of studies to try and quantify

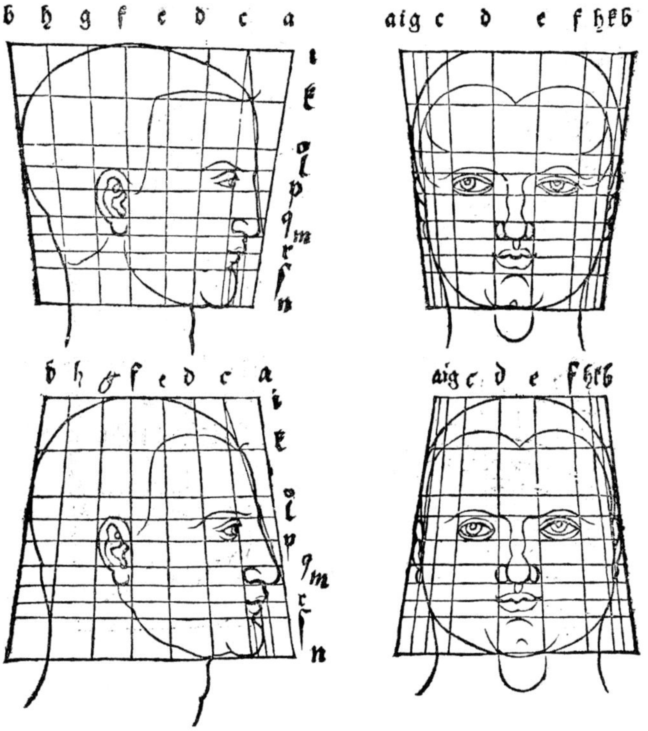

Figure 1. Albrecht Dürer's coordinate system constructed according to the location of landmarks and facial features. "Face Transformations" by Albrecht Dürer CC BY 4.0.

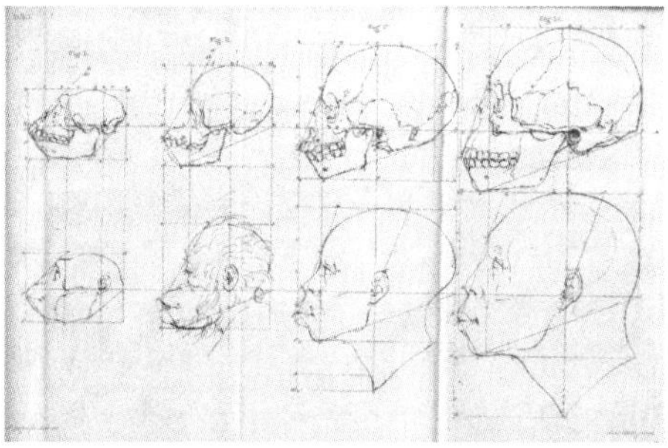

Figure 2. Petrus Camper's analysis of the facial form and facial angle in a tail monkey, a young orangutan, a native African, and a Kalmuk person. "Petrus Camper Facial Angles: Orangutan to Moor" by Adriaan Gilles CC BY 4.0.

normal values in populations to create a standard to treat patients and to differentiate between ethnic (ancestry) backgrounds. In addition, these studies were then combined with various analyses to describe the pathology of malocclusion including those by Steiner, Ricketts, McNamara, Wits, Tweed, and Downs (McNamara and Brudon 2001; Moorrees 2006). There are advantages and disadvantages to each of these analyses and this can be overwhelming and paralyzing to a novice. This is a testament to the fact that these analyses are trying to characterize a very complex, dynamic process into quantifiable, objective numbers. However, the end result is to compare the relationships of various components of the skull to each other to understand how a harmonious balance creates optimal function and esthetics.

The sagittal positioning of the jaws in relationship to the cranial base is fundamental to describing malocclusions and the subsequent effect on the esthetics of the face. In addition, the positioning of the teeth in dentoalveolar complex can have a distinct effect on masticatory function and facial esthetics. Therefore, when analyzing these positions, it is important to consider the position of the tooth in the jaw, the relationship of the jaw to the cranial base, and the relationship of the jaws to each other. The terminology used to describe these conditions is dependent on the analysis used and has changed over the course of time from early comparative zoology in the 18th century, to later orthodontic study, and finally to 19th century anthropologists in Europe (1).

The jaw position in relation to the cranial base can be described as too far forward or prognathic; too far backward or retrognathic; and in the correct position or orthognathic. This often relates to a facial type, such as the famous example of the Hapsburg family in Europe, which due to their familial trait of a large mandible (Figure 3), they were known to be prognathic or having a Class III malocclusion (5). However, this description is not illustrative enough because it does not give us an indication of which jaw is the etiology of the malocclusion or any indication of where the teeth are that support the lip profile. Nature tends to camouflage skeletal discrepancies with dental compensations so that teeth are in the best position for function. For example, a similar concept was seen by Begg in his study of Australian aboriginals and by Kaifu in his study of prehistoric Japanese populations where the teeth wore down due to diet and the anterior teeth compensated by retroclining to close any spacing (Kaifu 2000; Margvelashvili et al. 2013). Therefore, describing the jaws in relation to the cranial base as protrusive, retrusive, or normal gives a better indication of the problem and relates better to an etiology such as maxillary hypoplasia (small upper jaw) or mandibular prognathism (large lower jaw) or a combination of the two (McNamara and Brudon 2001; Nguyen and Profitt 2017; Proffit et al. 2012).

Figure 3. Portrait of Phillip II of Spain, a member of the Hapsburg house displaying their characteristic mandibular prognathia or Class III skeletal relationship. "Portrait of Phillip II, King of Spain" by Antonis Mor CC BY 4.0.

This concept can also be applied to the teeth but with an additional complexity of rotation. The teeth can be too far forward or protruded; too far backward or retruded; and in the correct position or normal. Continuing, they can also be too flared or proclined; too upright or retroclined; and in the correct position or normal. This is a simplistic description of a complex cranial anatomy as the teeth and jaws can also be described in the vertical and transverse dimension with three dimensional interactions that can be described by tip, roll, and yaw (Nguyen and Profitt 2017; Proffit et al. 2012). However, the sagittal dimension for both teeth and jaws is a classic starting point for descriptions of facial types regardless of the system of analysis.

The analysis of cephalometric points in anthropology can be done using photographs, radiographs, or computed tomography (CT) scans (Kaifu 2000; Utsuno et al. 2018; Margvelashvili et al. 2013; Luthur 1993; Pajevic et al. 2019). There are a multitude of different cranial landmarks that have been developed to augment different analyses of the cranium. Two traditional reference planes that can be utilized to represent cranial base are the Frankfort horizontal plane and the Sella-Nasion (S-N) line. Frankfort horizontal was developed by anthropologists utilizing dried skulls to represent the natural head position but suffers from inconsistencies in identification on a radiograph (Proffit et al. 2012; Ghafari 2006; Caufield 2006). The Frankfort horizontal plane is defined by a line through Porion (Po), the most superiorly positioned point on the external auditory meatus, and Orbitale (Or), the lowest point on the inferior rim of the orbit (Caufield 2006). On the other hand, the S-N line is only identifiable with a radiograph and is noted to have a high variability between individuals (Ghafari 2006). Sella (S) is defined as the geometric center of the pituitary fossa and Nasion (N) is located on the anterior most aspect of the frontonasial suture (Caufield 2006). Whenever possible, it is pertinent to utilize both lines and correlate them through an average measurement of about 7-9° between each other, which helps to ensure validity of derived measurements (Proffit et al. 2012; Ghafari 2006).

The next landmarks utilized give a representation of the position of the jaws; they are Point A, subspinale, and Point B, supermentale (Figures 4 and 5). Point A is defined as the posterior most point on the concavity between the anterior nasal spine (ANS) and the most inferior point on the alveolar bone above the maxillary incisor. Point B is defined as the most posterior point in the concavity of the mandible between the alveolar bone covering the mandibular incisor and pogonion (Pog), the most anterior point on the chin (Caufield 2006). From these points, the angles SNA, SNB, and ANB can be calculated. Alternatively, the Frankfort horizontal can be used to calculate corresponding angles (using the intersection of N-A and N-B, respectively) and the geometric relationship with S-N used to relate them. The ANB angle relates the jaws to each other, with a large value indicating a Class II relationship and a negative number indicating a Class III relationship. Continuing, a large value of SNA or SNB indicates protrusion or prognathia of the corresponding jaw, while a negative number indicates a retrusive or retrognathic jaw. Thus, utilizing all three of these angles can give an understanding of the relationship of both jaws to the cranium and the relative relationship to each other. True prognathism or bimaxillary protrusion, a forward or anterior jutting of the jaws, is when both the maxilla and mandible are prominent in relationship to the cranium. These relationships have a significant impact on facial esthetics and function however, with any of them the subject could still have a normal occlusion and harmonious facial esthetics. In addition, they should not be taken as absolutes because factors involved in the vertical growth of the face can affect the geometry and resulting angles (McNamara and Brudon 2001; Nguyen and Profitt 2017; Caufield 2006; Jacobson 2006).

Next, the positions of the teeth need to be evaluated in both relative position and angulation to the jaw. A line can be created from N-A and N-B, which act as references to the incisors for both jaws, respectably. The long axis of the tooth from incisal edge to root apex can be created and the relationship of this line to the N-A or N-B line indicates the relative proclination or retroclination of the tooth in degrees (Fig. 4 and 5). Continuing, a measurement from the long axis measured from the facial surface of the tooth to the respective N-A or N-B line is an indication of the relative protrusion or retrusion of the tooth in millimeters Caufield 2006). When both the maxillary and mandibular teeth procline and protrude, the resulting condition is described as bimaxillary dentoalveolar protrusion often mistakenly called bimaxillary protrusion. Bimaxillary protrusion is a term used by anthropologists to describe faces in which both jaws are prominent (protruded) in relation to the cranium (Proffit et al. 2012; Ghafari 2006). Another term often used by anthropologists is midfacial alveolar prognathism which, when described using cephalometric terminology, would be maxillary protusion (Fig. 4 and 5). This would result in a Class II skeletal relationship due to the maxilla being too far forward in comparison to the cranial base and mandible. However, the teeth

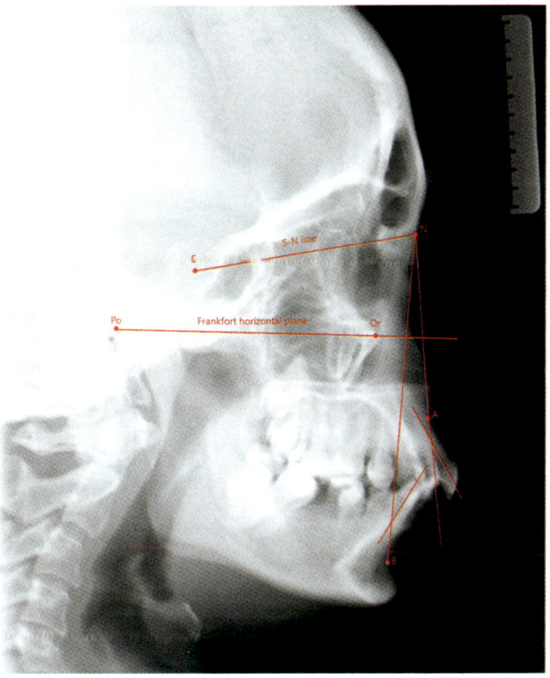

Figure 4. A lateral cephalogram of a Class II patient due to maxillary protrusion or midfacial alveolar prognathism. Frankfort horizontal, S-N line, A-point, B-Point have been traced to create angles and measurements used for analysis of the dentoalveolar complex. The upper and lower teeth are protruded and proclined. Private practice, Dr. J. DeMeo.

Figure 5. The skull of a Class II Pacific Islander male (JABSOM) due to maxillary protrusion or midfacial alveolar prognathism. Frankfort horizontal, A-point, B-Point have been traced to create angles and measurements used for analysis of the dentoalveolar complex. The upper and lower teeth are protruded and proclined. Mann-Labrash Osteological Collection, John A. Burns School of Medicine, University of Hawaii Manoa.

may be retroclined, proclined, or normal and can, therefore, lead to a variety of facial appearances despite the commonality of skeletal growth.

This measurement of the relationship between the jaws and the cranium can be powerful indications of function and esthetics. They also play a role in forensics in trying to establish an identification of an individual through ethnic, ancestral, or age-based comparative normal values (Albert et al. 2007; Machado et al. 2019; Cericato et al. 2016). In addition, they can be used for reconstruction or reproduction of the face to help identify a subject (Utsuno et al. 2018). Numerous groups have had normal values created including South African Bantus and Whites, African Americans, people of Northern European descent, Puerto Ricans, Koreans, Chinese, and Japanese (McNamara and Brudon 2001; Nguyen and Proffit 2017; Proffit et al. 2012; Hayashi et al. 2012; Ghafari 2006). It is important, though, to realize when applying this analysis to forensics that the differences between individuals is greater than the difference between populations. It is, thus, important when trying to assign an individual to a race or ancestry that the definition of race is not precise considering population overlap and genetic drift (Jorde and Wooding 2004). Consequently, understanding the characteristics of that individual or population is more important (Cericato et al. 2016). Therefore, it is imperative to utilize these analyses appropriately and with an aim of inclusion rather than exclusion when performing a forensic analysis using cephalometrics.

II. TRAITS AND FEATURES USED IN THIS ATLAS

TABLE 2.
Partial List of Non-metric Traits (by trait) Generally Used to Assess Ancestry*

Non-metric Trait	**European**	**African**	**Asian**
Nose form	Narrow	Wide	Medium
Nasal form	High	Low	Low
Interorbital distance	Narrow	Wide	Medium
Nasal spine	Large	Small	Medium
Nasal sill	Sill	Guttering	Guttering
Alveolar prognathism	None	Large	Medium
Orbit shape	Rhomboid	Round	Rhomboid
Malar	Small, retreating	Small, horizontal	Flaring

While the traits in this list, as in other non-metric trait lists, are often found in these ancestral groups, they may occur in varying combinations in any ancestral group. For example, the combination of a wide interorbital distance (usually Asian and African), presence of nasal guttering (usually Asian and African), and a large anterior nasal spine (usually European/White) are often present in Asians. Incipient or mild expression of nasal guttering (usually Asian and African) is sometimes present in Europeans. The combination of these features, however, is typically more important in ancestry determination than the presence or reliance of a single feature.

*Adapted from Bass (2005), Gill (1995), Hefner (2003), Hefner and Linde (2028), Rhine (1990), Vitek, 2012.

TABLE 3.
Partial List of Non-metric Traits to Assess Ancestry
(by Ancestral Group; adapted from Wheat 2009)*

European	**African**	**Asian**
Tower nasals	Quonset hut nasals	Tented nasals
Large nasal spine	Small nasal spine	Small nasal spine
Deep nasal sill	Guttered nasal border	Blurred nasal sill
Retreating zygomatic	Vertical zygomatic	Projecting (flat) zygomatic
S-shaped zygomaticomaxillary suture	S-shaped zygomaticomaxillary suture	Angled zygomaticomaxillary suture
Sloping orbits	Rectangular orbits	Rounded orbits
Long Base Chord	Long base chord	Short base chord
Round ext. auditory meatus	Round ext. auditory meatus	Elliptical ext. auditory meatus
Oval window visible	Oval window visible	Oval window not visible
Parabolic dental arcade	Hyperbolic dental arcade	Elliptical dental arcade
Bulging palatine suture	Bulging palatine suture	Straight palatine suture
Carabelli's trait	Molar crenulations	Enamel molar extensions
Bilobate and prominent chin	Blunt and vertical chin	Blunt and vertical chin
Undulating mandible	Straight mandibular border	Straight mandibular border
Pinched ramus	Pinched ramus	Wide ramus
Slanted ramus	Slanted ramus	Vertical ramus
Straight gonial angle	Straight gonial angle	Everted gonial angle
		Nasal overgrowth
		Buccal pits/shovel-shaped incisors

*Orbit shape, visualization of the oval window, and shape of the dental arch are often difficult to classify.

TABLE 4.
Cranial Traits Often Used to Assess Sex (adapted from Krogman 1962).

Trait	**Male**	**Female**
Overall size	Large	Small
Architecture	Rugged	Smooth
Supraorbital ridge	Medium to large	Small to medium
Mastoid process	Medium to large	Small to medium
Occipital area	Developed muscle markings	Not developed
Frontal eminence	Small	Large
Parietal eminence	Small	Large
Orbit	Squared, blunt rim	Rounded sharp rim
Forehead	Steep, less rounded	Rounded, full
Cheek bone	Heavy, laterally arched	Light, more compressed
Mandible	Large, broad	Small, rounded ramus
Palate	Large, broad, U-shape	Small, parabolic
Occipital condyle	Large	Small
Teeth	Large, lower M1 often 5-cusped	Small, molars often 4-cusped

TABLE 5.
Traits, Features, and Variants Used to Assess Ancestry and Sex in this Atlas*

Alveolar prognathism (small, medium, prominent)
Antegonial notch (present, absent, small, medium, large, prominent)
Anterior nasal spine (short, rounded; dull, medium, long and sharp, curved; small, medium, large)
Bathrocephaly (occipital "bun") (present, absent)
Buccal exostoses (present, absent; symmetrical, asymmetrical)
Chin shape (narrow, wide, rounded, square, bilobate, V-shaped)
Crenulated molars (present, absent)
Depression at nasion (small, medium, large)
Frontal bone (sloping, vertical)
Frontal grooves (present, absent)
Frontal process of the maxilla (short, moderate, long, prominent)
Inferior nasal morphology (incipient, guttered, partial sill, complete sill)
Inion hook (small, medium, large)
Interorbital breadth (narrow, intermediate/medium, wide)
Lingual exostoses (present, absent; symmetrical, asymmetrical)
Malar tubercle (absent, incipient, trace, present)
Mastoid process size (small, medium, large)
Metopic suture (complete, partial, trace)
Nasal aperture width (narrow, medium, wide)

*From Bass (2005), Brues (1990), Buikstra and Ubelaker (1994), France (1998), Gill (1995; 2016), Gill and Deeds (2008), Hefner (2003), Hefner and Linde (2018), Krogman (1955, 1962), Langley et al. (2017), Mann (2016), Martin and Saller (1957), Ossenberg (2013), Rhine (1990), Rogers (2008), Ubelaker (1989), Wheat (2009), Vitek (2012), White et al. (2012).
+Frontal process of the maxilla, when viewed laterally, is a feature deserving of further research as a possible ancestry trait.

TABLE 5—*Continued.*

Nasal bone structure/shape (Quonset-hut, tented, oval, semi-triangular, steepled, hourglass, pinched, waisted) (in this atlas, "pinched" and "waisted" are used synonymously and refers to medial curvature of the bones, usually near nasion)
Nasal guttering (absent, mild, pronounced)
Nasal overgrowth (present, absent)
Nasal sill (dull, incipient, moderate, pronounced/large; absent, dull, medium, sharp, and everted margin)
Nuchal crest (small, medium, large)
Occipital flattening (present, absent)
Orbit shape (rounded, rectangular, square; aviator
Palate depth (shallow, deep)
Palate size and shape (small, large; wide, shallow, rounded; V-shaped, elliptical, parabolic, hyperparabolic
Postbregmatic depression (present, absent)
Posterior border of mandible (straight, concave)
Posterior zygomatic tubercle (absent, weak, medium, strong/pronounced
Ramus posterior border (straight, concave)
Ramus width (narrow, wide)
Shovel-shaped incisors (present, absent)
Spatulate incisors (present, absent)
Stafne defect (present, absent)
Suprainion depression (present, absent)
Supramastoid crest (small, medium, large)
Supranasal suture (present, absent)

continued

TABLE 5–*Continued.*

Supraorbital margin (rounded/blunt, sharp/thin)
Supraorbital ridge (browridge/ glabella) (small, medium, large)
Torus palatinus (small, medium, large)
Transverse palatine suture shape (straight symmetrical, anterior bulging at midline symmetrical, anterior bulging right or left side, posterior bulging at midline symmetrical)
Vault length (short, intermediate, long)
Vault height (short, intermediate, high)
Vault shape (low, medium, high; rounded)
Vermiculate pattern in the frontal bone (present, absent)
Zygomatic arch (flat, concave, convex, curved)
Zygomatic bone morphology (anteriorly projecting, flat, retreating)
Zygomatic extension
Zygomatic suture shape (angled, smooth, S-shaped)

III. AFRICAN AMERICAN CRANIA

(5 Male and 3 Female)

1. AFRICAN AMERICAN MALE

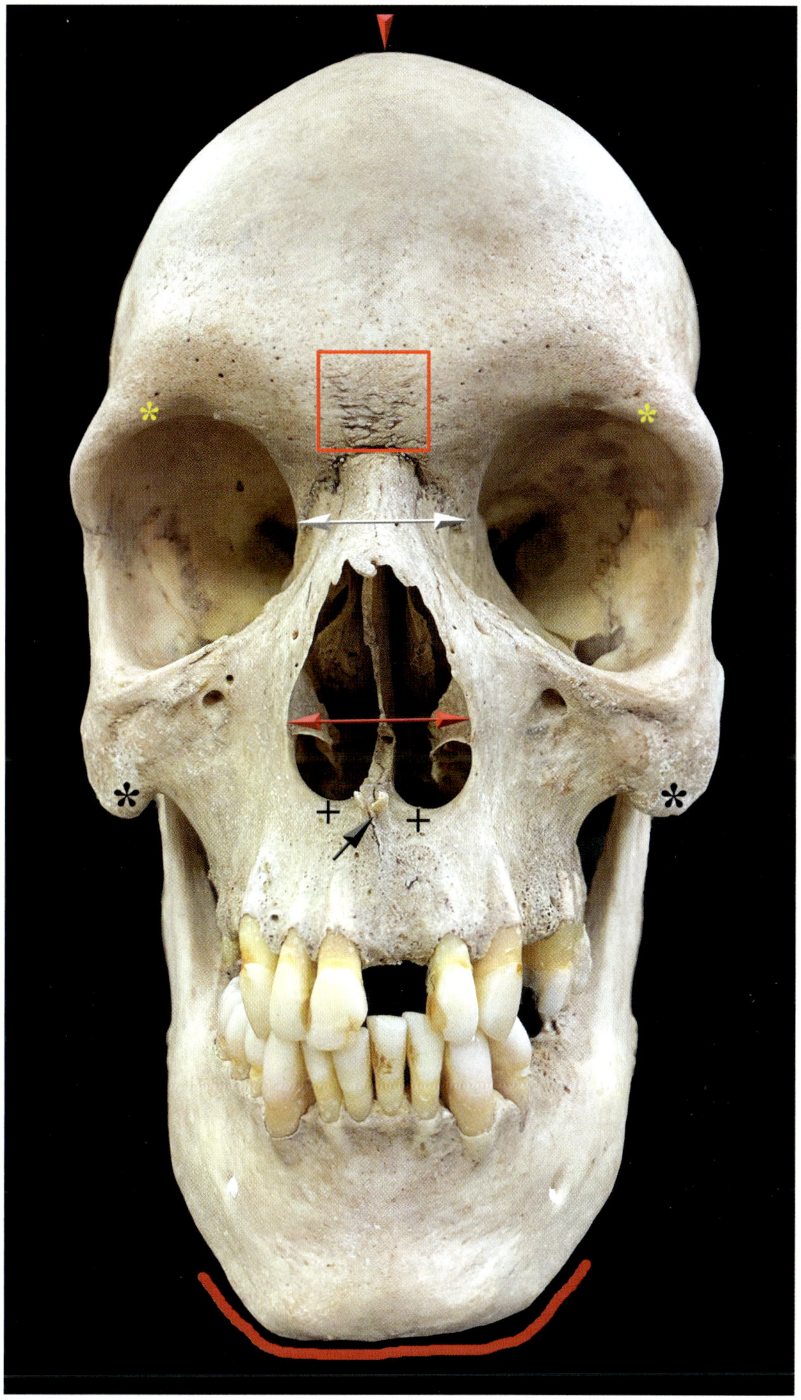

Figure 6. Anterior view 70-year-old African American male. Note the tall and narrow vertex (red arrowhead), trace supranasal suture (square), blunt/rounded superior margin of the orbit (yellow*), wide interorbital breadth (double white arrow), prominent malar tubercle (black*) ("zygomatic hook" Rhine 1990; see Hefner and Linde 2018), intermediate breadth of the nasal aperture (double red arrow), mild nasal guttering (+), and a broad chin (red line). The entire skull appears as if it were stretched superior-inferior, which it is not.

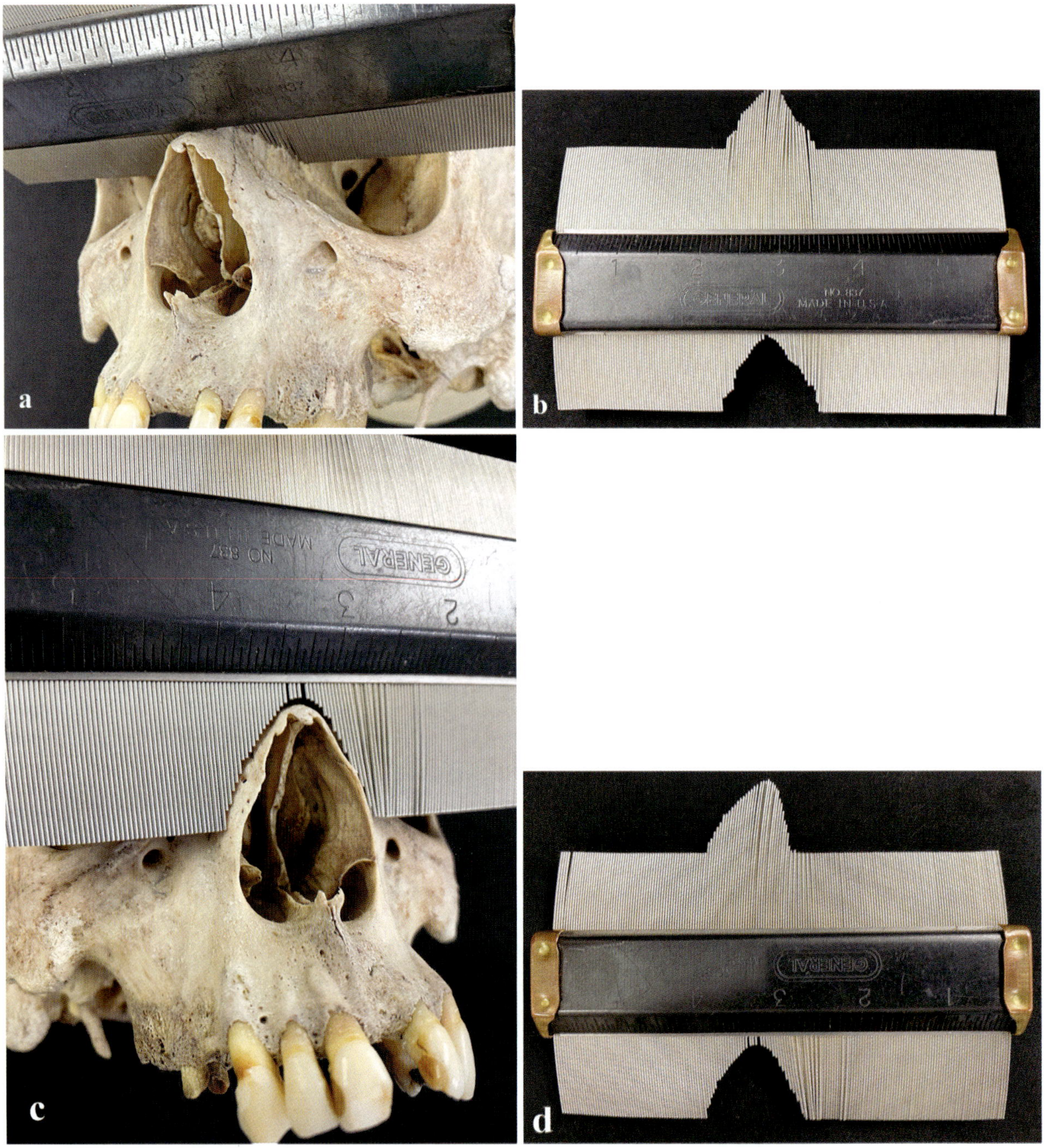

Figure 7a-d. Two methods for visualizing the size and contour of the nasal bones with a contour gauge in a 70-year-old African American male. (a-b) The first method reveals the contour of the nasal bones by placing the gauge perpendicular to the nasal bones, approximately 1 cm from nasion, and into the orbits. (c-d) The second method reveals the contour of the nasal bones and frontal process of the maxilla by placing the gauge perpendicular to the hard palate and orbits. Although both methods have been used, the second method (c-d) is typically preferred as it includes the frontal process of the maxilla and is used in this atlas (see Hefner and Linde 2018).

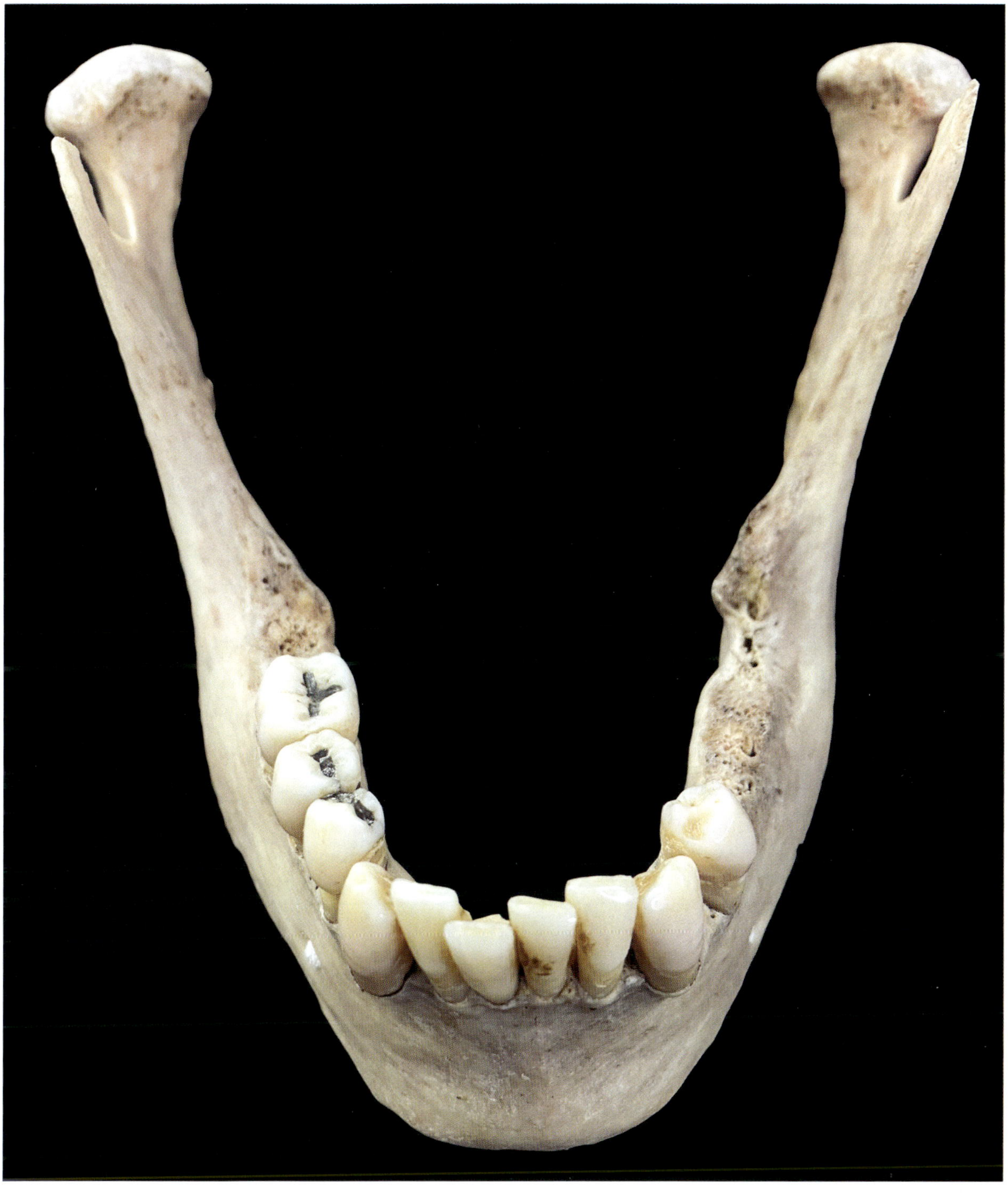

Figure 8. Mandible of 70-year-old African American male with a broad chin.

TABLE 6.
Cranial Measurements in a 70-year-old African American Male

Measurement	(mm)	Right	Left
Maximum Cranial Length (g-op, GOL)	204		
Maximum Cranial Breadth (eu-eu, XCB)	134		
Bizygomatic Breadth (zy-zy, ZYB)	129		
Basion-Bregma Height (ba-b, BBH)	150		
Cranial Base Length (ba-n, BNL)	110		
Basion-Prosthion (ba-pr, BPL)	106		
Maxillo-Alveolar Breadth (ecm-ecm, MAB)	60		
Maxillo-Alveolar Length (pr-alv, MAL)	60		
Biauricular Breadth (au-au, AUB)	121		
Upper Facial Height (n-pr)	72		
Minimum Frontal Breadth (ft-ft, WFB)	102		
Upper Facial Breadth (fmt-fmt)	112		
Nasal Height (n-ns, NLH)	52		
Nasal Breadth (al-al, NLB)	28		
Orbital Breadth (d-ec, OBB)	48		L
Orbital Height (OBH)	36		L
Biorbital Breadth (ec-ec, EKB)	101		
Interorbital Breadth (d-d, DKB)	28		
Frontal Chord (n-b, FRC)	127		
Parietal Chord (b-l, PAC)	129		
Occipital Chord (l-o, OCC)	100		
Foramen Magnum Length (ba-o, FOL)	36		
Foramen Magnum Breadth (FOB)	25		
Mastoid Length (MDH)		28	29

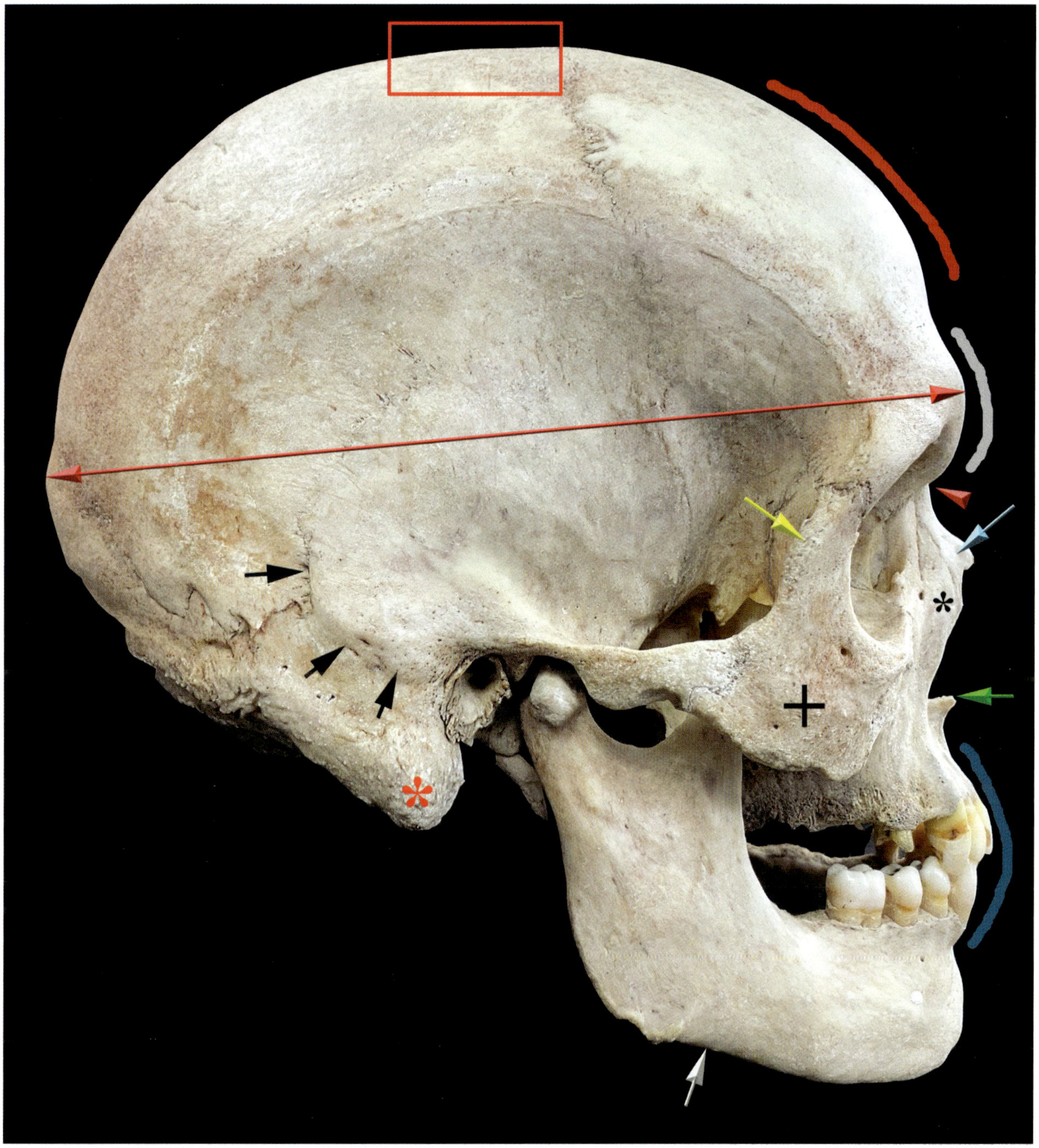

Figure 9. Right lateral view of 70-year-old African American male with a long vault (double red arrow), a prominent brow ridge/glabella region (curved gray line), nasion depression (red arrowhead), small "contoured" nasal bones (blue arrow), a short frontal process of the maxilla (black*) that usually accompanies short or small nasal bones, an anterior nasal spine (green arrow), absence of a posterior zygomatic tubercle (yellow arrow), mild alveolar prognathism (curved blue line), large malar bone (+), a mound-like supramastoid crest (black arrow), moderate-size mastoid process (red*), and small antegonial notch (white arrow). Note the absence of a postbregmatic depression (rectangle), a feature that is often present in Africans and African Americans. The presence and amount of alveolar prognathism, a forward projection/protrusion of the jaws, can be estimated or illustrated using several methods that include or exclude the teeth. (See Section I, "Cephalometric Analysis of the Dentoalveolar Complex: A Clinical Perspective on Alveolar Prognathism.")

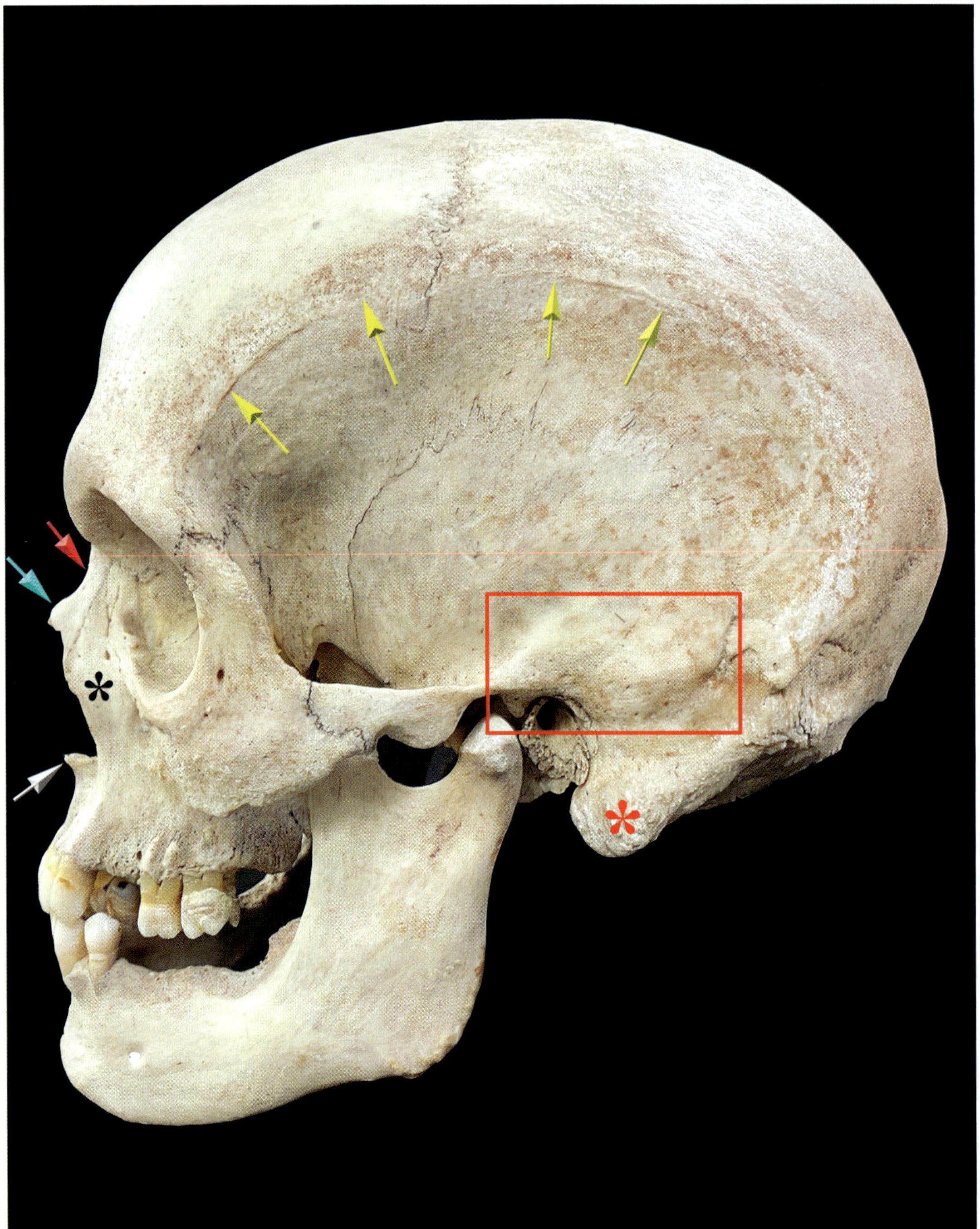

Figure 10. Left lateral view of 70-year-old African American male with a long vault, nasion depression, and small "contoured" nasal bones that are both concave (red arrow) and convex (blue arrow). The anterior nasal spine (white arrow) is small as is the frontal process of the maxilla (black*) situated anterior to the orbit and forming the lateral walls of the nasal aperture. Mild alveolar prognathism is evident. Note the prominent malar bones, a large brow ridge, the developed superior temporal line (yellow arrow), the highly developed supramastoid (rectangle) and nuchal crests, and a medium-size mastoid process (red*). The supramastoid crest in this individual is pronounced and more of a wide mound than a narrow crest.

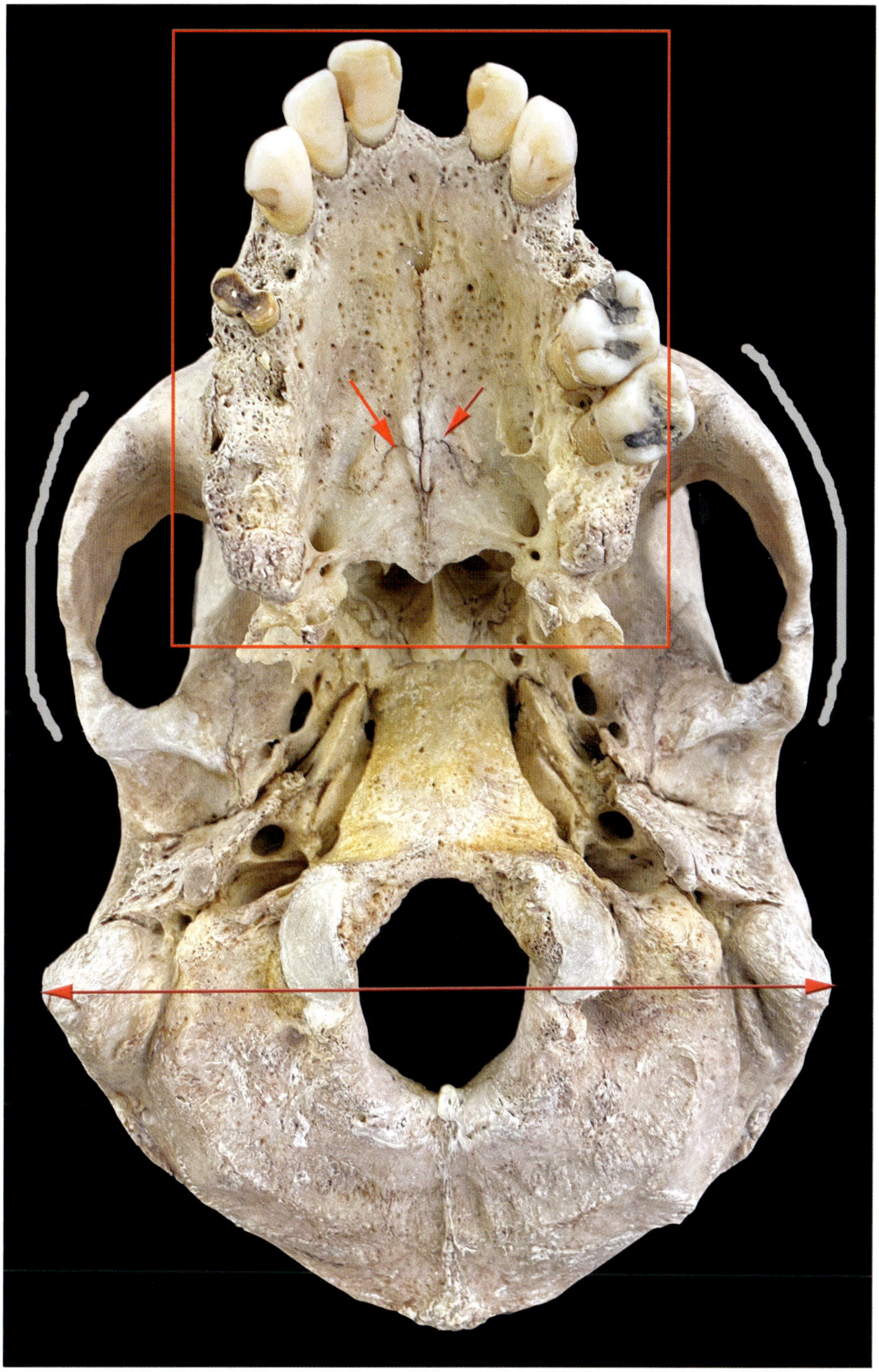

Figure 11. Inferior view of 70-year-old African American male. Note the anterior bulging of the transverse palatine suture (arrow) (Hauser and DeStefano 1989), convex zygomatic arches (curved gray line), long and narrow palate (rectangle), and wide distance between the mastoid processes (double arrow). The convex zygomatic arches, in comparison to those that are flat or more rarely concave is a feature that, with further research, might be of use in estimating ancestry and/or sex.

Figure 12. Superior view of 70-year-old African American male. Note that the oval-shaped skull is long (anterior-posterior) and narrow, a feature often seen in Africans and African Americans.

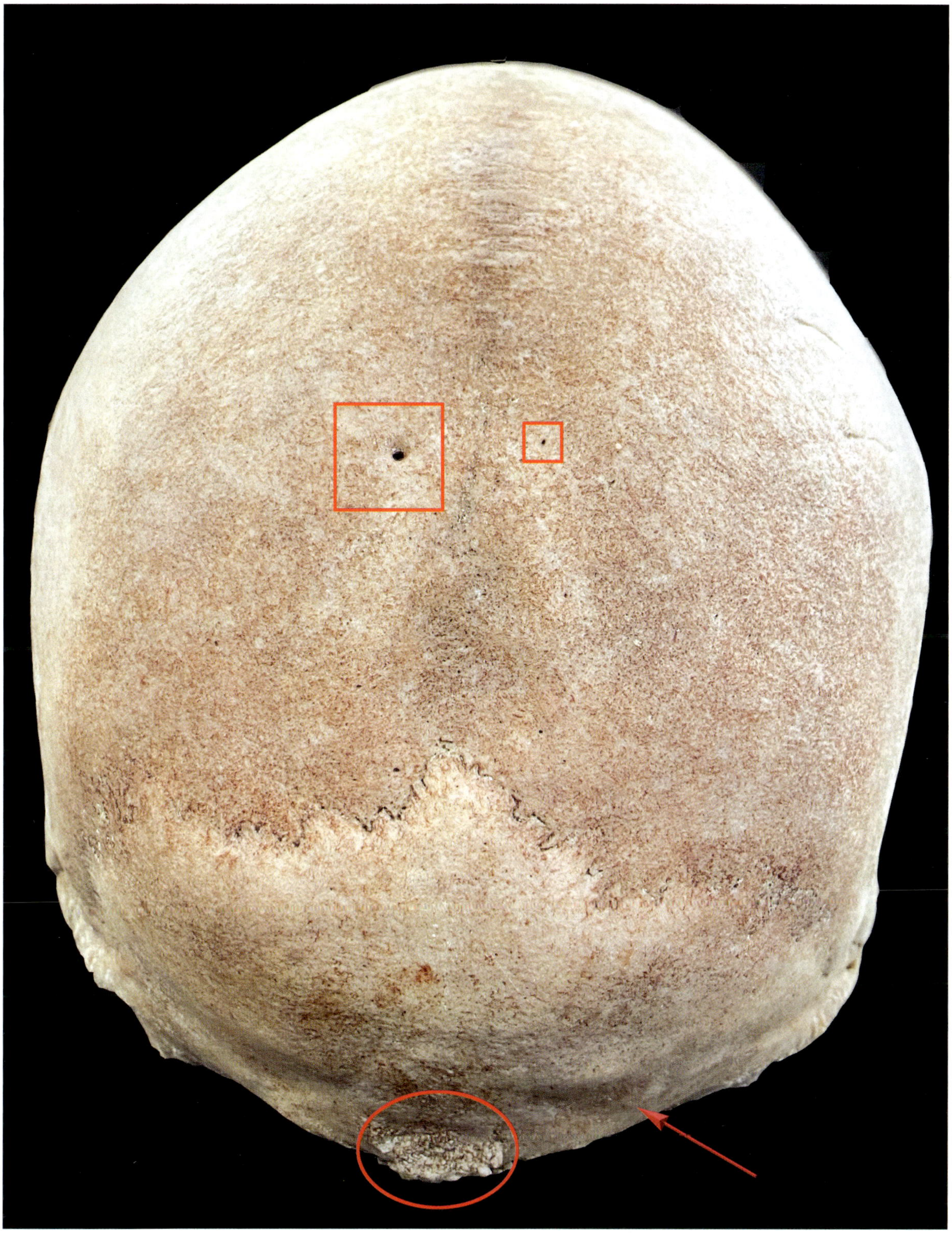

Figure 13. Posterior view of 70-year-old African American male. The sagittal suture is totally obliterated, the lambdoidal suture is partially visible, and the parietal foramina (square) are asymmetrical in size (normal variation). Note the developed nuchal crest (arrow) and inion hook (oval), features often present in males and some robust females.

2. AFRICAN AMERICAN MALE

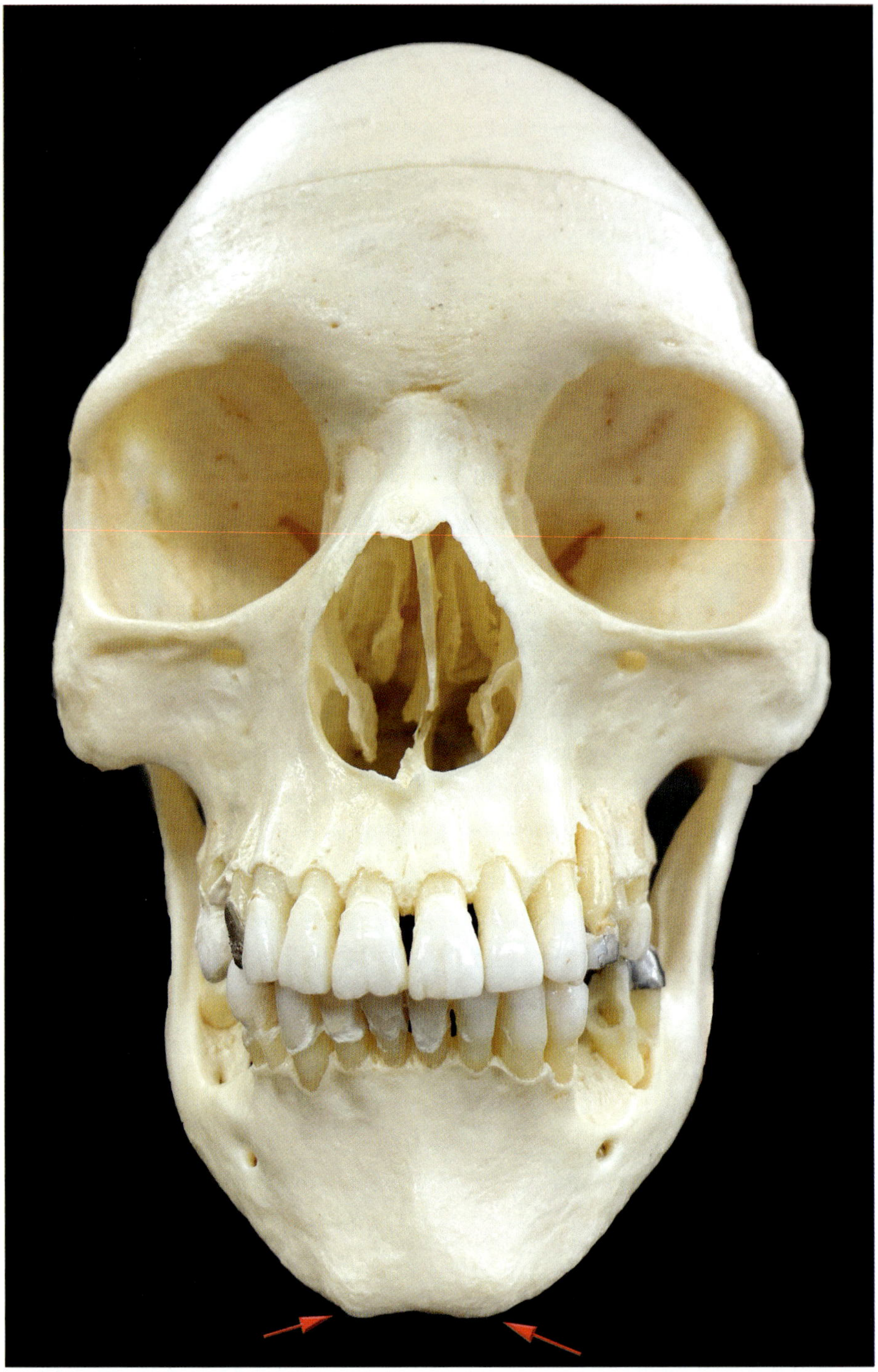

Figure 14. Anterior view 60-year-old African American male with a wide interorbital breadth, wide nasal aperture, and steepled (waist-like) nasal bones. Also present is a trace of the supranasal suture, small malar tubercle, and a square or slightly bilobate chin with prominent mandibular tubercles (arrow). Note that the malar (cheek) bones are wide in comparison to the cranial vault.

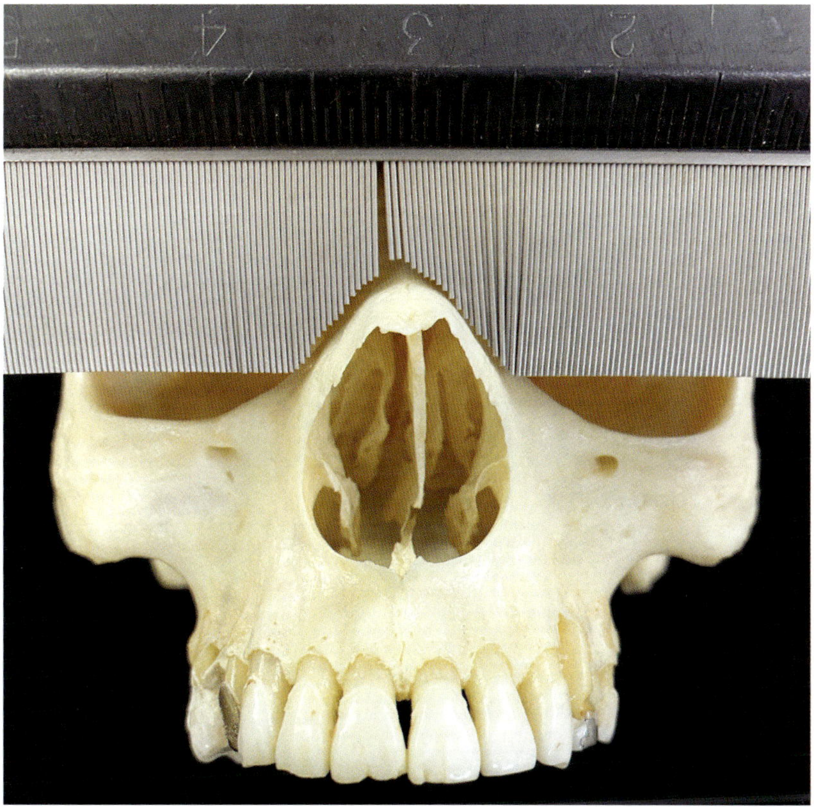

Figure 15. Nasal bone contour in a 60-year-old African American male.

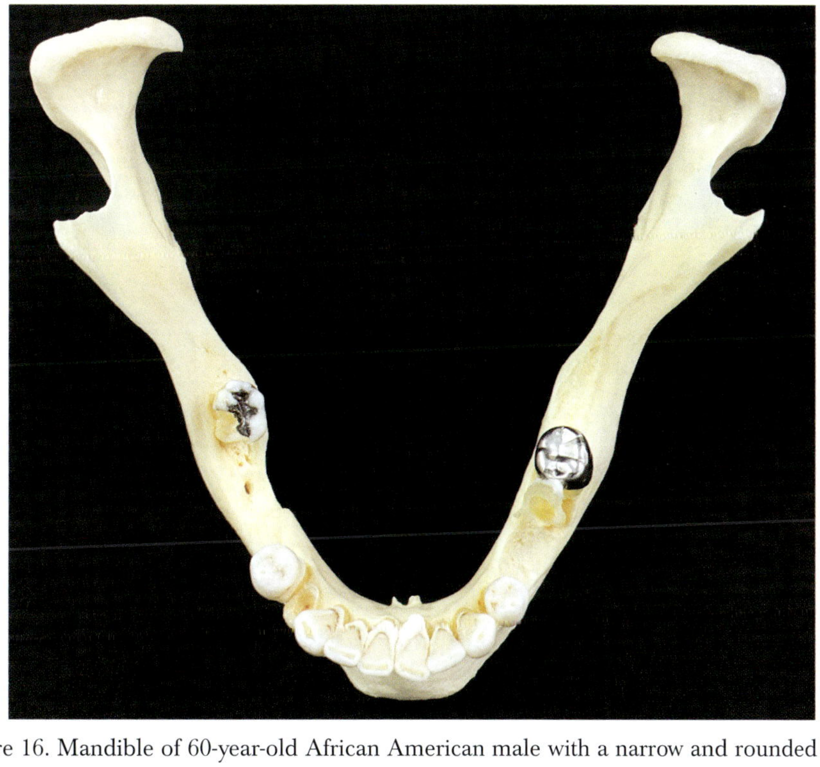

Figure 16. Mandible of 60-year-old African American male with a narrow and rounded chin.

TABLE 7.
Cranial Measurements in a 60-year-old African American Male

Measurement	(mm)	Right	Left
Maximum Cranial Length (g-op, GOL)	184		
Maximum Cranial Breadth (eu-eu, XCB)	136		
Bizygomatic Breadth (zy-zy, ZYB)	139		
Basion-Bregma Height (ba-b, BBH)	141		
Cranial Base Length (ba-n, BNL)	111		
Basion-Prosthion (ba-pr, BPL)	106		
Maxillo-Alveolar Breadth (ecm-ecm, MAB)	67		
Maxillo-Alveolar Length (pr-alv, MAL)	57		
Biauricular Breadth (au-au, AUB)	102		
Upper Facial Height (n-pr)	65		
Minimum Frontal Breadth (ft-ft, WFB)	93		
Upper Facial Breadth (fmt-fmt)	112		
Nasal Height (n-ns, NLH)	51		
Nasal Breadth (al-al, NLB)	30		
Orbital Breadth (d-ec, OBB)	43		L
Orbital Height (OBH)	34		L
Biorbital Breadth (ec-ec, EKB)	104		
Interorbital Breadth (d-d, DKB)	27		
Frontal Chord (n-b, FRC)	102		
Parietal Chord (b-l, PAC)	110		
Occipital Chord (l-o, OCC)	109		
Foramen Magnum Length (ba-o, FOL)	41		
Foramen Magnum Breadth (FOB)	31		
Mastoid Length (MDH)		25	26

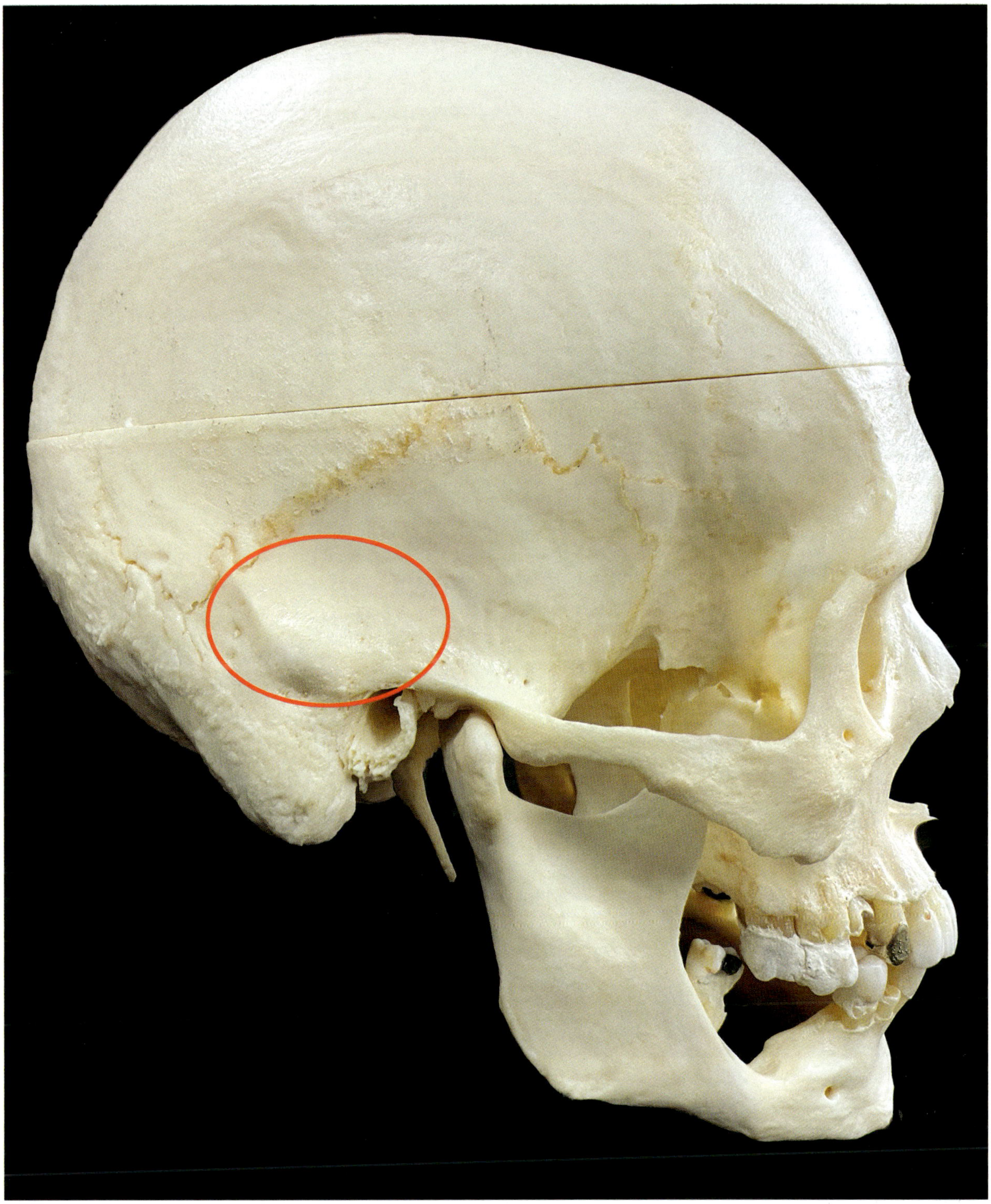

Figure 17. Right lateral view 60-year-old African American male with a sloping frontal bone, a small brow ridge, low/rounded nasal bones, anterior nasal spine, and alveolar prognathism. Note the developed supramastoid crest (oval) and the absence of a postbregmatic depression. The body of the mandible has resorbed dramatically due to the antemortem loss of teeth.

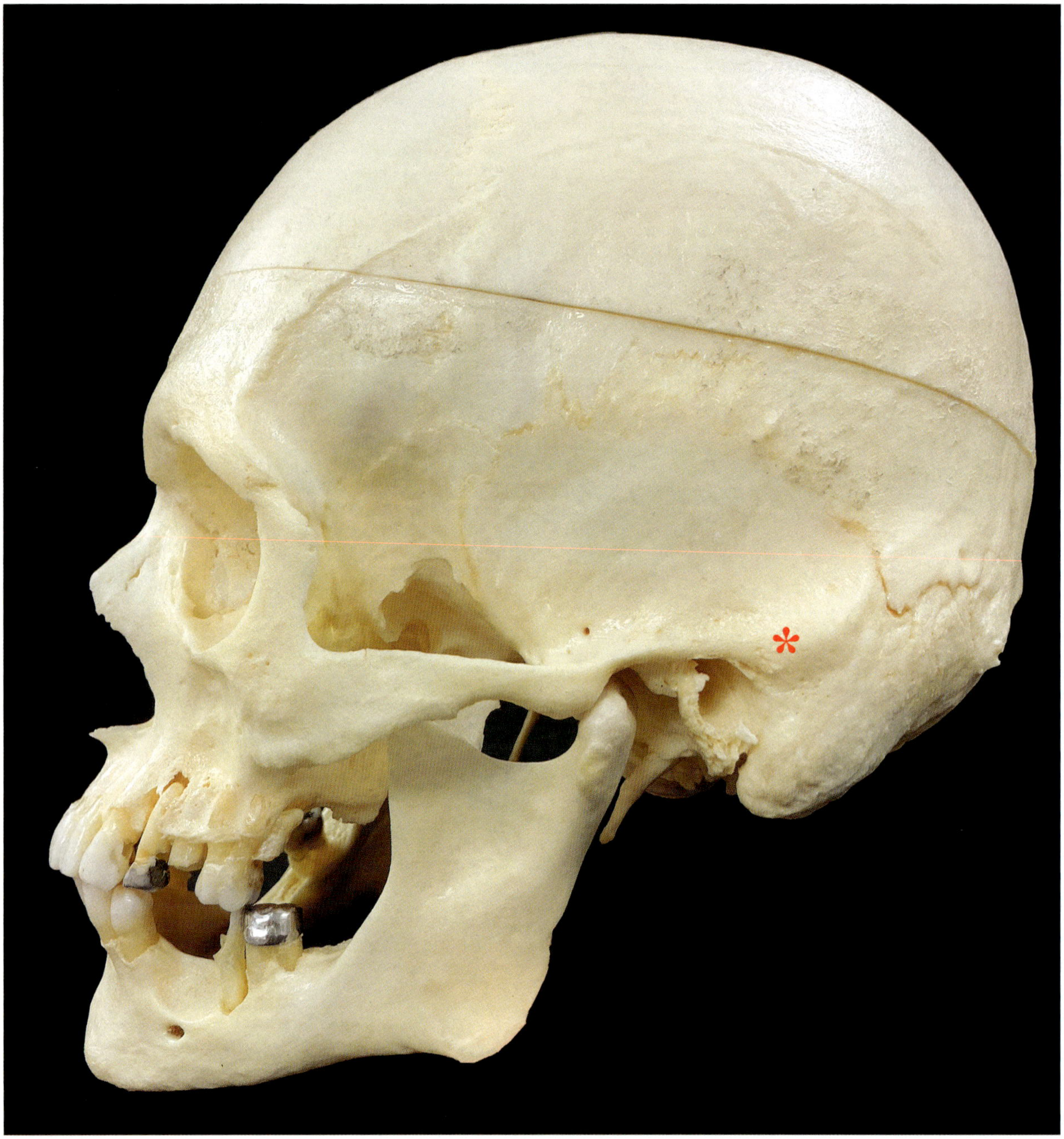

Figure 18. Left lateral view 60-year-old African American male with a pronounced supramastoid crest (*).

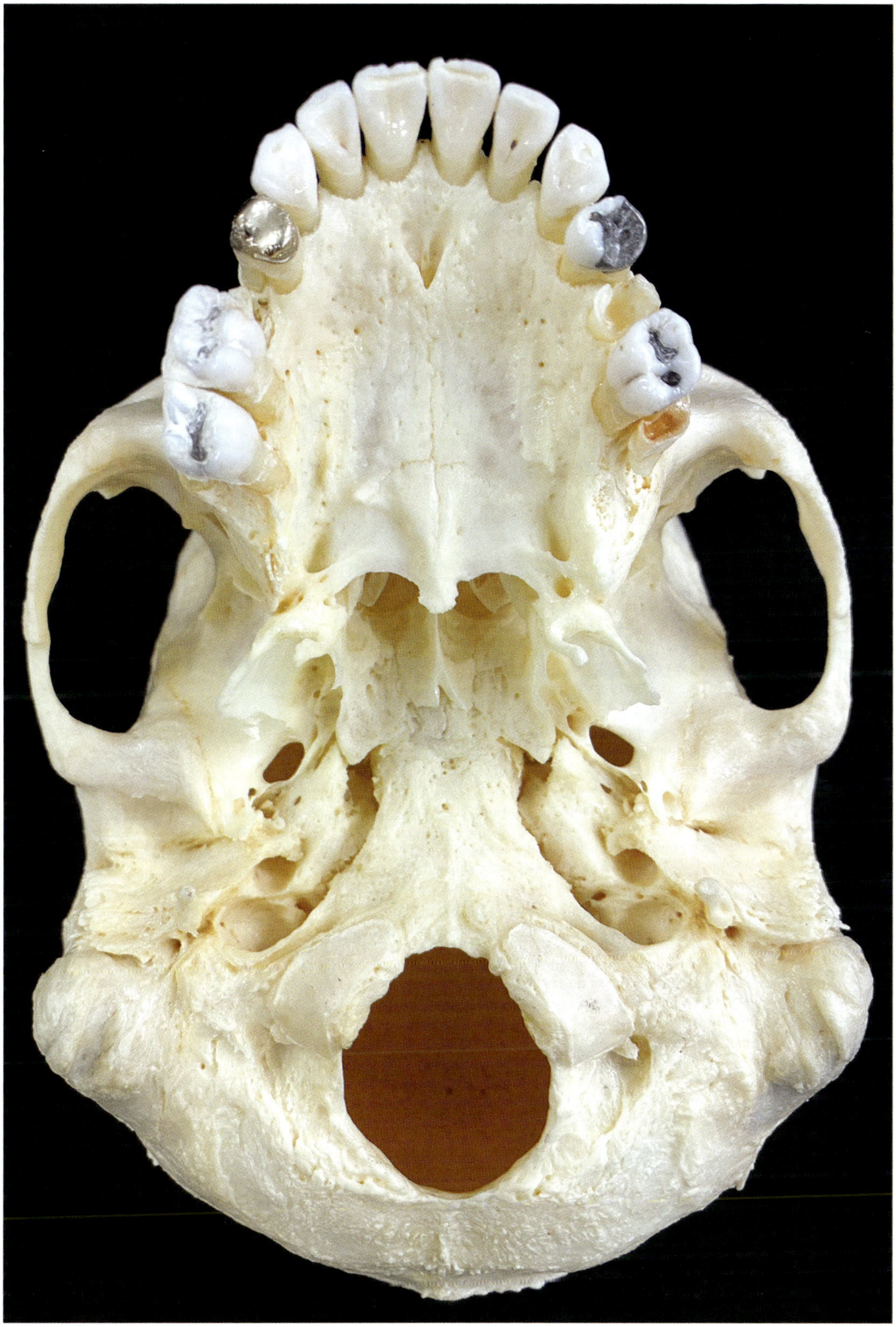

Figure 19. Inferior view 60-year-old African American male with a wide, shallow palate and a short occipital region.

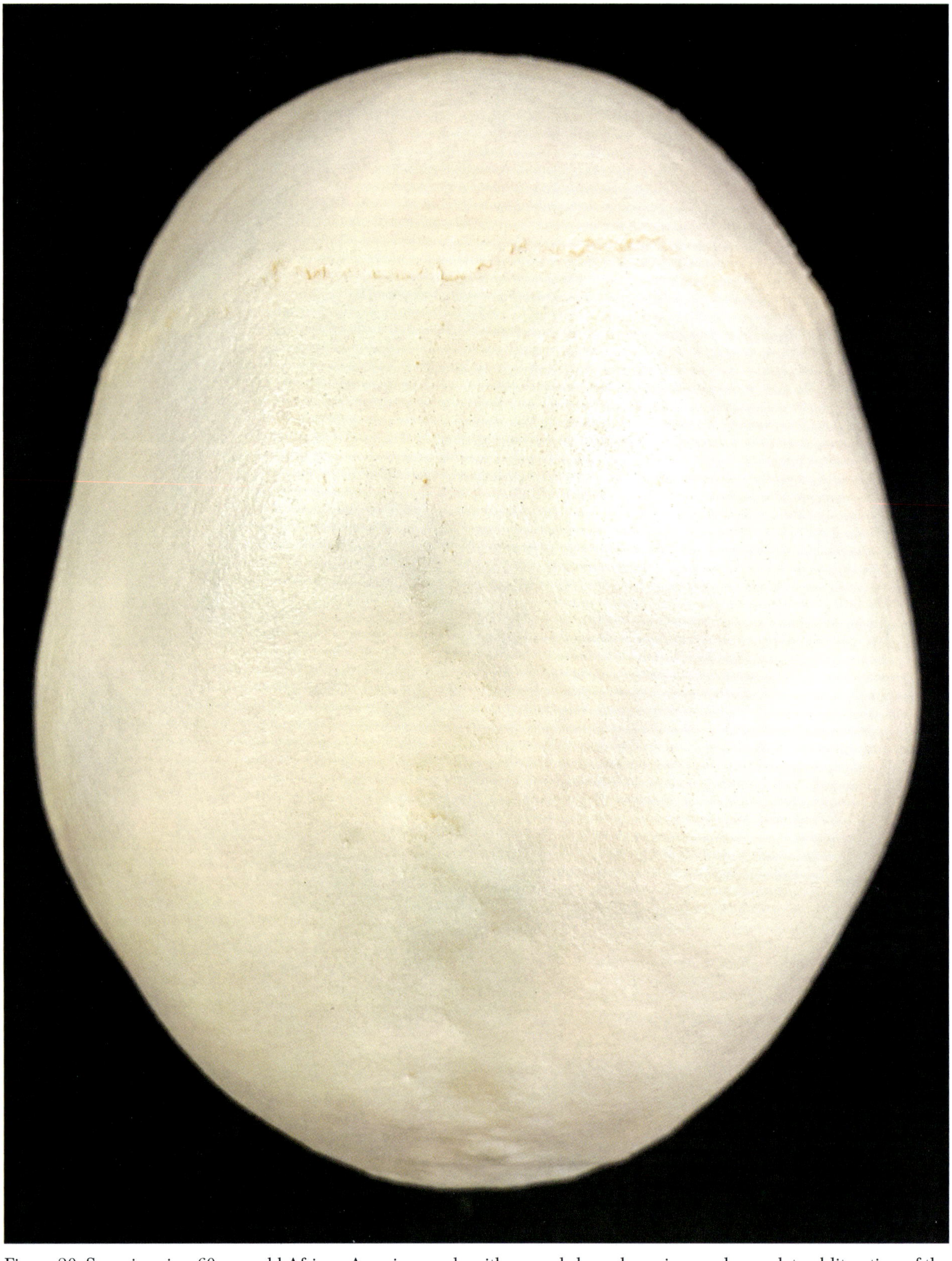

Figure 20. Superior view 60-year-old African American male with an oval-shaped cranium and complete obliteration of the sagittal and lambdoidal sutures.

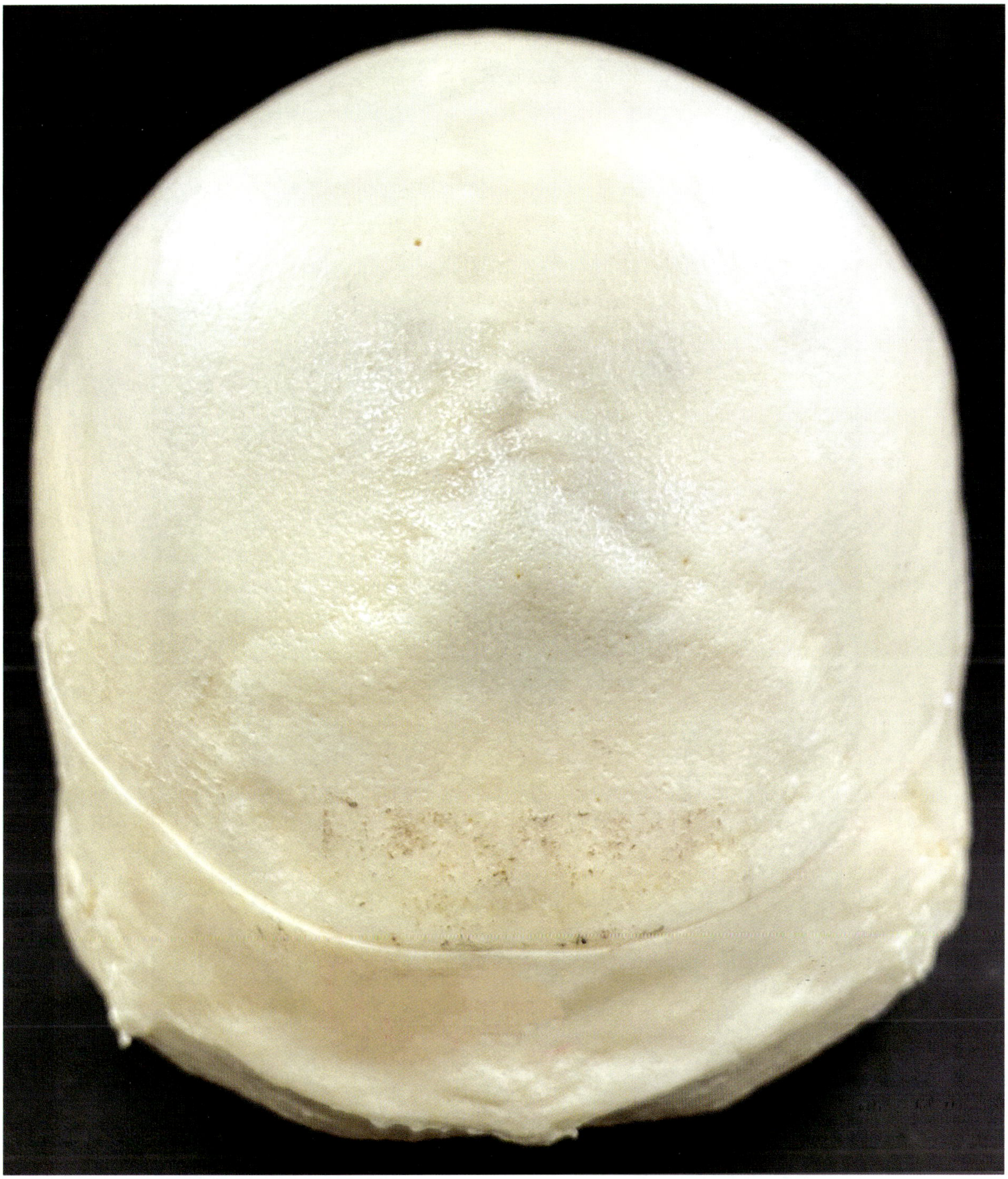

Figure 21. Posterior view 60-year-old African American male with a prominent nuchal crest and inion hook, and complete obliteration of the sutures.

3. AFRICAN AMERICAN MALE

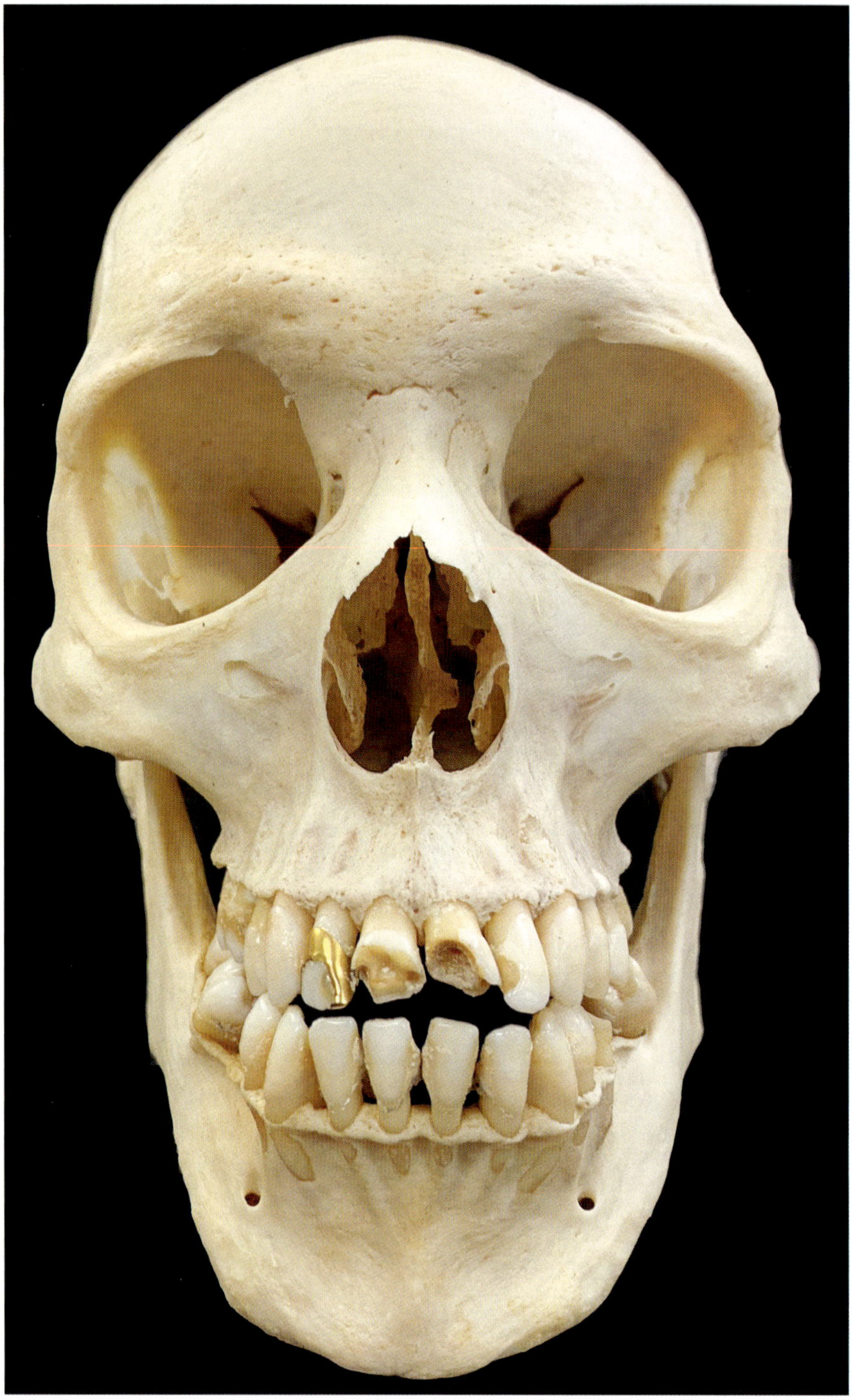

Figure 22. Anterior view of 59-year-old African American male with a prominent brow ridge, trace supranasal suture, waisted (narrowed) nasal bones, and wide zygomas. Note the broad interorbital breadth and nasal aperture and slight nasal guttering.

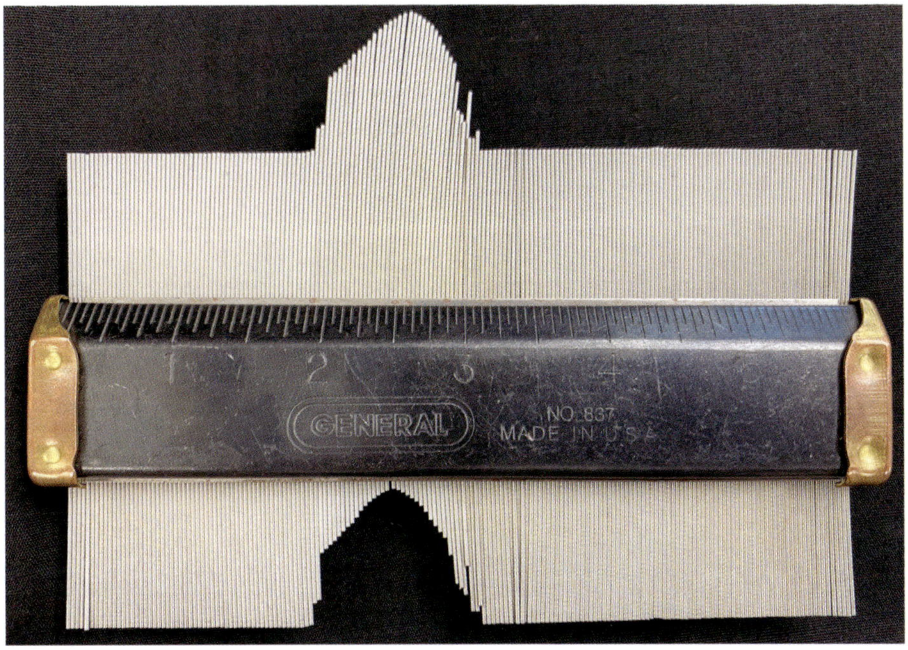

Figure 23. Nasal bone contour in 59-year-old African American male.

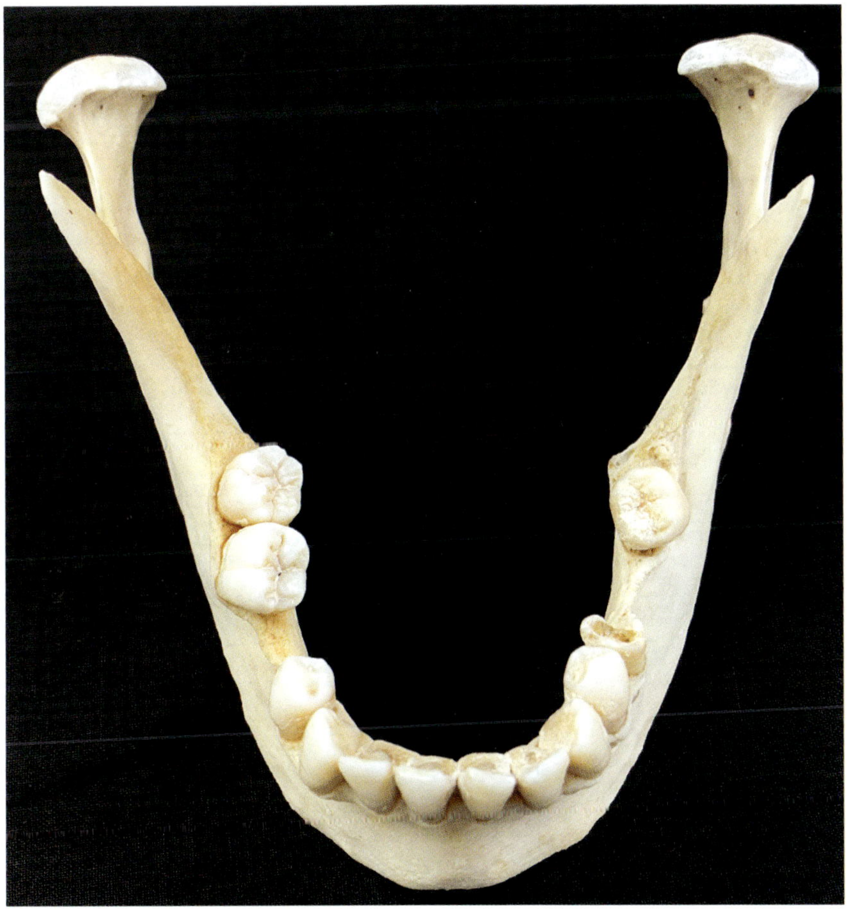

Figure 24. Mandible of 59-year-old African American male with rounded chin.

TABLE 8.
Cranial Measurements in a 59-year-old African American Male

Measurement	(mm)	Right	Left
Maximum Cranial Length (g-op, GOL)	185		
Maximum Cranial Breadth (eu-eu, XCB)	129		
Bizygomatic Breadth (zy-zy, ZYB)	128		
Basion-Bregma Height (ba-b, BBH)	123		
Cranial Base Length (ba-n, BNL)	95		
Basion-Prosthion (ba-pr, BPL)	91		
Maxillo-Alveolar Breadth (ecm-ecm, MAB)	61		
Maxillo-Alveolar Length (pr-alv, MAL)	52		
Biauricular Breadth (au-au, AUB)	114		
Upper Facial Height (n-pr)	65		
Minimum Frontal Breadth (ft-ft, WFB)	90		
Upper Facial Breadth (fmt-fmt)	103		
Nasal Height (n-ns, NLH)	49		
Nasal Breadth (al-al, NLB)	24		
Orbital Breadth (d-ec, OBB)	41		L
Orbital Height (OBH)	31		L
Biorbital Breadth (ec-ec, EKB)	93		
Interorbital Breadth (d-d, DKB)	25		
Frontal Chord (n-b, FRC)	106		
Parietal Chord (b-l, PAC)	107		
Occipital Chord (l-o, OCC)	97		
Foramen Magnum Length (ba-o, FOL)	37		
Foramen Magnum Breadth (FOB)	29		
Mastoid Length (MDH)		24	24

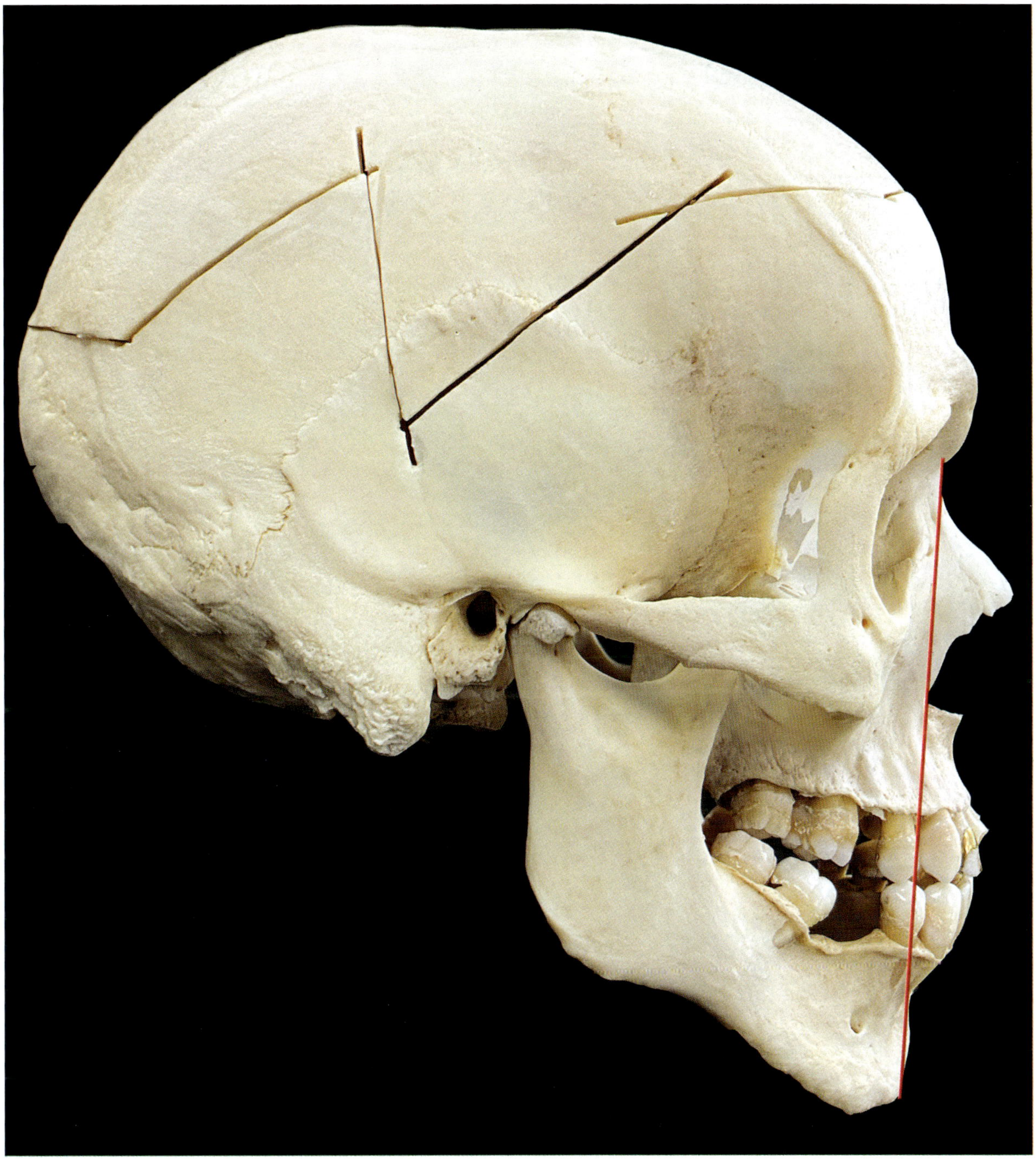

Figure 25. Right lateral view of 59-year-old African American male with a postbregmatic depression and a long vault. Note the small anterior nasal spine, an antegonial notch, and alveolar prognathism.

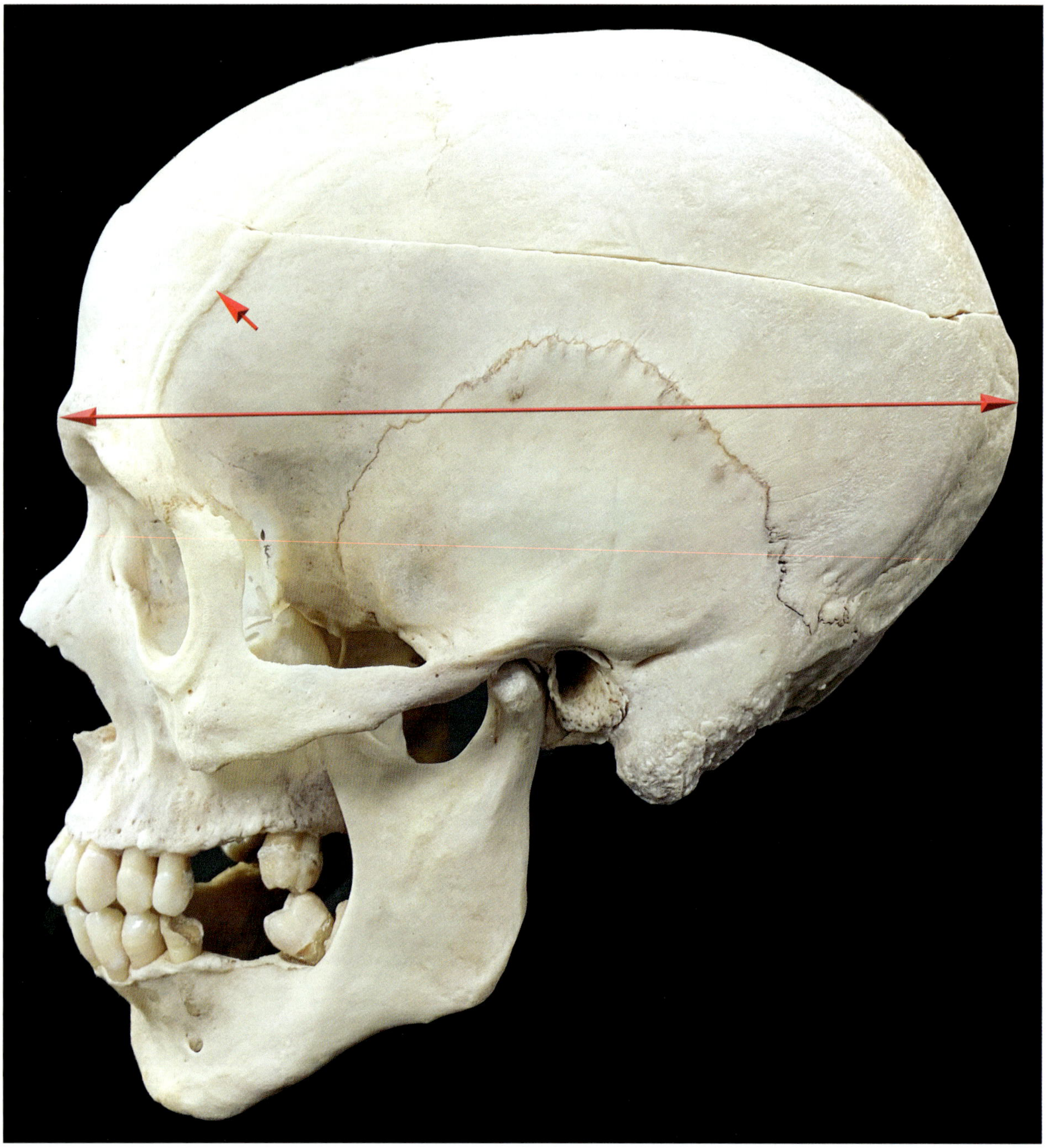

Figure 26. Left lateral view of 59-year-old African American male with a developed temporal line (arrow) and a long (anterior-posterior) vault (double arrow).

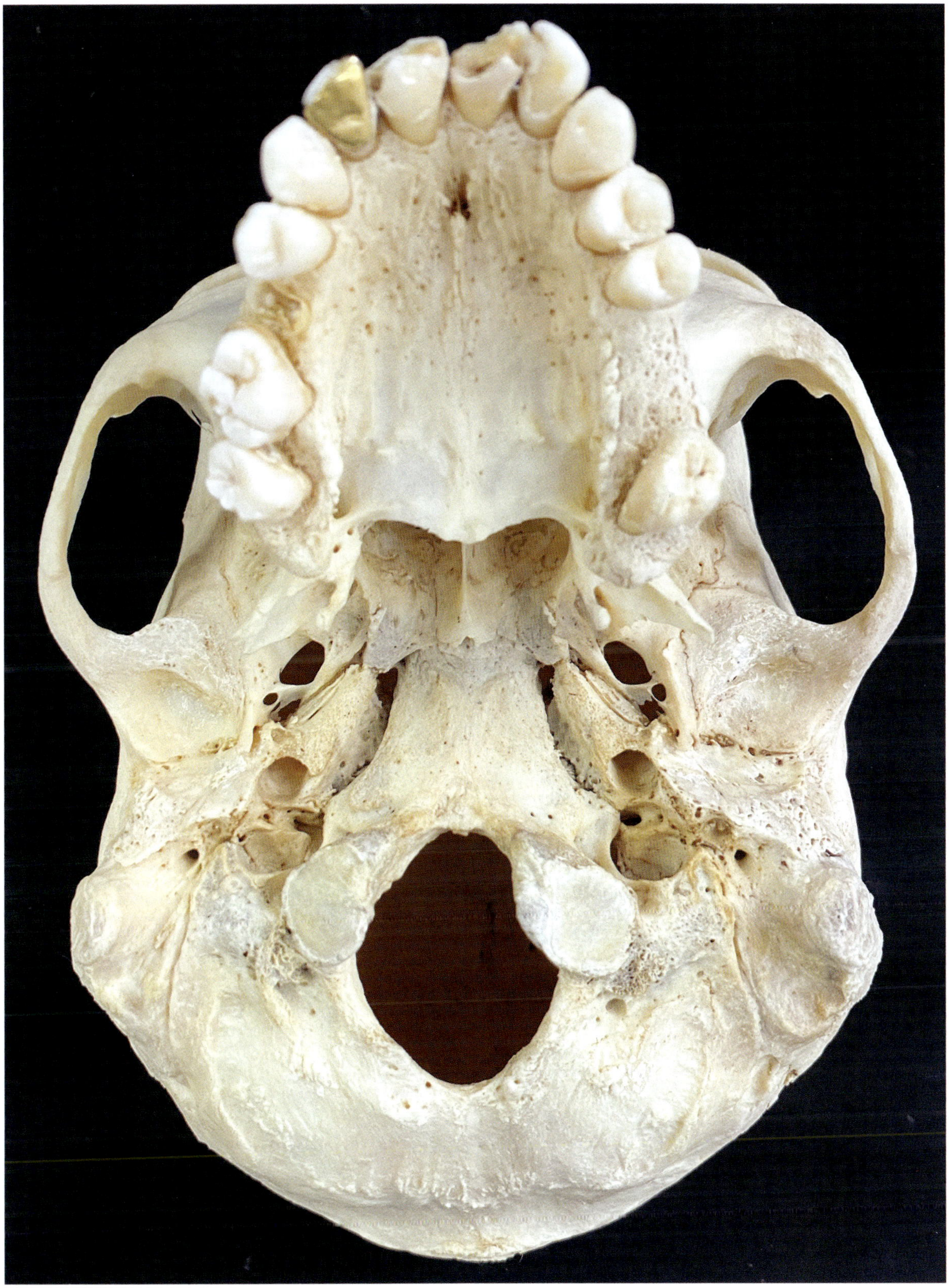

Figure 27. Inferior view of 59-year-old African American male.

Figure 28. Superior view of 59-year-old African American male with a long, oval-shaped cranium and nearly complete obliteration of all sutures.

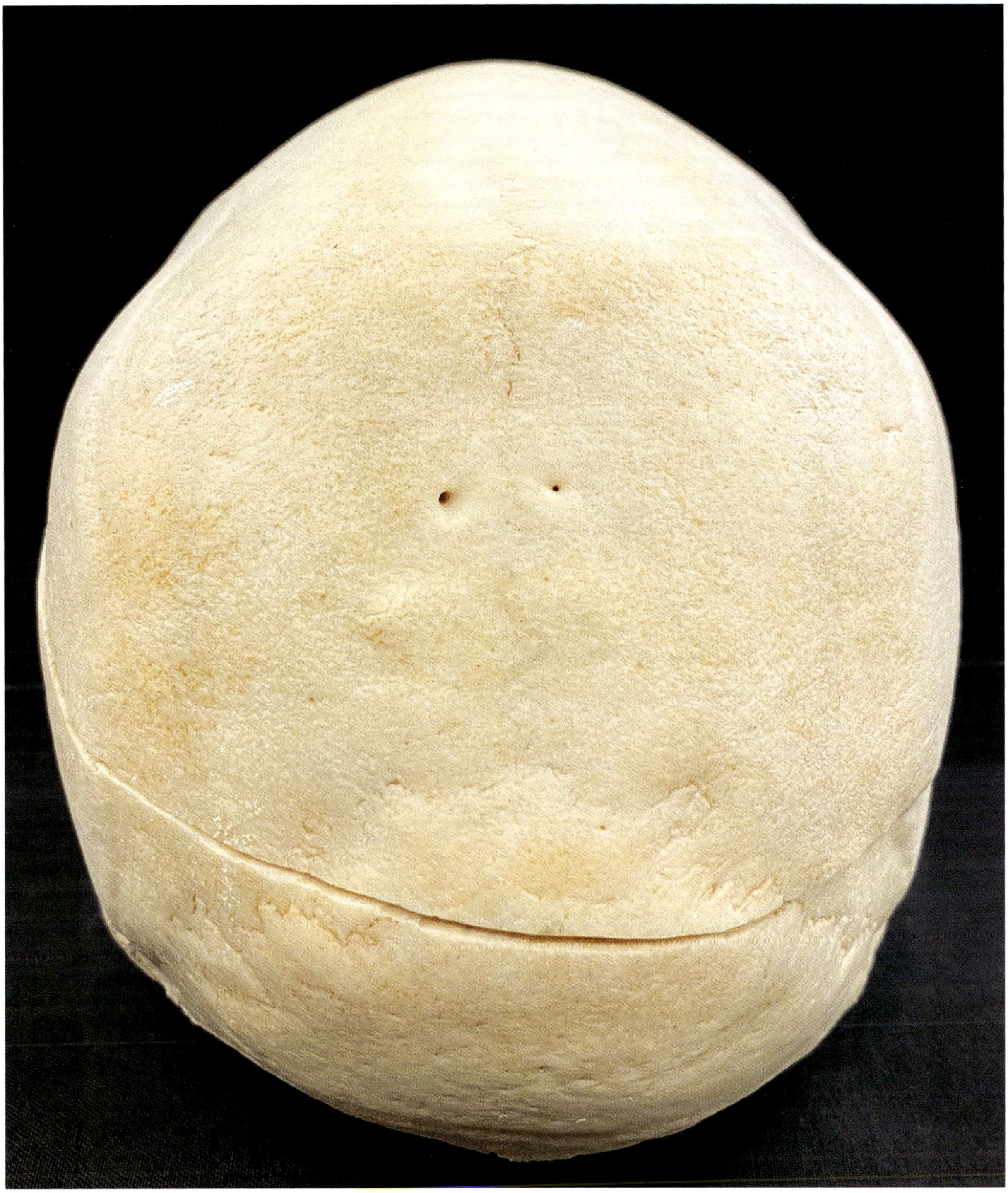

Figure 29. Posterior view of 59-year-old African American male with two parietal foramina (that in the living, transmit emissary veins) and nearly complete obliteration of the sutures.

4. AFRICAN AMERICAN MALE

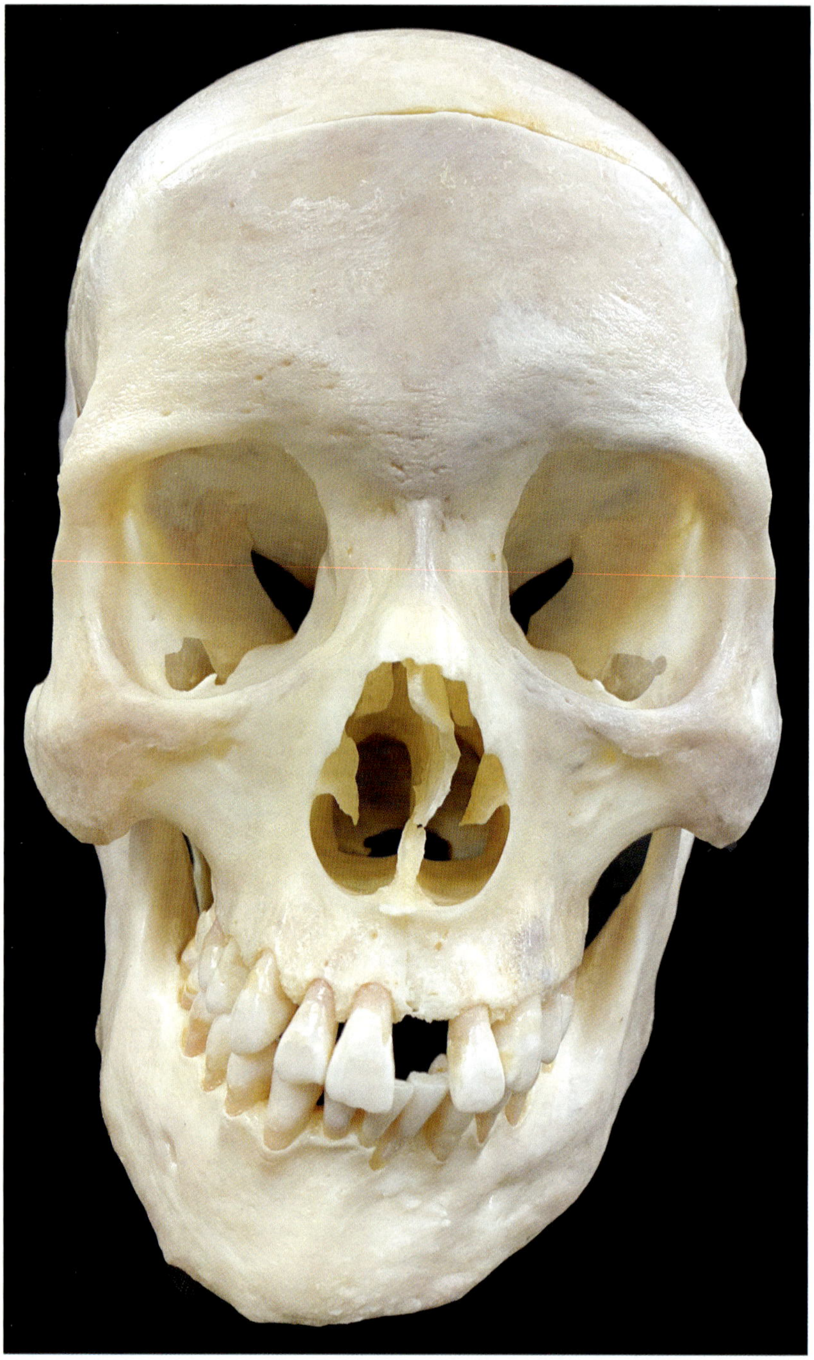

Figure 30. Anterior view of 38-year-old African American male with broad interorbital breadth and wide nasal aperture, and prominent malar tubercles. The nasal bones are 'pinched' (Saul and Saul 2010) near nasion, the nasal aperture is asymmetrical, and the nasal septum is deviated to the left. The mandible is asymmetrical likely due to osteosarcoma (Personal Communication Dr. J. DeMeo) and the maxillary left central incisor is missing and remodeled antemortem. (Numerous descriptive terms have been used to describe the shape of nasal bones such as these including: pinched, tented, narrowed, hourglass, waisted, and steepled.)

Figure 31. Nasal bone contour in a 38-year-old African American male.

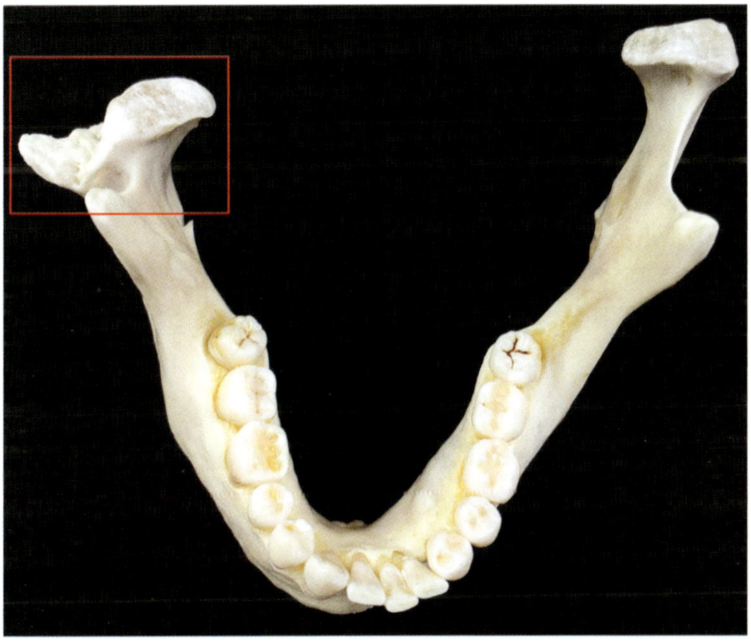

Figure 32. Mandible of 38-year-old African American male with an asymmetrical mandible, primarily of the right mandibular condyle and neck (rectangle) and the temporomandibular joint (not shown). The likely cause of the asymmetry is osteosarcoma of the temporomandibular joint resulting in asymmetry of the entire mandible (Personal Communication Dr. J. DeMeo).

TABLE 9.
Cranial Measurements in a 38-year-old African American Male

Measurement	(mm)	Right	Left
Maximum Cranial Length (g-op, GOL)	185		
Maximum Cranial Breadth (eu-eu, XCB)	129		
Bizygomatic Breadth (zy-zy, ZYB)	123		
Basion-Bregma Height (ba-b, BBH)	136		
Cranial Base Length (ba-n, BNL)	104		
Basion-Prosthion (ba-pr, BPL)	96		
Maxillo-Alveolar Breadth (ecm-ecm, MAB)	55		
Maxillo-Alveolar Length (pr-alv, MAL)	48		
Biauricular Breadth (au-au, AUB)	115		
Upper Facial Height (n-pr)	61		
Minimum Frontal Breadth (ft-ft, WFB)	87		
Upper Facial Breadth (fmt-fmt)	99		
Nasal Height (n-ns, NLH)	50		
Nasal Breadth (al-al, NLB)	26		
Orbital Breadth (d-ec, OBB)	36		L
Orbital Height (OBH)	32		L
Biorbital Breadth (ec-ec, EKB)	87		
Interorbital Breadth (d-d, DKB)	23		
Frontal Chord (n-b, FRC)	105		
Parietal Chord (b-l, PAC)	112		
Occipital Chord (l-o, OCC)	94		
Foramen Magnum Length (ba-o, FOL)	36		
Foramen Magnum Breadth (FOB)	28		
Mastoid Length (MDH)		22	24

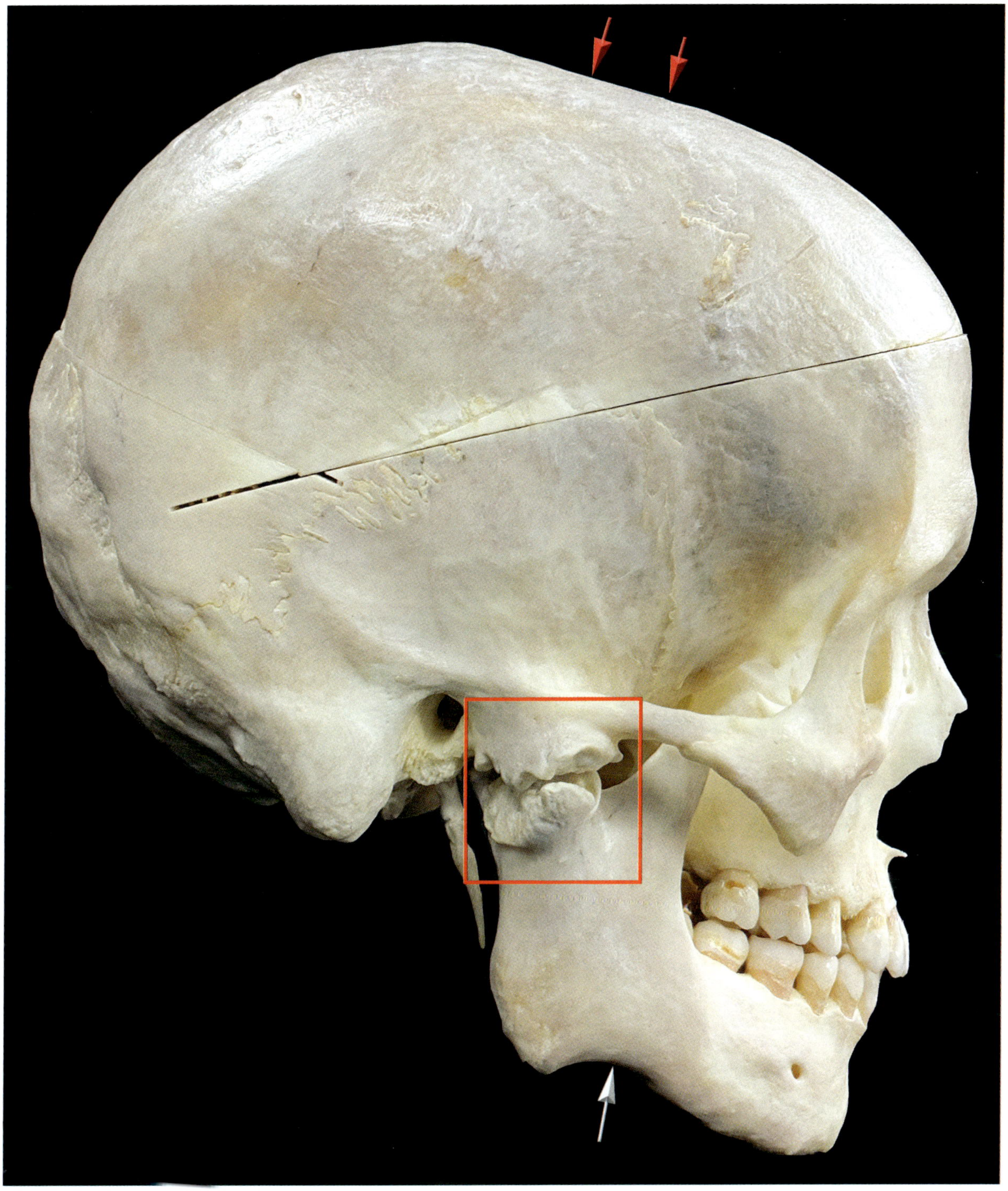

Figure 33. Right lateral view 38-year-old African American male with an asymmetrical mandible and severe degenerative joint disease of right mandibular condyle and temporomandibular joint (square), likely due to osteosarcoma (Personal Communication Dr. J. DeMeo). Note also the postbregmatic depression (red arrow), low and mildly concave nasal bones, the anterior nasal spine, absence of alveolar prognathism, and a prominent antegonial notch (white arrow).

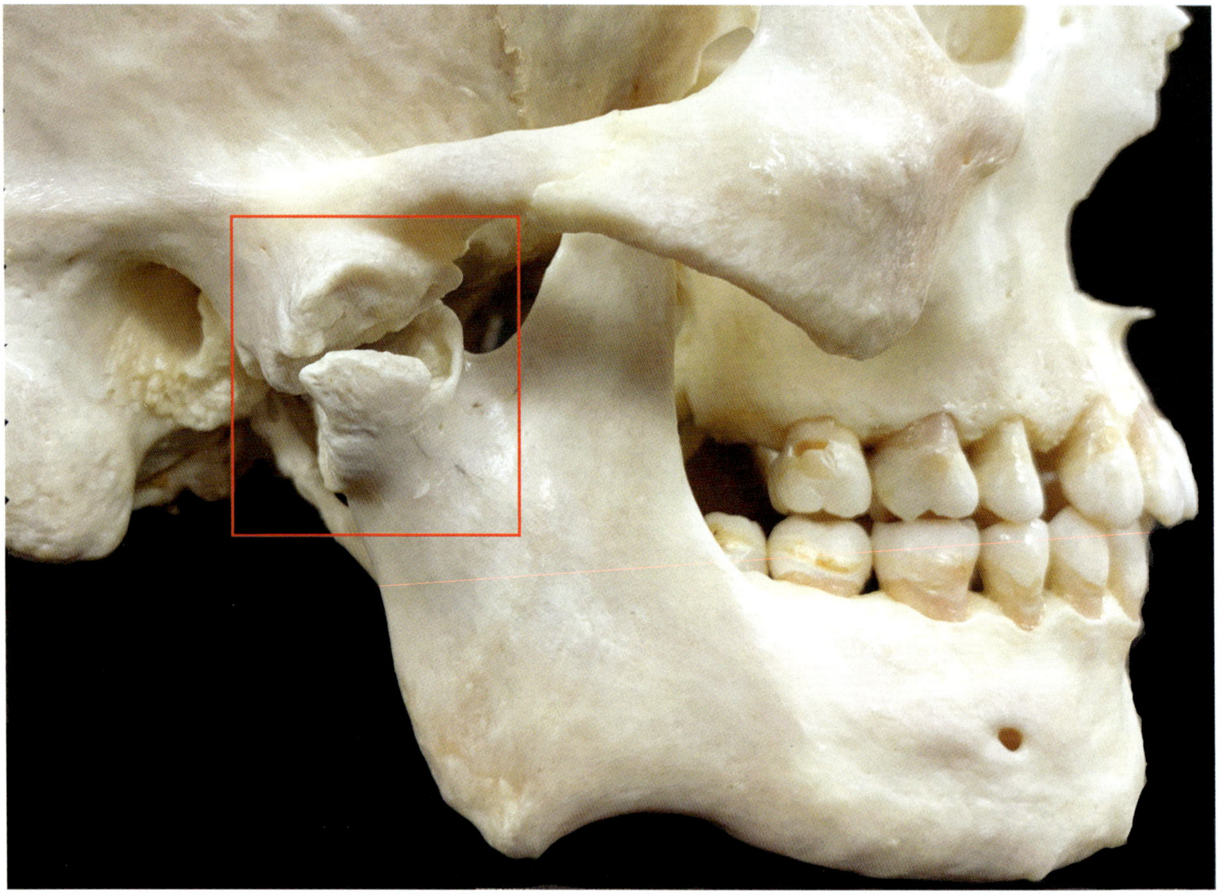

Figure 34. Close-up of right lateral view 38-year-old African American male with an asymmetrical mandible and severe degenerative joint disease of right mandibular condyle and temporomandibular joint (square) likely caused by osteosarcoma (Personal Communication Dr. J. DeMeo).

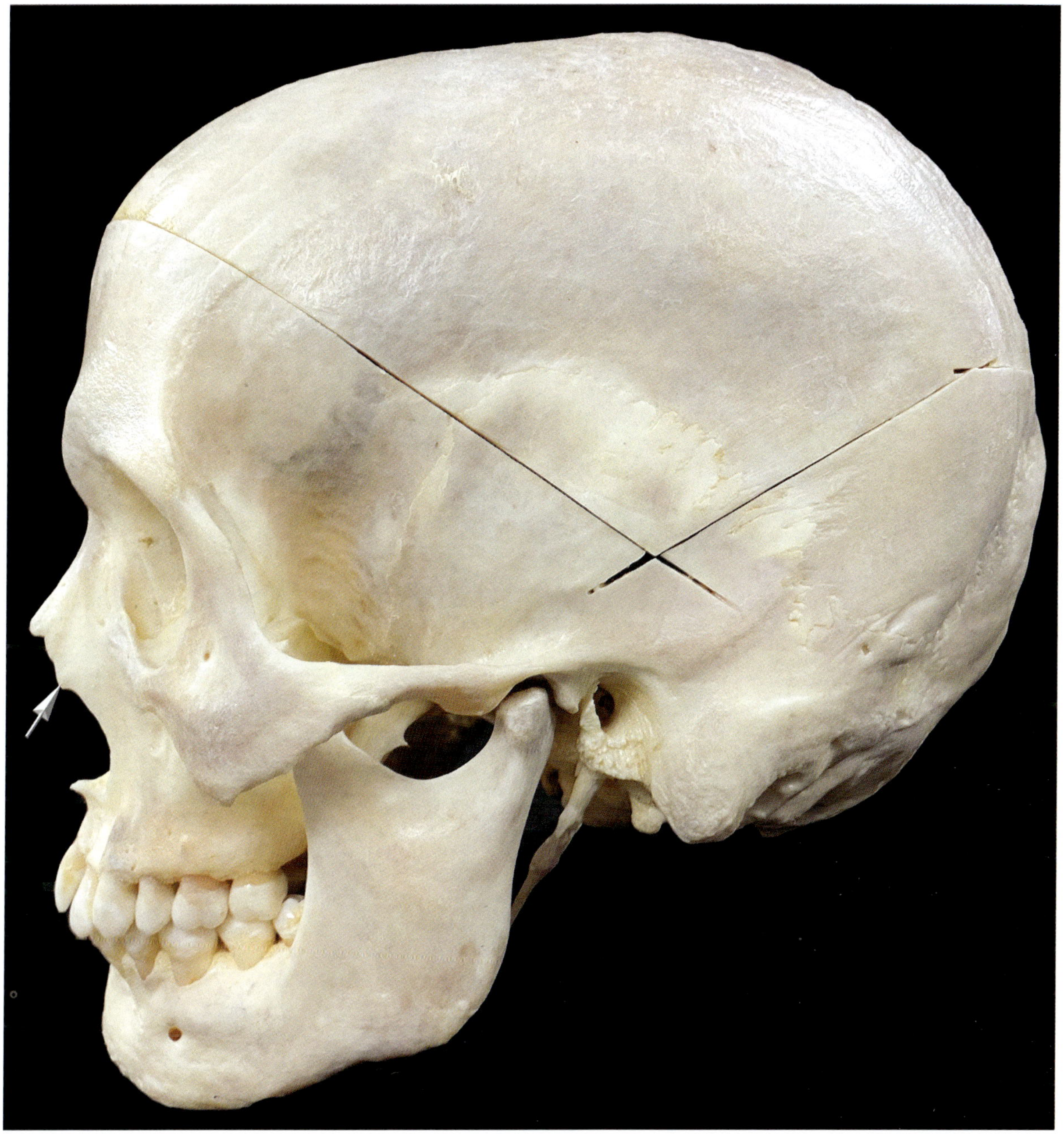

Figure 35. Left lateral view 38-year-old African American male with an asymmetrical mandible. Note the projection (arrow) that is an extension of the frontal process and not the nasal bones; this projection is not part of the feature known as over-hanging nasal bones.

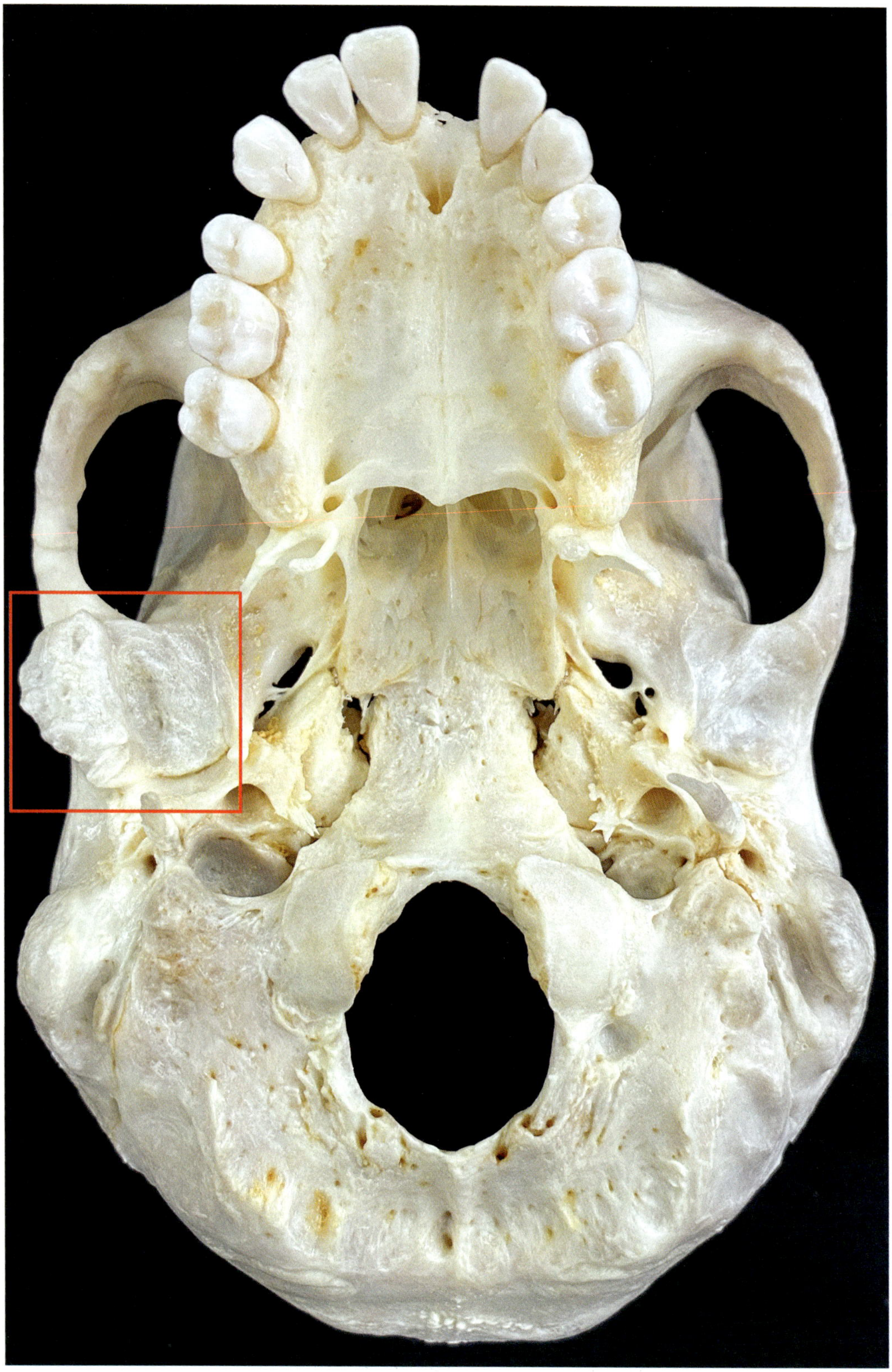

Figure 36. Inferior view 38-year-old African American male with asymmetrical mandible and alteration of the right temporomandibular joint (rectangle) likely due to osteosarcoma (Personal Communication Dr. J. DeMeo). The teeth are spatulate, the maxillary left central incisor is missing antemortem, and the right mastoid process is larger than the left mastoid process.

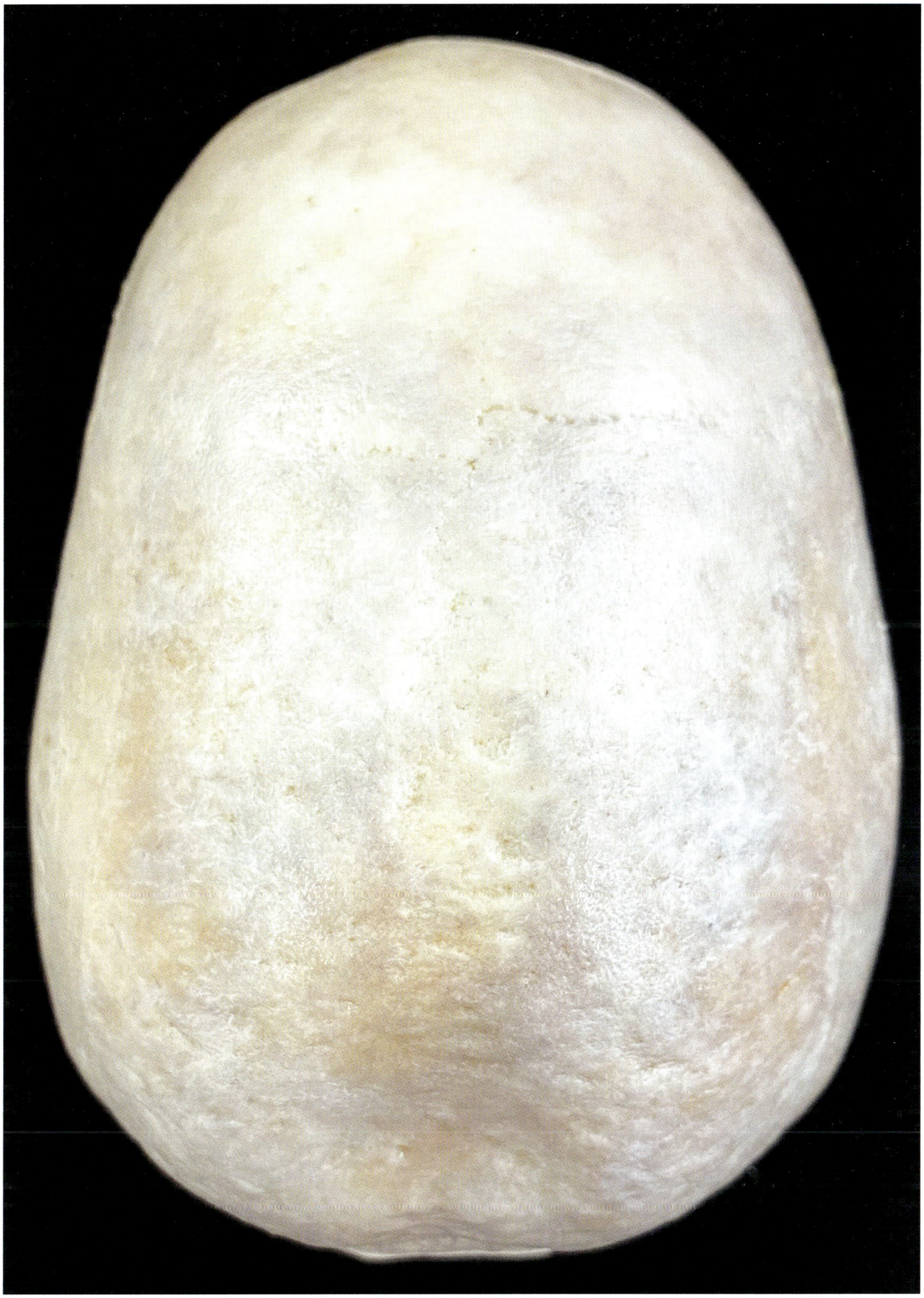

Figure 37. Superior view 38-year-old African American male with a long (anterior-posterior), oval-shaped vault and complete obliteration of all sutures.

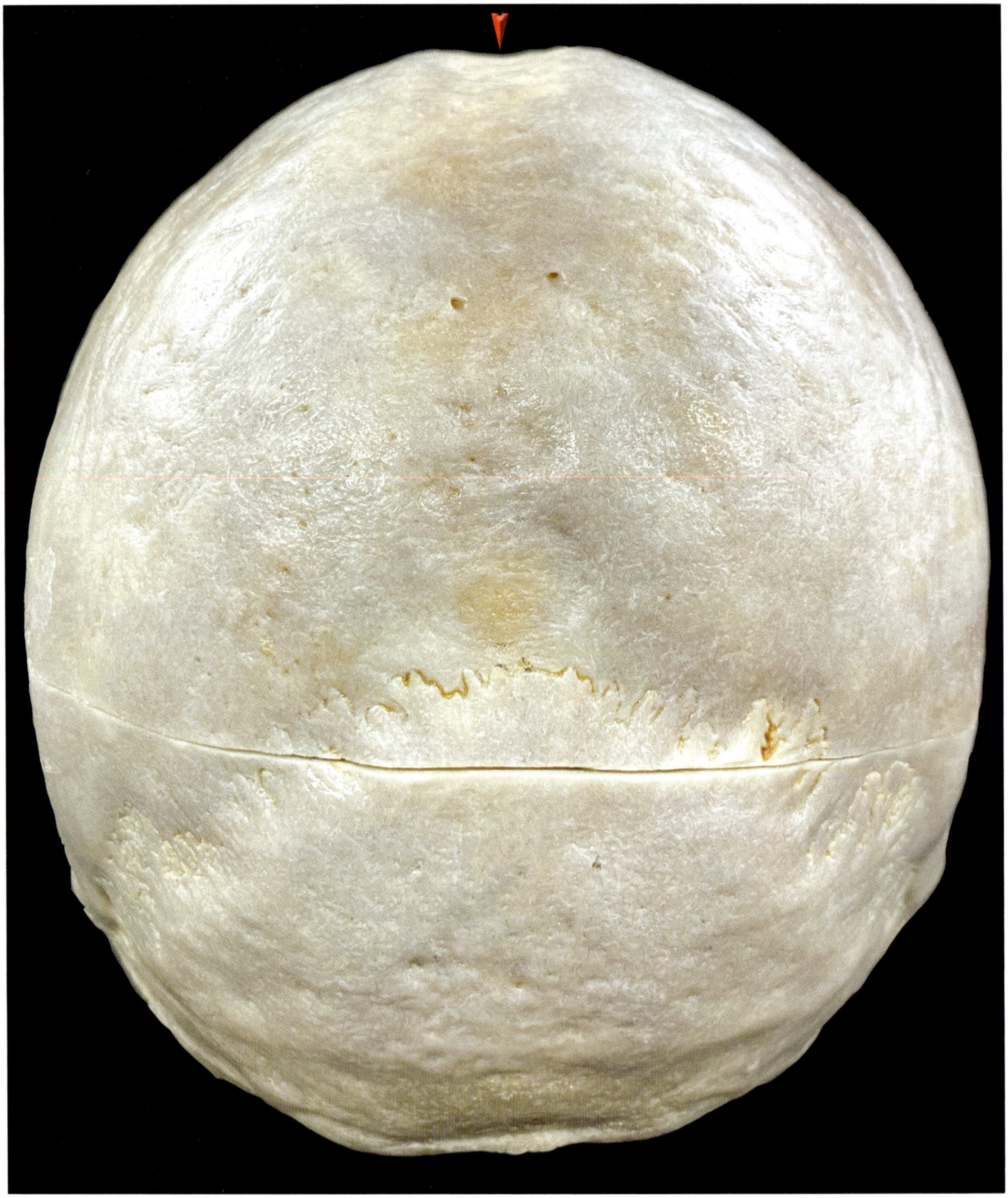

Figure 38. Posterior view 38-year-old African American male with complete obliteration of the sagittal suture and a slight depression (arrowhead) along the sagittal suture.

5. AFRICAN AMERICAN HISPANIC (PUERTO RICAN) MALE

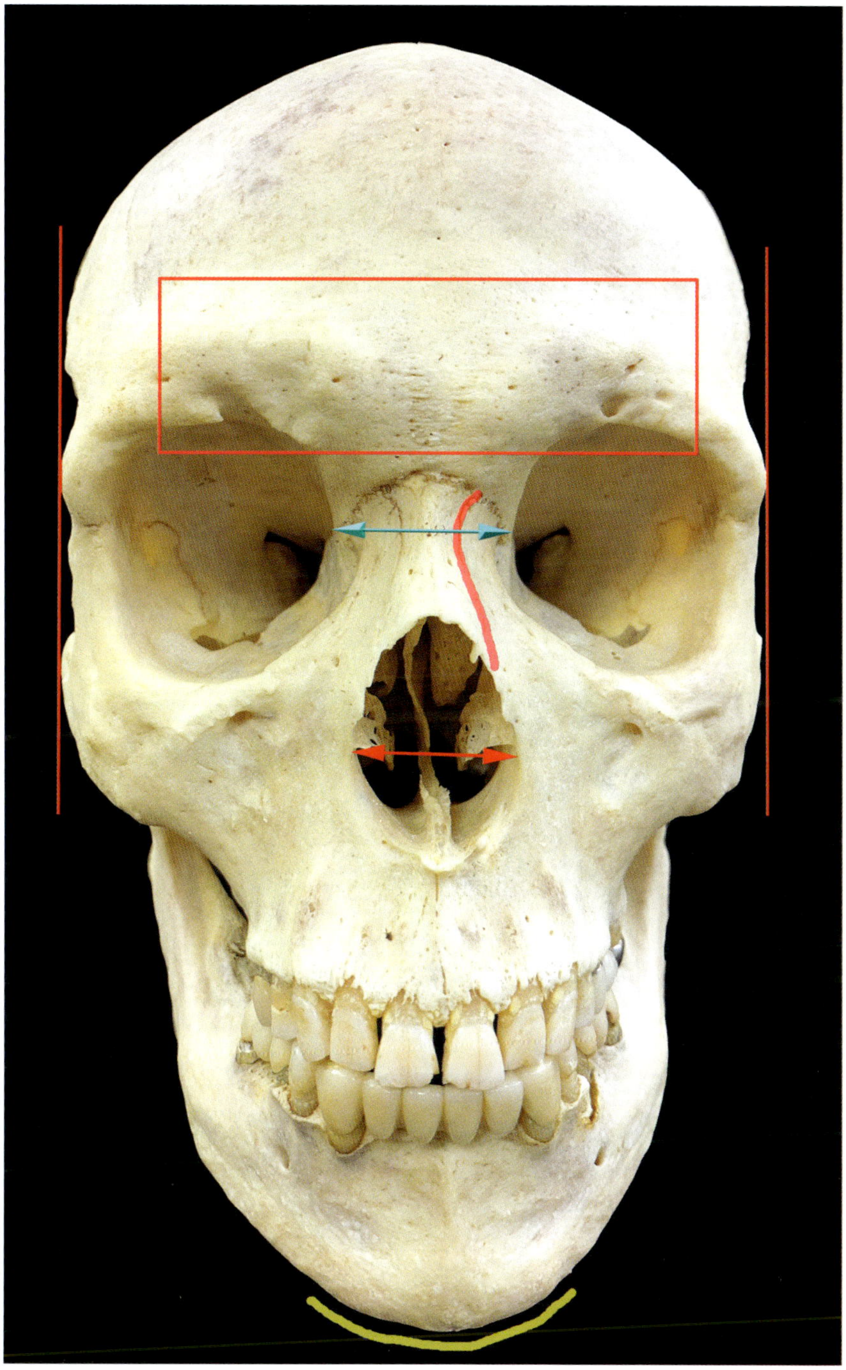

Figure 39. Anterior view of 96-year-old African American Hispanic (Puerto Rico) male with a large brow ridge (rectangle), narrow nasal aperture (red double arrow), a wide interorbital breadth (blue double arrow), contoured/waisted nasal bones (curved red line), and a rounded chin (curved yellow line). The vertical red lines emphasize the narrow facial region in contrast to the vault.

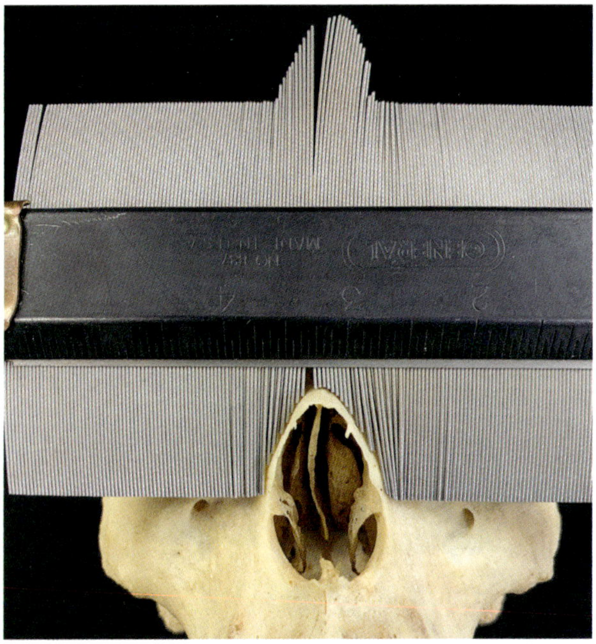

Figure 40. Contour of the nasal bones in a 96-year-old African American Hispanic (Puerto Rico) male.

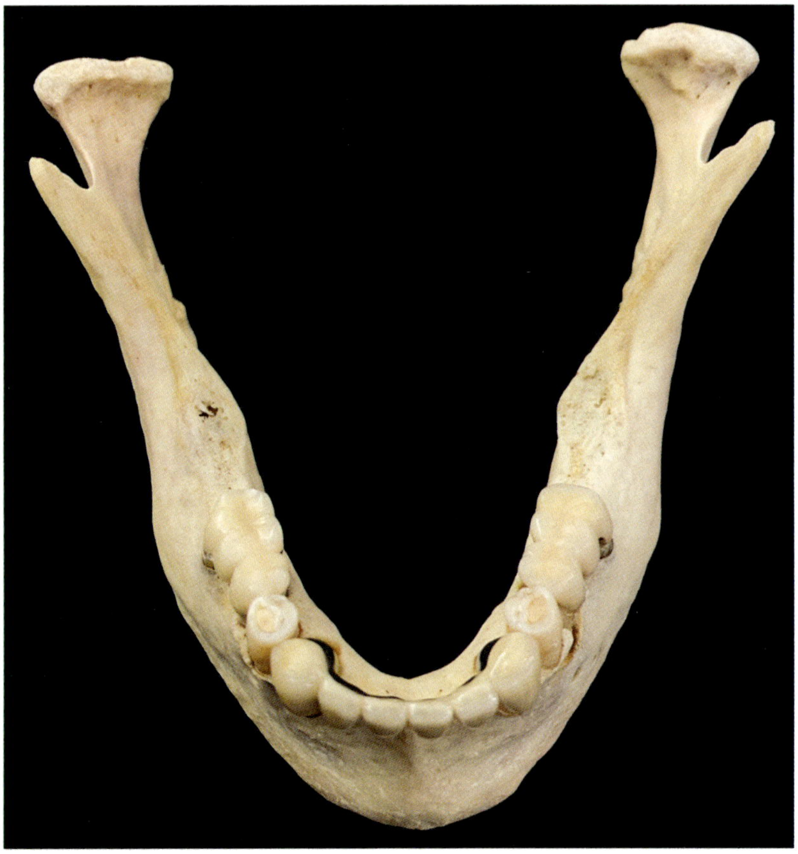

Figure 41. Mandible of 96-year-old African American Hispanic (Puerto Rico) male with a dental appliance and implants of the anterior teeth.

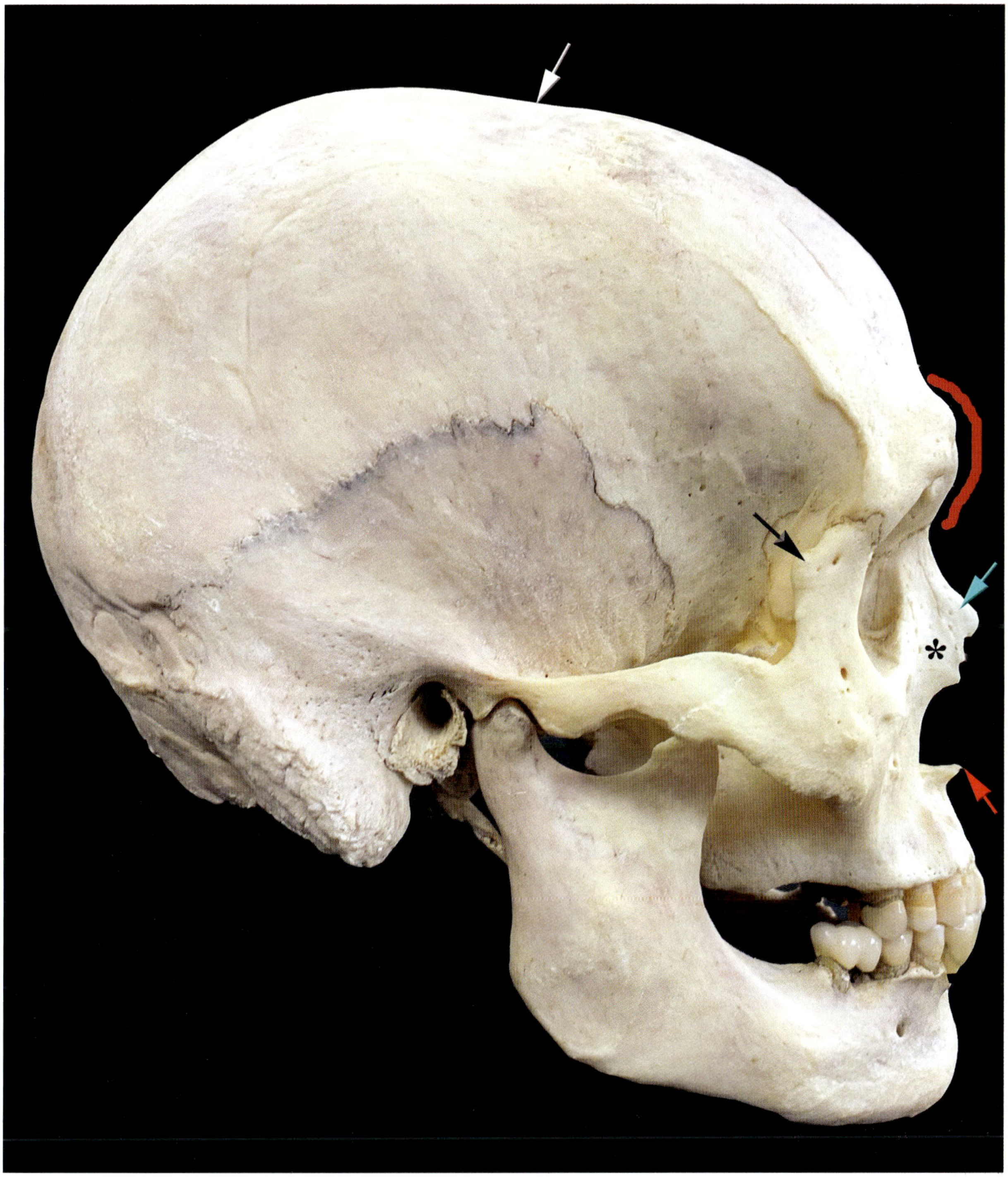

Figure 42. Right lateral view of 96-year-old African American Hispanic (Puerto Rico) male with a postbregmatic depression (white arrow). Note the prominent brow ridge (curved red line), small nasal bone (blue arrow) moderate projection of the frontal process of the maxilla (*), an anterior nasal spine (red arrow), and posterior zygomatic tubercle (black arrow). The frontal process of the maxilla refers to the forward projection/protrusion of the bone inferior to the nasal bone and that forms the lateral walls of the nasal aperture.

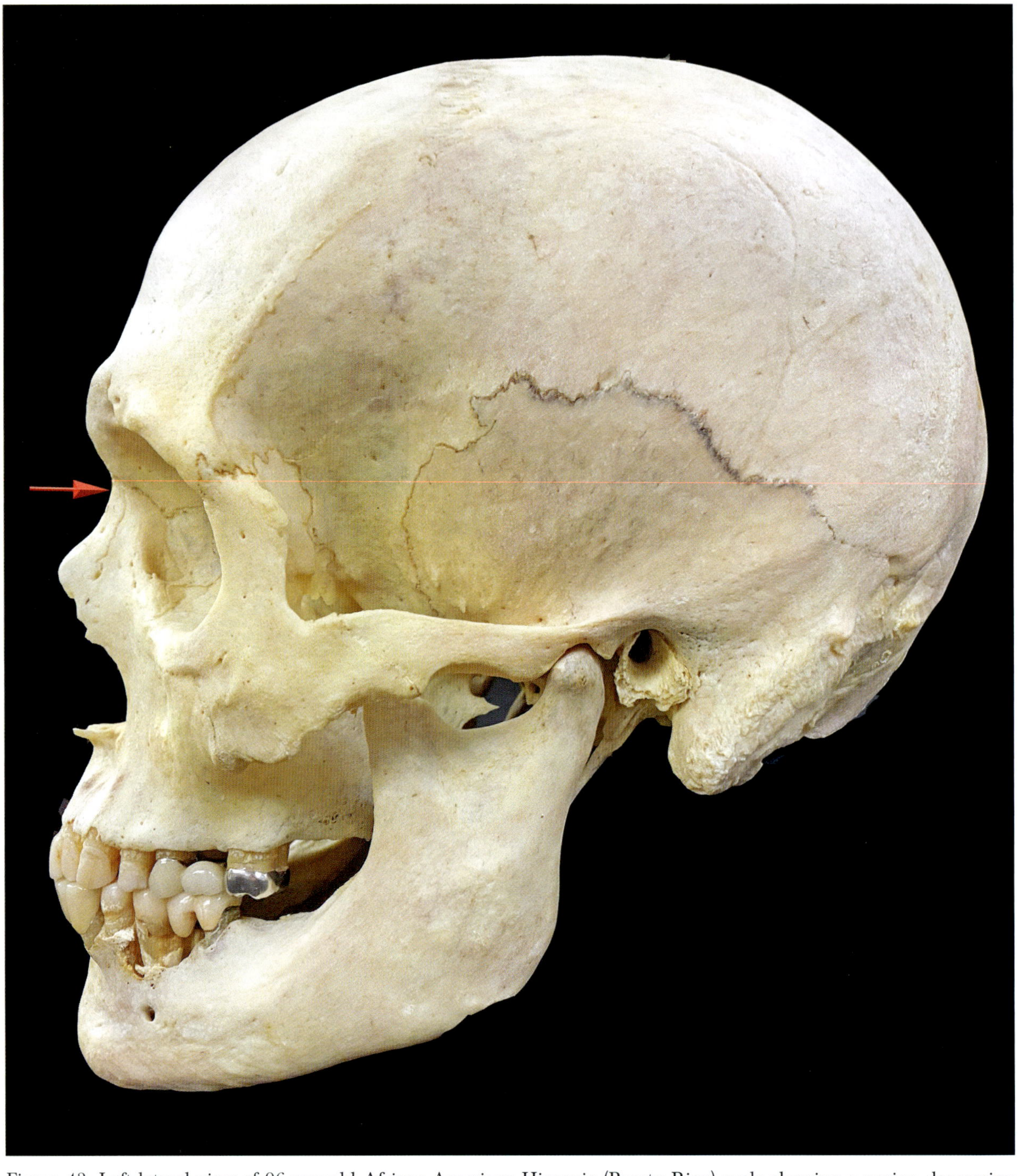

Figure 43. Left lateral view of 96-year-old African American Hispanic (Puerto Rico) male showing a nasion depression (arrow).

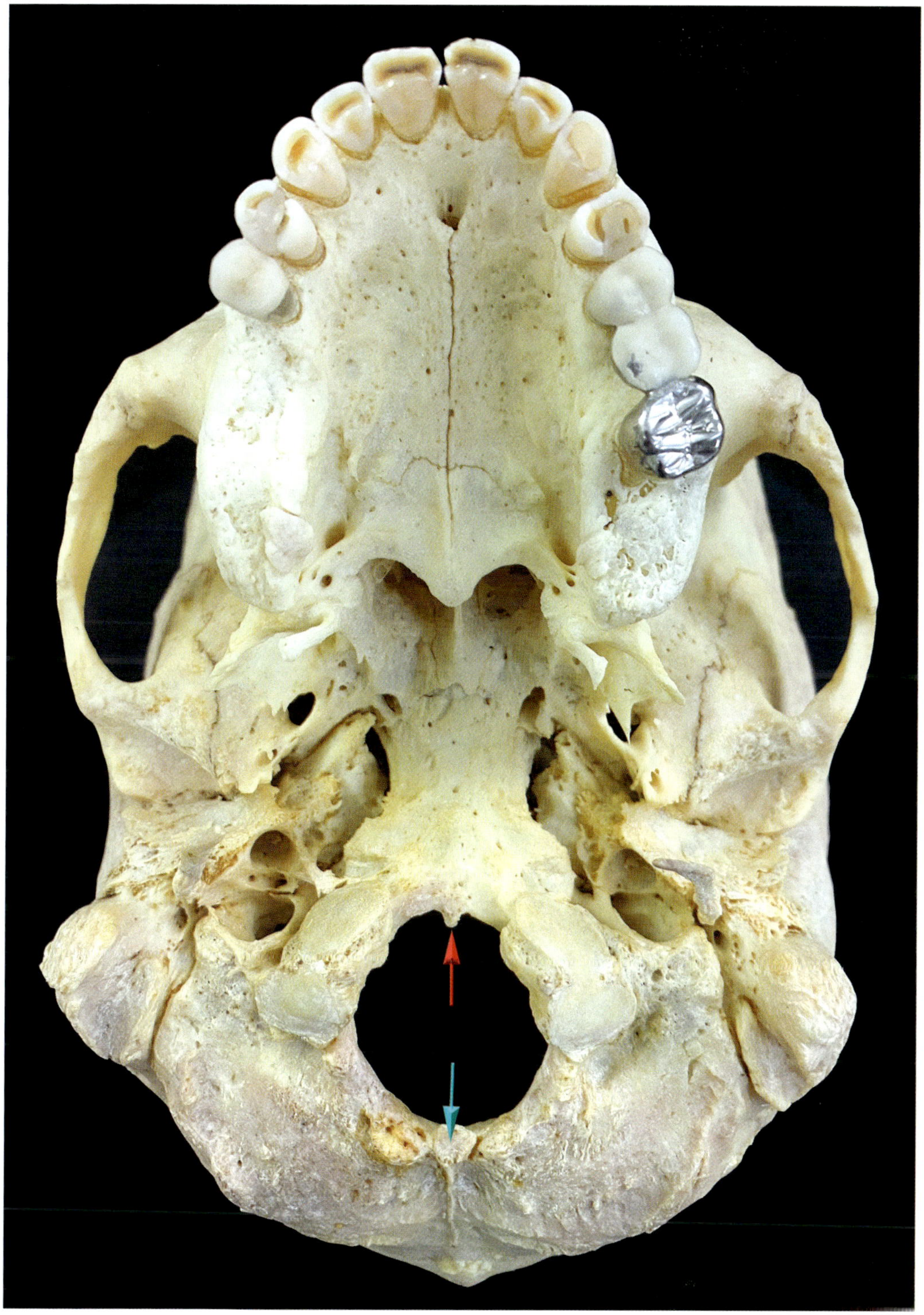

Figure 44. Inferior view of 96-year-old African American Hispanic (Puerto Rico) male with a small osteophyte (calcified apical ligament) at basion (red arrow) and a Kerckring's ossicle (blue arrow) along the posterior margin of the foramen magnum. Both of these bony growths may obscure osteometric landmarks. The anterior teeth exhibit advanced dental attrition/wear possibly due to bruxism (grinding of the teeth), dietary factors, anatomy, and a natural sequela of aging among other causes.

Figure 45. Superior view of 96-year-old African American Hispanic (Puerto Rico) male with an oblong-shaped vault and complete obliteration of the sagittal suture.

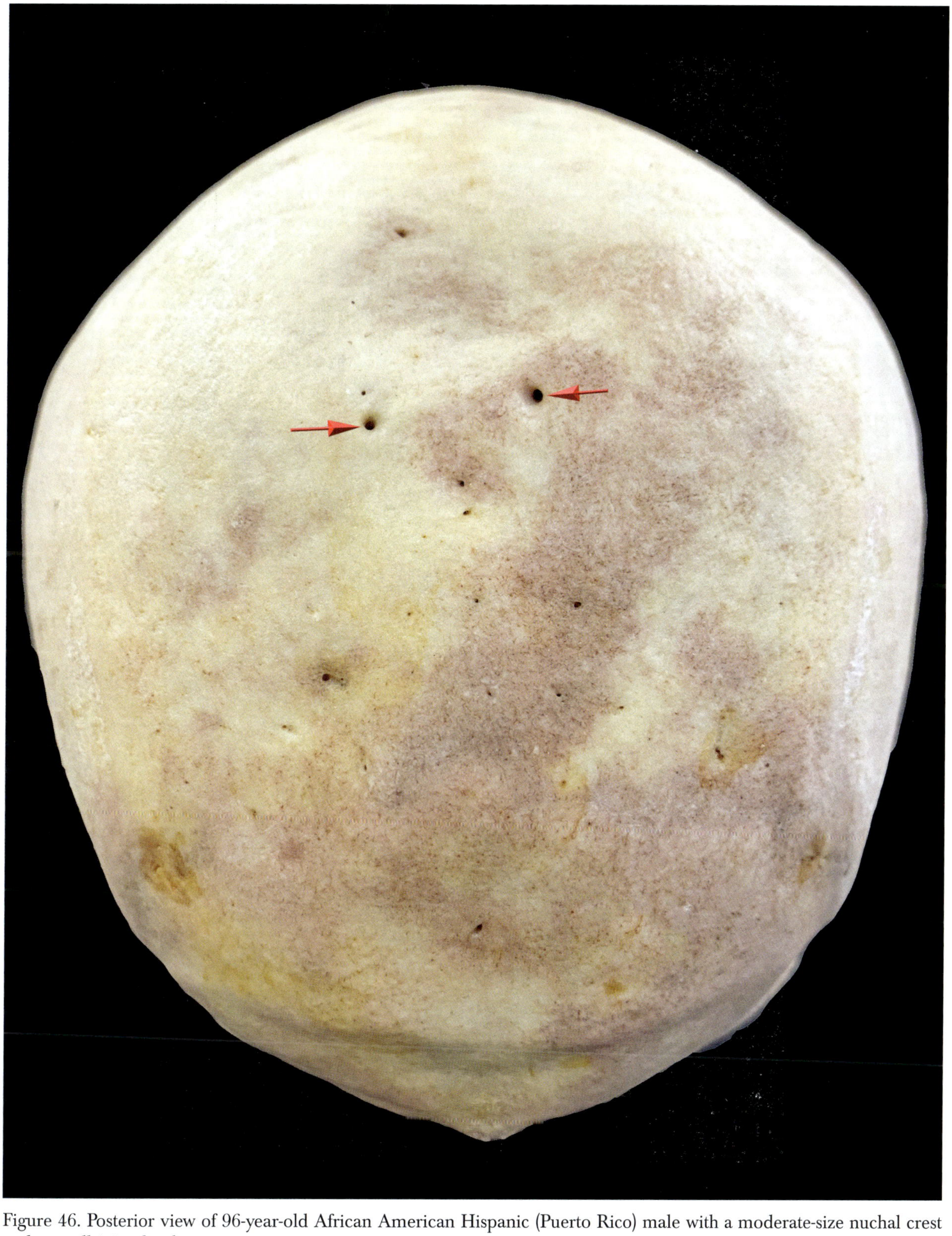

Figure 46. Posterior view of 96-year-old African American Hispanic (Puerto Rico) male with a moderate-size nuchal crest and a small inion hook.

6. AFRICAN AMERICAN FEMALE

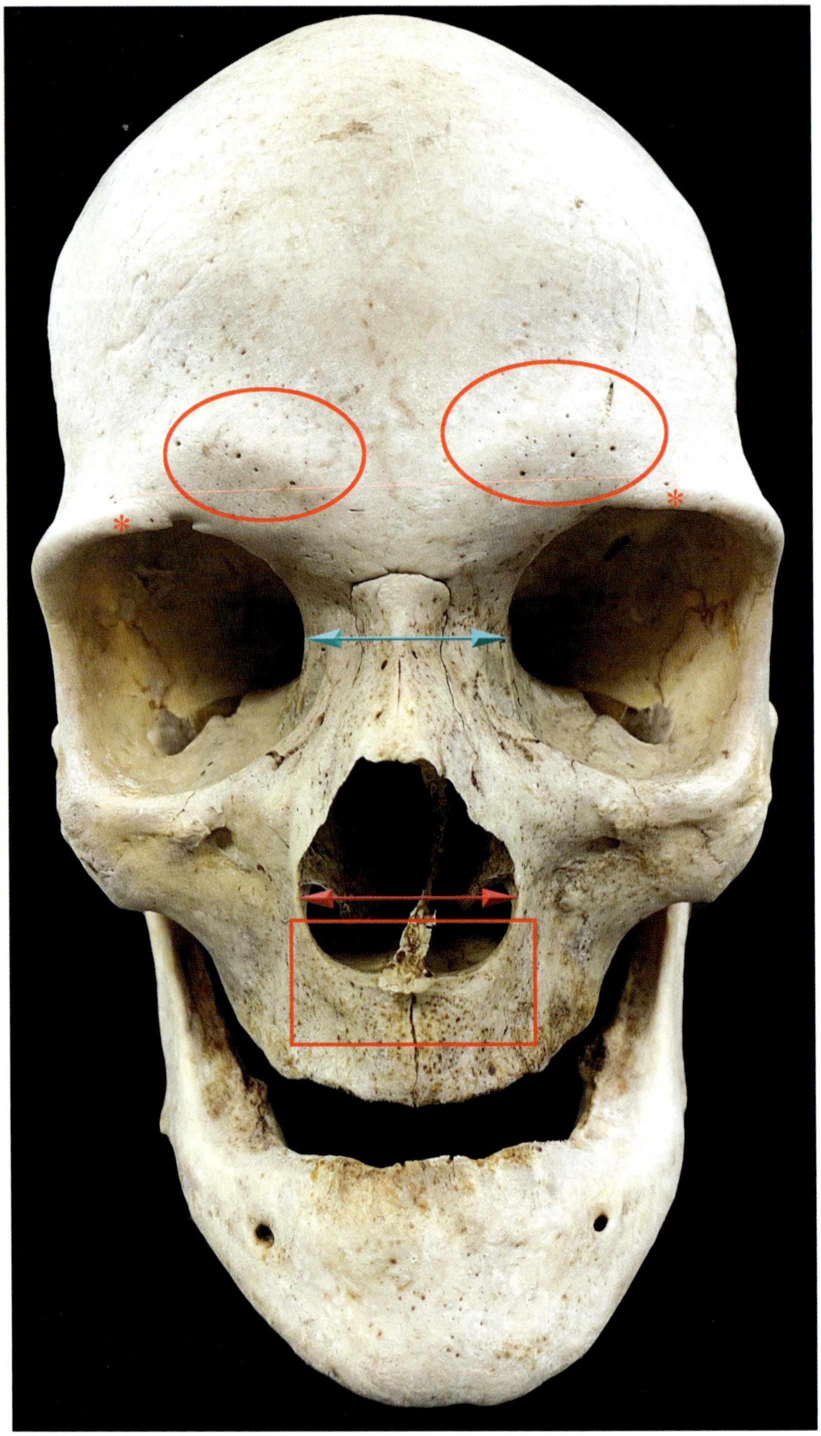

Figure 47. Anterior view of an edentulous 84-year-old African American female with a small and separate brow ridge (oval), blunt orbital rims (*), wide interorbital breadth (double blue arrow) and nasal aperture (double red arrow), and no nasal guttering (rectangle) (see Figure 48).

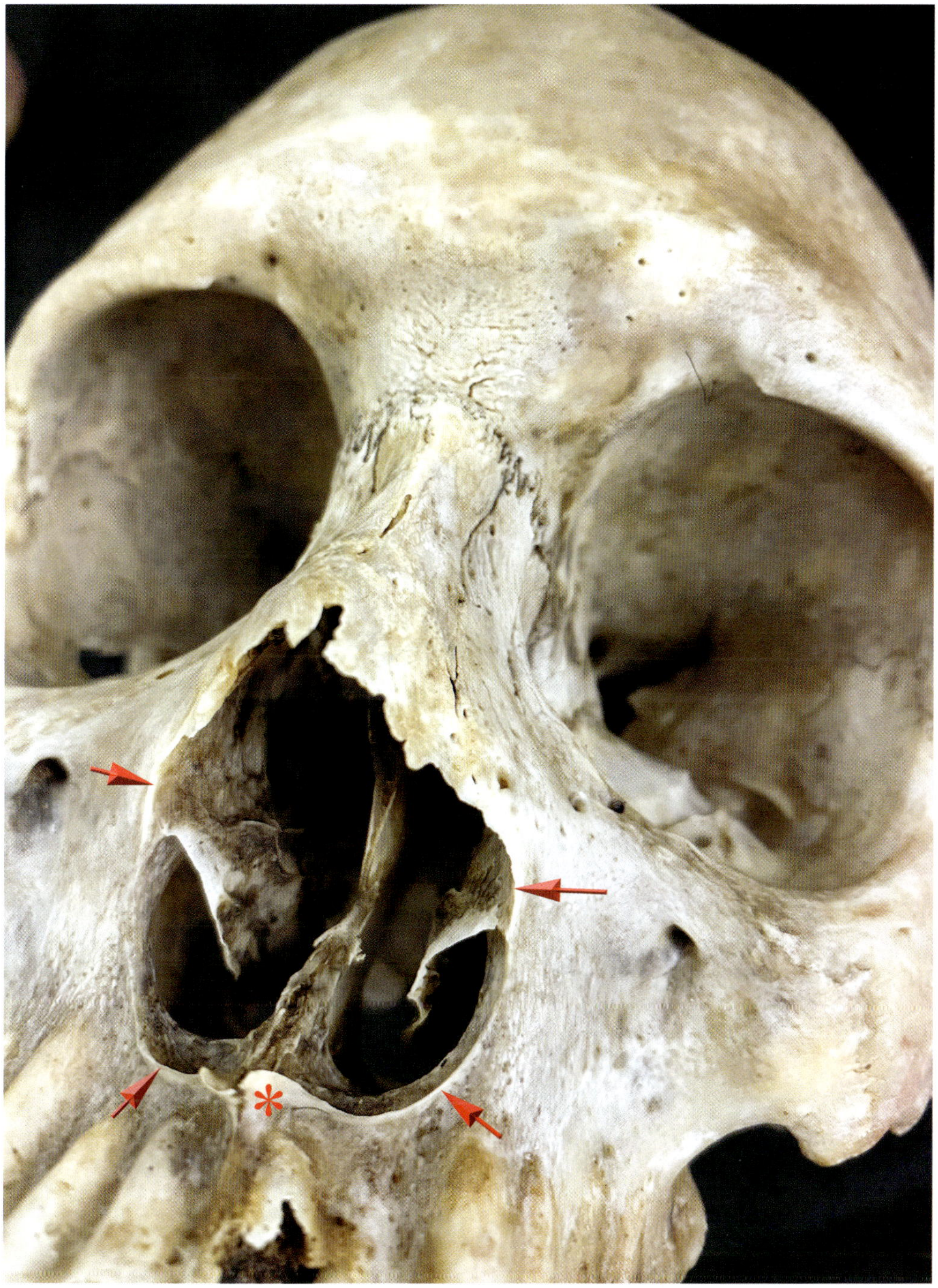

Figure 48. Anterior-lateral view of 84-year-old African American female showing a sharp outline of the nasal aperture (arrow) and a prominent anterior nasal spine (*). Immersion in acetone or a mild bleach and water solution to degrease bones often causes a color differential in bone such as seen in this photograph. In this individual, the bony rim and anterior nasal spine, because they are so thin, are much lighter in color than the surrounding bone. Chemical immersion may also cause accessory bones and osteophytes to become whiter/lighter in color than surrounding bone, possibly because of their developmental composition. To some researchers, the presence of a small or large anterior nasal spine (ANS) is not as diagnostic of ancestry as the presence/absence of nasal guttering, especially in individuals with medium or large guttering.

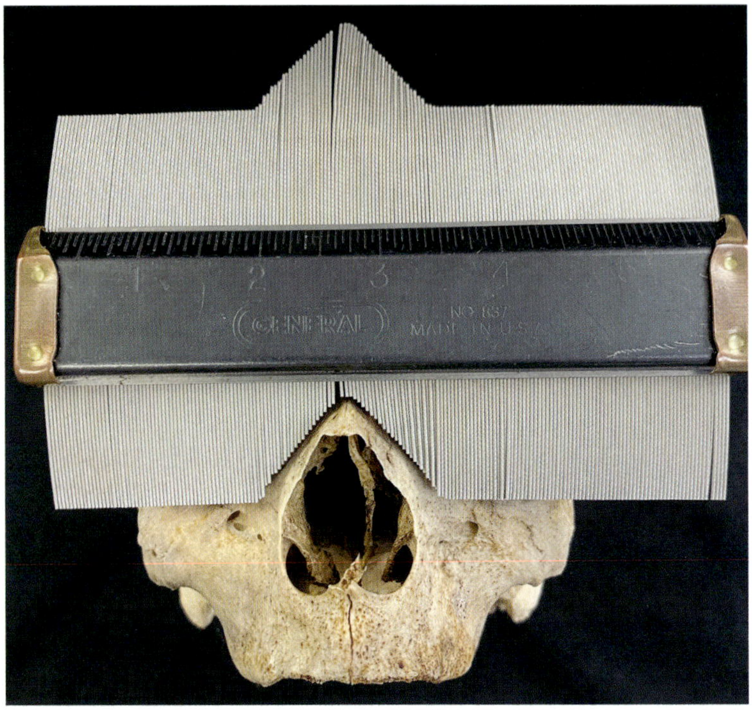

Figure 49. Nasal bone contour in an 84-year-old African American female.

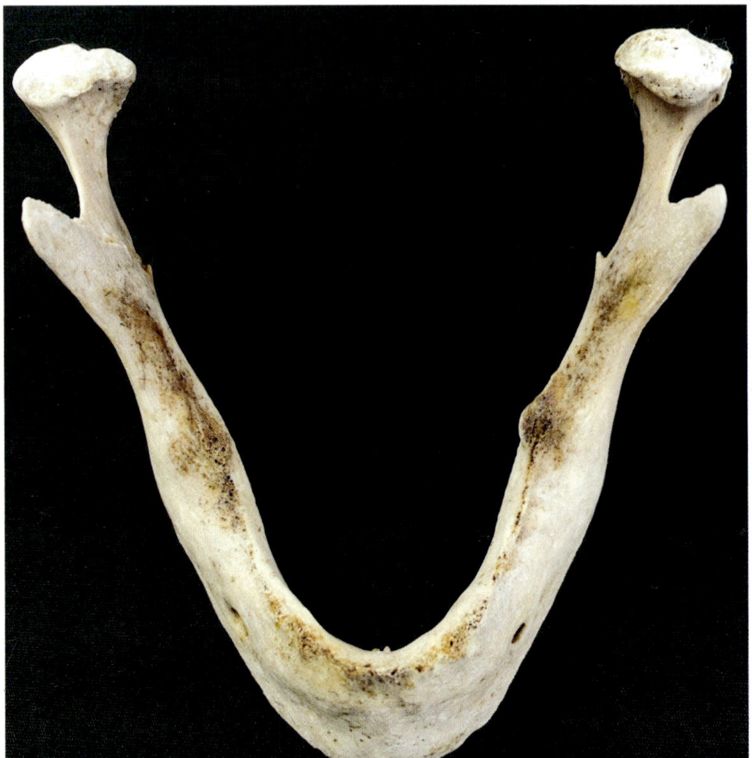

Figure 50. Edentulous mandible of 84-year-old African American female with a rounded chin.

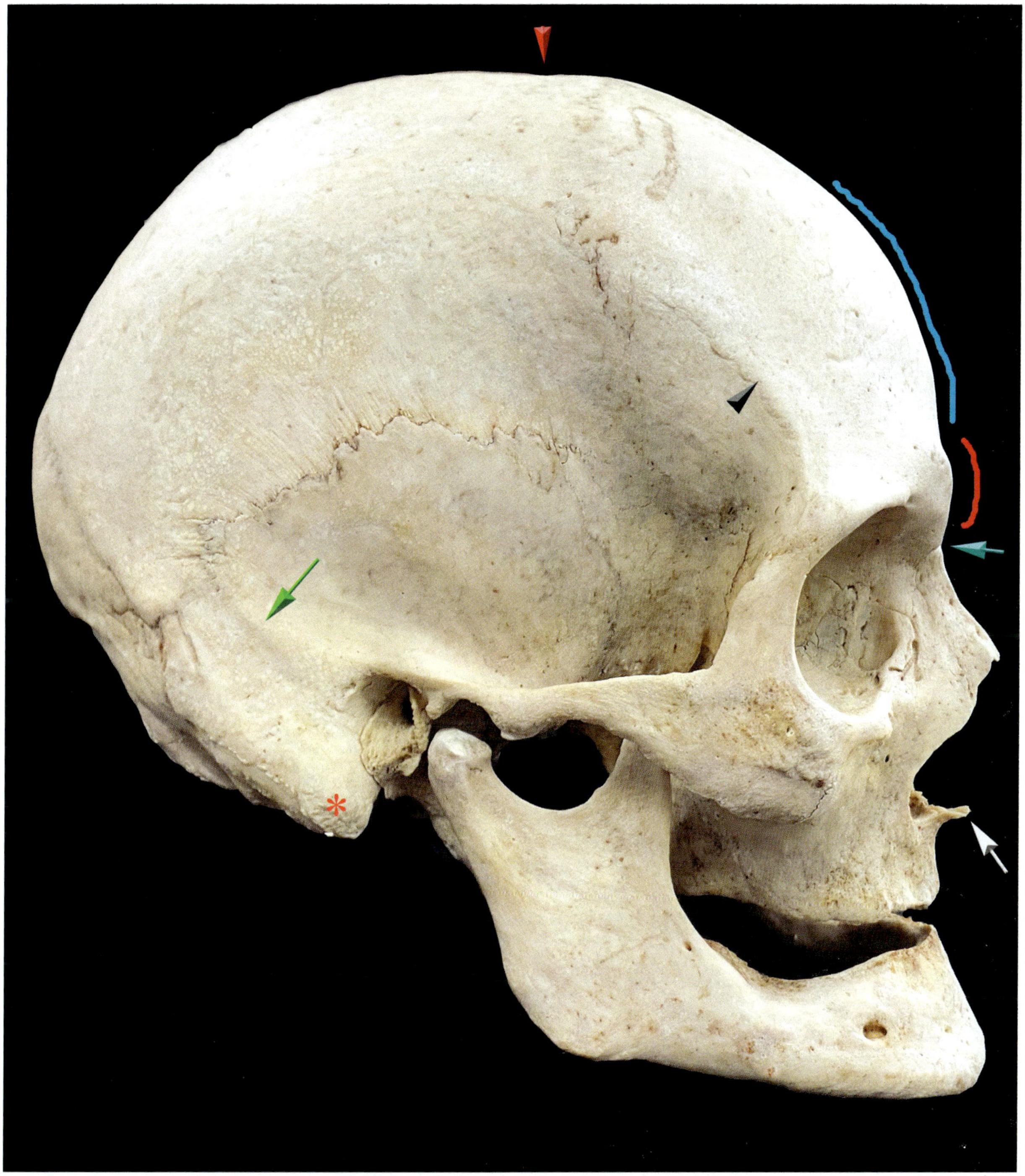

Figure 51. Right lateral view of 84-year-old African American female with a postbregmatic depression (red arrowhead), a small brow ridge (curved red line), curved frontal bone (curved blue line), absent depression at nasion (blue arrow), a prominent anterior nasal spine (white arrow), mild development of the temporal crest (black arrowhead), a supramastoid crest (green arrow), and a small mastoid process (*).

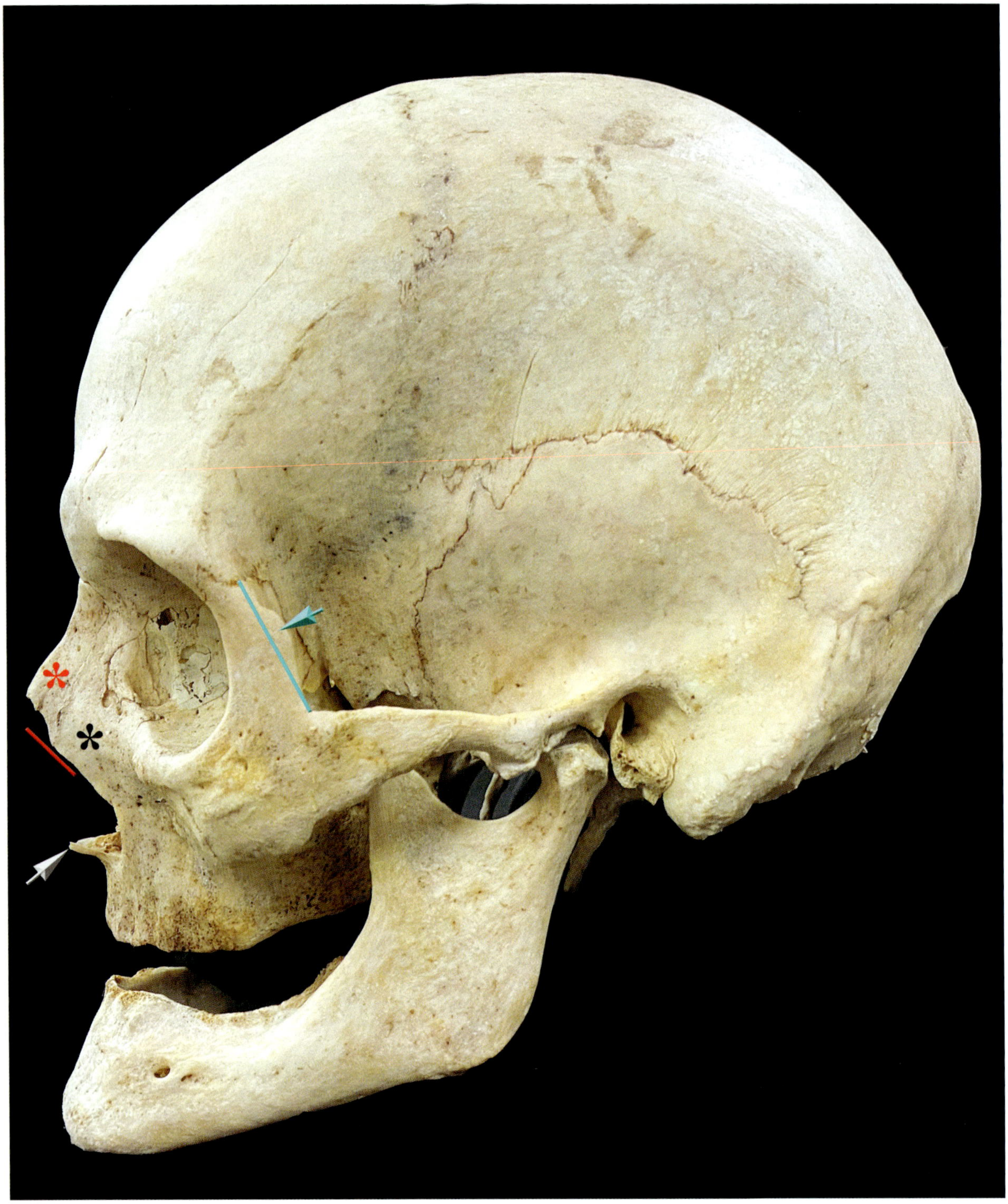

Figure 52. Left lateral view of 84-year-old edentulous African American female with small nasal bones (red*), absence of overhanging nasal bones (red line), moderate-size frontal process of the maxilla (black*), prominent anterior nasal spine (white arrow), posterior zygomatic tubercle (blue line and arrow), and marked resorption of the maxilla and mandible.

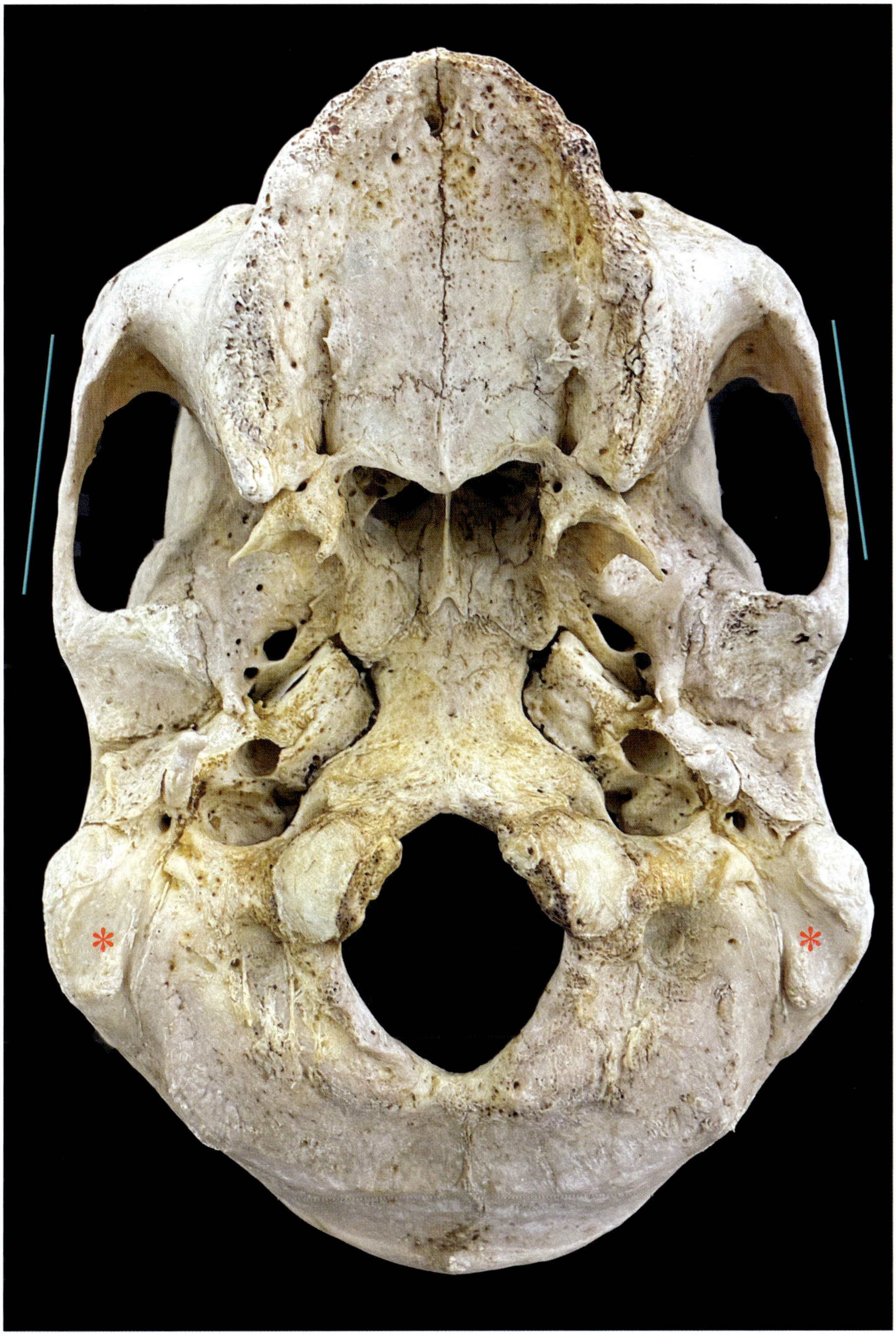

Figure 53. Inferior view of an edentulous 84-year-old African American female with flat zygomatic arches (blue line) and a wide and shallow hard palate. The shape of the zygomatic arches, whether flat or convex, is not a feature commonly used in assessing ancestry or sex, but is deserving of further research.

Figure 54. Superior view of 84-year-old African American female with partial obliteration of the sutures.

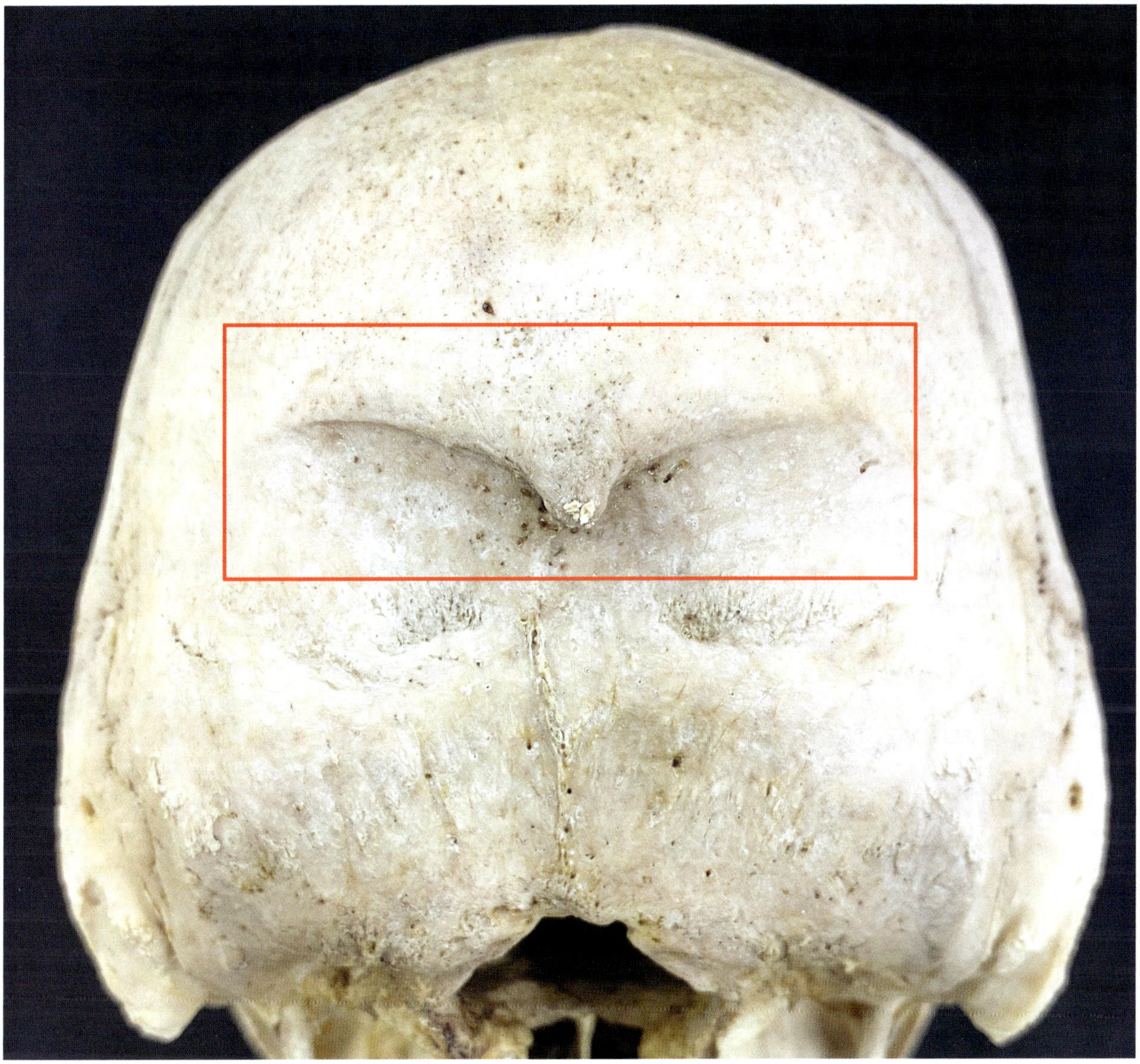

Figure 55. Posterior view of 84-year-old African American female with an unusually developed nuchal crest and inion hook (rectangle).

7. AFRICAN AMERICAN (ETHIOPIAN) SPANISH DOMINICAN FEMALE

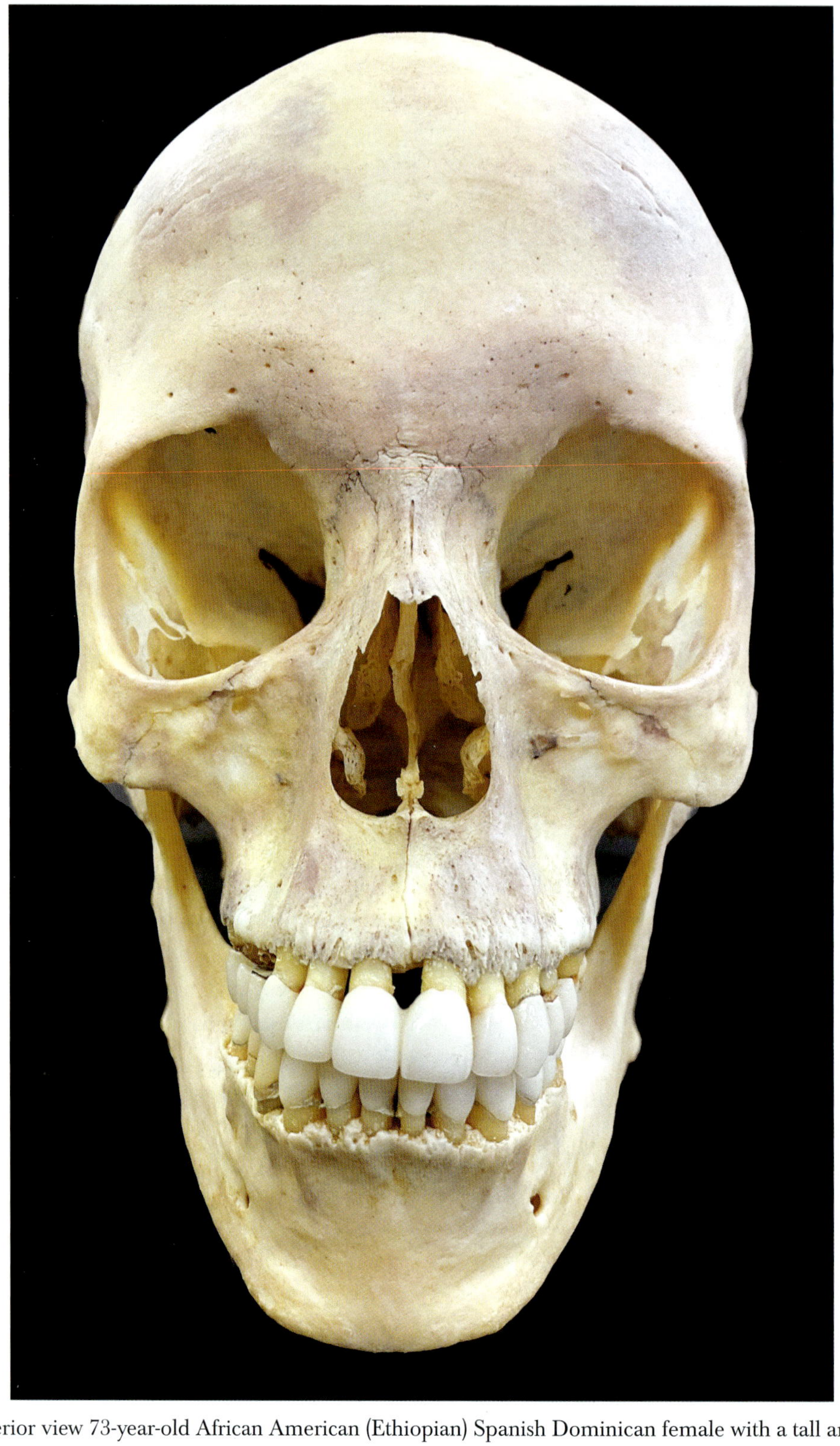

Figure 56. Anterior view 73-year-old African American (Ethiopian) Spanish Dominican female with a tall and narrow skull.

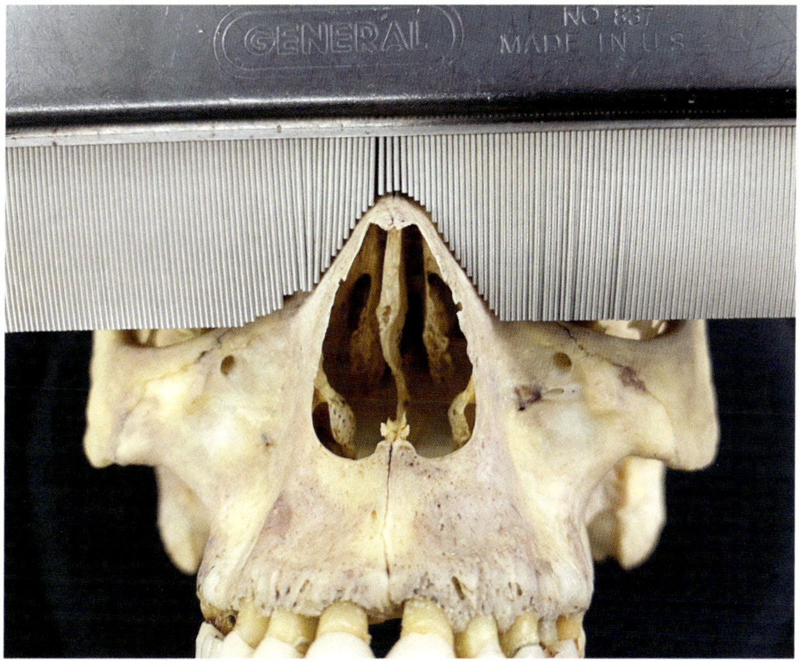

Figure 57. Nasal bone contour in a 73-year-old African American (Ethiopian) Spanish Dominican female.

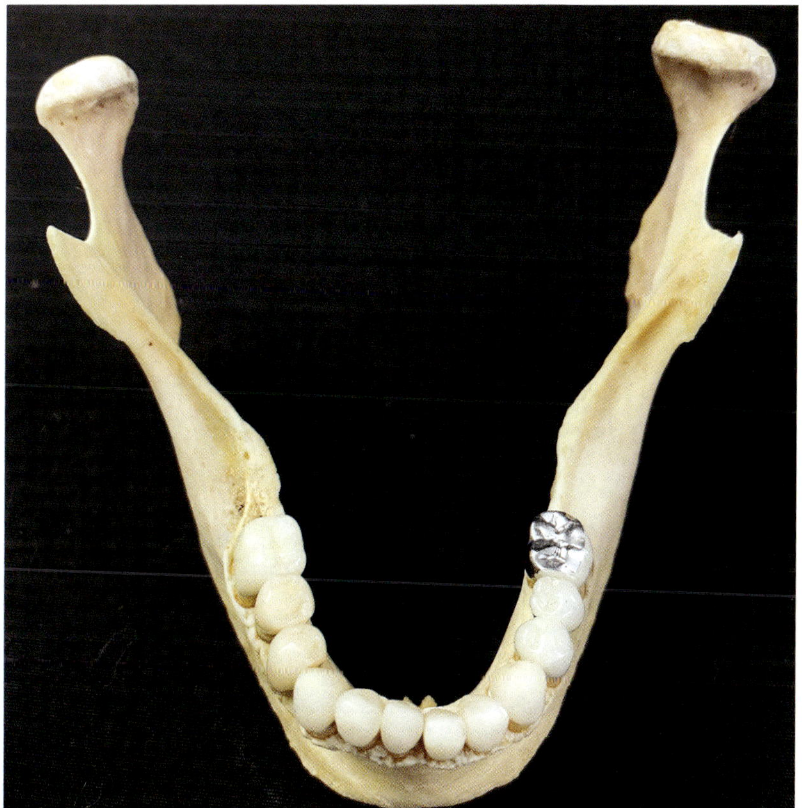

Figure 58. Mandible of 73-year-old African American (Ethiopian) Spanish Dominican female.

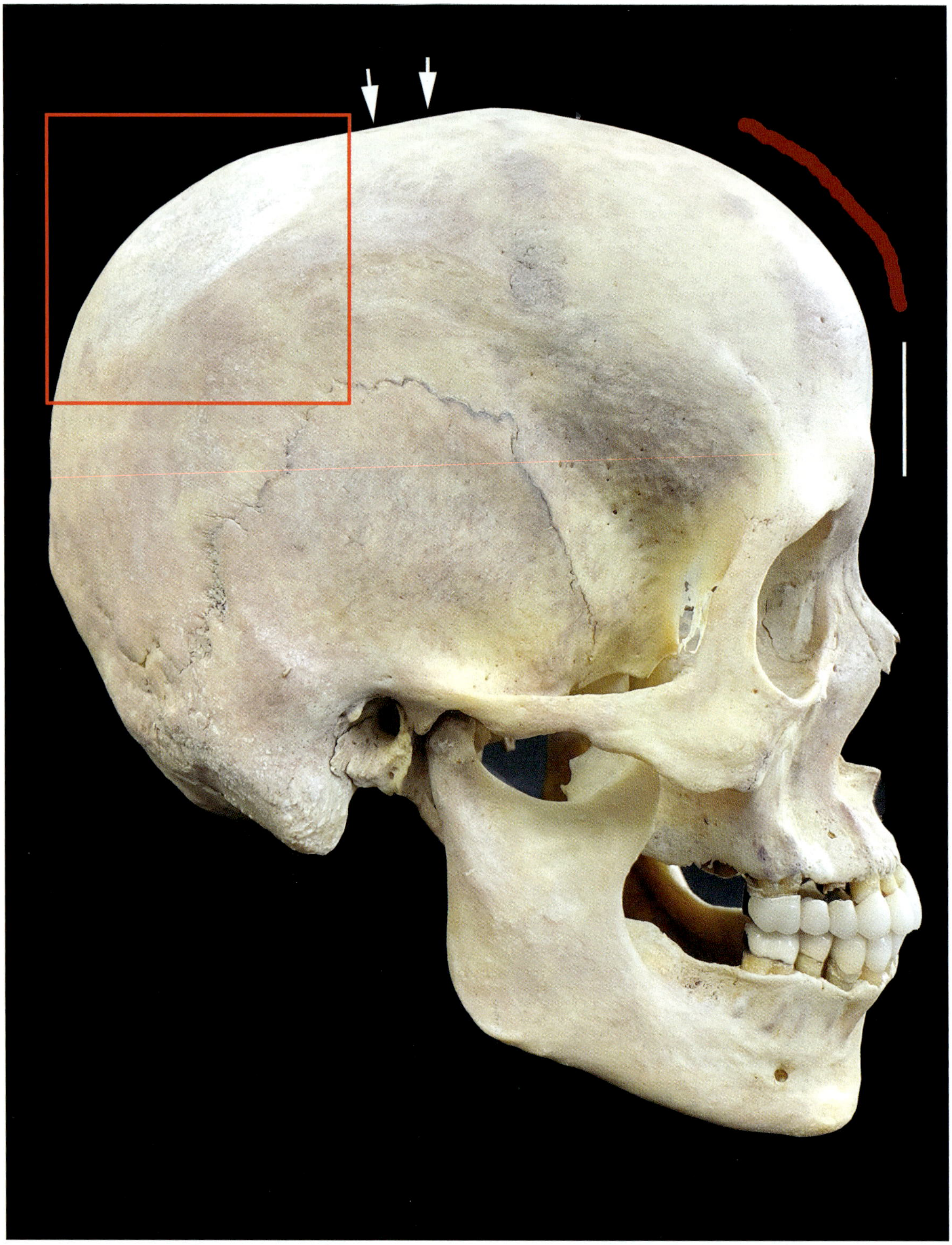

Figure 59. Right lateral view of 73-year-old African American (Ethiopian) Spanish Dominican female with a postbregmatic depression (white arrow), sloping frontal bone (curved red line), vertical brow ridge and inferior portion of the frontal bone (white line), and biparietal thinning (square). Parietal thinning results from resorption of the outer table and diploe in elderly individuals between the sagittal suture and temporal line in one (parietal) or both (biparietal thinning) parietal bones resulting in flat or oval-shaped depressions.

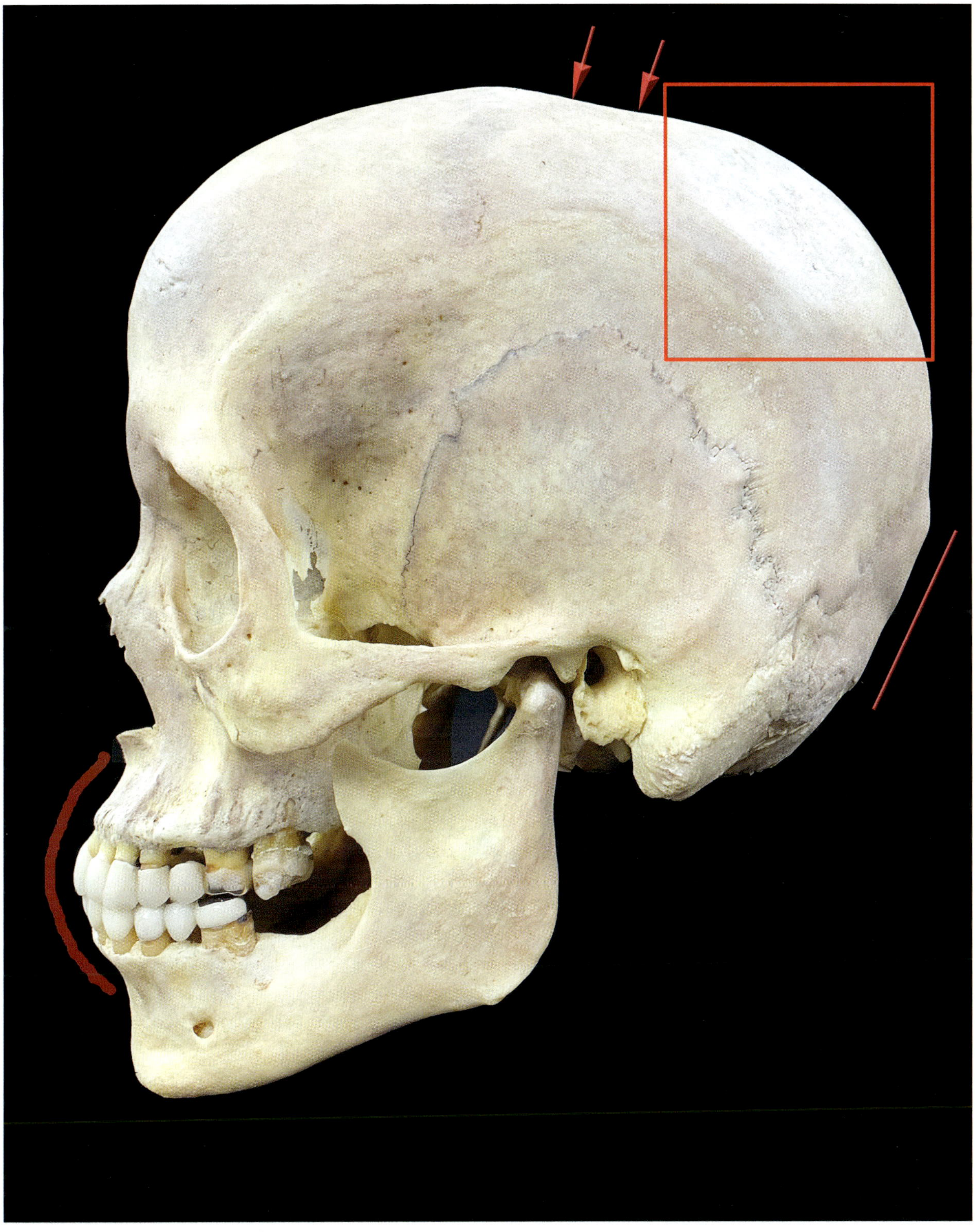

Figure 60. Left lateral view 73-year-old African American (Ethiopian) Spanish Dominican female with a postbregmatic depression (arrow), biparietal thinning (square), flat occipital region (straight red line), and alveolar prognathism (curved red line).

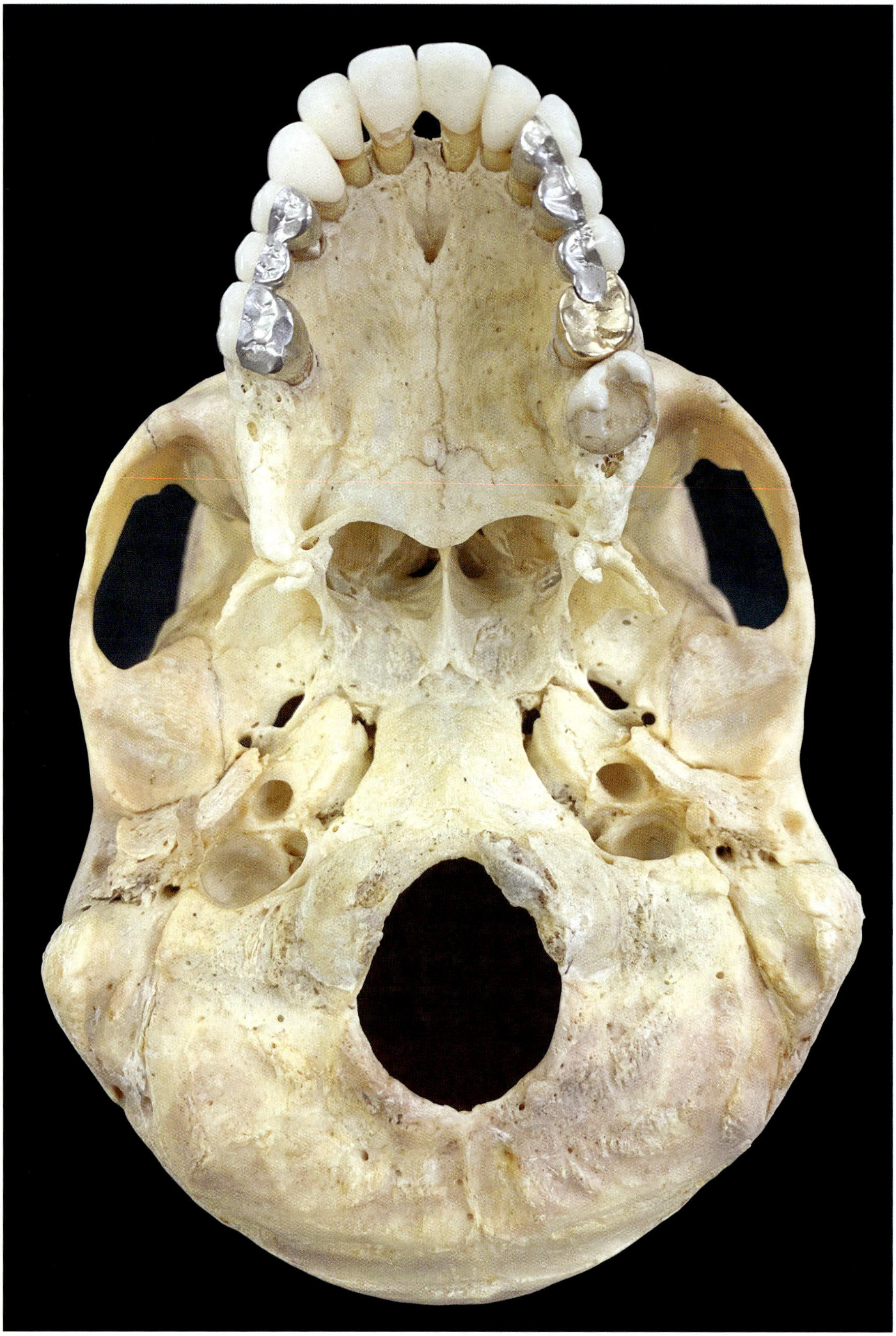

Figure 61. Inferior view 73-year-old African American (Ethiopian) Spanish Dominican female.

Figure 62. Superior view 73-year-old African American (Ethiopian) Spanish Dominican female with an oval-shape cranium and complete obliteration of the sutures.

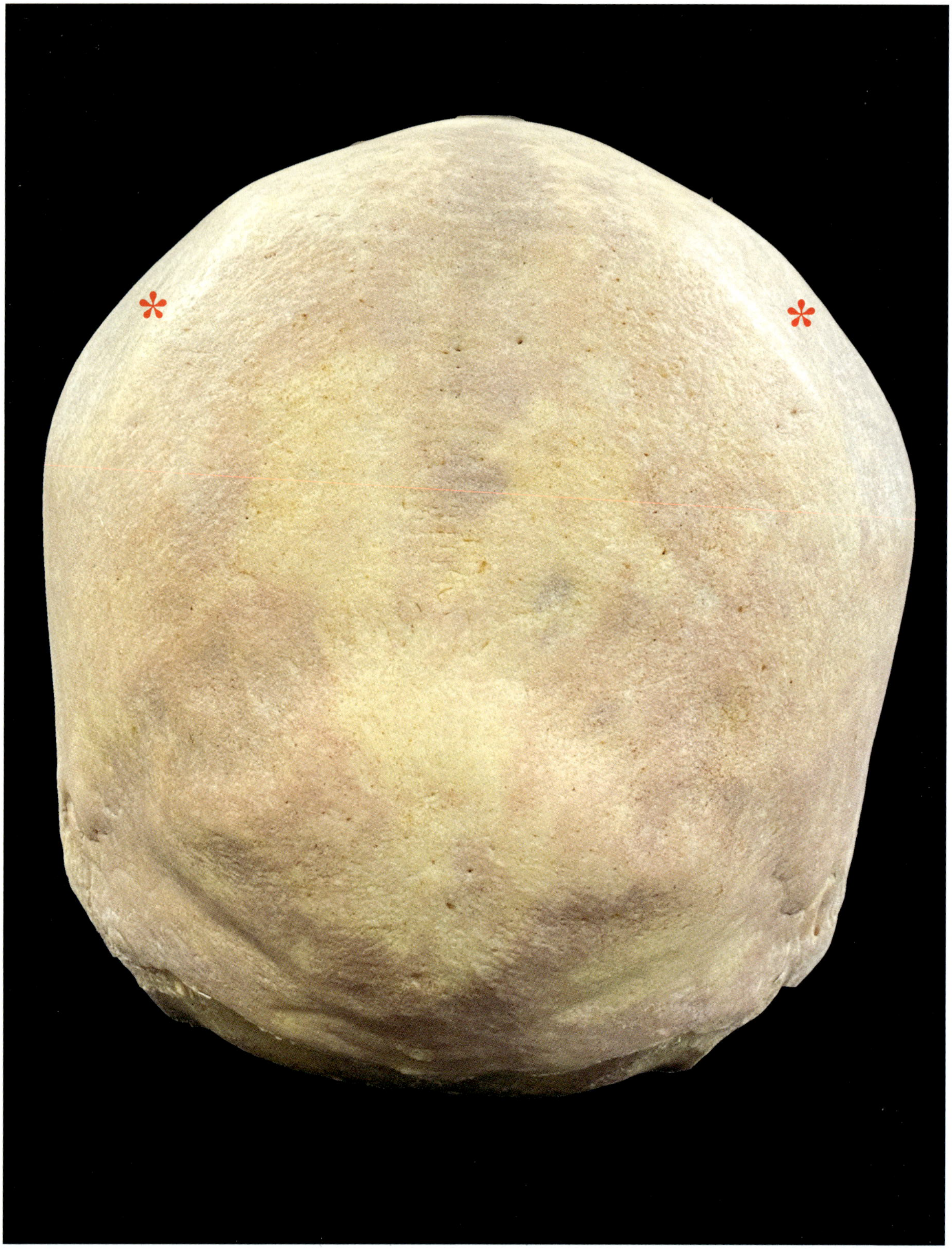

Figure 63. Posterior view 73-year-old African American (Ethiopian) Spanish Dominican female with biparietal thinning (*) and complete obliteration of the sutures.

8. AFRICAN AMERICAN NATIVE AMERICAN FEMALE

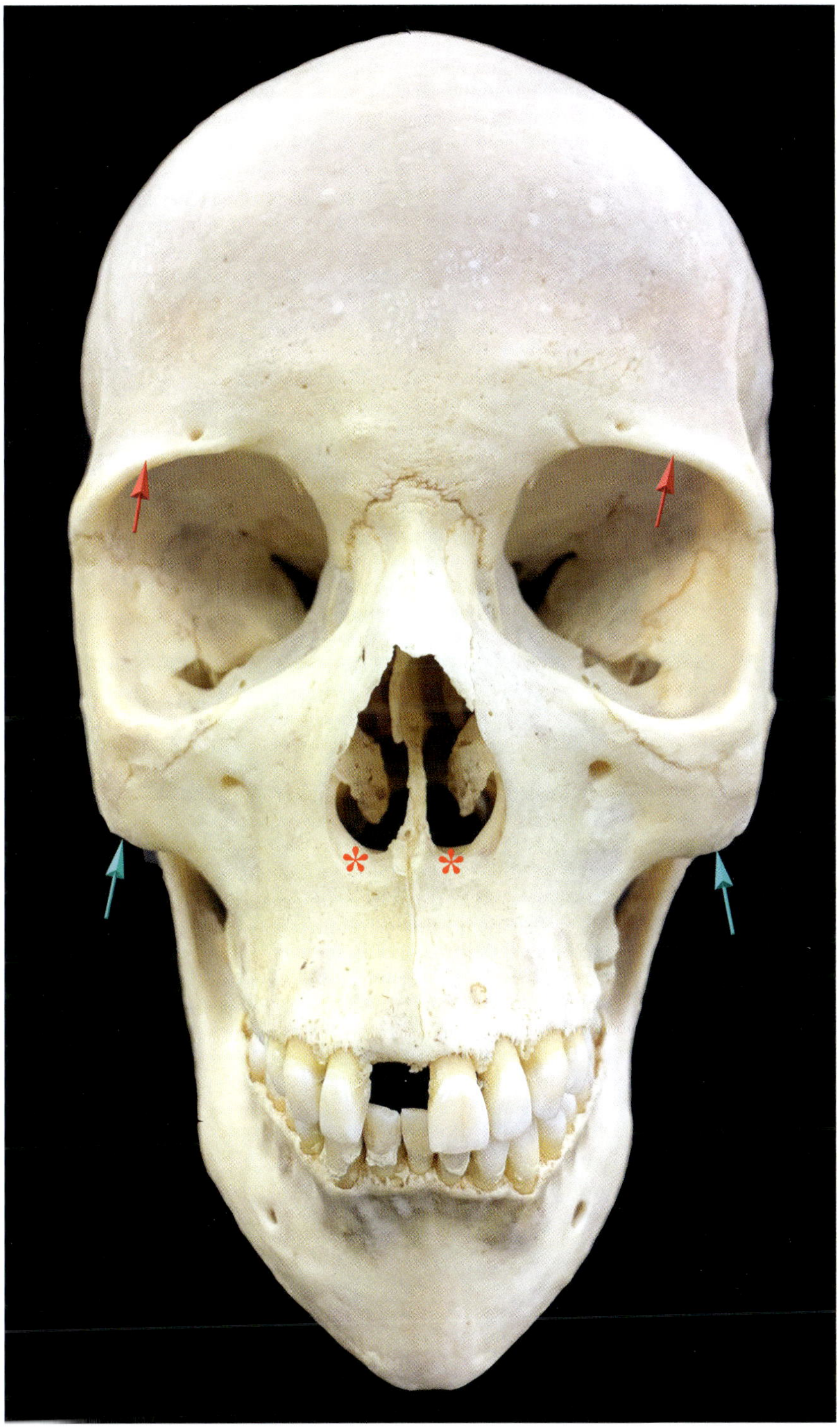

Figure 64. Anterior view of 63-year-old African American Native American female with a sharp superior orbital margin (red arrow), nasal guttering (*), a narrow and pointed chin, and absent malar tubercle (blue arrow). The absent maxillary right central incisor is antemortem. Fordisc (3.0) analysis (Jantz and Ousley 2005) using 24 variables classified this skull as a Black male with a 0.288 posterior probability.

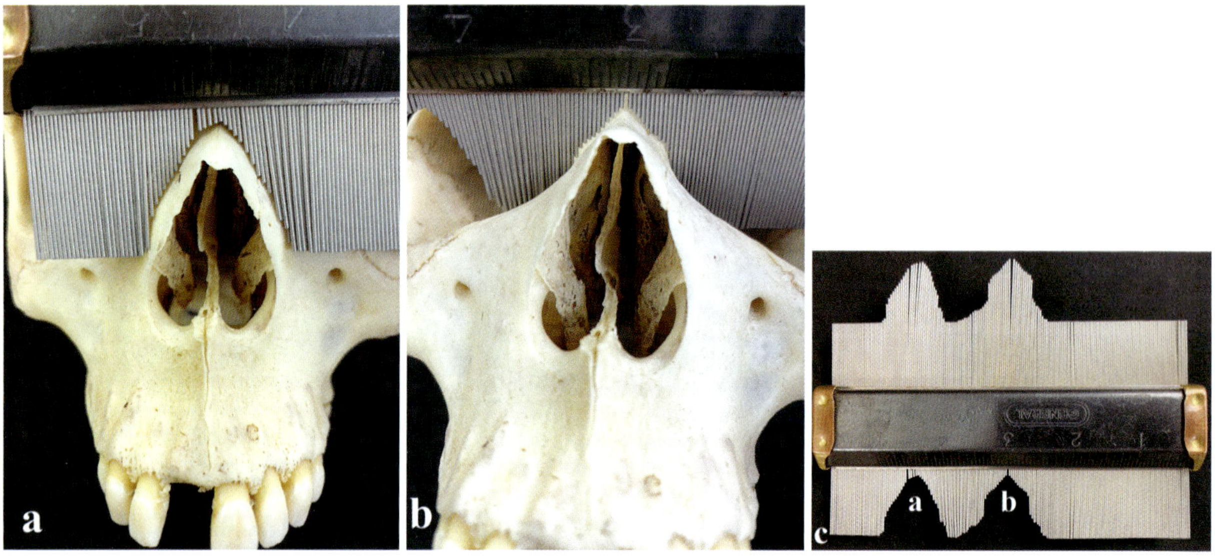

Figure 65a-c. Nasal bone contour in a 63-year-old African American Native American female. (a) Contour gauge including the frontal process of the maxilla, (b) contour gauge perpendicular to the nasal bones, and (c) a comparison of the two methods showing differences in nasal bone contour shown in (a) and (b) which is the more angular of the two.

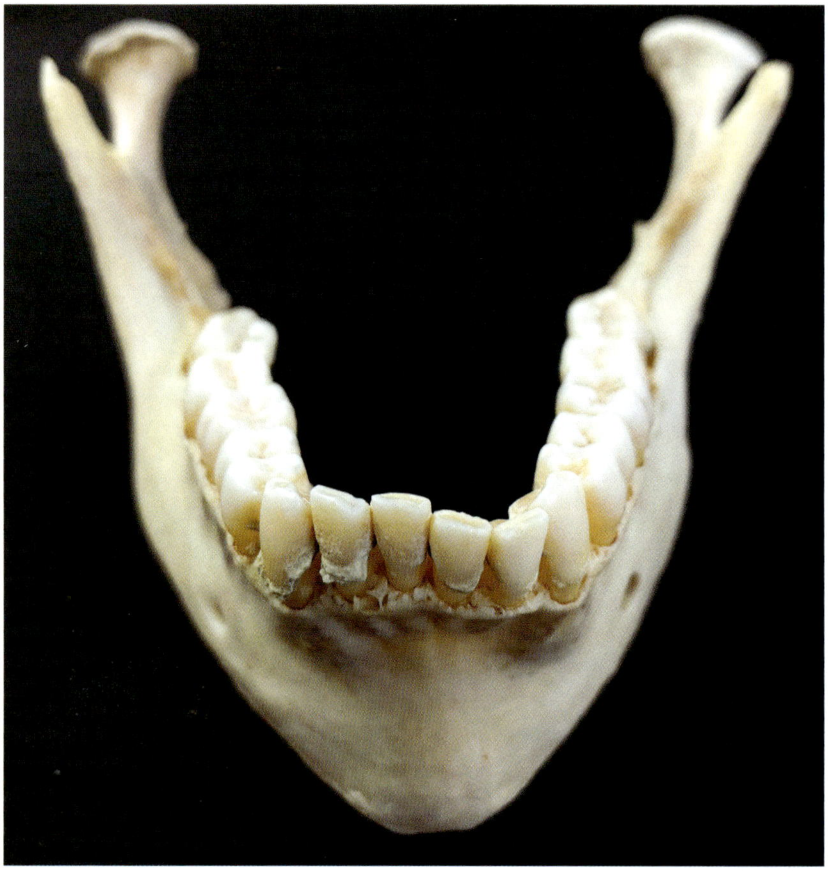

Figure 66. Mandible of 63-year-old African American Native American female showing mild crowding of the anterior teeth and a pointed chin.

TABLE 10.
Cranial Measurements in a 63-year-old African American Native American Female

Measurement	(mm)	Right	Left
Maximum Cranial Length (g-op, GOL)	178		
Maximum Cranial Breadth (eu-eu, XCB)	142		
Bizygomatic Breadth (zy-zy, ZYB)	121		
Basion-Bregma Height (ba-b, BBH)	136		
Cranial Base Length (ba-n, BNL)	100		
Basion-Prosthion (ba-pr, BPL)	103		
Maxillo-Alveolar Breadth (ecm-ecm, MAB)	64		
Maxillo-Alveolar Length (pr-alv, MAL)	60		
Biauricular Breadth (au-au, AUB)	115		
Upper Facial Height (n-pr)	77		
Minimum Frontal Breadth (ft-ft, WFB)	96		
Upper Facial Breadth (fmt-fmt)	108		
Nasal Height (n-ns, NLH)	53		
Nasal Breadth (al-al, NLB)	26		
Orbital Breadth (d-ec, OBB)	42		L
Orbital Height (OBH)	34		L
Biorbital Breadth (ec-ec, EKB)	99		
Interorbital Breadth (d-d, DKB)	23		
Frontal Chord (n-b, FRC)	114		
Parietal Chord (b-l, PAC)	115		
Occipital Chord (l-o, OCC)	92		
Foramen Magnum Length (ba-o, FOL)	34		
Foramen Magnum Breadth (FOB)	26		
Mastoid Length (MDH)			28

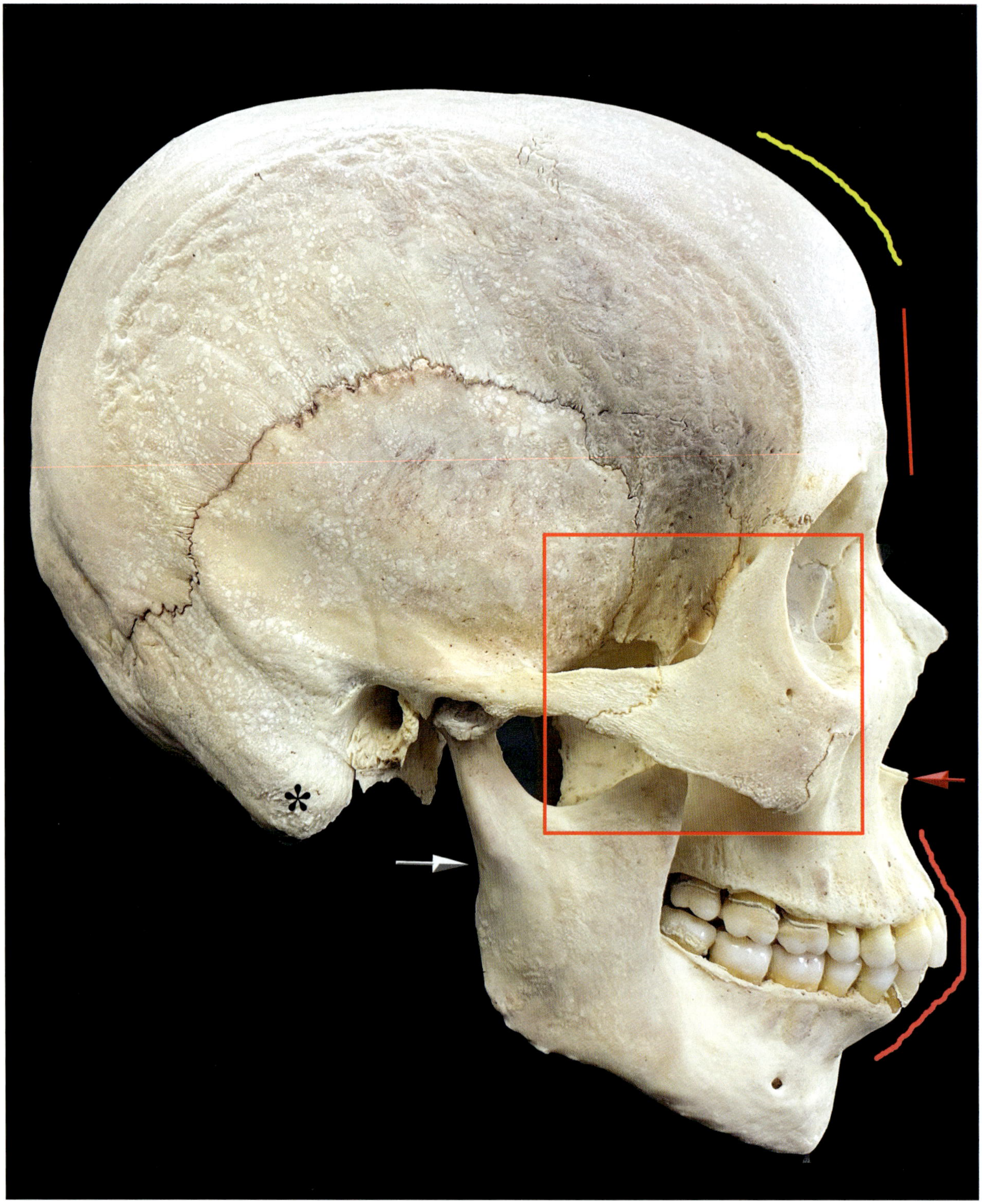

Figure 67. Right lateral view of 63-year-old African American Native American female showing a sloping frontal bone (curved yellow line), vertical glabellar region and absence of a brow ridge (straight red line), small anterior nasal spine (red arrow), pronounced alveolar prognathism (curved red line), flat zygoma (square), a concave posterior ramus (white arrow), and a small mastoid process (*).

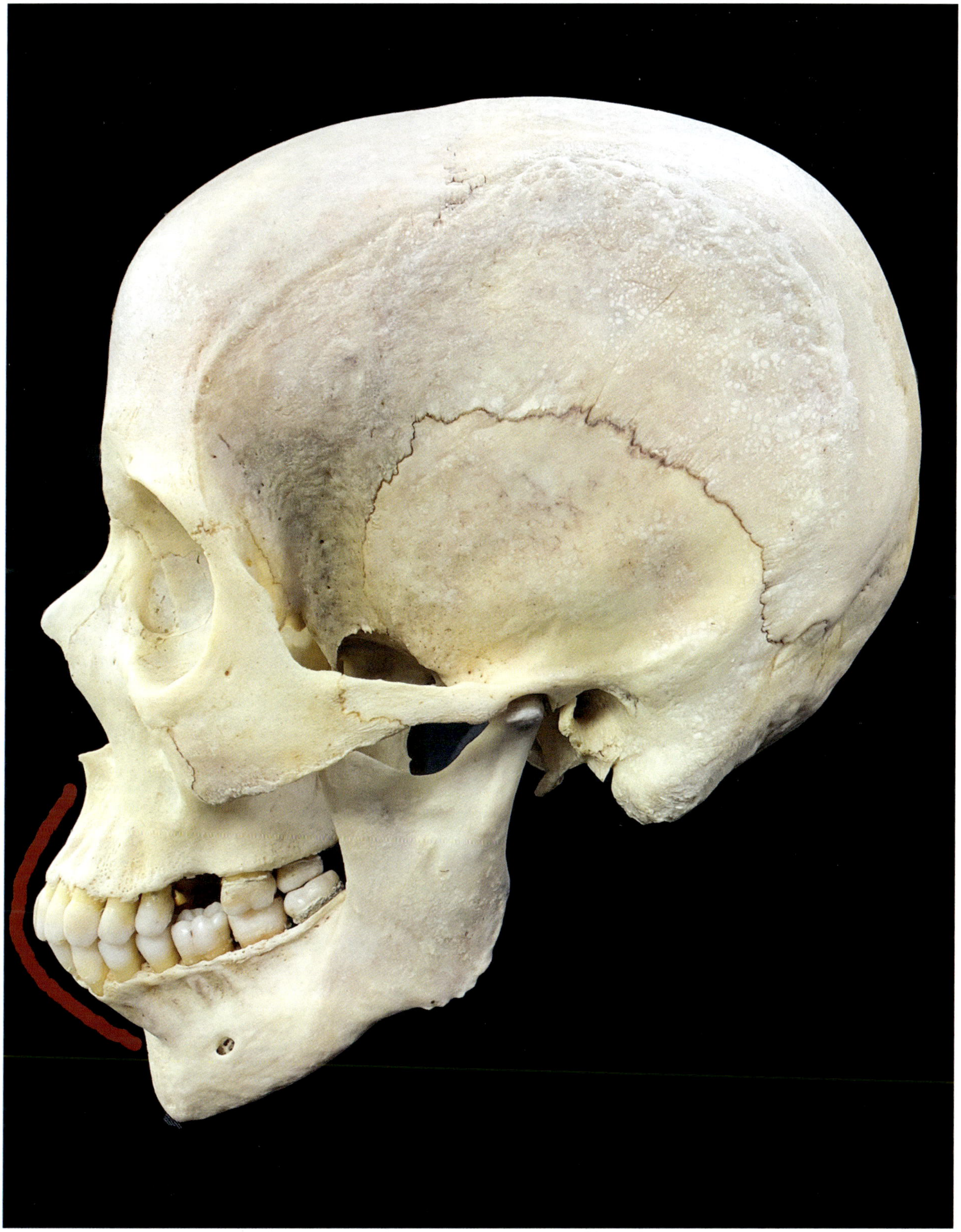

Figure 68. Left lateral view of 63-year-old African American Native American female with pronounced alveolar prognathism (curved red line).

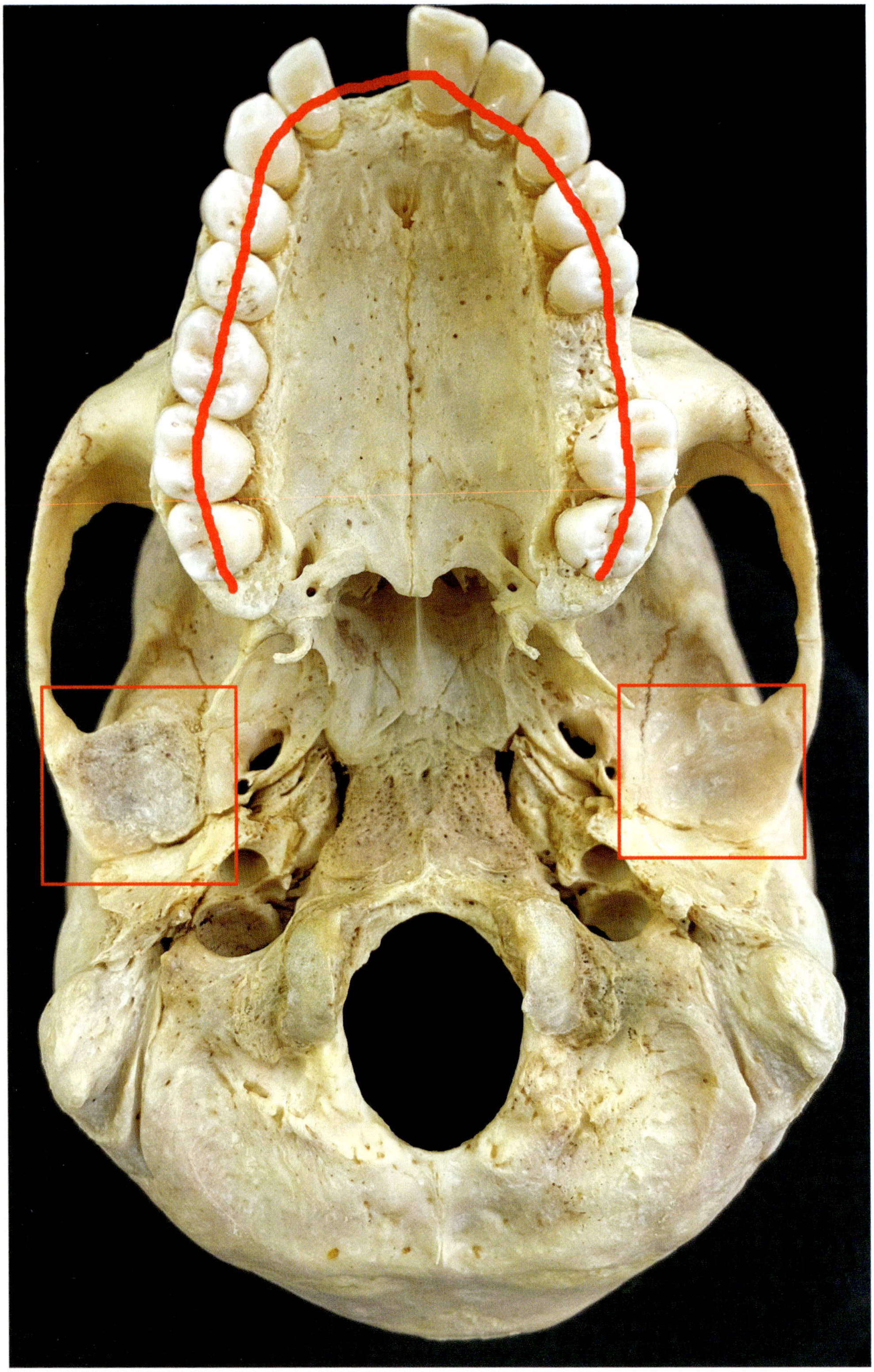

Figure 69. Inferior view of 63-year-old African American Native American female with spatulate-shaped incisors, narrow and long palate with the posterior portion of the palate curved medially (curved red line), and antemortem loss of the maxillary right central incisor and left first molar. The temporomandibular joints (square) exhibit degenerative joint disease that deepened and widened this articulation.

Figure 70. Superior view of 63-year-old African American Native American female with partial obliteration of the coronal suture and complete obliteration of the sagittal suture.

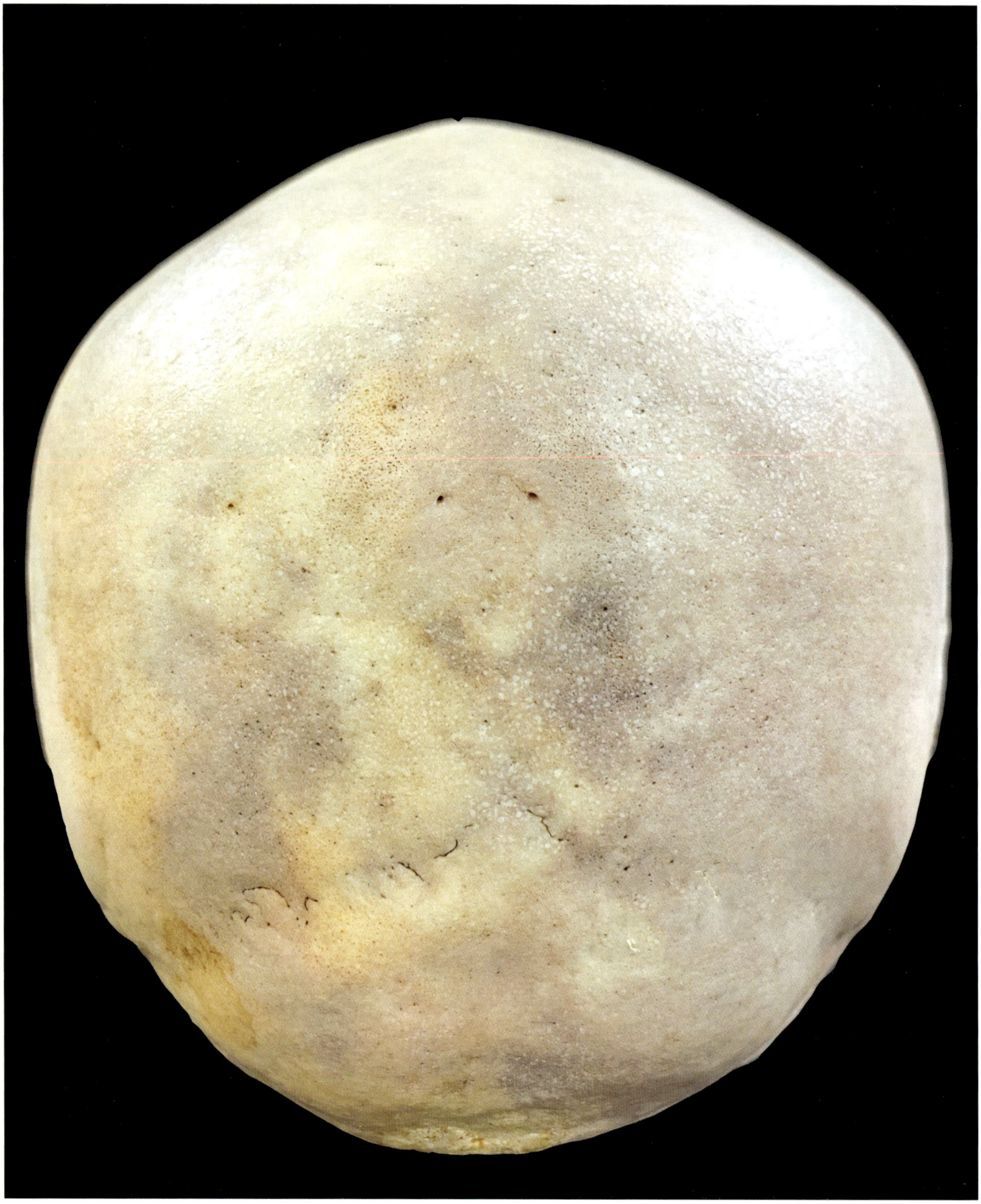

Figure 71. Posterior view of 63-year-old African American Native American female with a small nuchal crest and nearly complete obliteration of the lambdoidal suture.

IV. ASIAN CRANIA

(12 Male and 8 Female)

1. BURMESE (MYANMAR) MALE

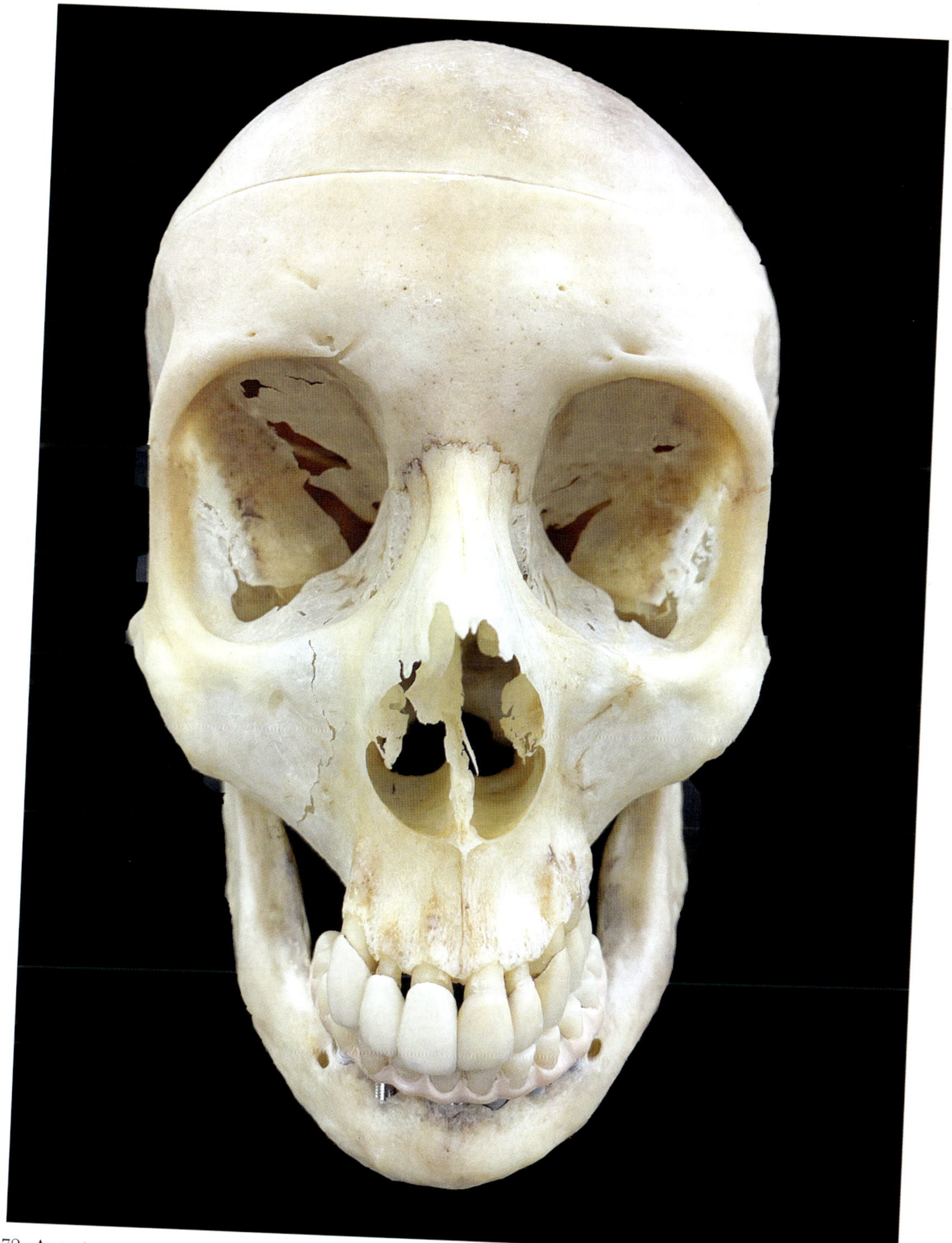

Figure 72. Anterior view of 84-year-old Burmese (Myanmar) male with a narrow maxilla due to the antemortem loss of teeth, and dental implants in the mandible. Note the nearly circular nasal aperture, large nasal bones, paper-thin (osteoporosis) nasal bones and maxilla, and absence of nasal guttering.

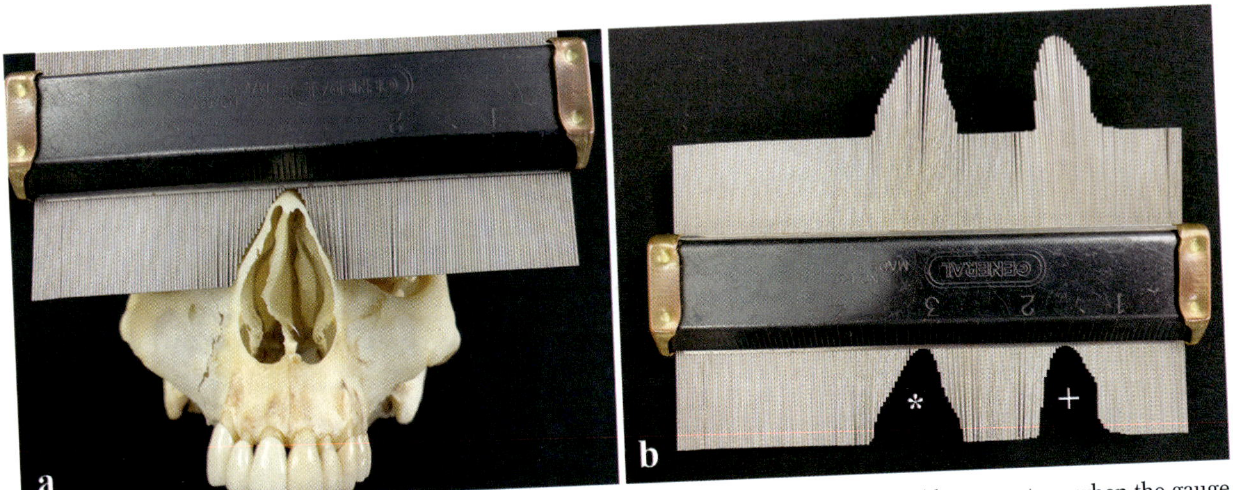

Figure 73a-b. Nasal bone contour in an 84-year-old Burmese (Myanmar) male. (a) The nasal bone contour when the gauge is held perpendicular to the hard palate and orbits and (b) a comparison of the contour as shown in (a) (*) and when the gauge is held perpendicular to the nasal bones (+). Note that the two contours differ greatly.

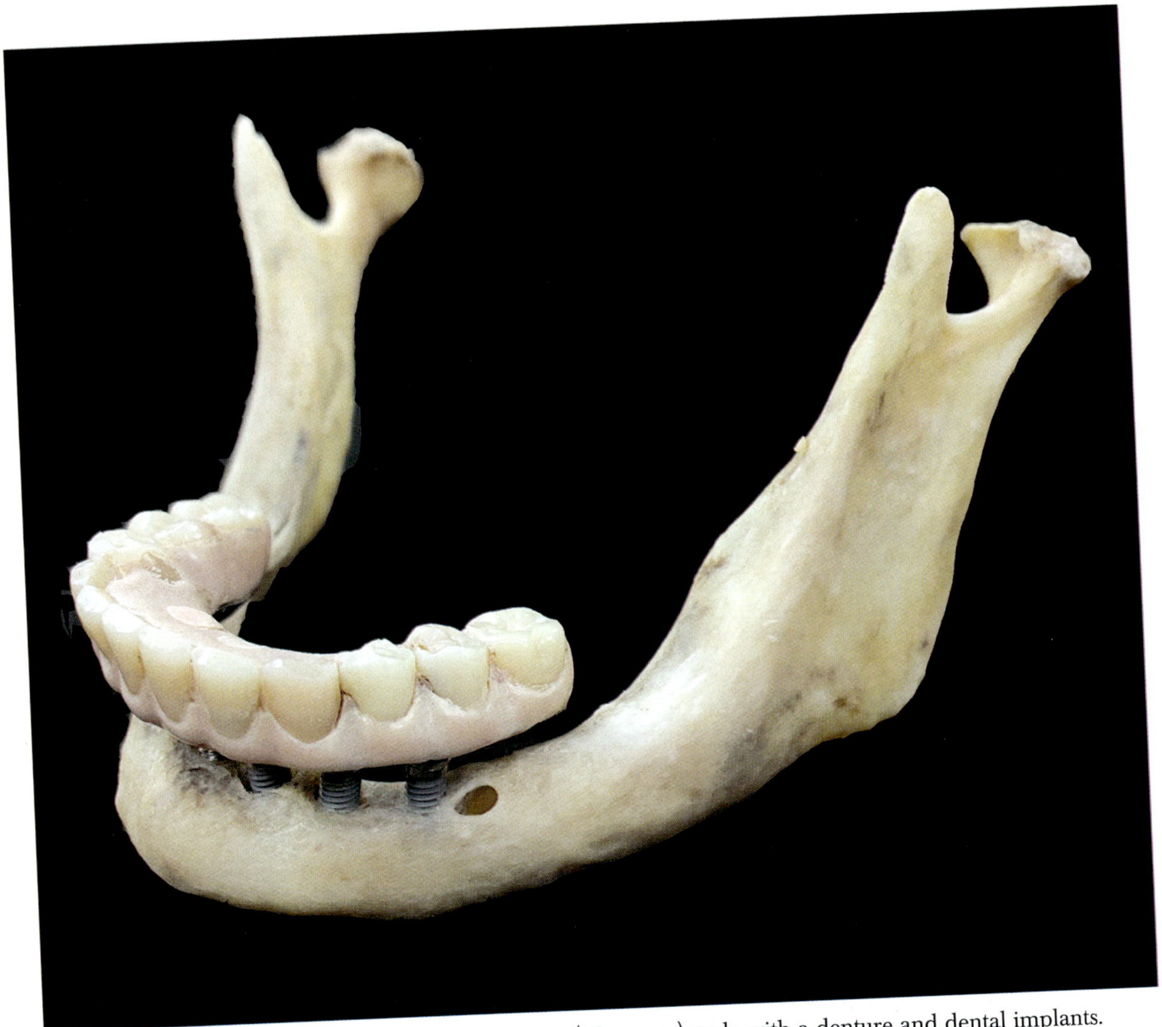

Figure 74. Mandible of an 84-year-old Burmese (Myanmar) male with a denture and dental implants.

TABLE 11.
Cranial Measurements in an 84-year-old Burmese (Myanmar) Male

Measurement	(mm)	Right	Left
Maximum Cranial Length (g-op, GOL)	171		
Maximum Cranial Breadth (eu-eu, XCB)	135		
Bizygomatic Breadth (zy-zy, ZYB)	123		
Basion-Bregma Height (ba-b, BBH)	135		
Cranial Base Length (ba-n, BNL)	100		
Basion-Prosthion (ba-pr, BPL)	93		
Maxillo-Alveolar Breadth (ecm-ecm, MAB)	49		
Maxillo-Alveolar Length (pr-alv, MAL)	53		
Biauricular Breadth (au-au, AUB)	119		
Upper Facial Height (n-pr)	69		
Minimum Frontal Breadth (ft-ft, WFB)	89		
Upper Facial Breadth (fmt-fmt)	98		
Nasal Height (n-ns, NLH)	53		
Nasal Breadth (al-al, NLB)	25		
Orbital Breadth (d-ec, OBB)	41		L
Orbital Height (OBH)	39		L
Biorbital Breadth (ec-ec, EKB)	91		
Interorbital Breadth (d-d, DKB)	20		
Frontal Chord (n-b, FRC)	114		
Parietal Chord (b-l, PAC)	107		
Occipital Chord (l-o, OCC)	97		
Foramen Magnum Length (ba-o, FOL)	33		
Foramen Magnum Breadth (FOB)	28		
Mastoid Length (MDH)		22	21

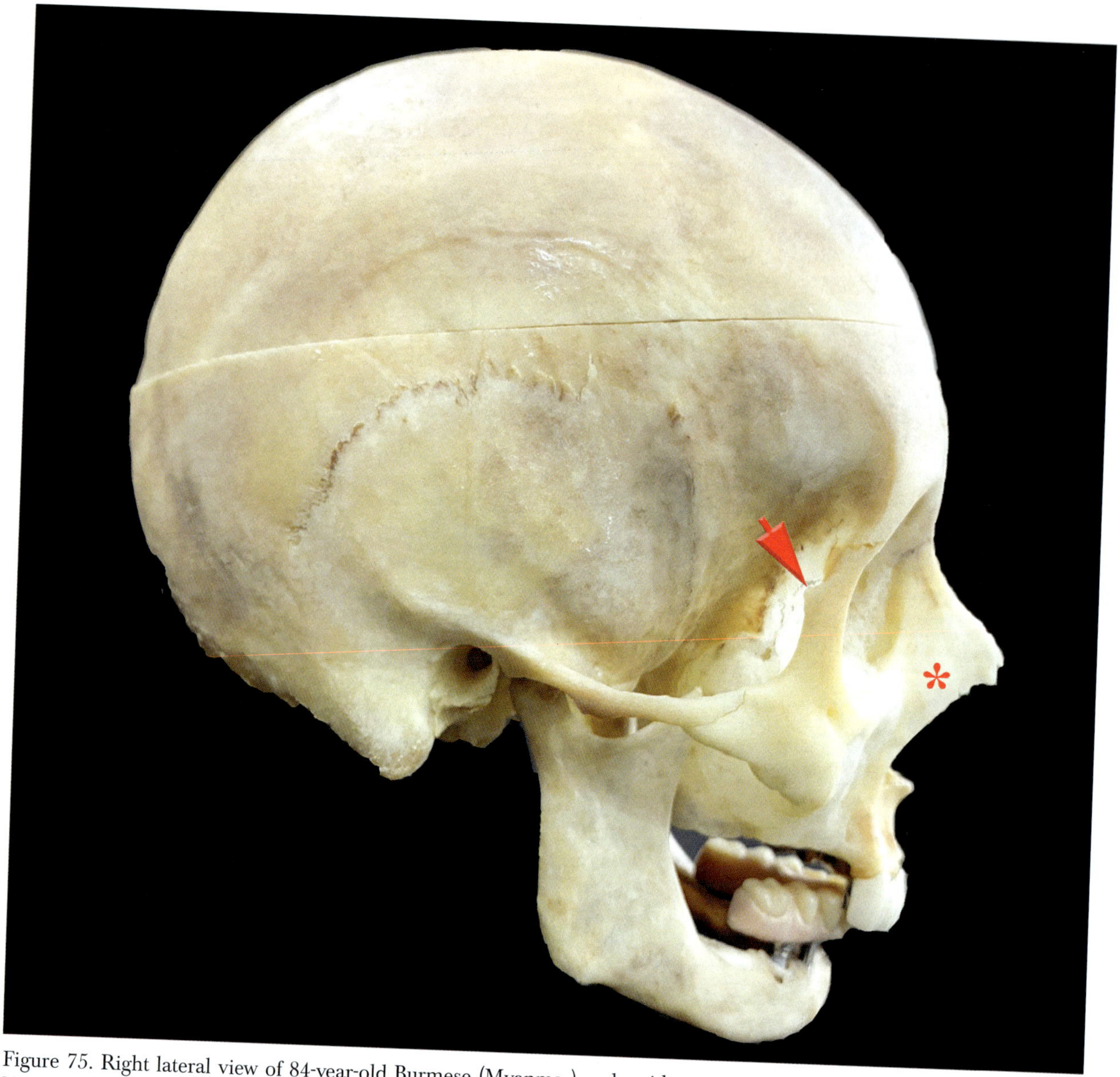

Figure 75. Right lateral view of 84-year-old Burmese (Myanmar) male with a small posterior zygomatic tubercle (arrow). Note the nasal overgrowth/overhanging nasal bones, and the prominent and projecting frontal process (*) of the maxilla. The mandible has a denture anchored by implants.

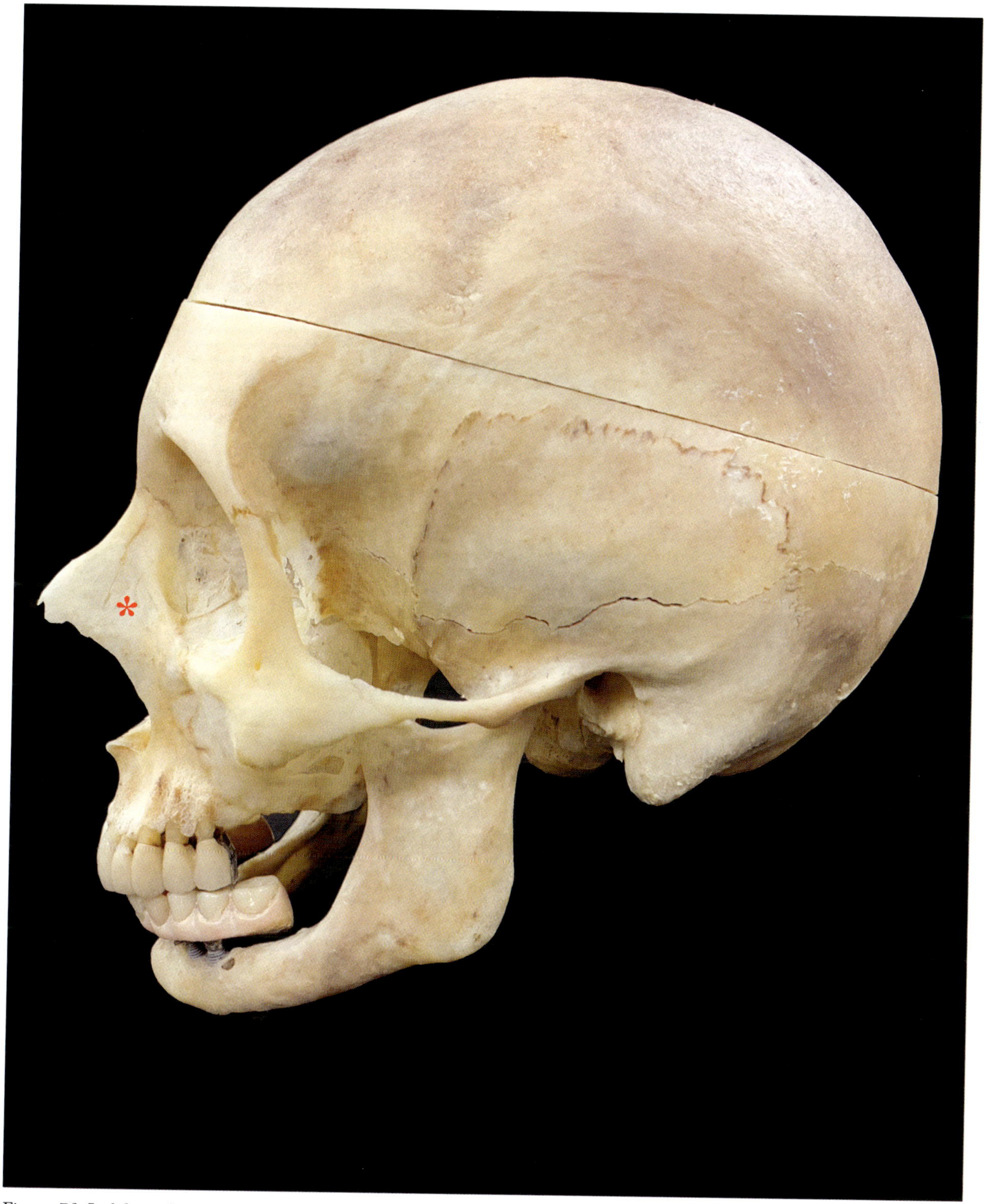

Figure 76. Left lateral view 84-year-old Burmese (Myanmar) male with projecting nasal bones and severe resorption of the mandible. Note the nasal overgrowth and the prominent and projecting frontal process (*) of the maxilla. The horizontal fracture in the left temporal bone occurred postmortem.

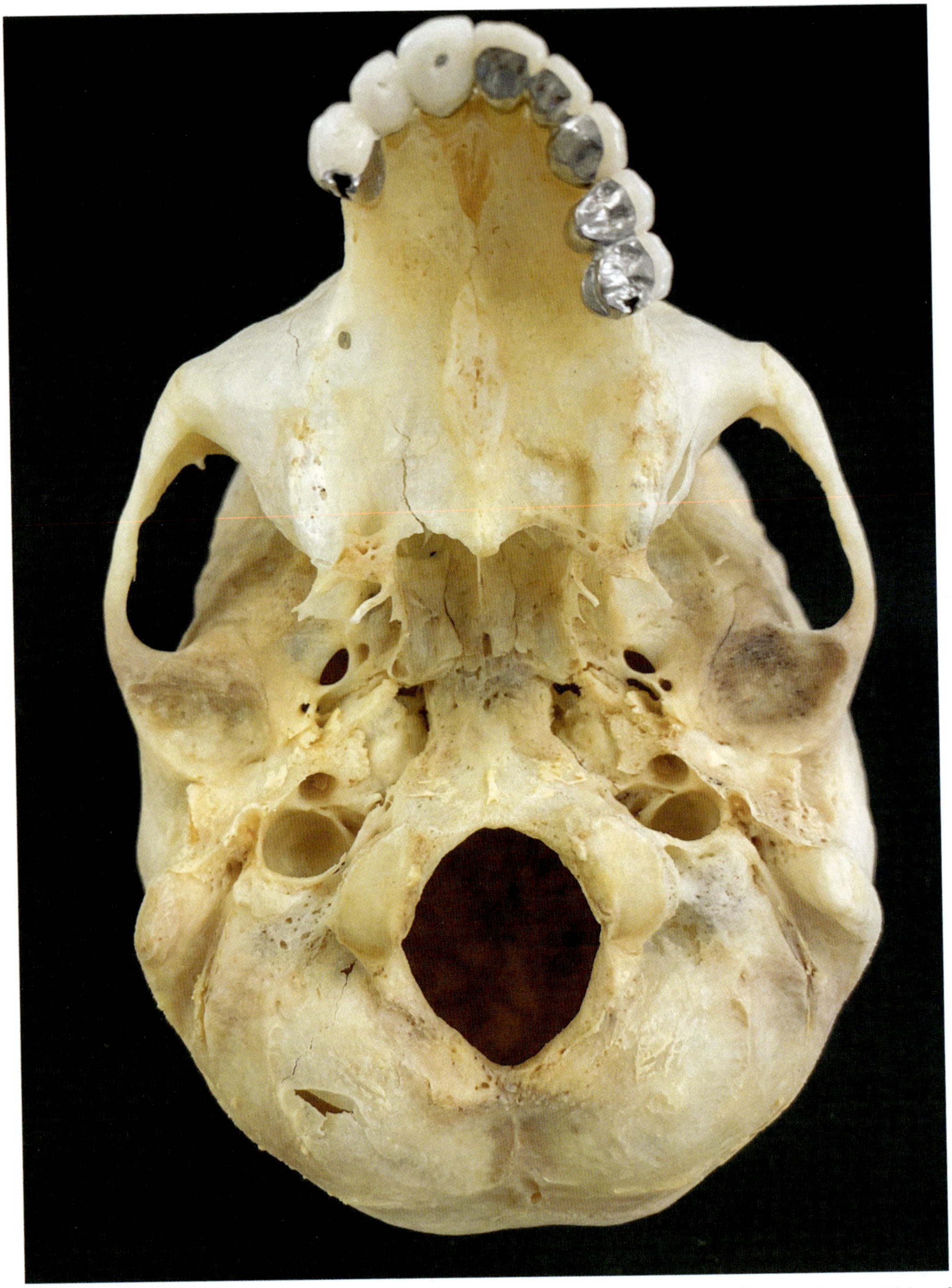

Figure 77. Inferior view 84-year-old Burmese (Myanmar) male with severe osteoporosis. Even with careful handling, the paper-thin bone resulted in postmortem breakage of the occipital bone and hard palate.

Figure 78. Superior view 84-year-old Burmese (Myanmar) male with complete obliteration of the coronal and sagittal sutures.

Figure 79. Posterior view 84-year-old Burmese (Myanmar) male with nearly complete obliteration of the sutures and only a trace of the lambdoidal suture.

2. CHINESE MALE

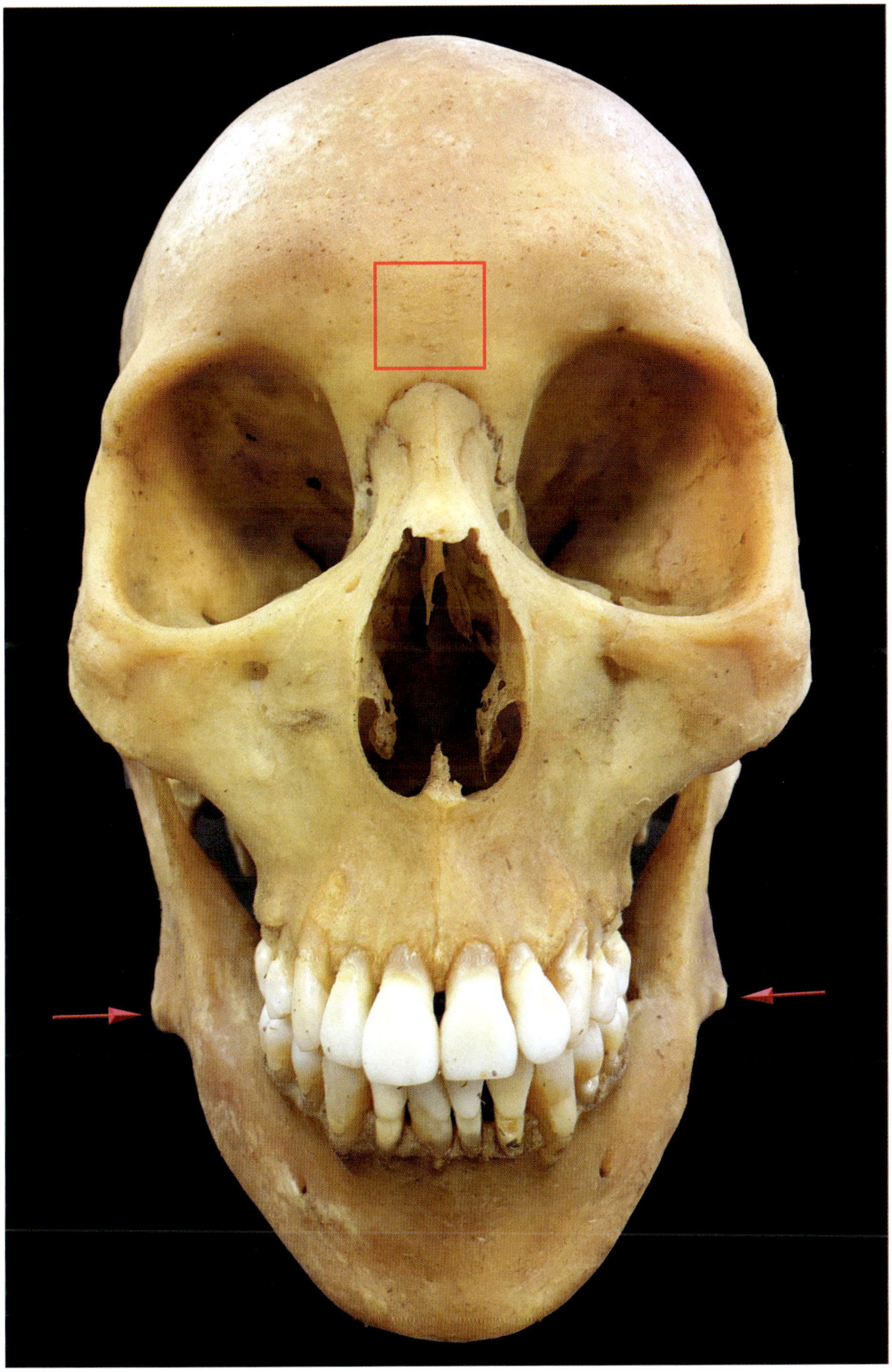

Figure 80. Anterior view 64-year-old Chinese male with a narrow and tall vault, trace supranasal suture (square), and gonial eversion (arrow). The bones are naturally dark before being immersed in acetone to degrease them.

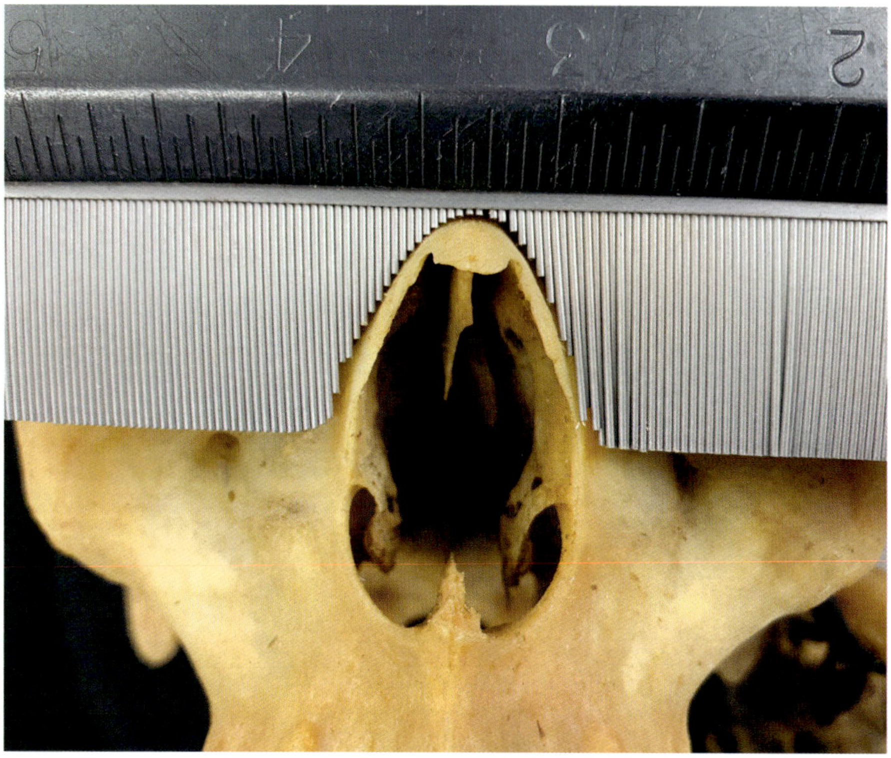

Figure 81. Nasal bone contour in a 64-year-old Chinese male.

Figure 82. Mandible of 64-year-old Chinese male with a rotated right central incisor.

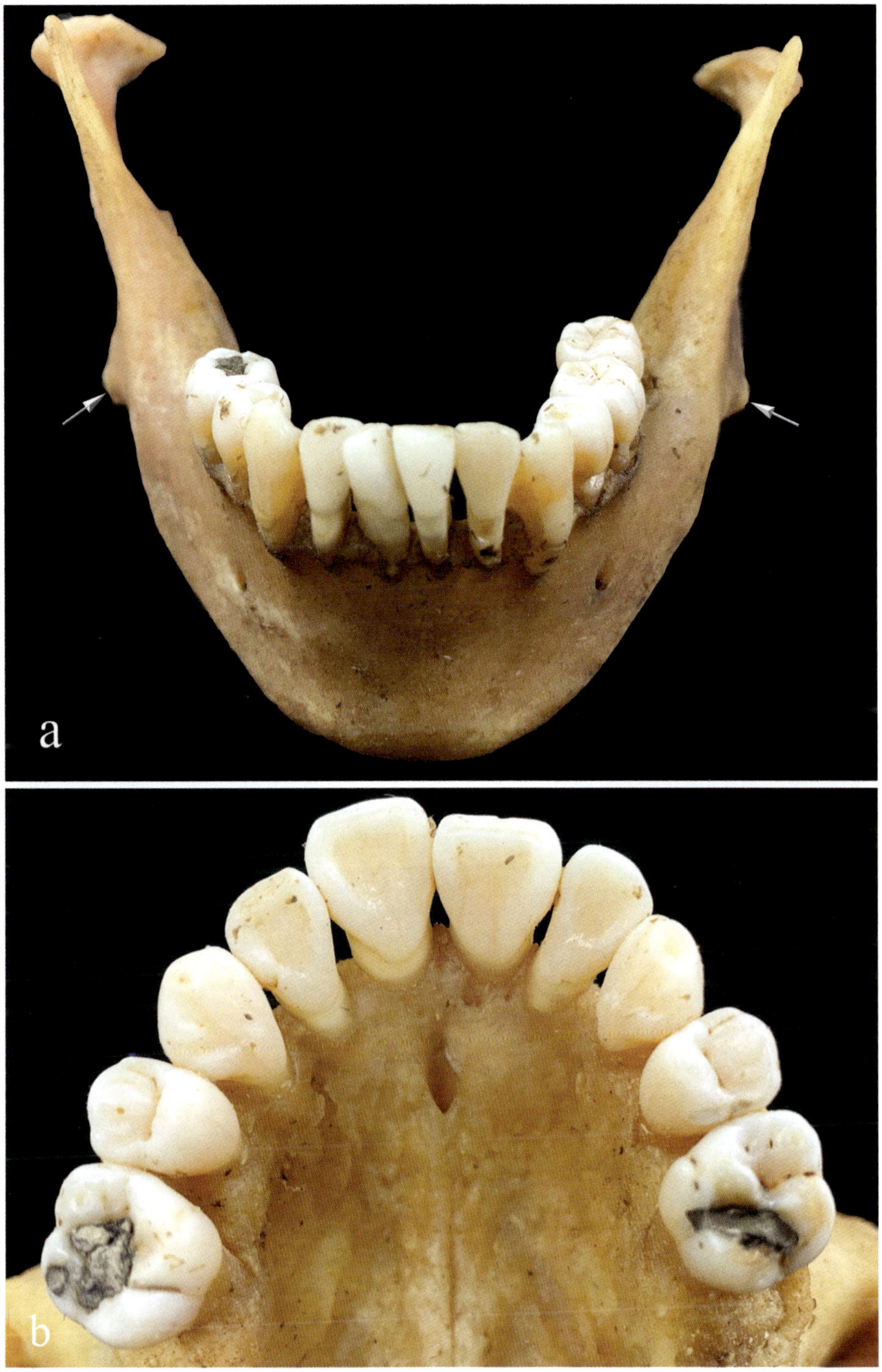

Figure 83a-b. (a) Anterior view of mandible of 64-year-old Chinese male showing gonial eversion (arrow). Gonial eversion is often used as a sex indicator, but has also been shown to be related to functional and adaptive responses of the mandible (Kemkes-Grottenthaler et al. 2002). (b) Inferior view of maxillary teeth showing shovel-shaped incisors.

TABLE 12.
Cranial Measurements in a 64-year-old Chinese Male

Measurement	(mm)	Right	Left
Maximum Cranial Length (g-op, GOL)	180		
Maximum Cranial Breadth (eu-eu, XCB)	142		
Bizygomatic Breadth (zy-zy, ZYB)	128		
Basion-Bregma Height (ba-b, BBH)	139		
Cranial Base Length (ba-n, BNL)	94		
Basion-Prosthion (ba-pr, BPL)	80		
Maxillo-Alveolar Breadth (ecm-ecm, MAB)	58		
Maxillo-Alveolar Length (pr-alv, MAL)	48		
Biauricular Breadth (au-au, AUB)	121		
Upper Facial Height (n-pr)	76		
Minimum Frontal Breadth (ft-ft, WFB)	89		
Upper Facial Breadth (fmt-fmt)	101		
Nasal Height (n-ns, NLH)	55		
Nasal Breadth (al-al, NLB)	23		
Orbital Breadth (d-ec, OBB)	40		L
Orbital Height (OBH)	37		L
Biorbital Breadth (ec-ec, EKB)	96		
Interorbital Breadth (d-d, DKB)	22		
Frontal Chord (n-b, FRC)	120		
Parietal Chord (b-l, PAC)	112		
Occipital Chord (l-o, OCC)	102		
Foramen Magnum Length (ba-o, FOL)	35		
Foramen Magnum Breadth (FOB)	29		
Mastoid Length (MDH)			26

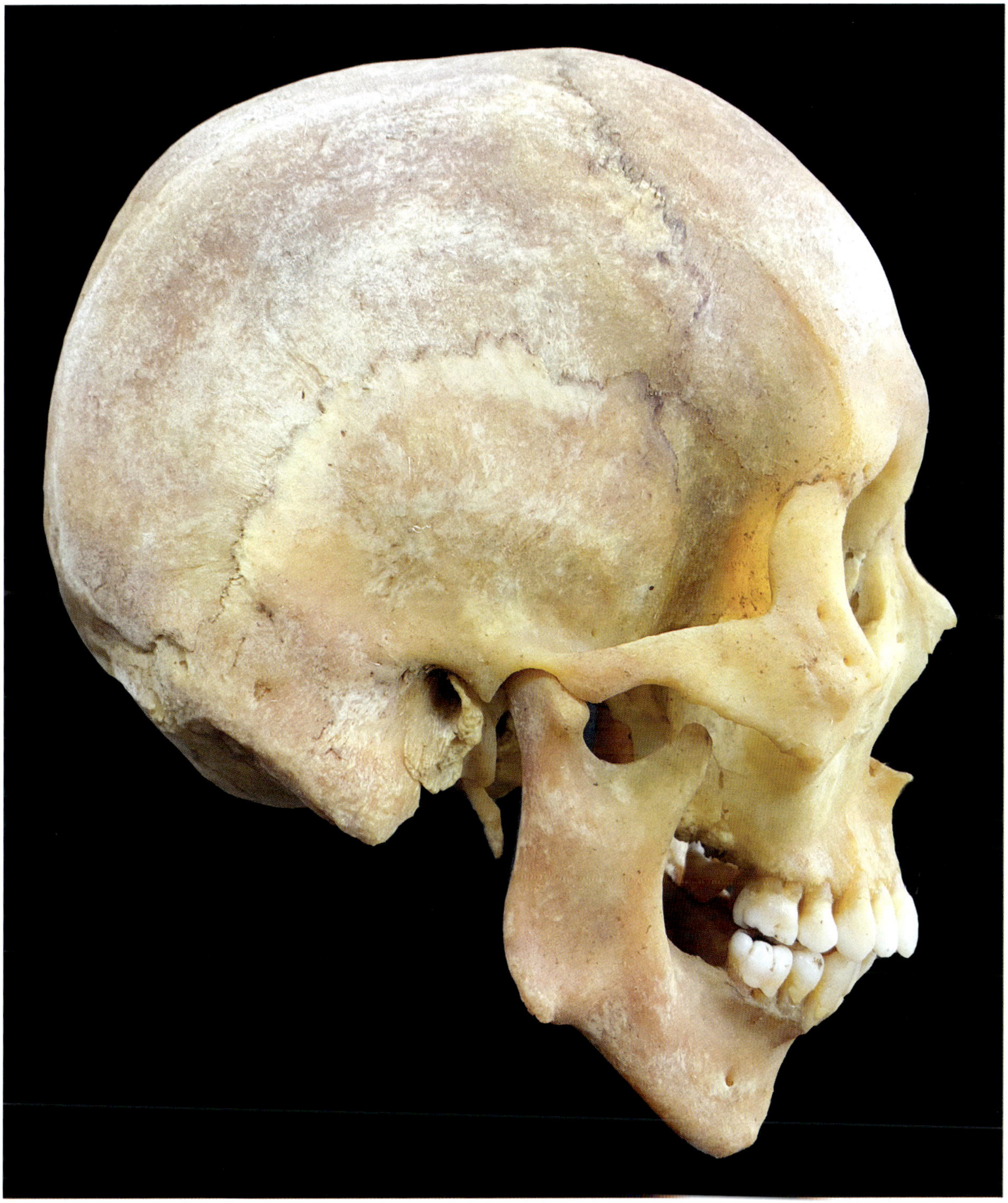

Figure 84. Right lateral view 64-year-old Chinese male with a retrognathic (posteriorly positioned) mandible.

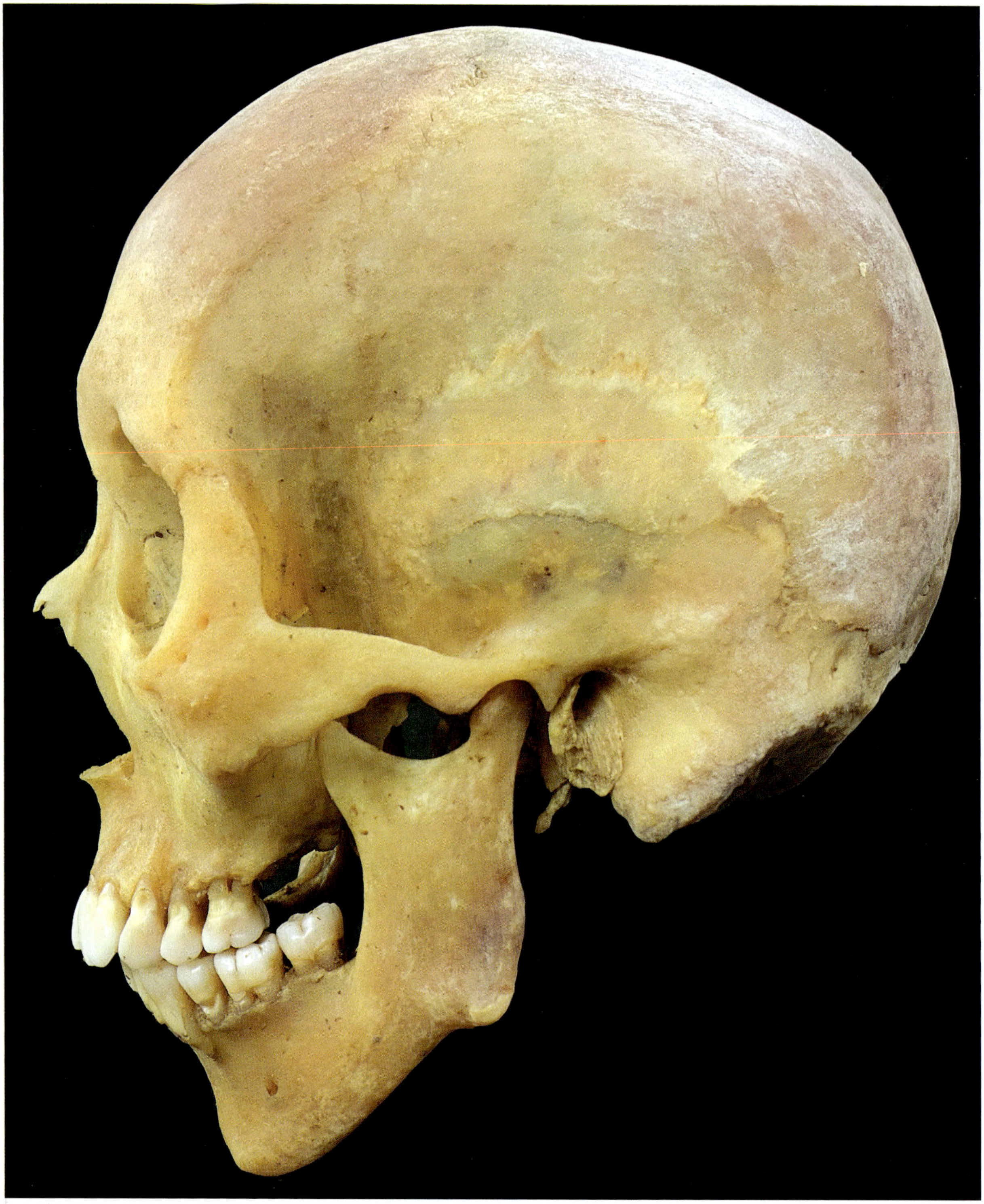

Figure 85. Left lateral view 64-year-old Chinese male.

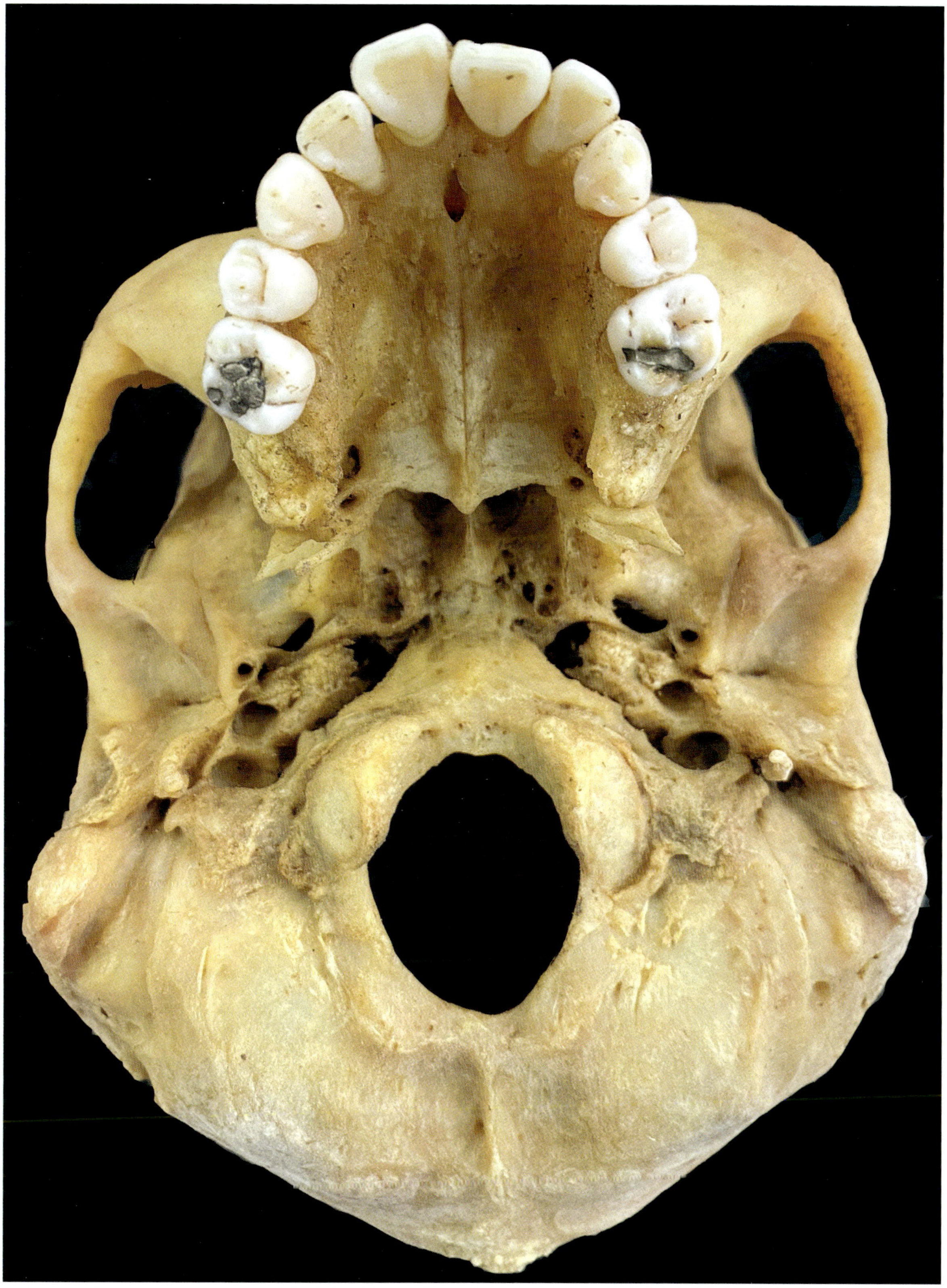

Figure 86. Inferior view 64-year-old Chinese male. Note the shovel-shaped central and lateral incisors.

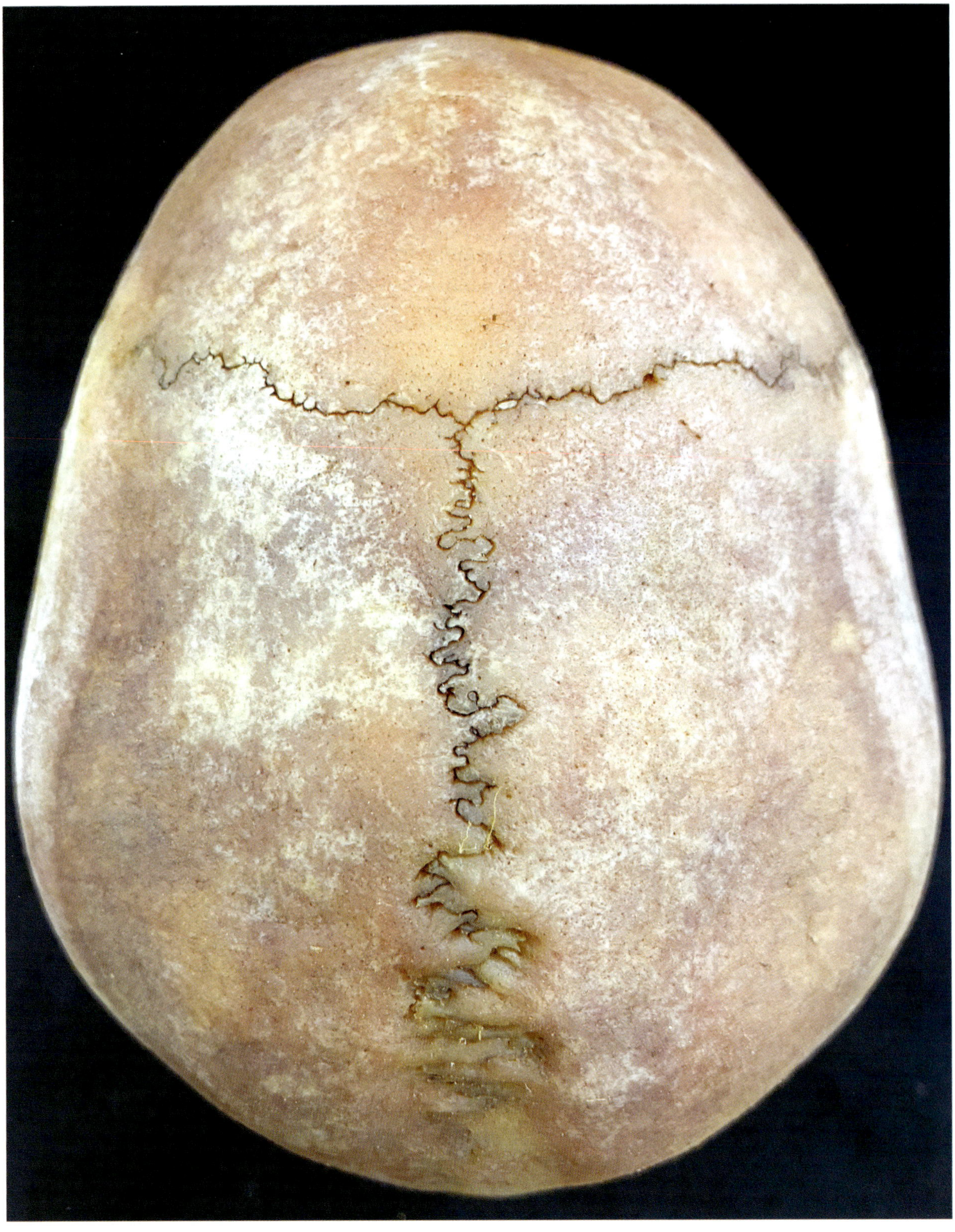

Figure 87. Superior view 64-year-old Chinese male with open sutures.

Figure 88. Posterior view 64-year-old Chinese male with open sutures and a small inion hook.

3. FILIPINO EUROPEAN MALE

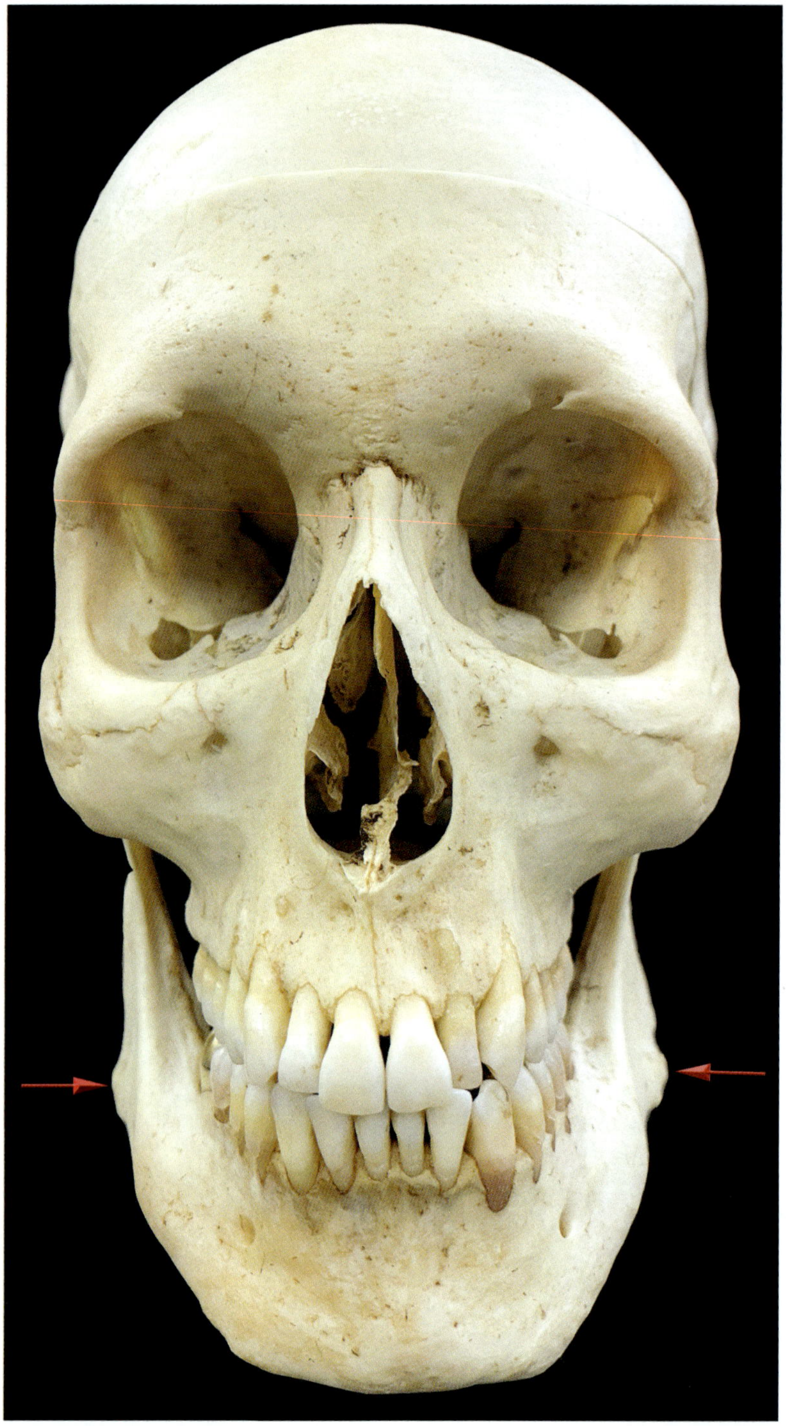

Figure 89. Anterior view 44-year-old Filipino European male with wide interorbital breadth, small round orbits, and blunt superior orbital rims. This male has a very narrow nasal aperture and nasal bones (see Figure 90), and an anterior nasal spine. Note the robust zygomatic bones, broad chin, and gonial eversion (arrow). This skull is a good example of "racial" admixture with a combination of Asian and European features.

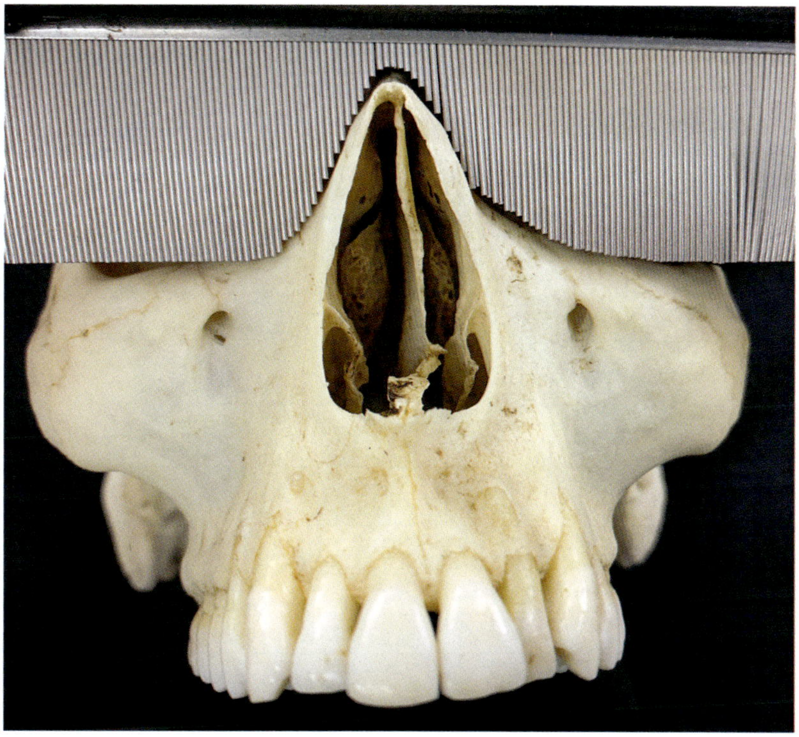

Figure 90. Nasal bone contour in a 44-year-old Filipino European male.

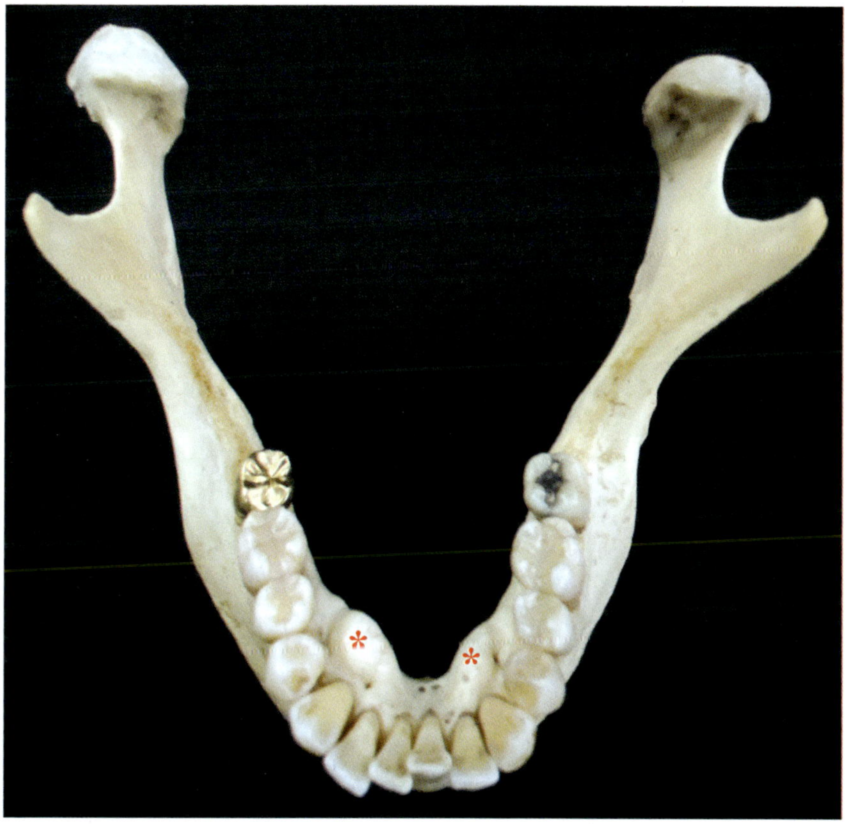

Figure 91. Mandible of 44-year-old Filipino European male with symmetrical lingual exostoses (*).

TABLE 13.
Cranial Measurements in a 44-year-old Filipino European Male

Measurement	(mm)	Right	Left
Maximum Cranial Length (g-op, GOL)	183		
Maximum Cranial Breadth (eu-eu, XCB)	136		
Bizygomatic Breadth (zy-zy, ZYB)	125		
Basion-Bregma Height (ba-b, BBH)	141		
Cranial Base Length (ba-n, BNL)	102		
Basion-Prosthion (ba-pr, BPL)	97		
Maxillo-Alveolar Breadth (ecm-ecm, MAB)	60		
Maxillo-Alveolar Length (pr-alv, MAL)	53		
Biauricular Breadth (au-au, AUB)	119		
Upper Facial Height (n-pr)	68		
Minimum Frontal Breadth (ft-ft, WFB)	86		
Upper Facial Breadth (fmt-fmt)	98		
Nasal Height (n-ns, NLH)	55		
Nasal Breadth (al-al, NLB)	20		
Orbital Breadth (d-ec, OBB)	40		L
Orbital Height (OBH)	32		L
Biorbital Breadth (ec-ec, EKB)	90		
Interorbital Breadth (d-d, DKB)	20		
Frontal Chord (n-b, FRC)	116		
Parietal Chord (b-l, PAC)	116		
Occipital Chord (l-o, OCC)	106		
Foramen Magnum Length (ba-o, FOL)	33		
Foramen Magnum Breadth (FOB)	29		
Mastoid Length (MDH)		27	31

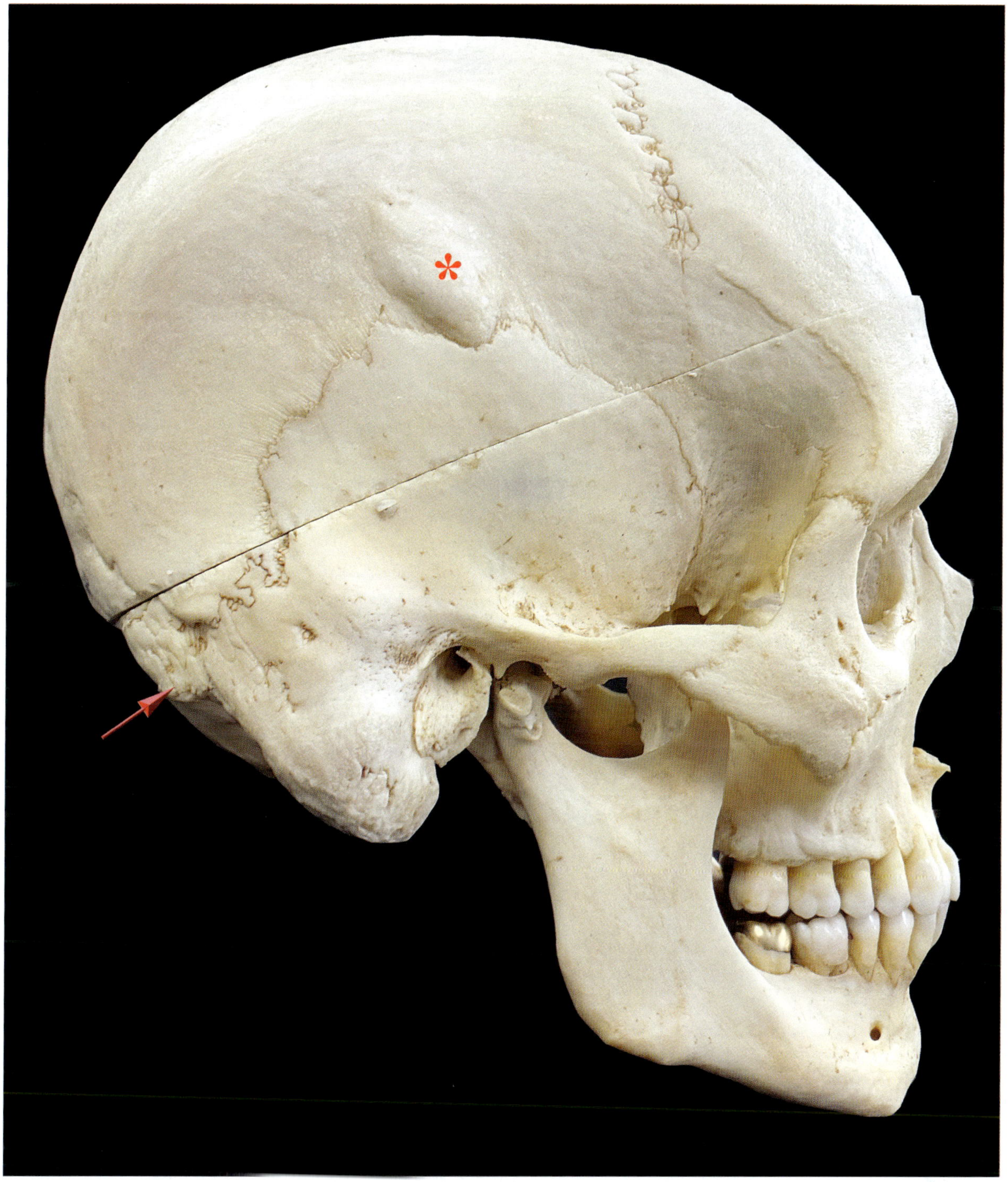

Figure 92. Right lateral view of 44-year-old Filipino European male with a possible benign tumor (*). Note that the lesion abruptly stops at the squamosal suture. This skull also exhibits mild alveolar prognathism and a developed nuchal crest (arrow).

Figure 93. Left lateral view of 44-year-old Filipino European male with a posteriorly sloping frontal bone, rounded vault, small brow ridge, nasion depression, and an anterior nasal sill. Note the large zygomatic bone and zygomatic arch.

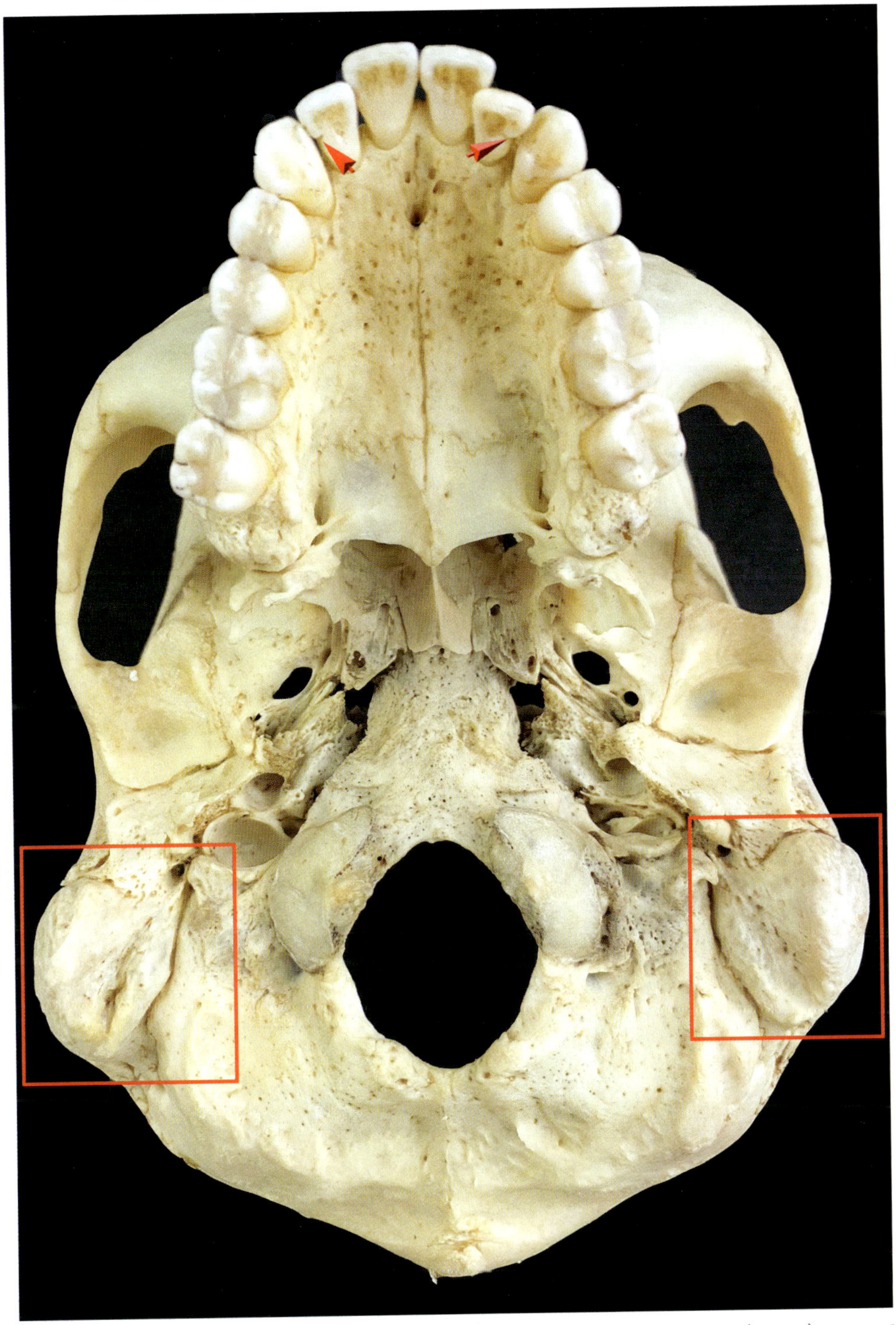

Figure 94. Inferior view of 44-year-old Filipino European male with large mastoid processes (square), a "V"-shaped palate with a small torus palatinus, and shovel-shaped incisors. The bilateral interruption grooves (palatoradicular grooves) in the distal margins of the lateral incisors (arrow) are a rare dental anomaly (Kim et al. 2017). This anomaly is commonly seen in East Asians at a frequency of 40-60% (Scott and Irish 2017). (See Figure 95.)

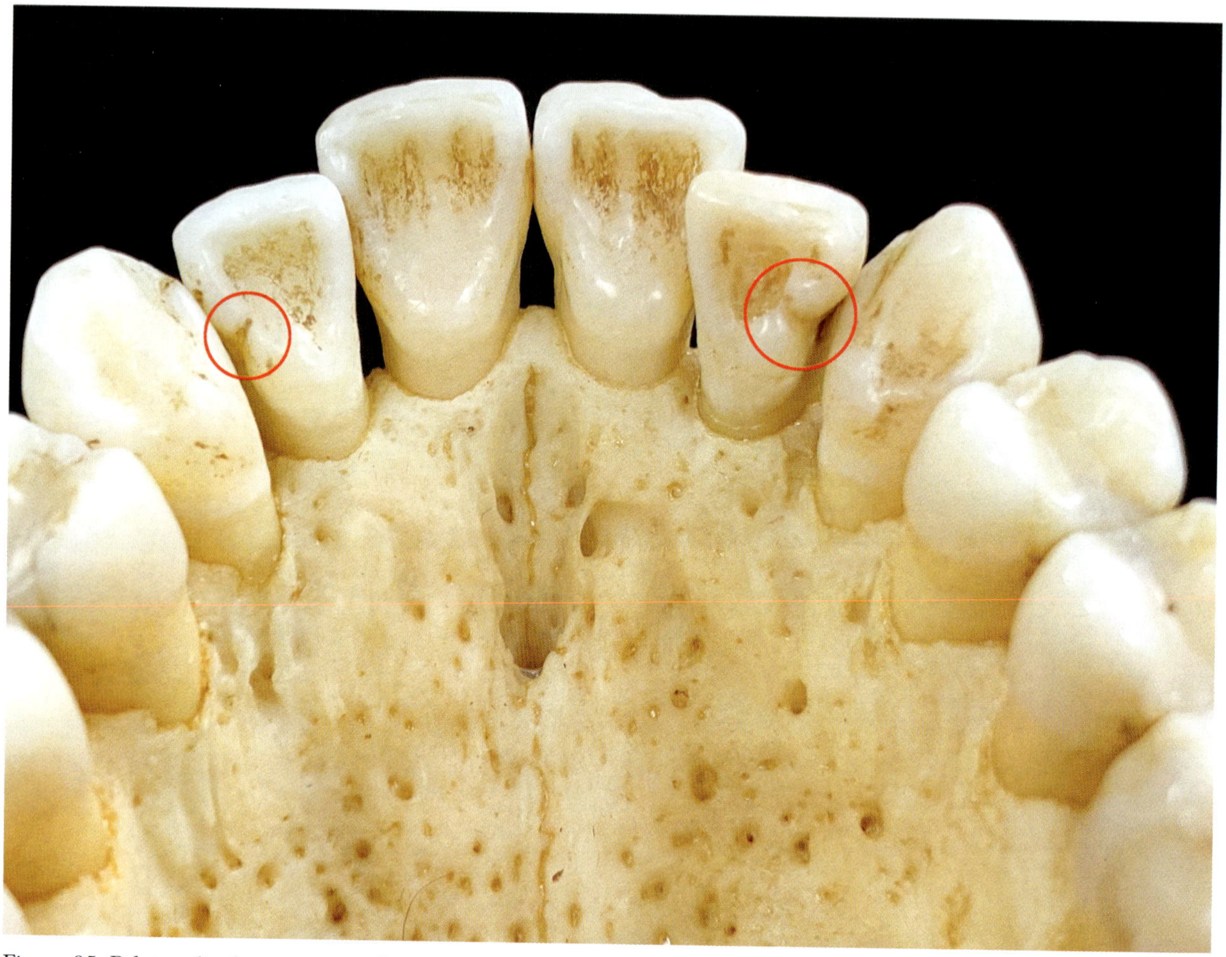

Figure 95. Palatoradicular grooves in the lingual-distal margins of the lateral incisors (circles) in a 44-year-old Filipino European male. Palatoradicular grooves, also known as interruption grooves in the anthropological literature, are rare developmental anomalies with a genetic component and most commonly present in maxillary lateral incisors (Kim et al. 2017; Suchetha et al. 2012). The frequency of palatoradicular grooves may be unilateral or bilateral and varies from 2.8 to 8.5% (Kogon 1986; Withers et al. 1981). Interruption grooves are more common in East Asians and derived groups than in Africans, African Americans, or Europeans (Scott and Irish 2017).

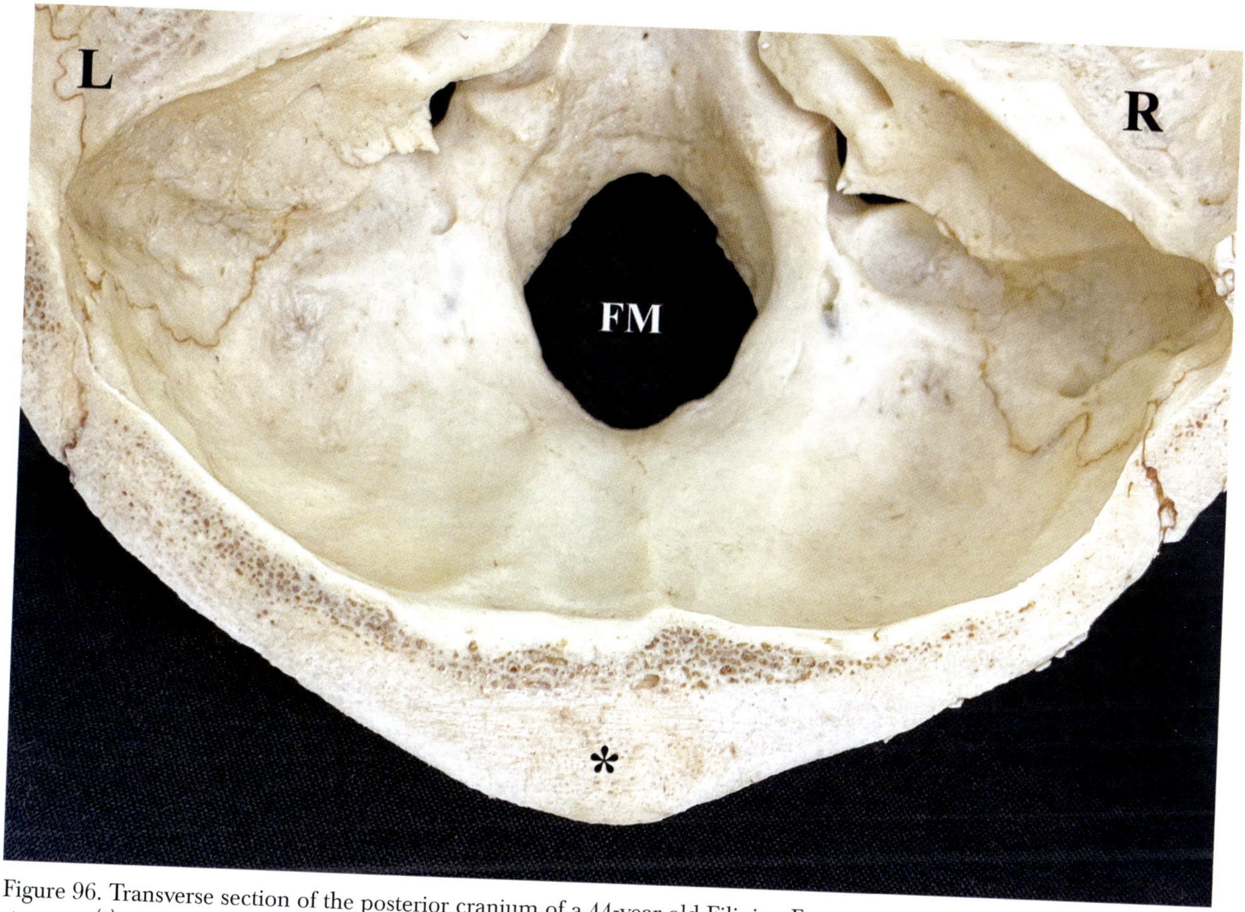

Figure 96. Transverse section of the posterior cranium of a 44-year-old Filipino European male showing the dense internal structure (*) of the occipital bone through the inion crest and hook (1880). "L"= left side, "R"=right side, and "FM" is foramen magnum.

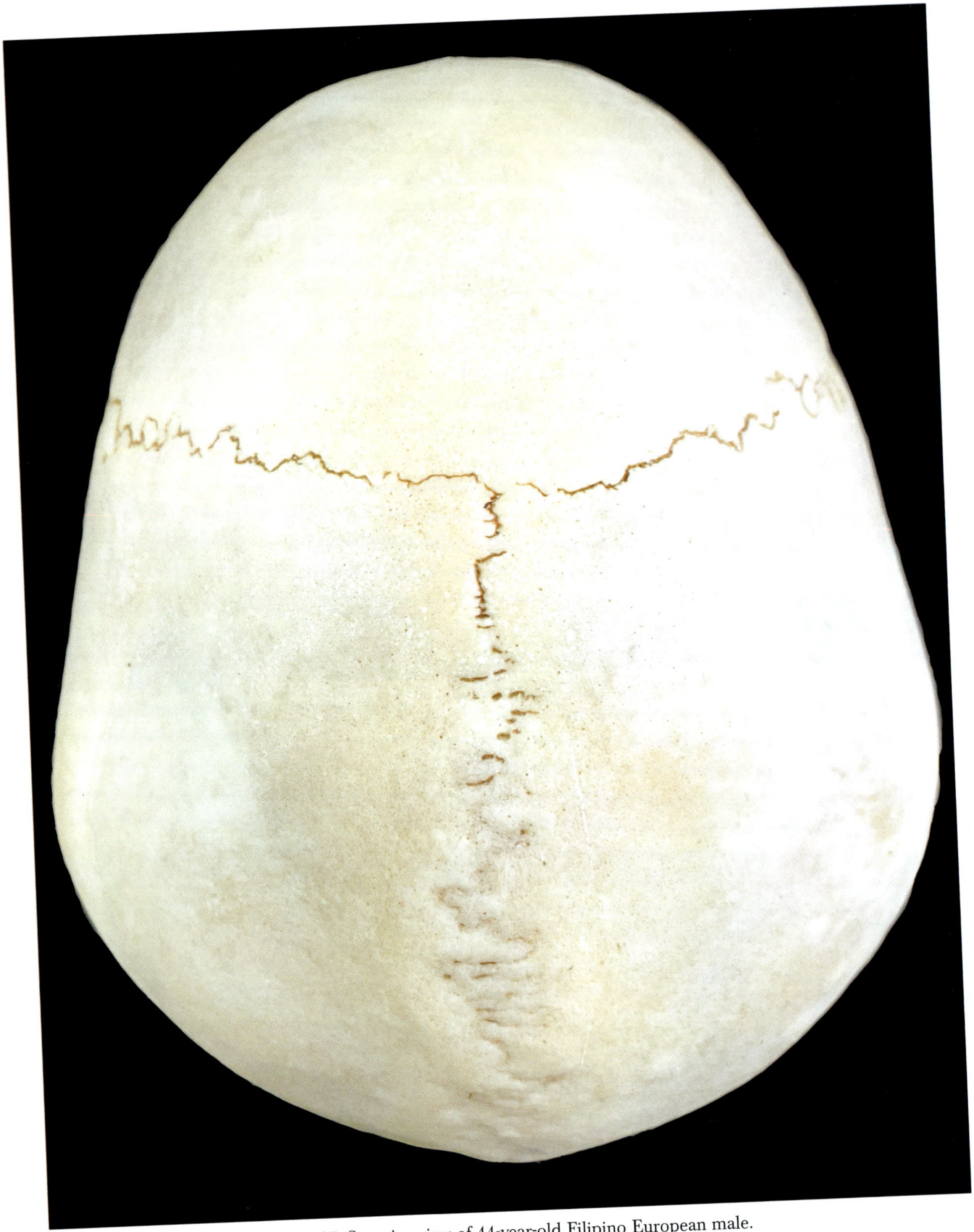

Figure 97. Superior view of 44-year-old Filipino European male.

Figure 98. Posterior view of 44-year-old Filipino European male.

4. CHINESE FEMALE

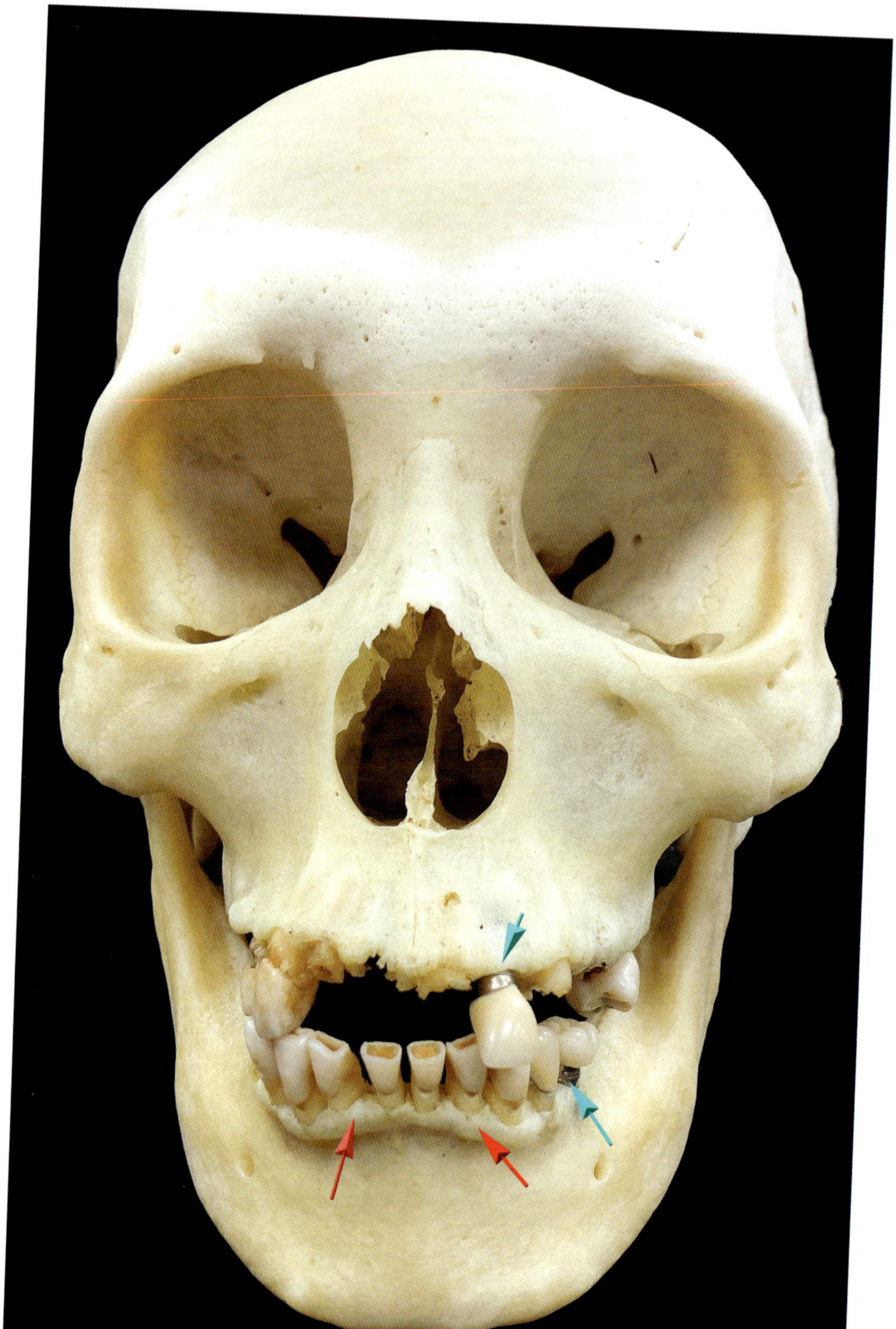

Figure 99. Anterior view 95-year-old Chinese female exhibiting blunt superior orbital margins (typically a male trait), a wide interorbital breadth, and a wide nasal aperture. Note the prominent brow ridge, buccal exostoses (red arrow) in the mandible, severe occlusal wear of the anterior teeth, and dental implant (blue arrow).

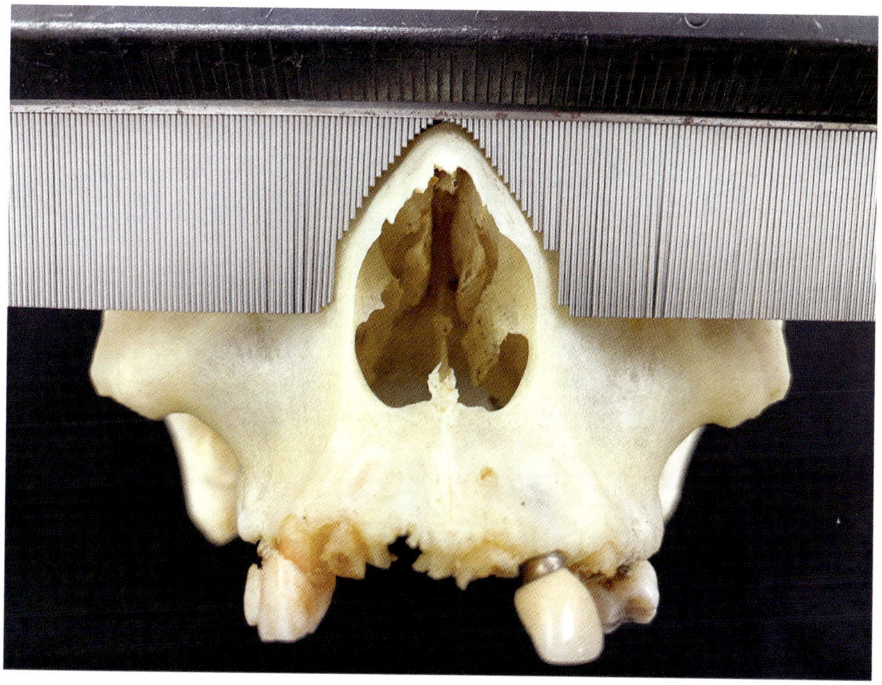

Figure 100. Nasal bone contour in a 95-year-old Chinese female.

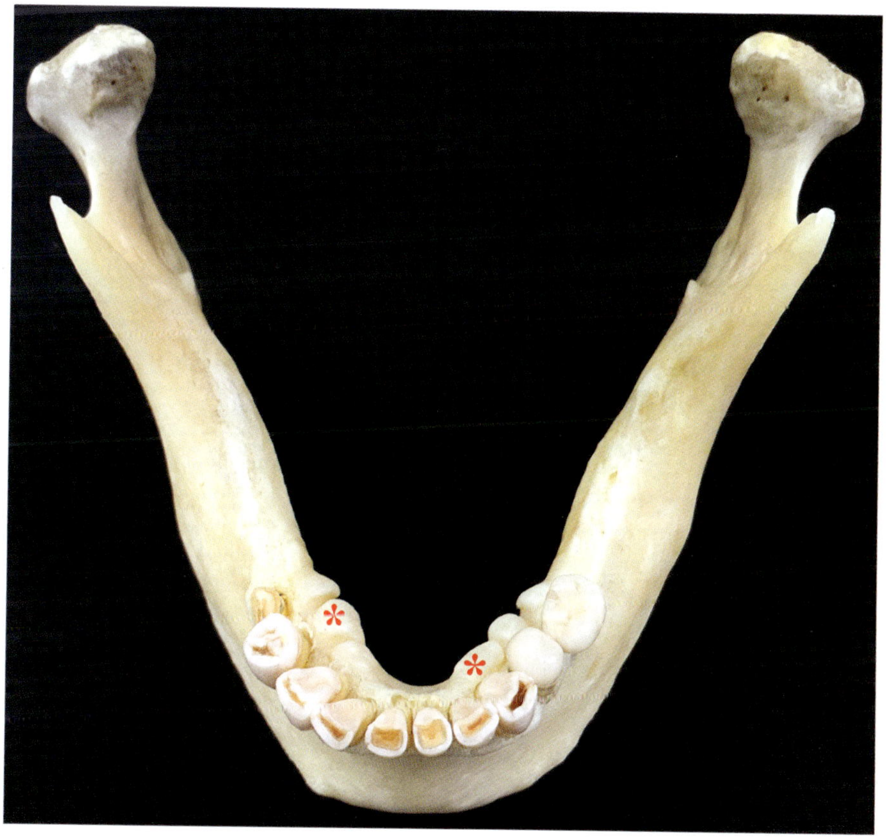

Figure 101. Mandible of 95-year-old Chinese female with a broad and rounded mental eminence, lingual exostoses (*) buttressing each of the existing teeth, and heavy wear (attrition) of the anterior teeth. The other mandibular tooth sockets (alveoli) did not have exostoses due to antemortem loss of teeth.

TABLE 14.
Cranial Measurements in a 95-year-old Chinese Female

Measurement	(mm)	Right	Left
Maximum Cranial Length (g-op, GOL)	175		
Maximum Cranial Breadth (eu-eu, XCB)	150		
Bizygomatic Breadth (zy-zy, ZYB)	131		
Basion-Bregma Height (ba-b, BBH)	130		
Cranial Base Length (ba-n, BNL)	95		
Basion-Prosthion (ba-pr, BPL)	94		
Maxillo-Alveolar Breadth (ecm-ecm, MAB)	64		
Maxillo-Alveolar Length (pr-alv, MAL)	53		
Biauricular Breadth (au-au, AUB)	133		
Upper Facial Height (n-pr)	67		
Minimum Frontal Breadth (ft-ft, WFB)	93		
Upper Facial Breadth (fmt-fmt)	105		
Nasal Height (n-ns, NLH)	51		
Nasal Breadth (al-al, NLB)	24		
Orbital Breadth (d-ec, OBB)	42		L
Orbital Height (OBH)	34		L
Biorbital Breadth (ec-ec, EKB)	96		
Interorbital Breadth (d-d, DKB)	23		
Frontal Chord (n-b, FRC)	111		
Parietal Chord (b-l, PAC)	112		
Occipital Chord (l-o, OCC)	98		
Foramen Magnum Length (ba-o, FOL)	36		
Foramen Magnum Breadth (FOB)	27		
Mastoid Length (MDH)		28	31

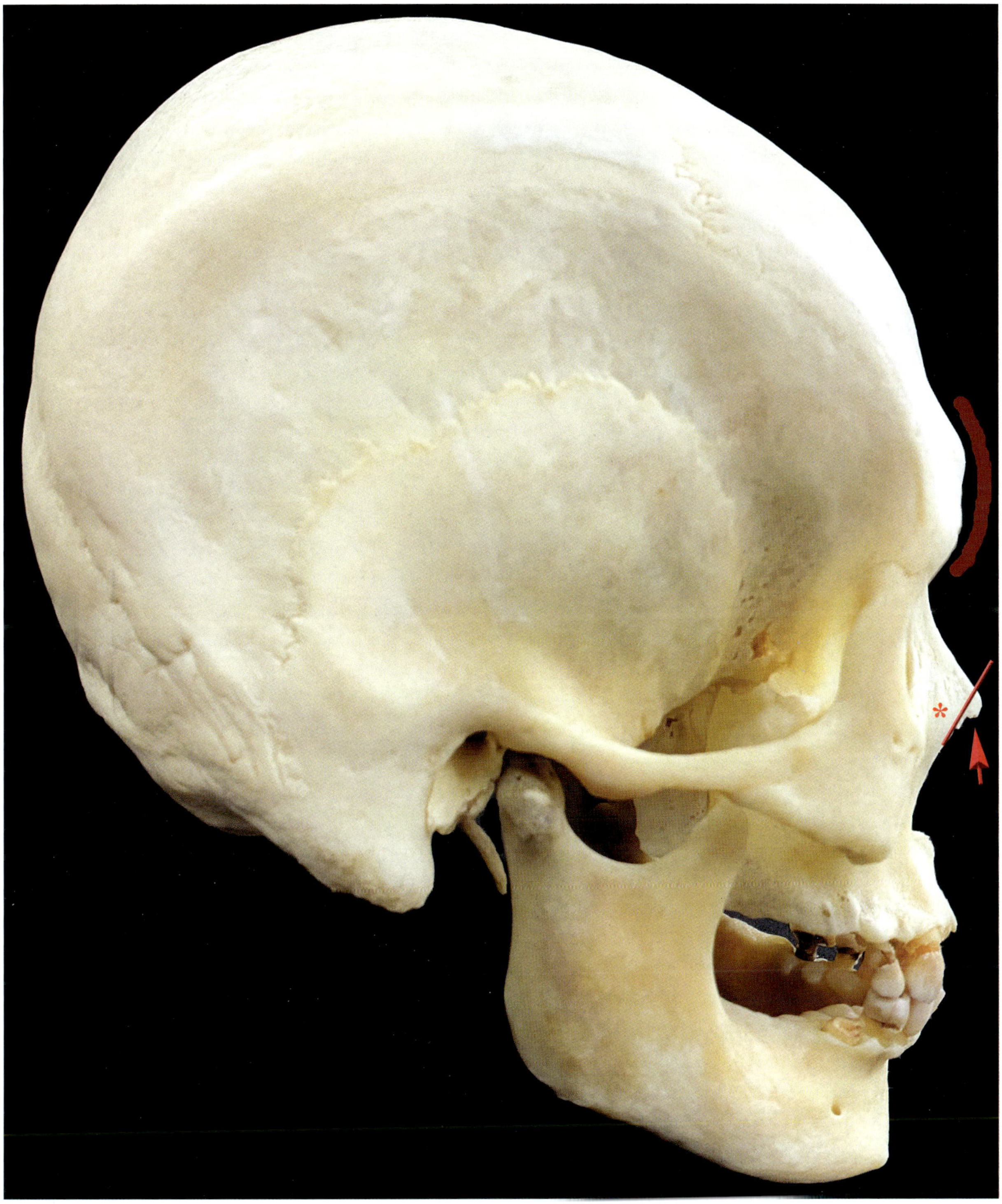

Figure 102. Right lateral view 95-year-old Chinese female with a large, protruding brow ridge (curved red line) and nasal overgrowth (red straight line and arrow), and a small, short (anterior-posterior) frontal process of the maxilla (*).

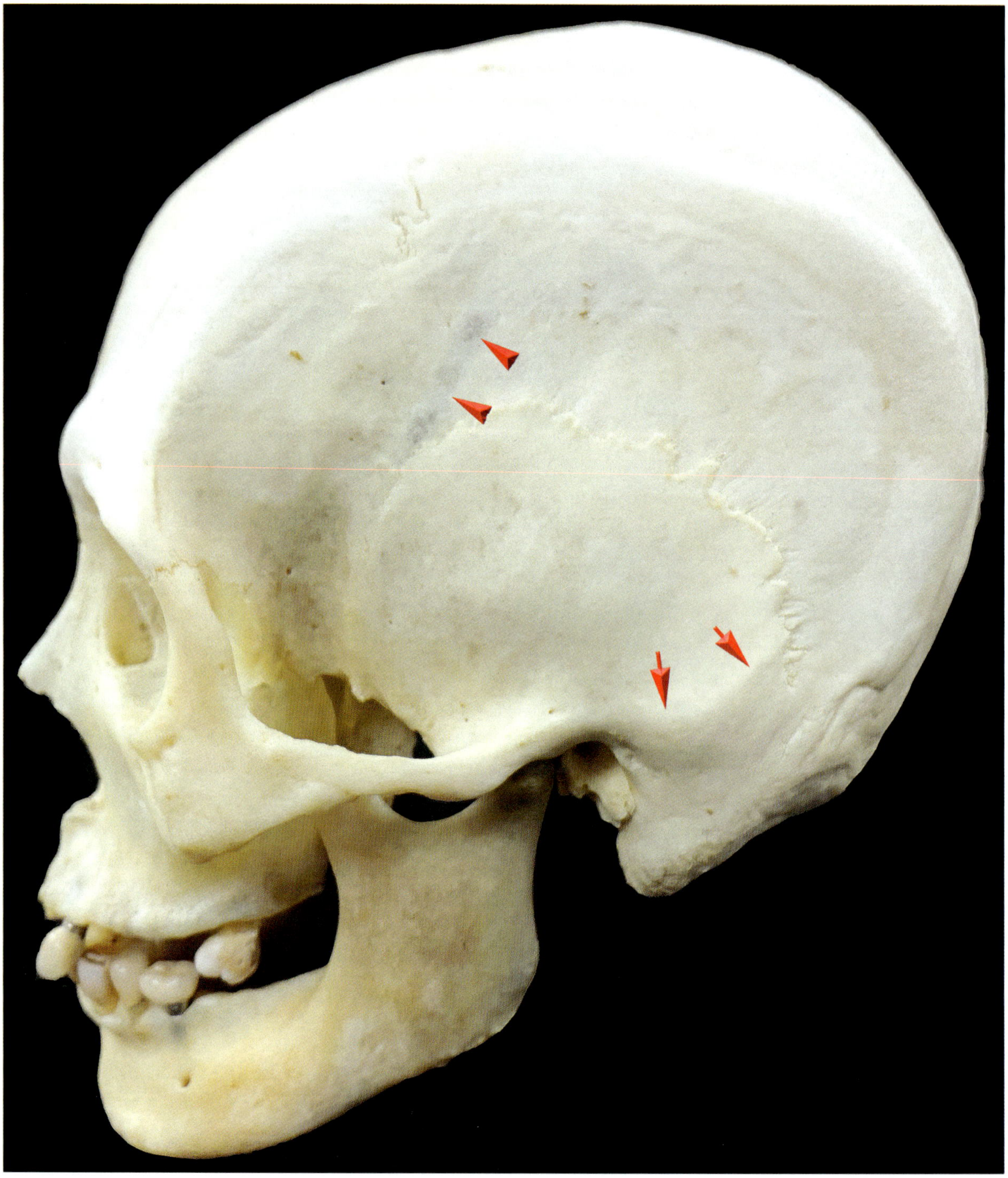

Figure 103. Left lateral view 95-year-old Chinese female with thin, translucent left parietal bone due to osteoporosis (arrowhead), and a developed supramastoid crest (zygomatic extension; Langley et al. [2017]) (arrow).

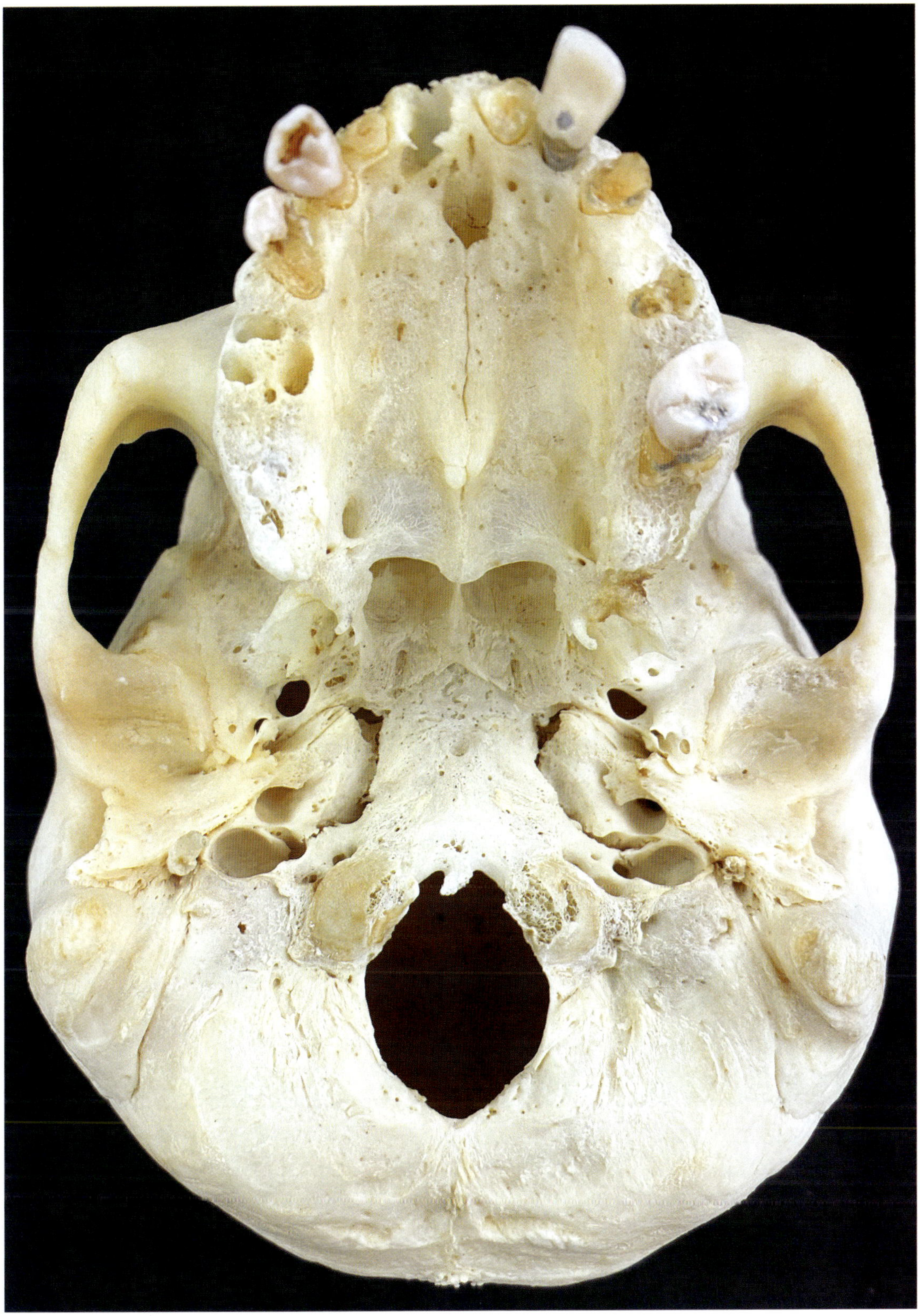

Figure 104. Inferior view 95-year-old Chinese female with an elliptical/horseshoe-shaped hard palate (Maier et al. 2015) and a small torus palatinus. Note the translucent (osteoporotic) bone in the hard palate.

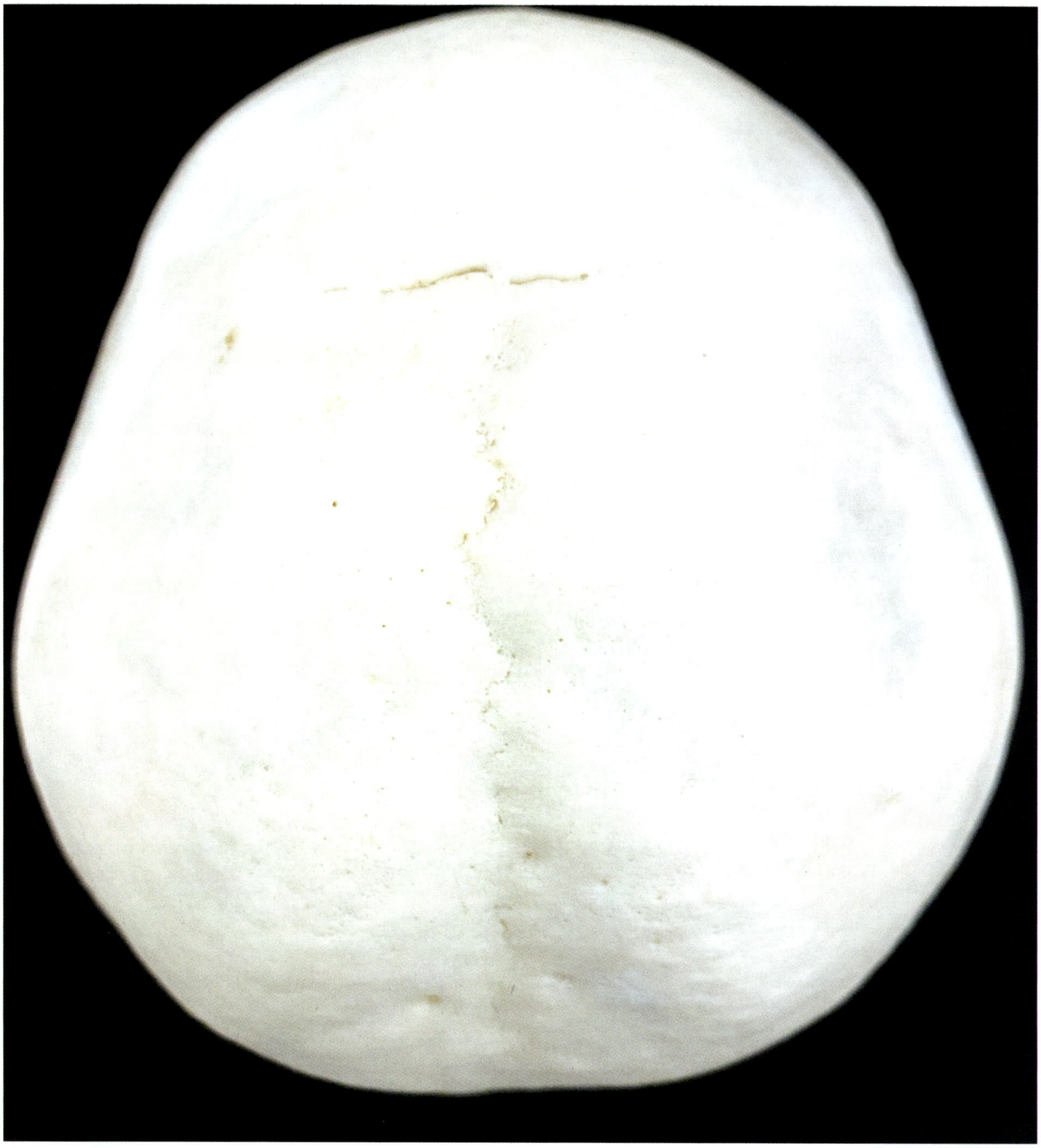

Figure 105. Superior view 95-year-old Chinese female with nearly complete obliteration of all sutures.

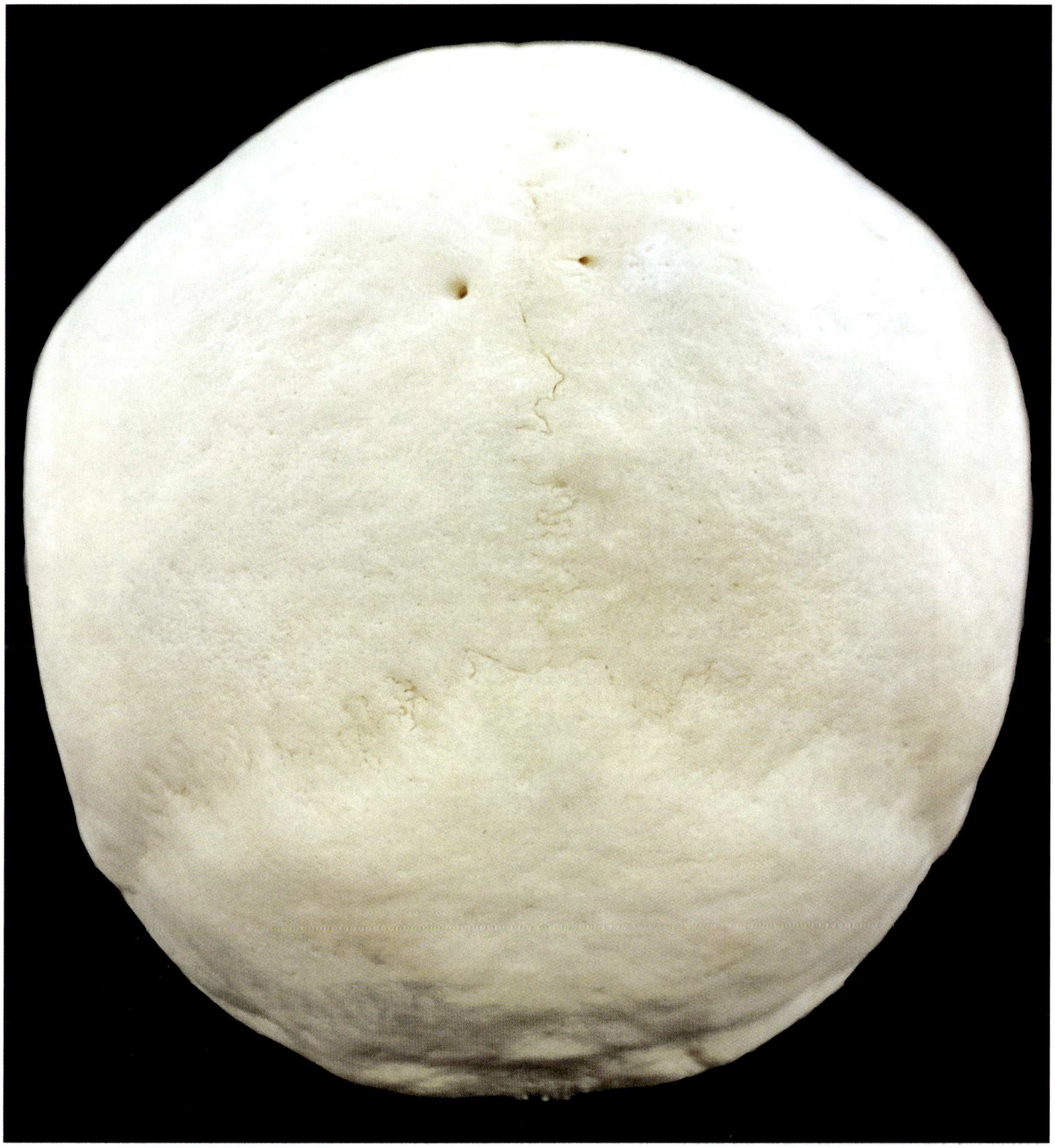

Figure 106. Posterior view 95-year-old Chinese female with bilateral parietal foramina. Parietal foramina are normal anatomical variants that transmit emissary veins.

5. FILIPINO FEMALE (SOUTHEAST ASIA)

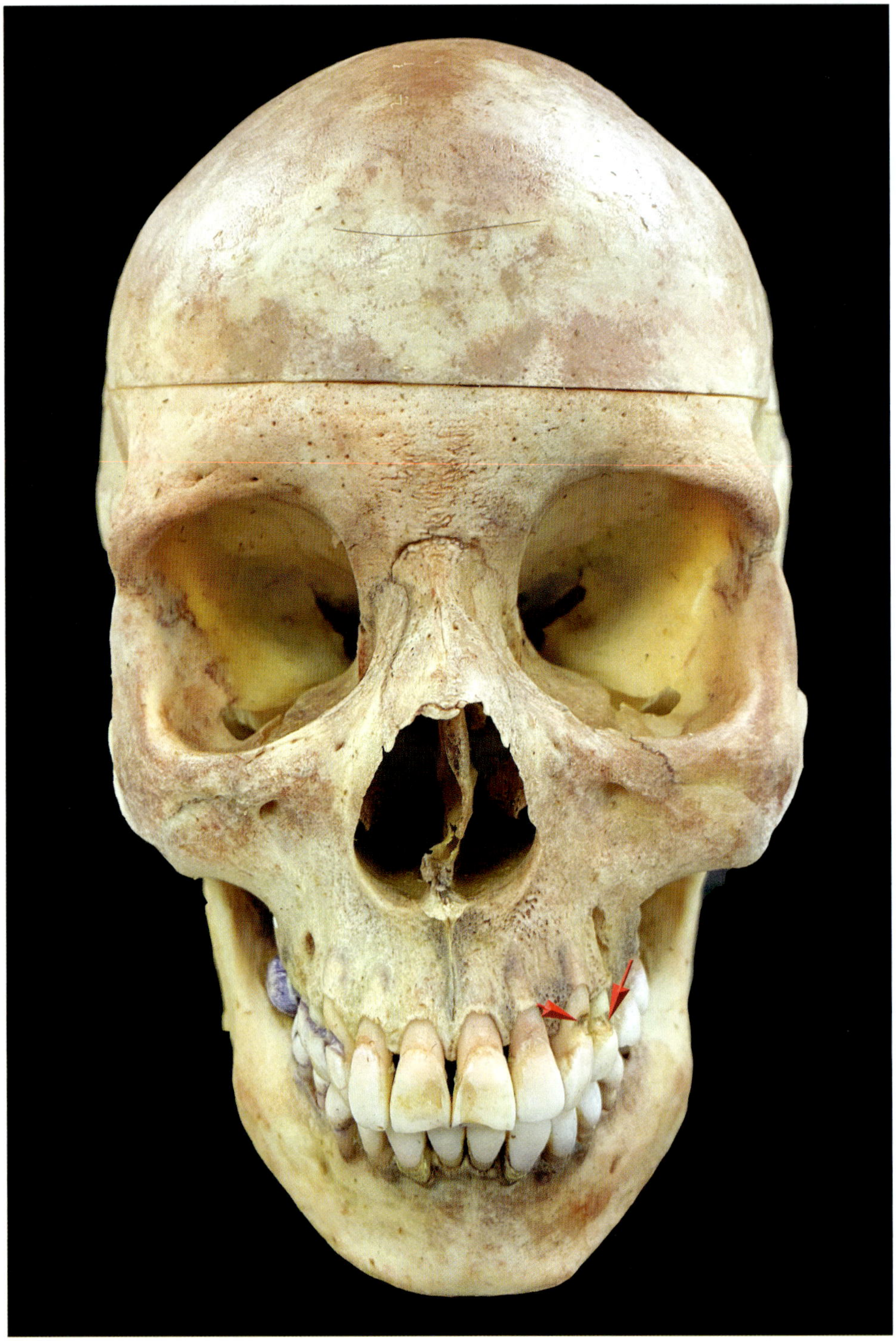

Figure 107. Anterior view of 44-year-old Filipino female (Southeast Asia) with blunt supraorbital margins, and intermediate/medium interorbital and nasal breadth. Note the waisted nasal bones, and a medium anterior nasal spine. Observe the toothbrush grooves (arrow) along many of the maxillary teeth (see Figure 109).

Figure 108. Nasal bone contour in a 44-year-old Filipino female (Southeast Asia).

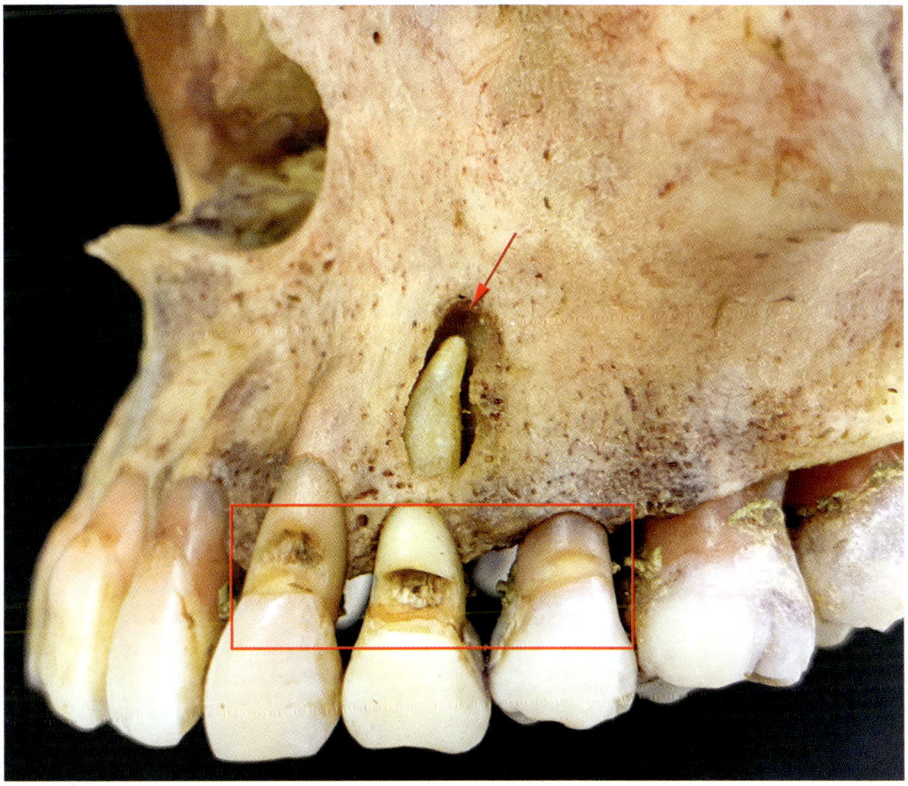

Figure 109. Periapical lesion (probably an abscess) (arrow) and toothbrush grooves (rectangle) in three maxillary left teeth. The toothbrush groove in the canine tooth exposed the pulp chamber, likely resulting in the periapical lesion.

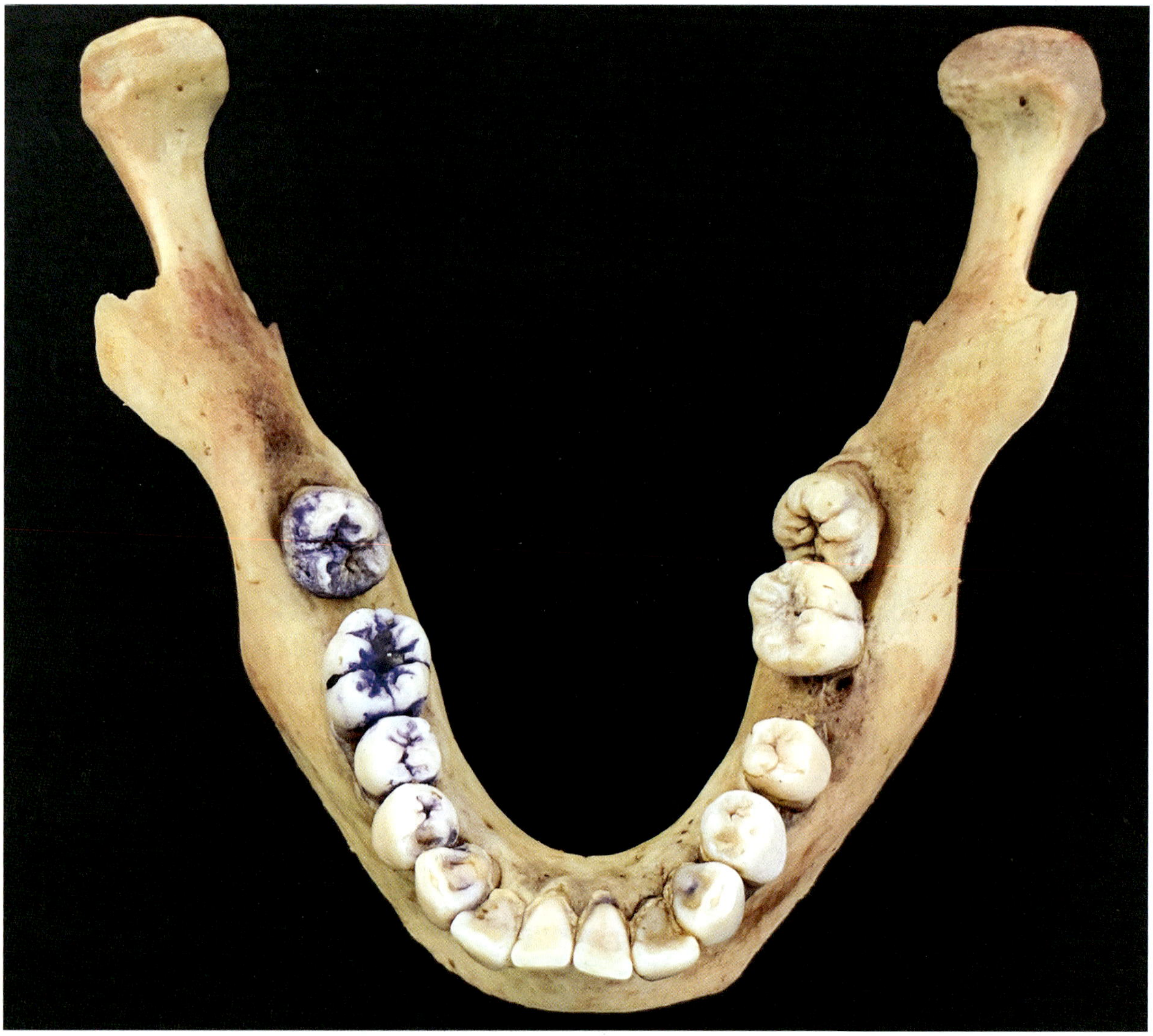

Figure 110. Mandible of 44-year-old Filipino female (Southeast Asia). Consultation with an oral surgeon revealed that the blue stain in some of the teeth is of unknown etiology.

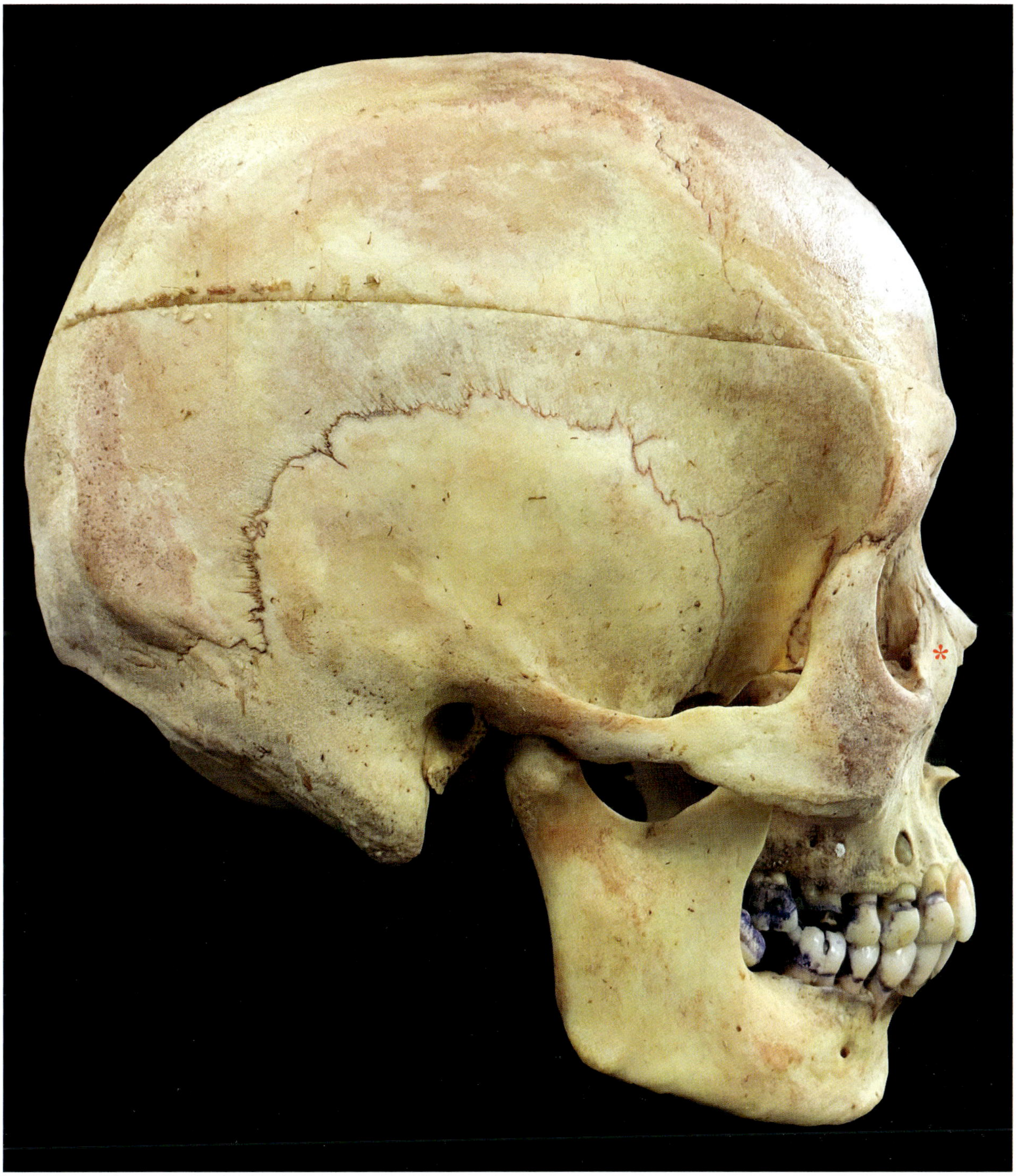

Figure 111. Right lateral view of 44-year-old Filipino female (Southeast Asia) with a small brow ridge and alveolar prognathism. Note the small and straight nasal bones, the anterior nasal spine, robust malar bone and short frontal process of the maxilla (*). There are several periapical lesions in the maxilla.

Figure 112. Left lateral view of 44-year-old Filipino female (Southeast Asia) with a vertical frontal bone, alveolar prognathism, small mastoid process, a periapical lesion (arrow) in the maxillary left first premolar, and a high vertex (peak of the skull near bregma).

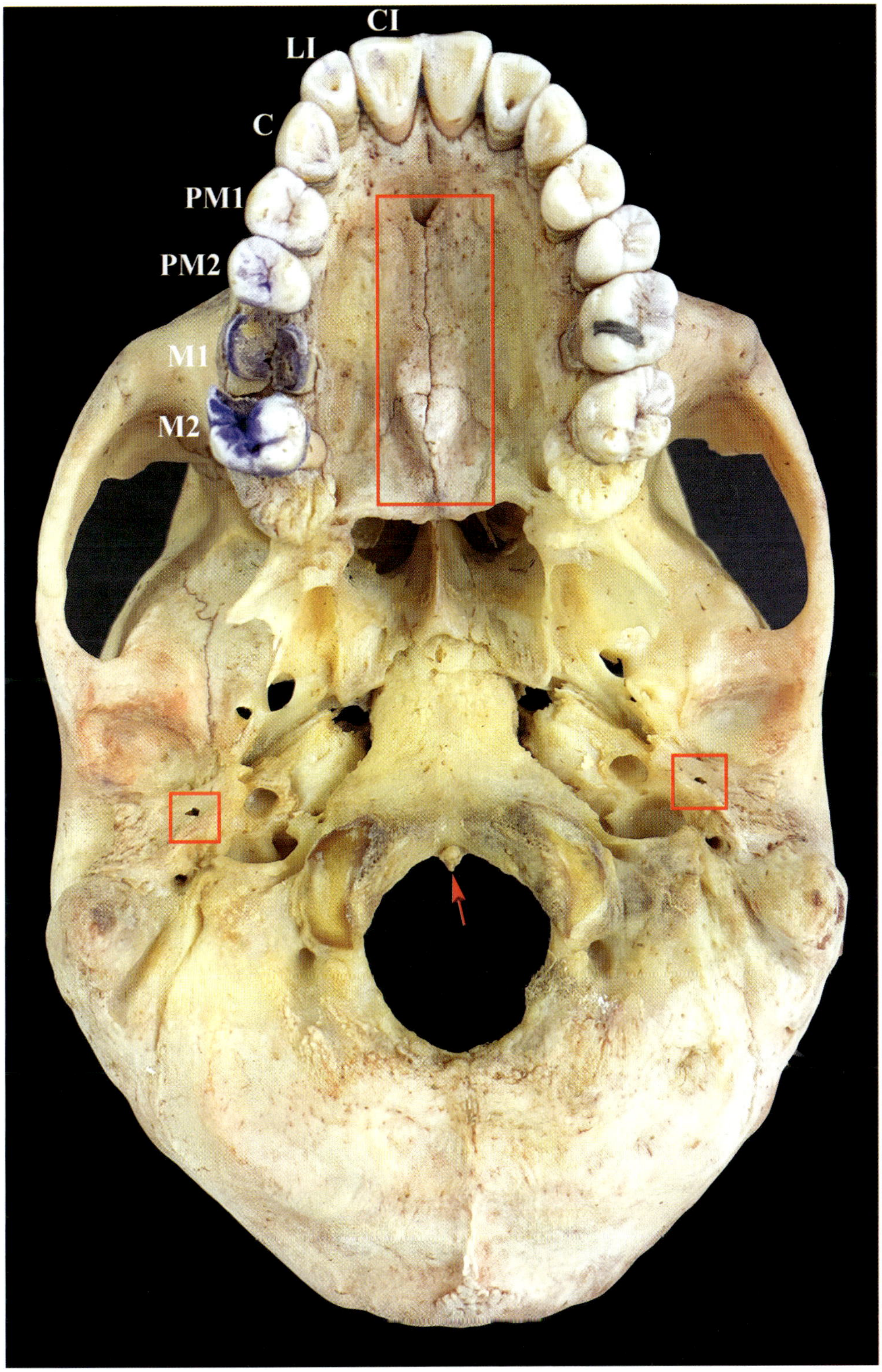

Figure 113. Inferior view 44-year-old Filipino female (Southeast Asia) with small torus palatinus (rectangle), possible calcified apical ligament (arrow), and tympanic dehiscence (square). Take note of the shovel-shaped incisors (CI and LI) and canines (C) consisting of raised marginal ridges that sometimes must be filed down ("enameloplasty") to properly occlude the teeth. Also note the "peg-lateral" incisor that can be associated with congenitally missing teeth (Personal Communication Dr. J. DeMeo).

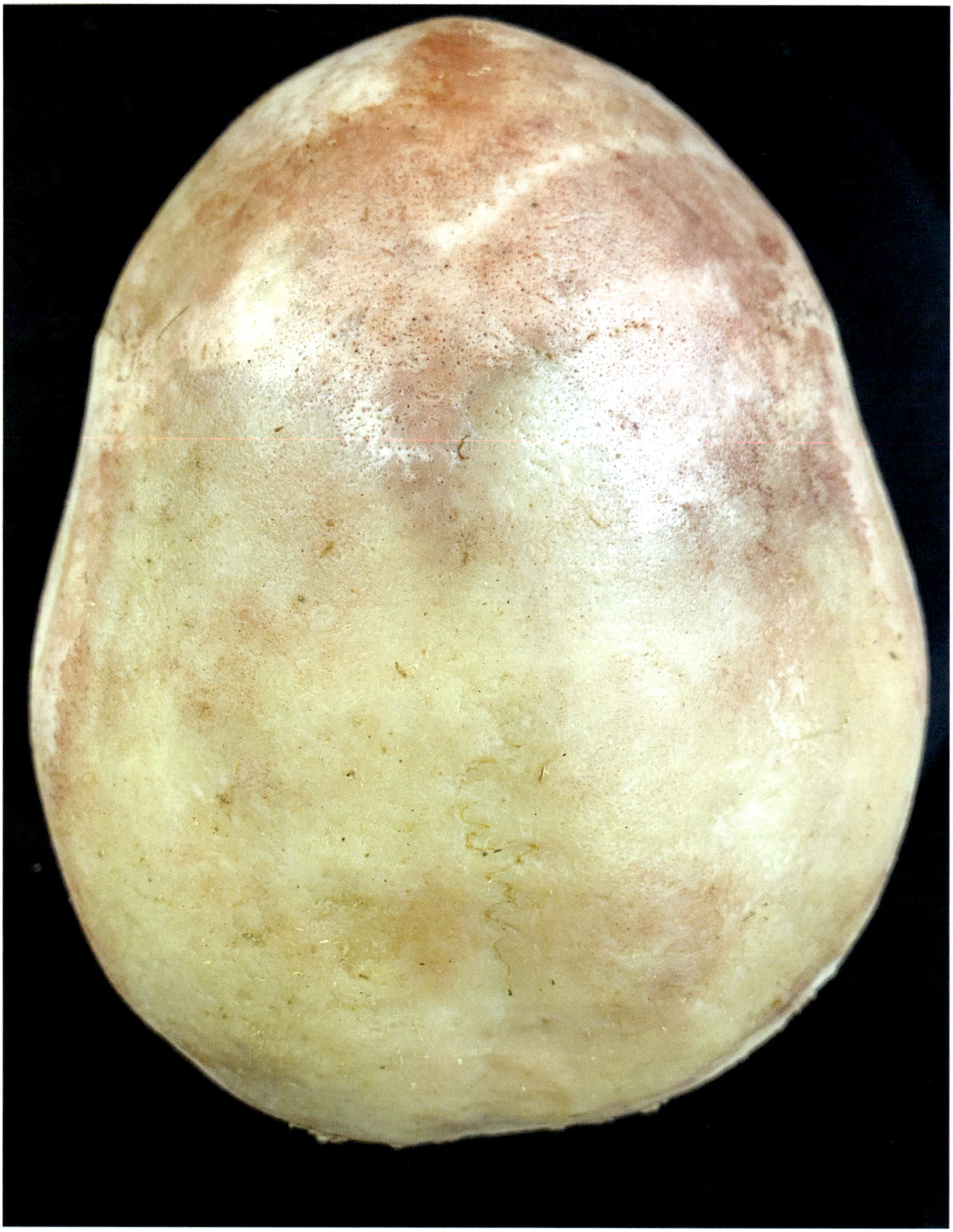

Figure 114. Superior view of 44-year-old Filipino female (Southeast Asia) with complete obliteration of the sutures.

Figure 115. Posterior view of 44-year-old Filipino female (Southeast Asia) with an angular skull and bossing (squaring) of the parietal eminences.

6. FILIPINO MALE

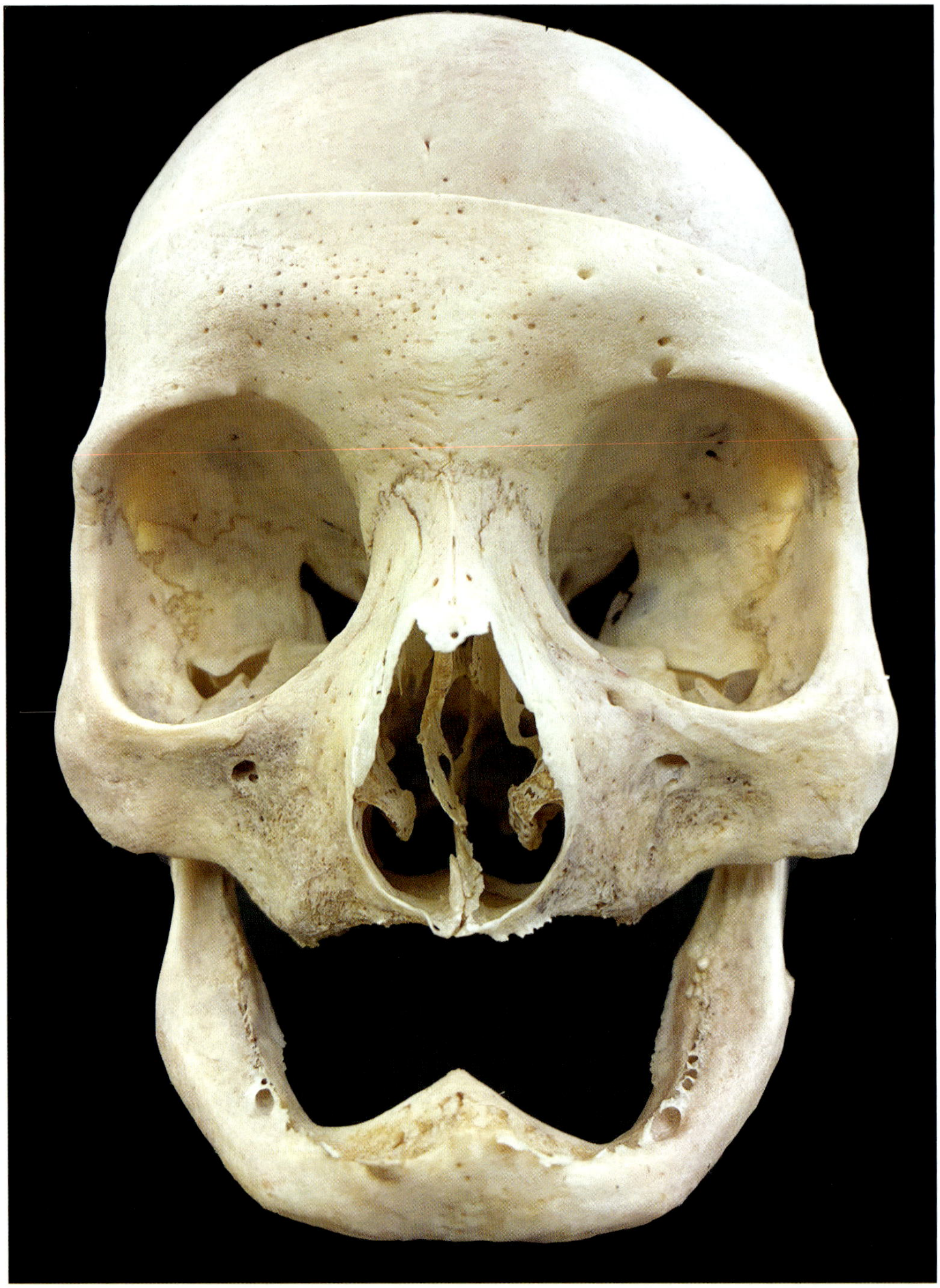

Figure 116. Anterior view 71-year-old Filipino male with severe resorption of the maxilla and body of the mandible. This small/gracile skull demonstrates the difficulty and limitation of utilizing metrics and visual examination (morphology) in establishing ancestry and sex due to antemortem loss of teeth. Bony resorption has resulted in a paper-thin maxilla/nasal floor. Due to extreme alveolar resorption, the maxilla and mandible can only be approximated. Note the peak in the mandible due to greater resorption of bone lateral to the midline.

Figure 117. Nasal bone contour in a 71-year-old Filipino male.

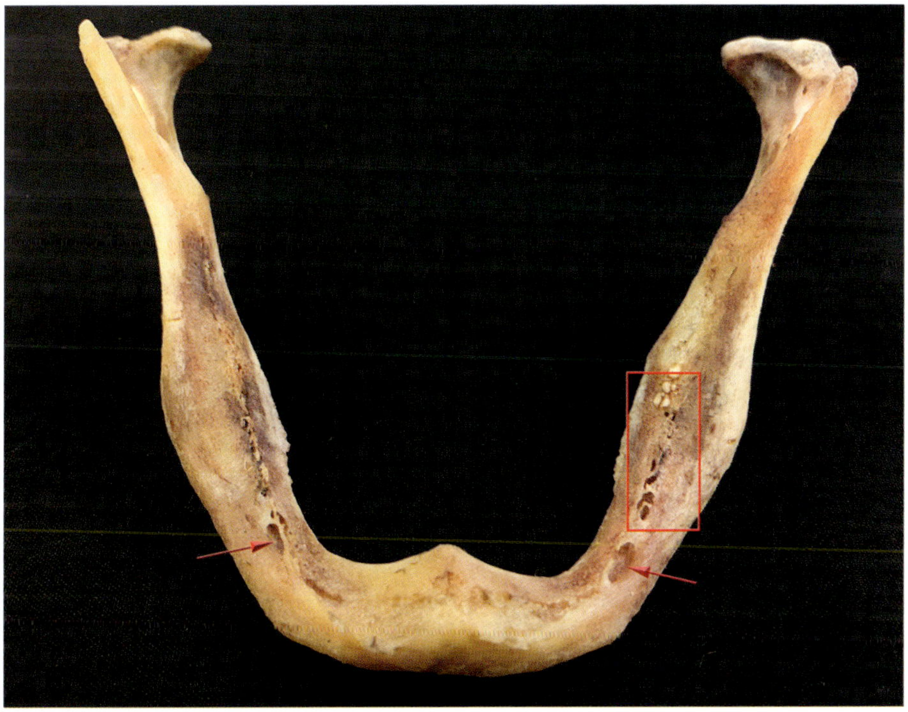

Figure 118. Superior view of mandible in a 71-year-old Filipino male showing extreme resorption of the condyles due to osteoarthritis. Note the altered size and shape of the mental foramina (arrow) due to antemortem tooth loss. The mental foramina transmit neurovascular bundles and, as in this individual, can lie along the alveolar crest due to bone resorption. This condition is known as foramina transposition (Charalampakis et al. 2017) and results from antemortem tooth loss and bone resorption. Note multiple perforations into the mental canal (rectangle) in the left hemi-mandible.

TABLE 15.
Cranial Measurements in a 71-year-old Filipino Male

Measurement	(mm)	Right	Left
Maximum Cranial Length (g-op, GOL)	174		
Maximum Cranial Breadth (eu-eu, XCB)	132		
Bizygomatic Breadth (zy-zy, ZYB)	131		
Basion-Bregma Height (ba-b, BBH)	132		
Cranial Base Length (ba-n, BNL)	99		
Basion-Prosthion (ba-pr, BPL)	*		
Maxillo-Alveolar Breadth (ecm-ecm, MAB)	49		
Maxillo-Alveolar Length (pr-alv, MAL)	*		
Biauricular Breadth (au-au, AUB)	123		
Upper Facial Height (n-pr)	*		
Minimum Frontal Breadth (ft-ft, WFB)	93		
Upper Facial Breadth (fmt-fmt)	102		
Nasal Height (n-ns, NLH)	48		
Nasal Breadth (al-al, NLB)	24		
Orbital Breadth (d-ec, OBB)	39		L
Orbital Height (OBH)	36		L
Biorbital Breadth (ec-ec, EKB)	94		
Interorbital Breadth (d-d, DKB)	21		
Frontal Chord (n-b, FRC)	112		
Parietal Chord (b-l, PAC)	114		
Occipital Chord (l-o, OCC)	88		
Foramen Magnum Length (ba-o, FOL)	33		
Foramen Magnum Breadth (FOB)	26		
Mastoid Length (MDH)		26	24

*Not recordable due to resorption of maxilla

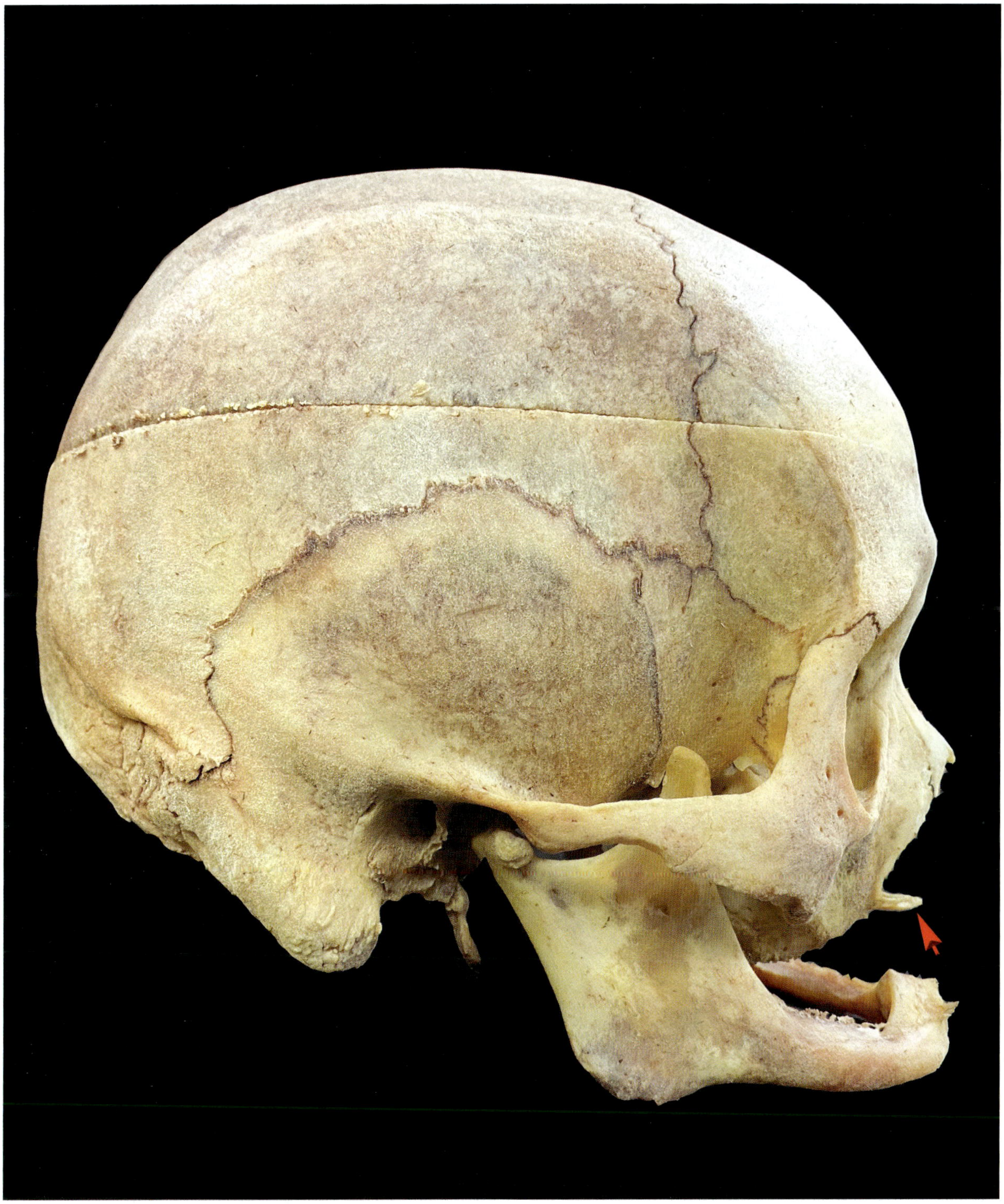

Figure 119. Right lateral view of 71-year-old Filipino male with severe resorption of the maxilla and body of the mandible. Note that resorption of the maxilla has exaggerated the length of the anterior nasal spine (arrow). Although the small size of this skull is suggestive of female, the developed nuchal crest, supramastoid crest, and robust mandibular ramus are strong indicators of male.

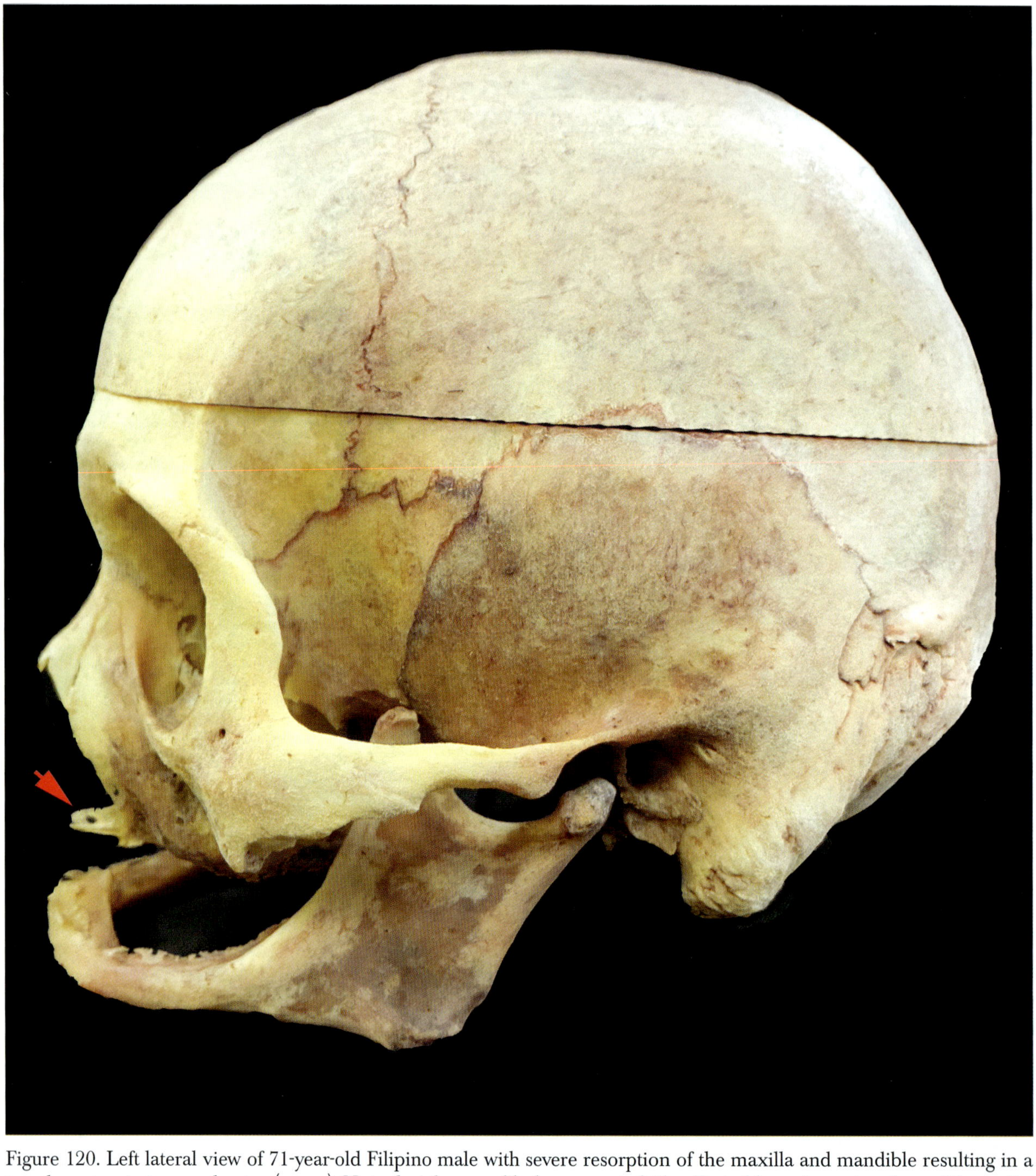

Figure 120. Left lateral view of 71-year-old Filipino male with severe resorption of the maxilla and mandible resulting in a translucent anterior nasal spine (arrow). Note that the mandibular ramus is little affected by antemortem loss of the teeth.

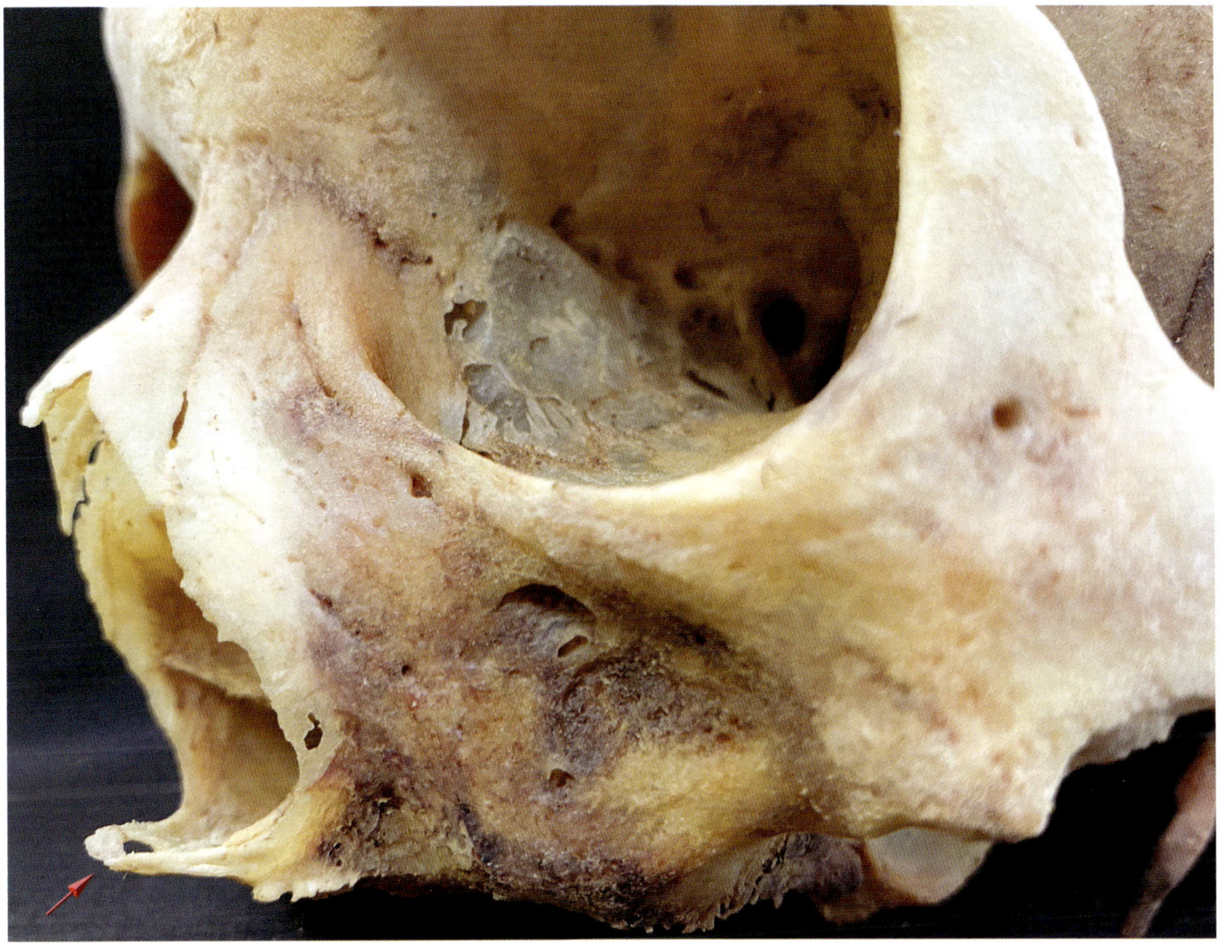

Figure 121. Left oblique view of 71-year-old Filipino male with severe resorption of the maxilla that appears to have exaggerated the length of the anterior nasal spine (arrow).

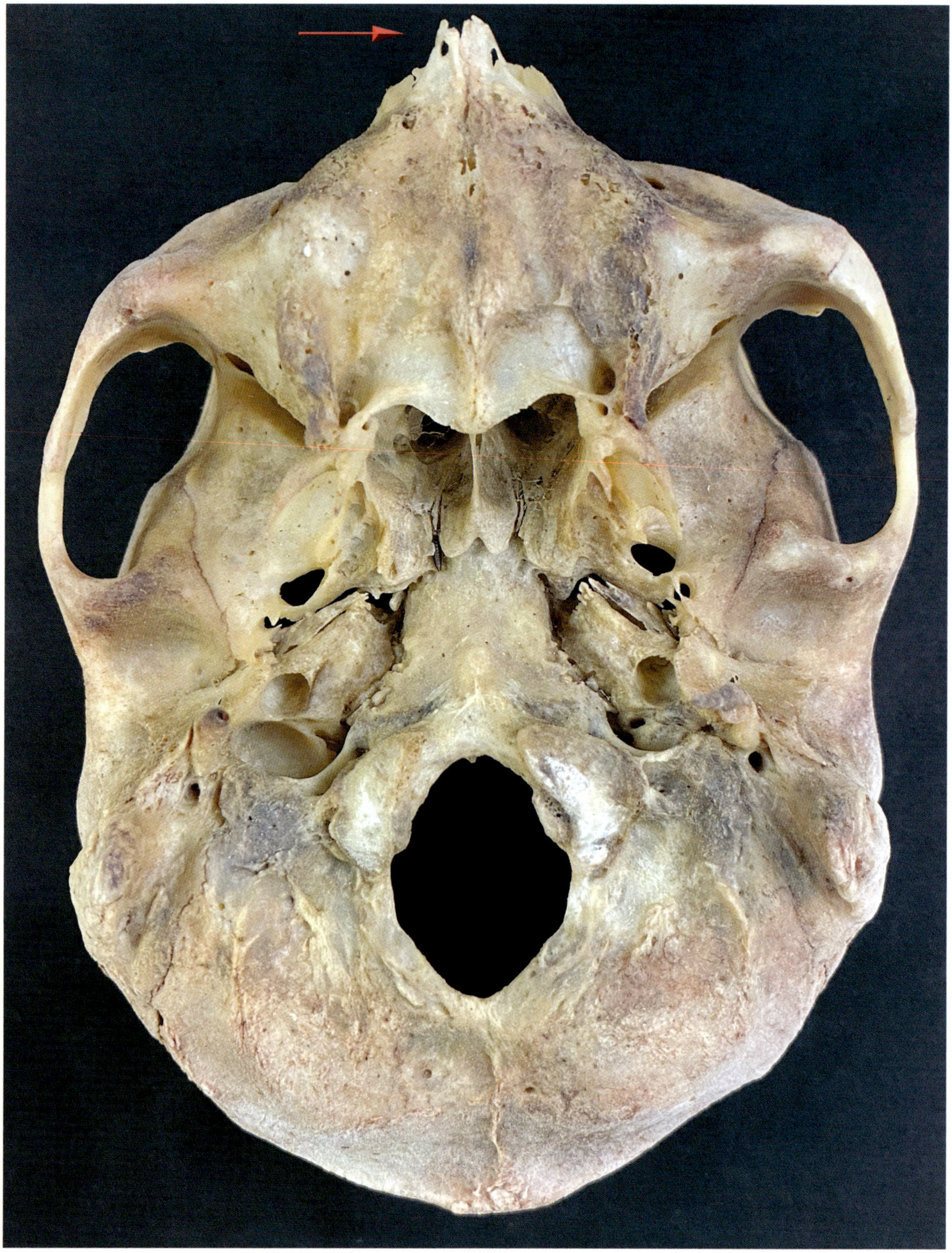

Figure 122. Inferior view of 71-year-old Filipino male with extreme resorption of the maxilla resulting in exaggeration of the anterior nasal spine (arrow).

Figure 123. Superior view of the cranium in a 71-year-old Filipino male with minimal obliteration of the sutures.

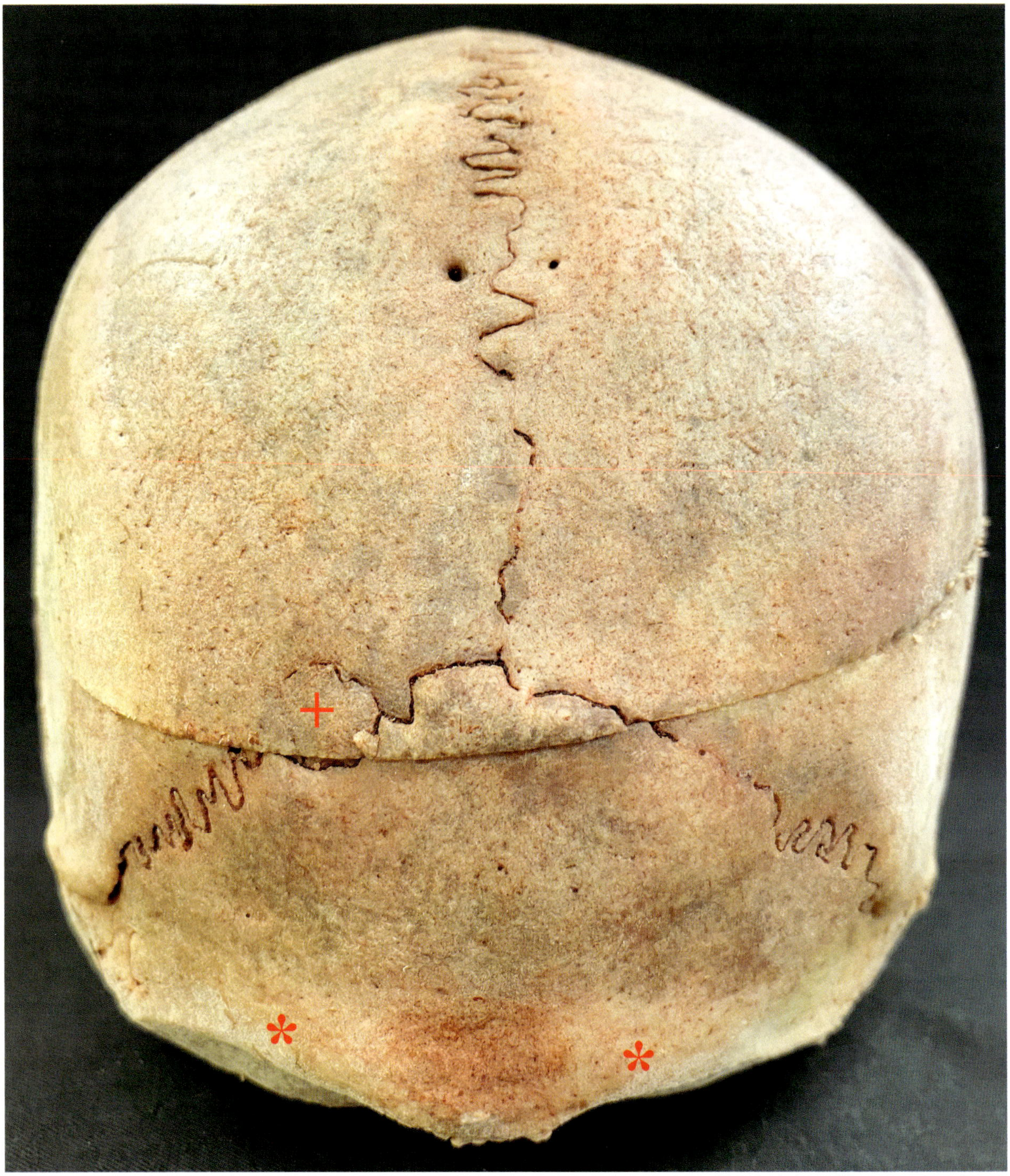

Figure 124. Posterior view of 71-year-old Filipino male with a well-developed nuchal crest (*) and a small lambdoidal ossicle (+). Although numerous lambdoidal ossicles (Wormian bones) are often present in Asians, they can be found in other ancestral groups as well.

7. JAPANESE MALE

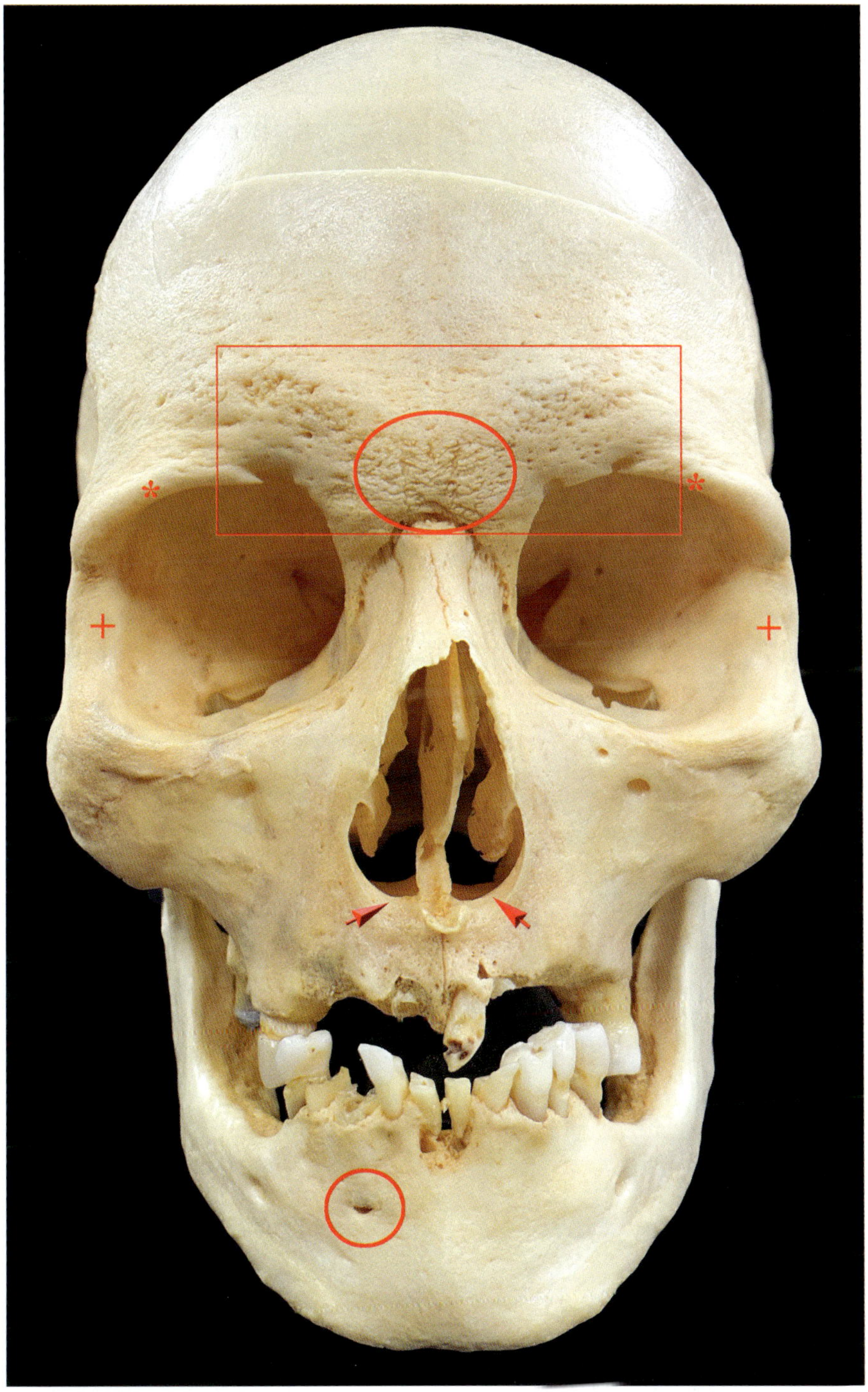

Figure 125. Anterior view of 96-year-old Japanese male with a vermiculate pattern of pits and grooves in the brow ridge (rectangle) and a trace supranasal suture (oval). Note the blunt supraorbital margin (*) and thickened lateral margin of the orbits (+). Also present is mild nasal guttering (arrow) and a small periapical abscess of the mandibular right canine (circle) that could be mistaken for a foramen that transmits vessels.

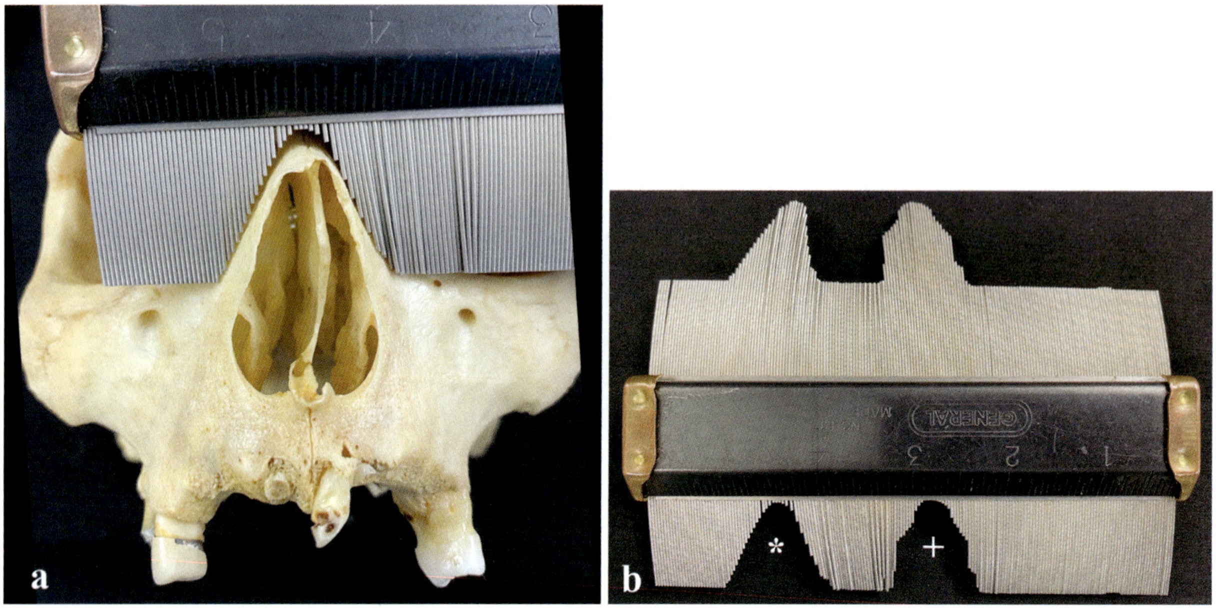

Figure 126a-b. Nasal bone contour in a 96-year-old Japanese male showing (a) the nasal contour with the gauge held perpendicular to the hard palate and (b) a comparison of the nasal contour when the contour gauge is held perpendicular (*) to the hard palate, and perpendicular to the nasal bones (+). The gauge was held at the same location along the nasal bones. Note that the contour of the nasal bones differs greatly depending on how the gauge is oriented.

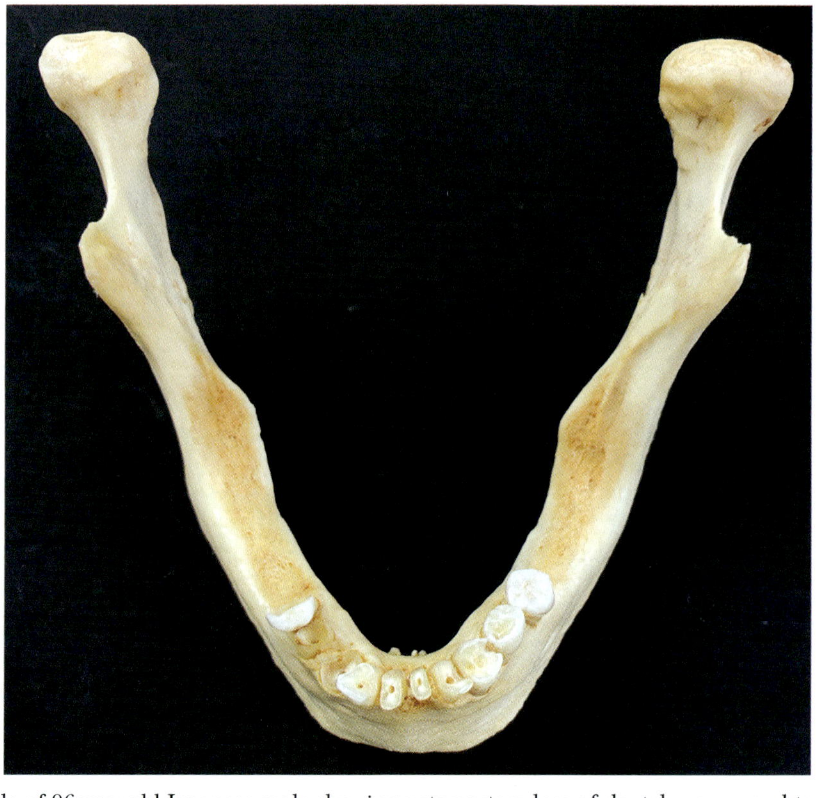

Figure 127. Mandible of 96-year-old Japanese male showing antemortem loss of dental crowns and teeth with subsequent alveolar resorption.

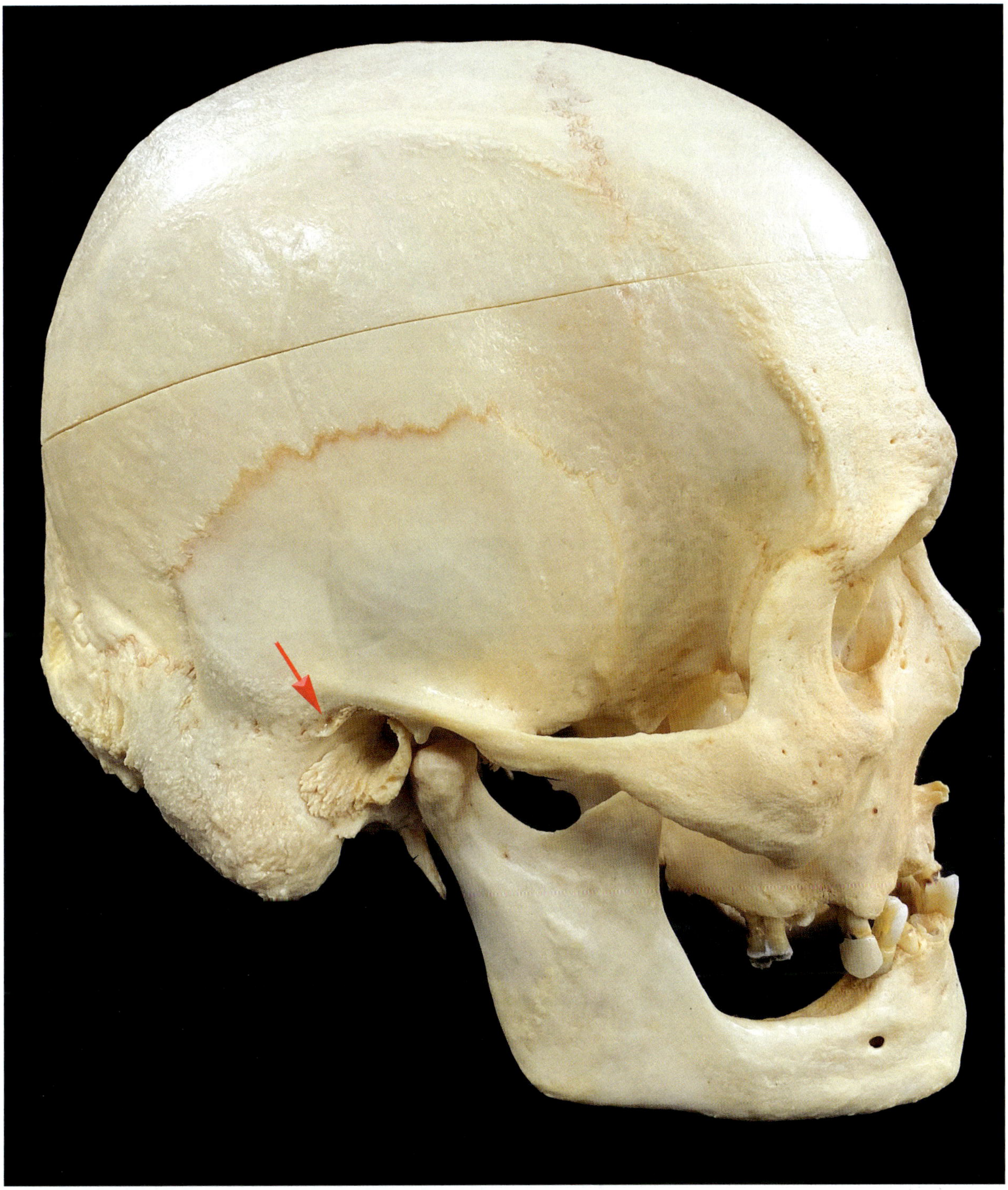

Figure 128. Right lateral view of 96-year-old Japanese male with resorption of the mandible due to antemortem loss of teeth and a flattened occipital region due in part to a developed nuchal crest. Also present are an anterior nasal spine, prominent zygoma, large suprameatal pit (arrow) (a male trait), a sloping frontal bone, depression at nasion, and a short anterior-posterior (brachycephalic) cranium.

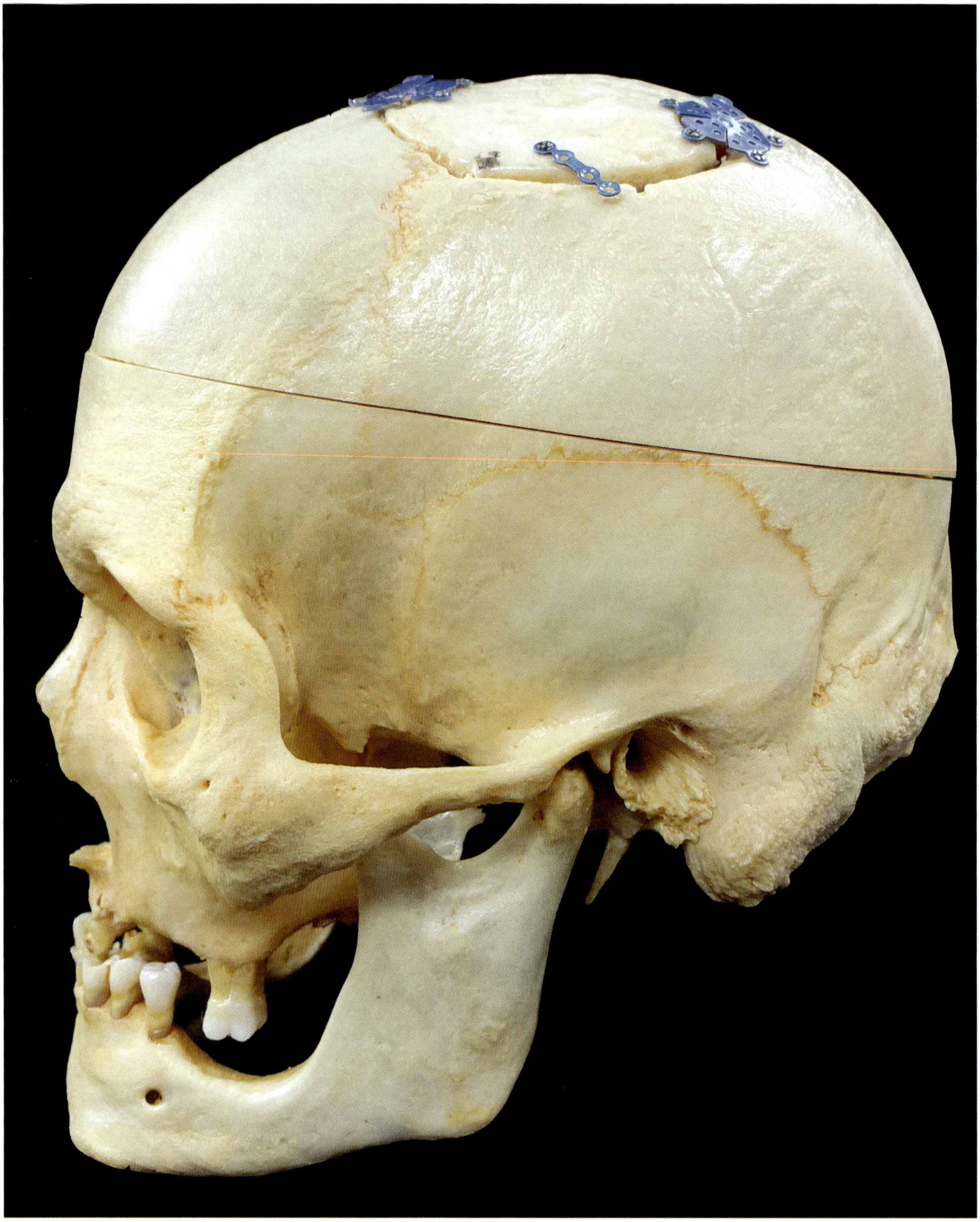

Figure 129. Left lateral view of 96-year-old Japanese male with a healed cranioplasty. Note the surgical hardware in the left parietal bone. Cranioplasty is the surgical repair of a defect in the cranium and the replacement of the "bone flap" with other material such as an acrylic or metallic plate.

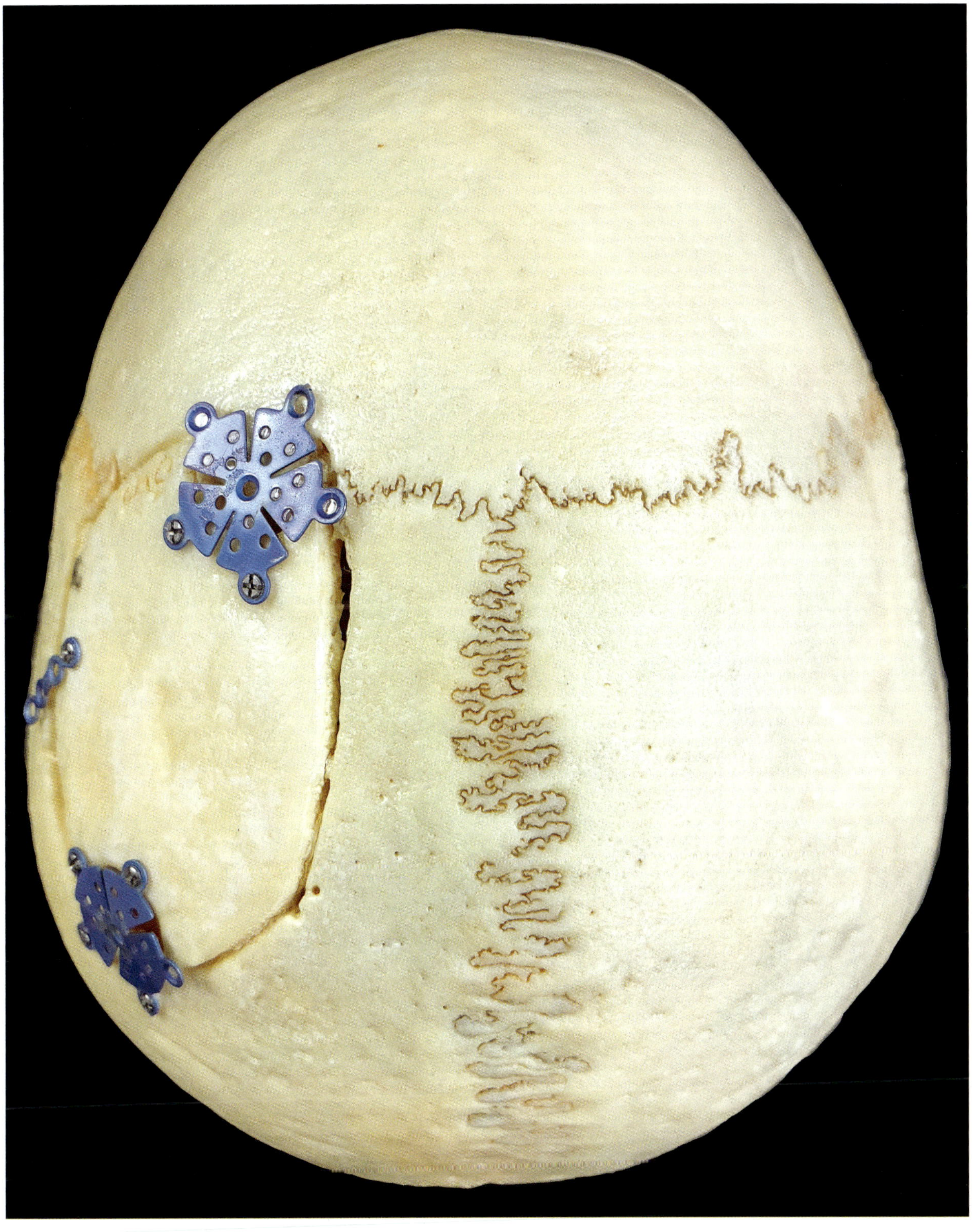

Figure 130. Superior view of 96-year-old Japanese male with healed cranioplasty using plates and screws.

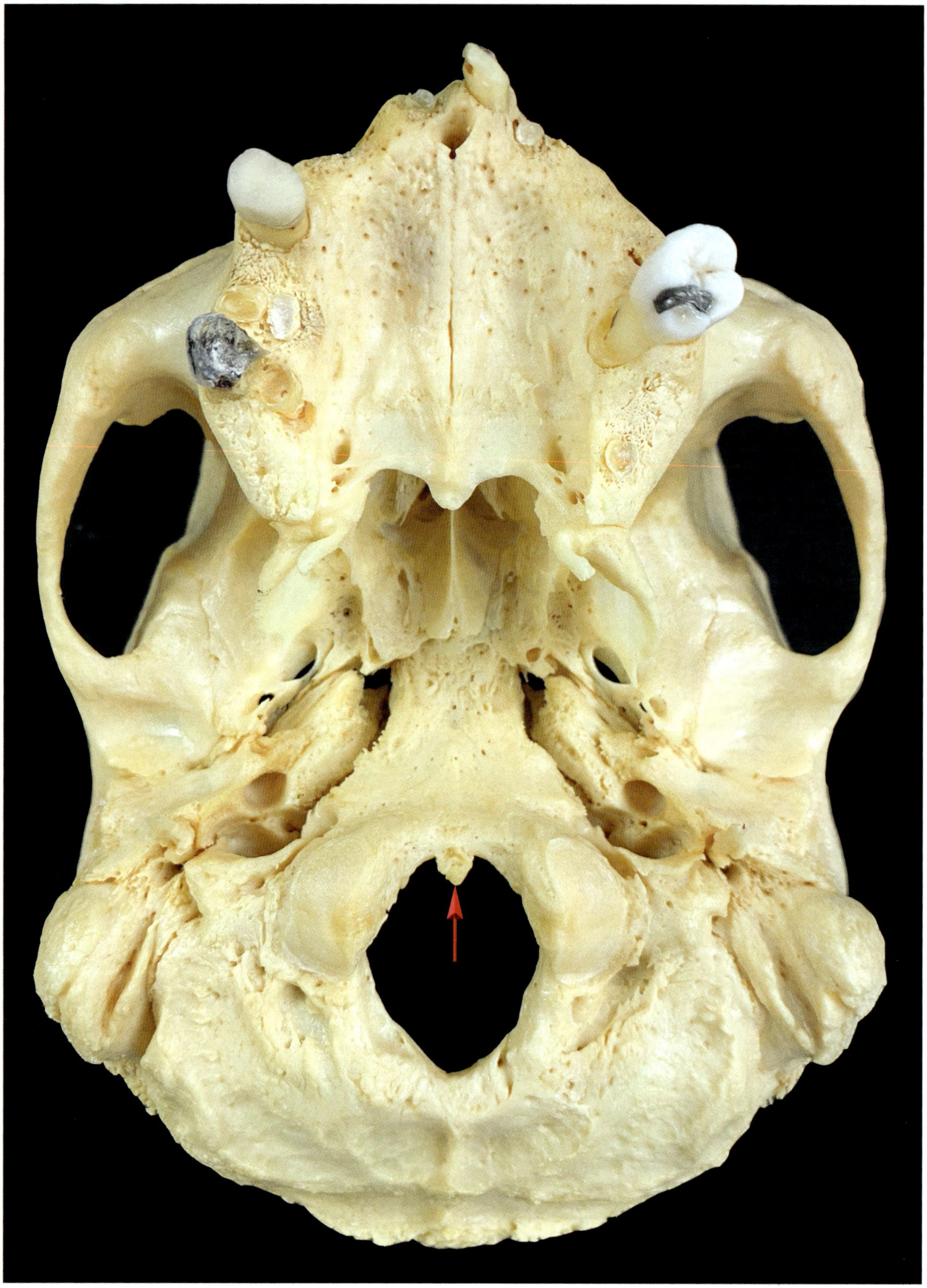

Figure 131. Inferior view of 96-year-old Japanese male. Note the small bony exostosis/calcified apical ligament (arrow) along the anterior margin of the foramen magnum, and the loss of several dental crowns with retention of tooth roots.

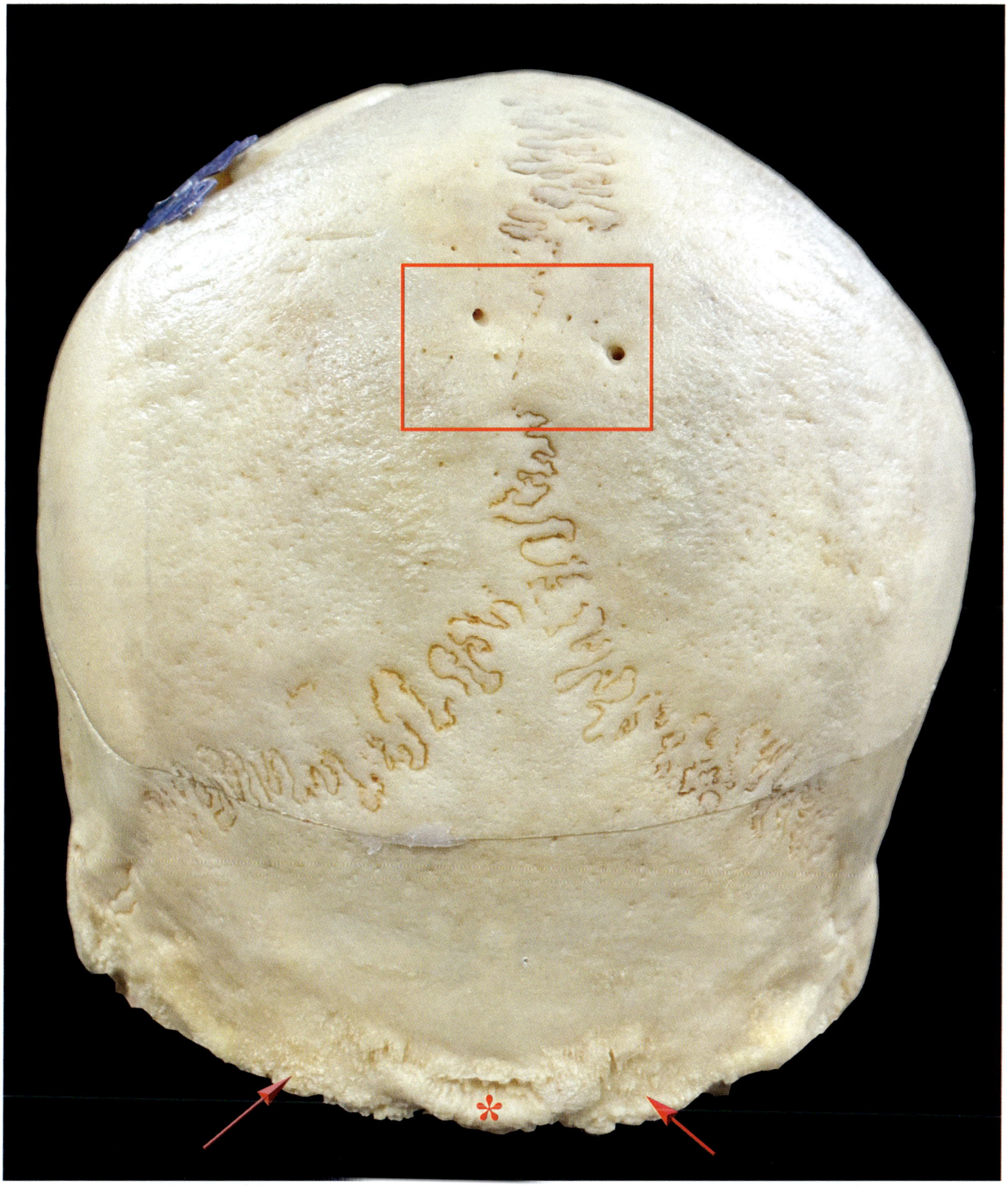

Figure 132. Posterior view of 96-year-old Japanese male with a highly developed and rugged nuchal crest (arrow) and inion hook (*), and multiple small and large parietal foramina in the area of obelion. Obelion (square) is that portion of the sagittal suture and parietal foramina that forms the division symbol (÷) obelos. Obelion is usually the simplest portion of the sagittal suture.

8. JAPANESE MALE

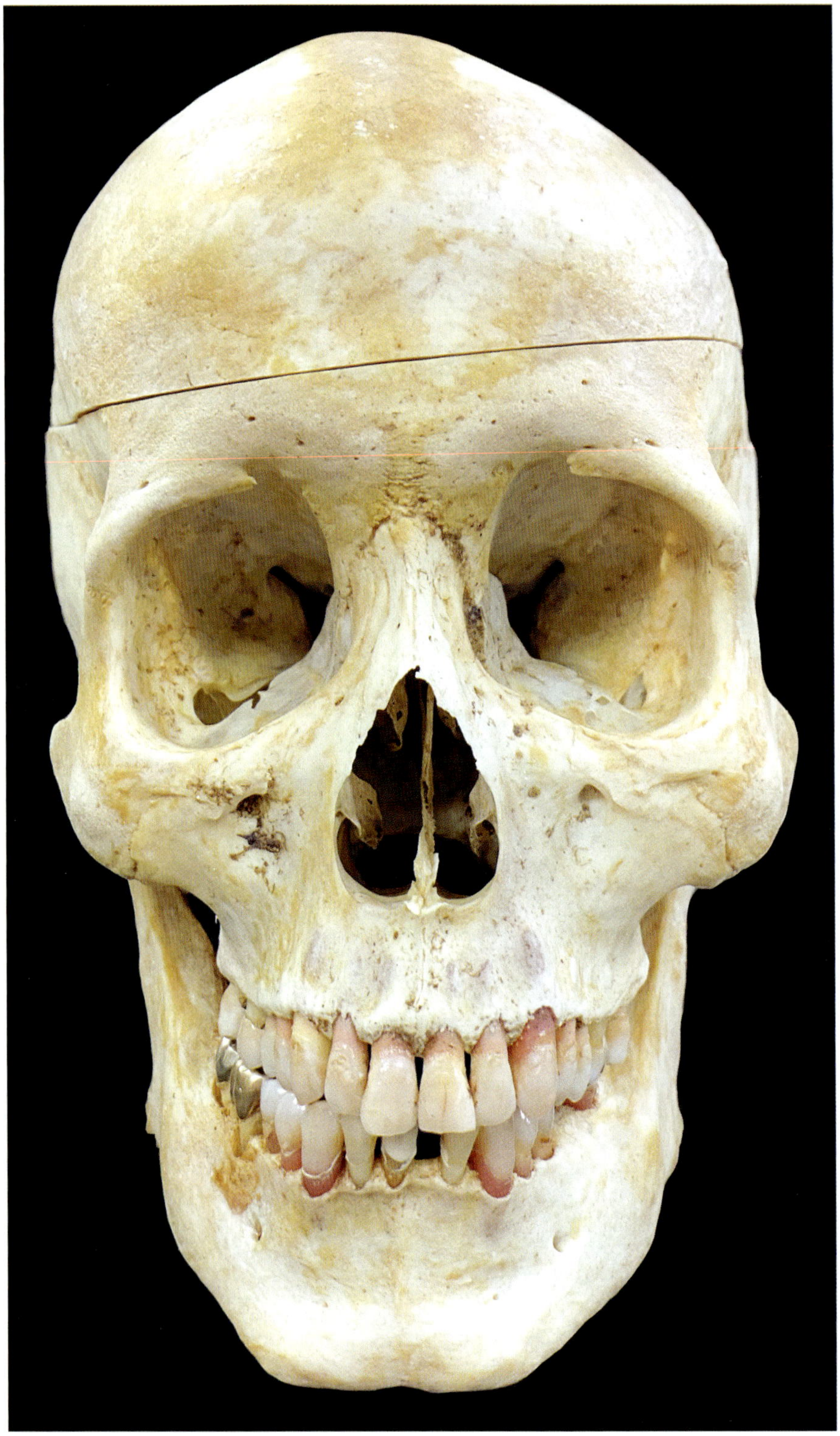

Figure 133. Anterior view 64-year-old Japanese male with "pink teeth" affecting the tooth roots (see Figure 134).

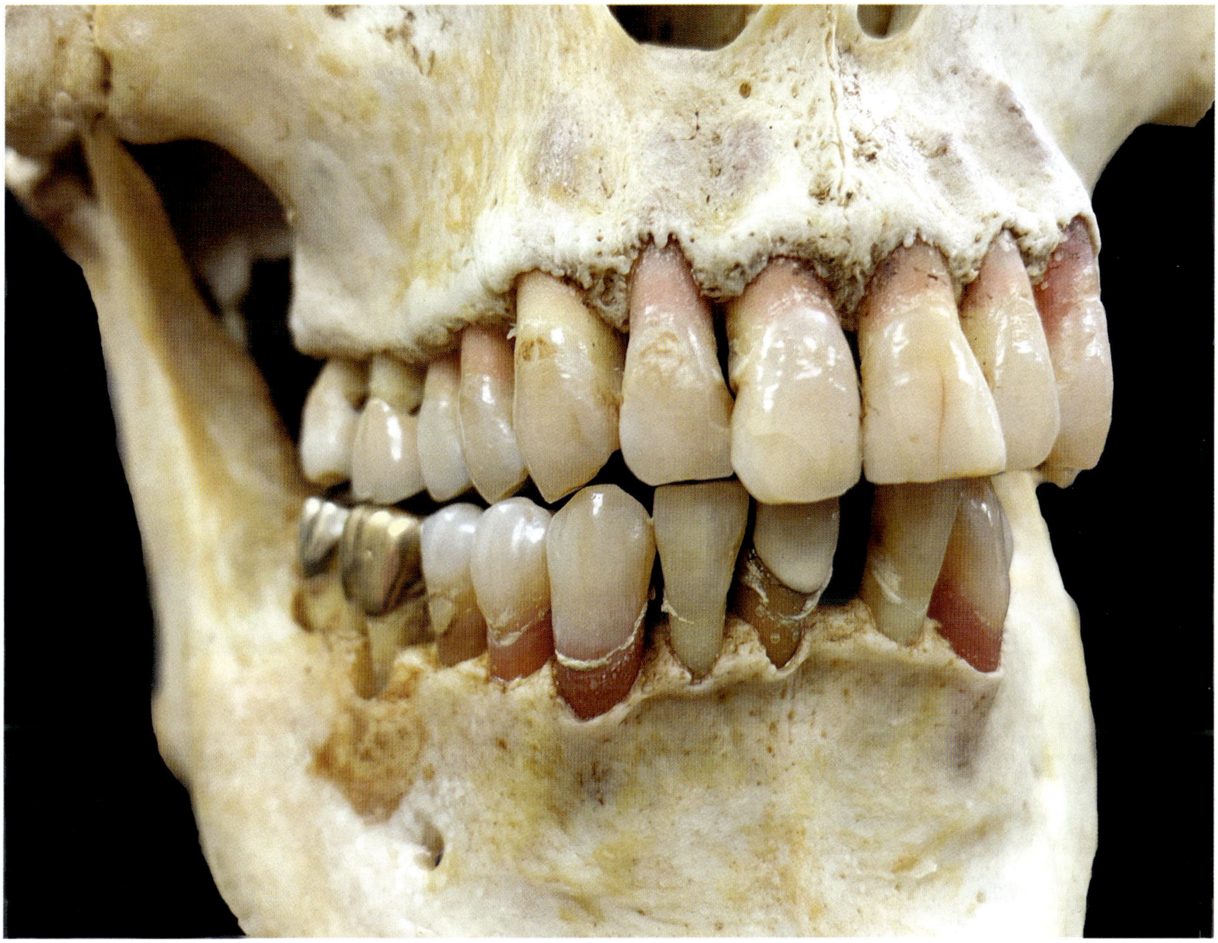

Figure 134. Right oblique view of 64-year-old Japanese male with pink teeth. The etiology of pink teeth is controversial and has been associated with asphyxia due to hanging, strangulation and drowning, severe trauma, and a moist environment postmortem (Charan Gowda et al. 2015).

Figure 135. Nasal bone contour in a 64-year-old Japanese male.

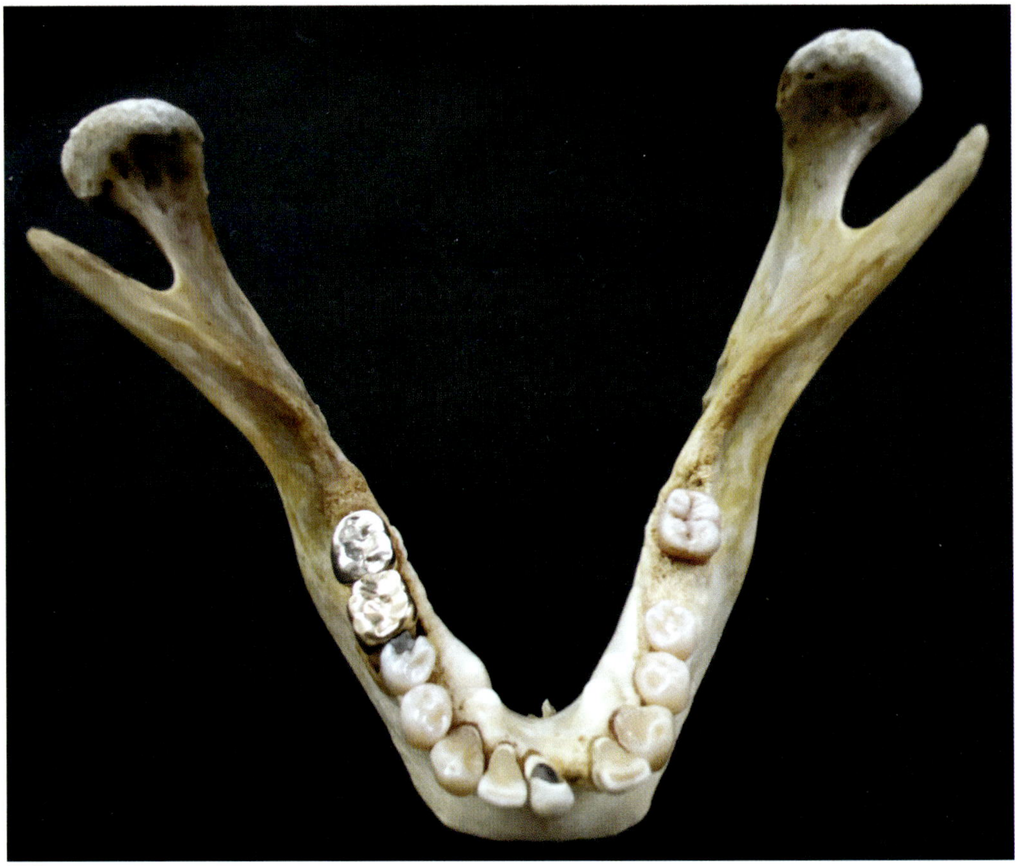

Figure 136. Mandible of 64-year-old Japanese male with lingual exostoses and a wide and square chin. Note that the rami are asymmetrical.

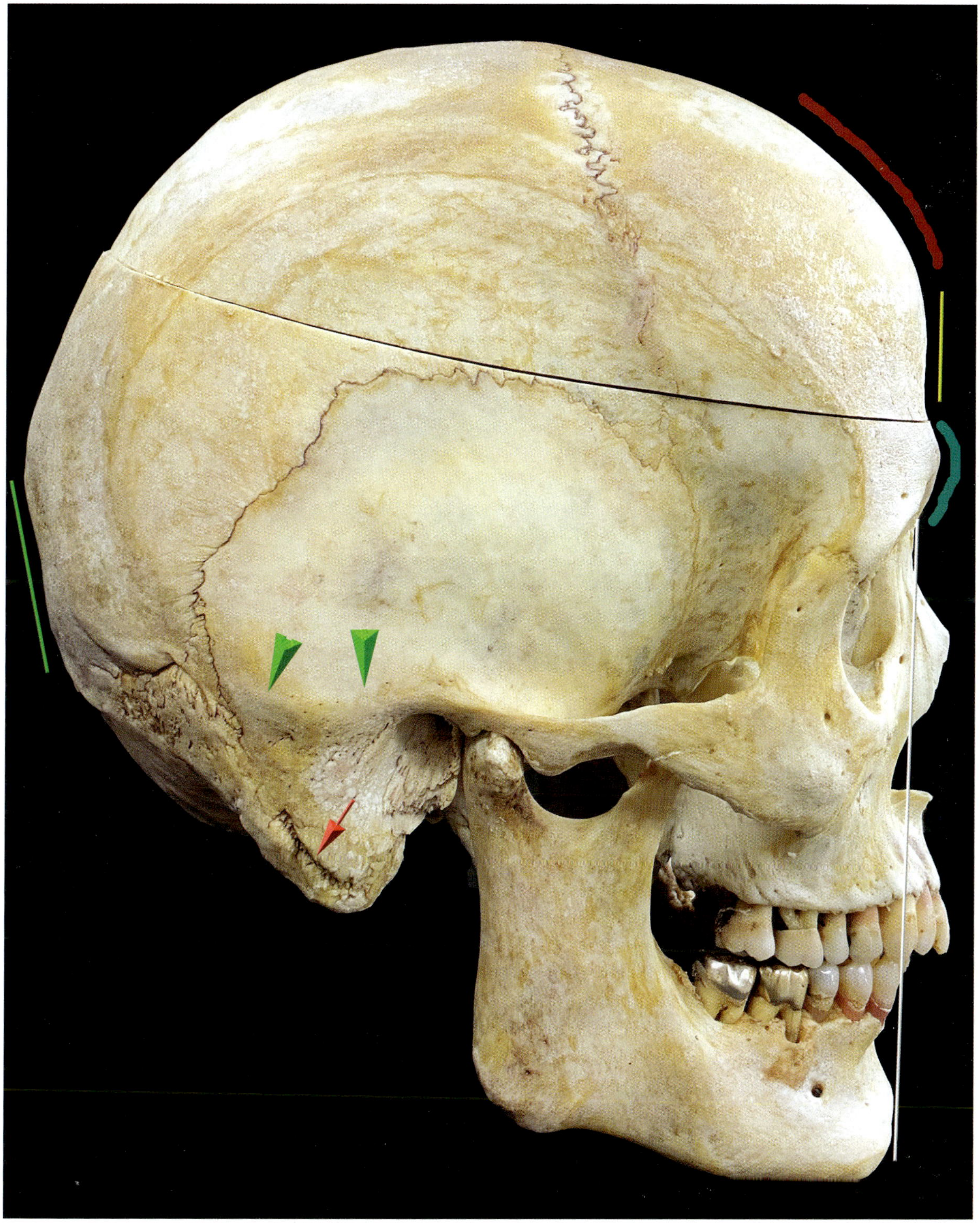

Figure 137. Right lateral view of 64-year-old Japanese male with a petrosquamousal suture (red arrow), also known in the radiological literature as Körner's septum, and minimal alveolar prognathism (straight white line). The petrosquamousal suture is a common anatomical variant most often found in males (Wojciechowski et al. 2019). Note the brow ridge (curved green line), short vertical frontal bone (yellow line) that becomes sloping (curved red line), flat occipital region (straight green line), and supramastoid crest (green arrowhead).

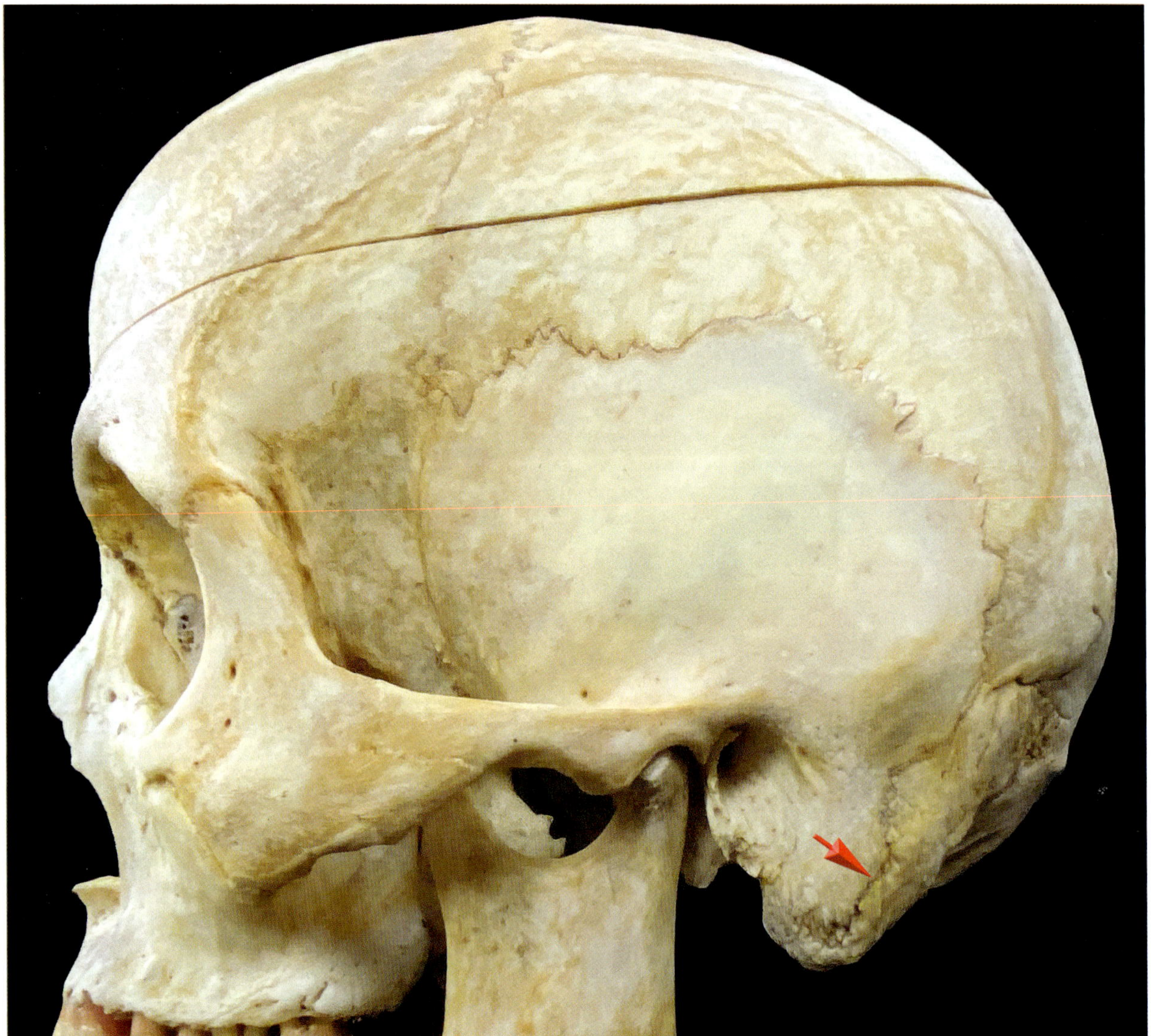

Figure 138. Left lateral view of 64-year-old Japanese male with petrosquamousal suture (arrow) (Mann et al. 2016), small nasal bones, an anterior nasal spine, robust malar bone, and minimal alveolar prognathism.

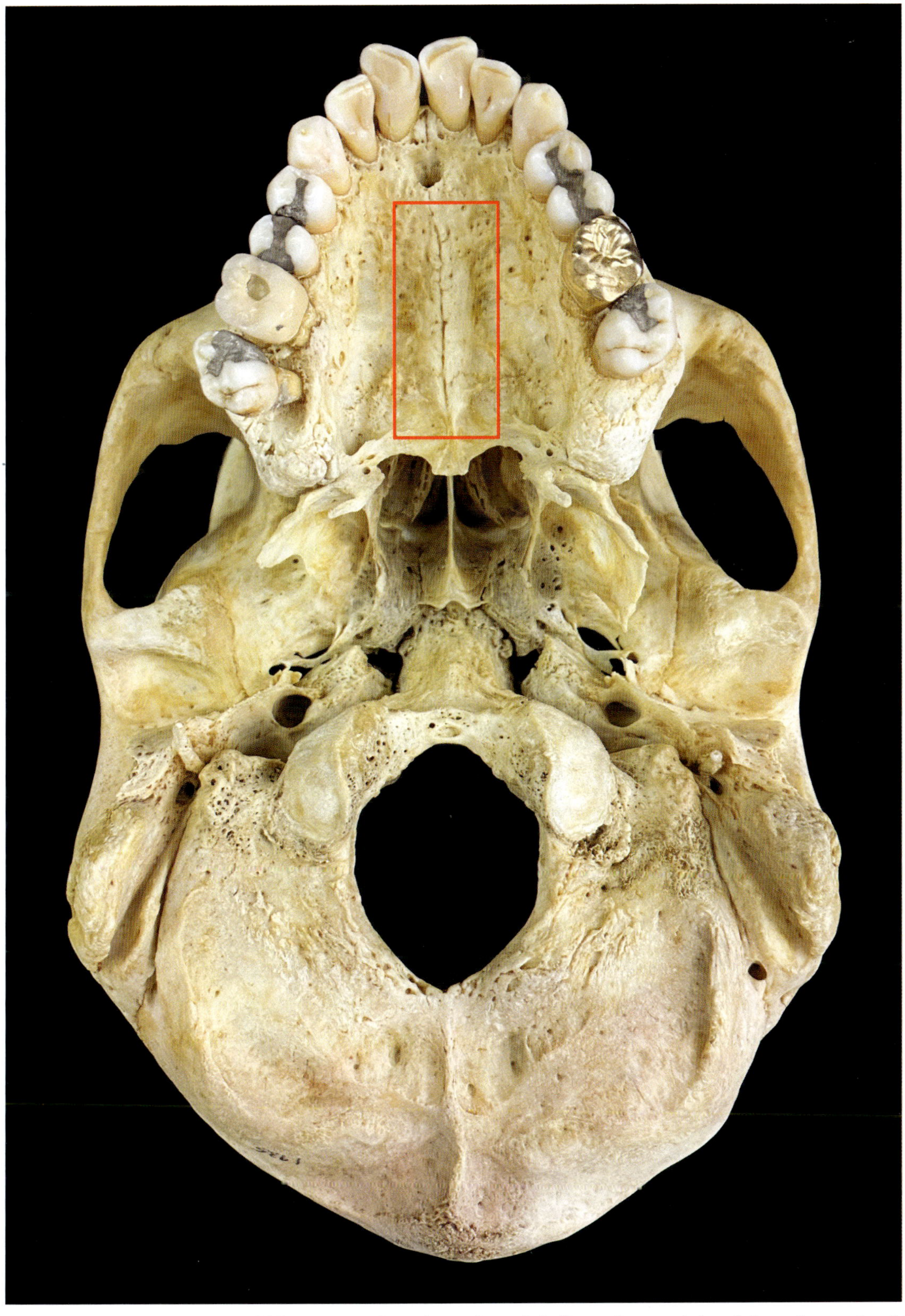

Figure 139. Inferior view of 64-year-old Japanese male with a wide and shallow palate, and shovel-shaped central and lateral incisors. Additionally, the central incisors are winged/angled and there is a small torus palatinus (rectangle).

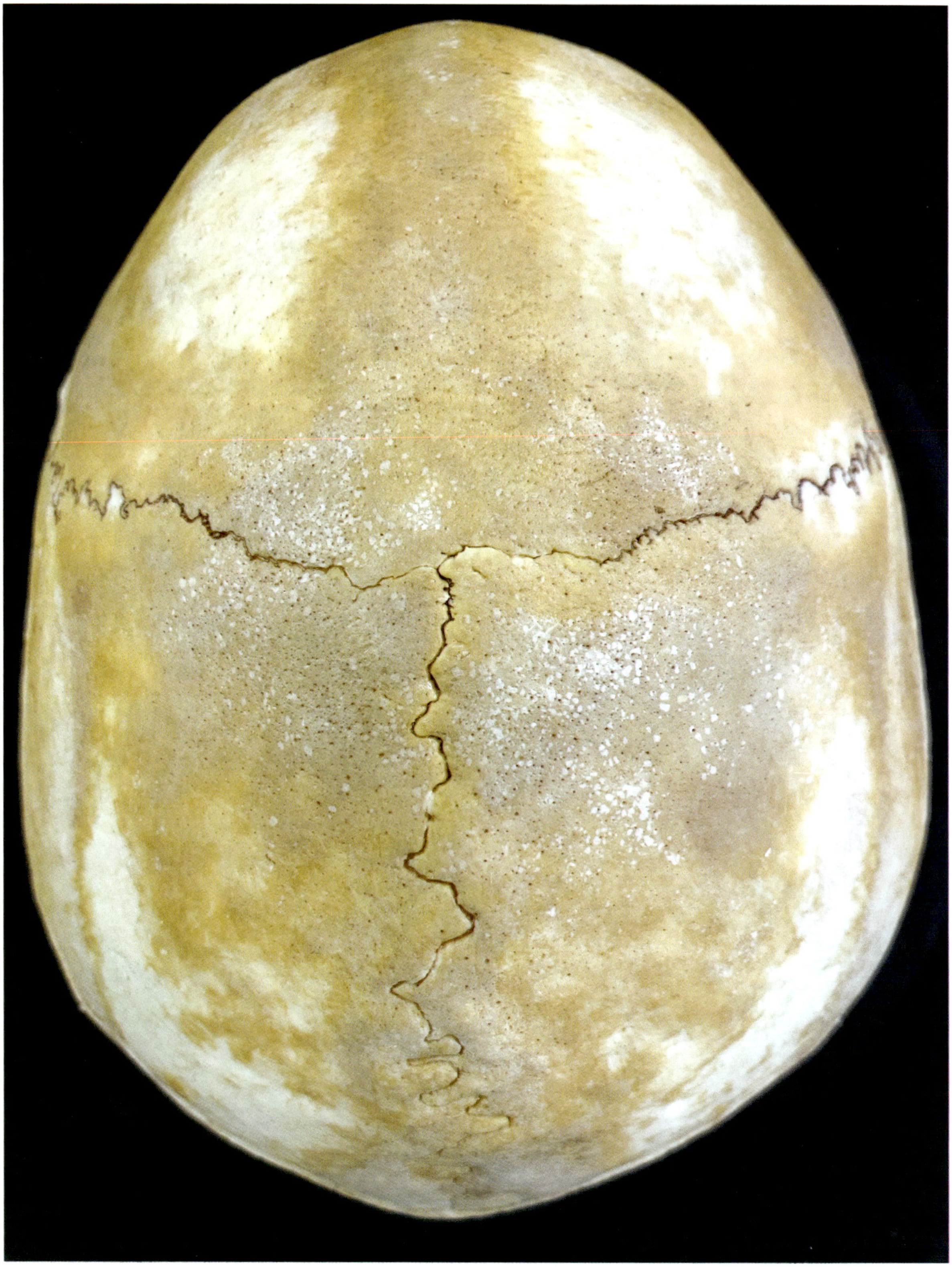

Figure 140. Superior view of 64-year-old Japanese male with an oval-shaped cranium and minimal obliteration of the sutures.

Figure 141. Posterior view of 64-year-old Japanese male with sagittal keeling (arrow) in the parietal bones, minimal obliteration of the sagittal and lambdoidal sutures, a moderately-sized nuchal crest, and a single right parietal foramen (circle).

9. JAPANESE MALE

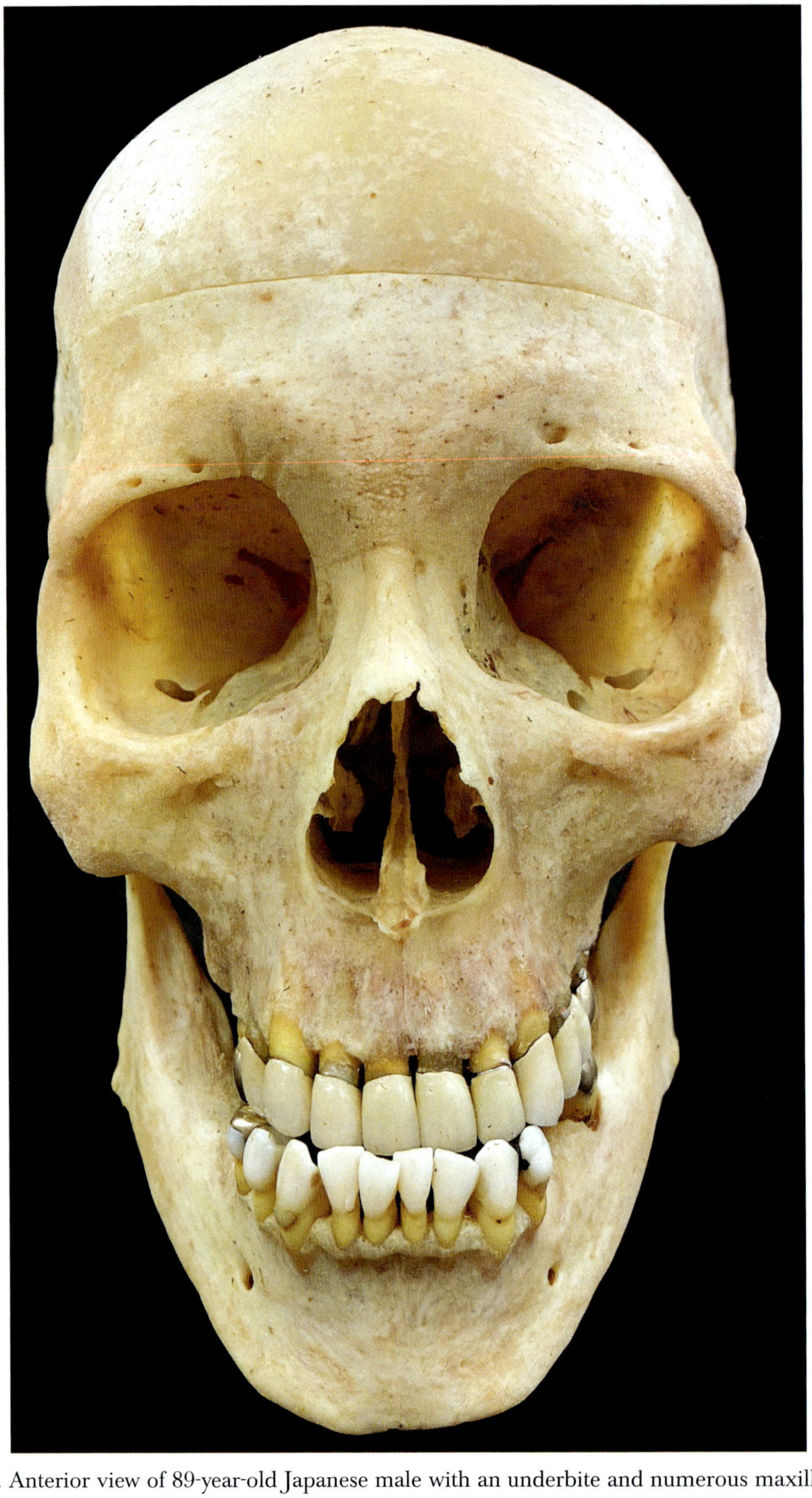

Figure 142. Anterior view of 89-year-old Japanese male with an underbite and numerous maxillary crowns.

Figure 143. Nasal bone contour in an 89-year-old Japanese male.

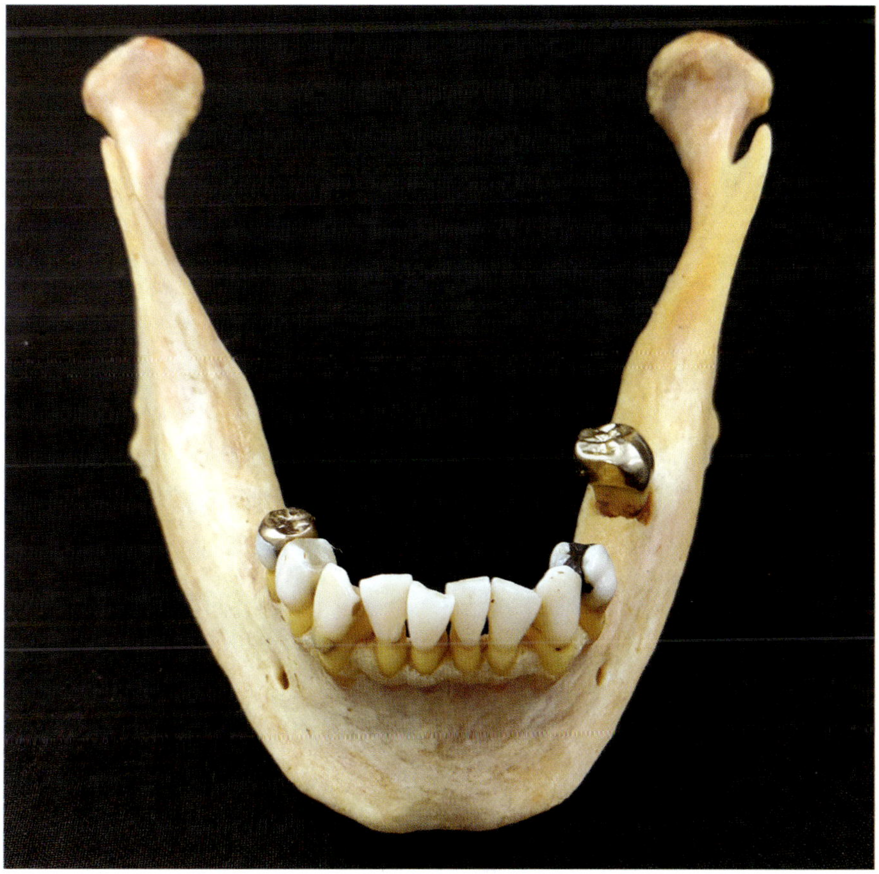

Figure 144. Mandible in an 89-year-old Japanese male.

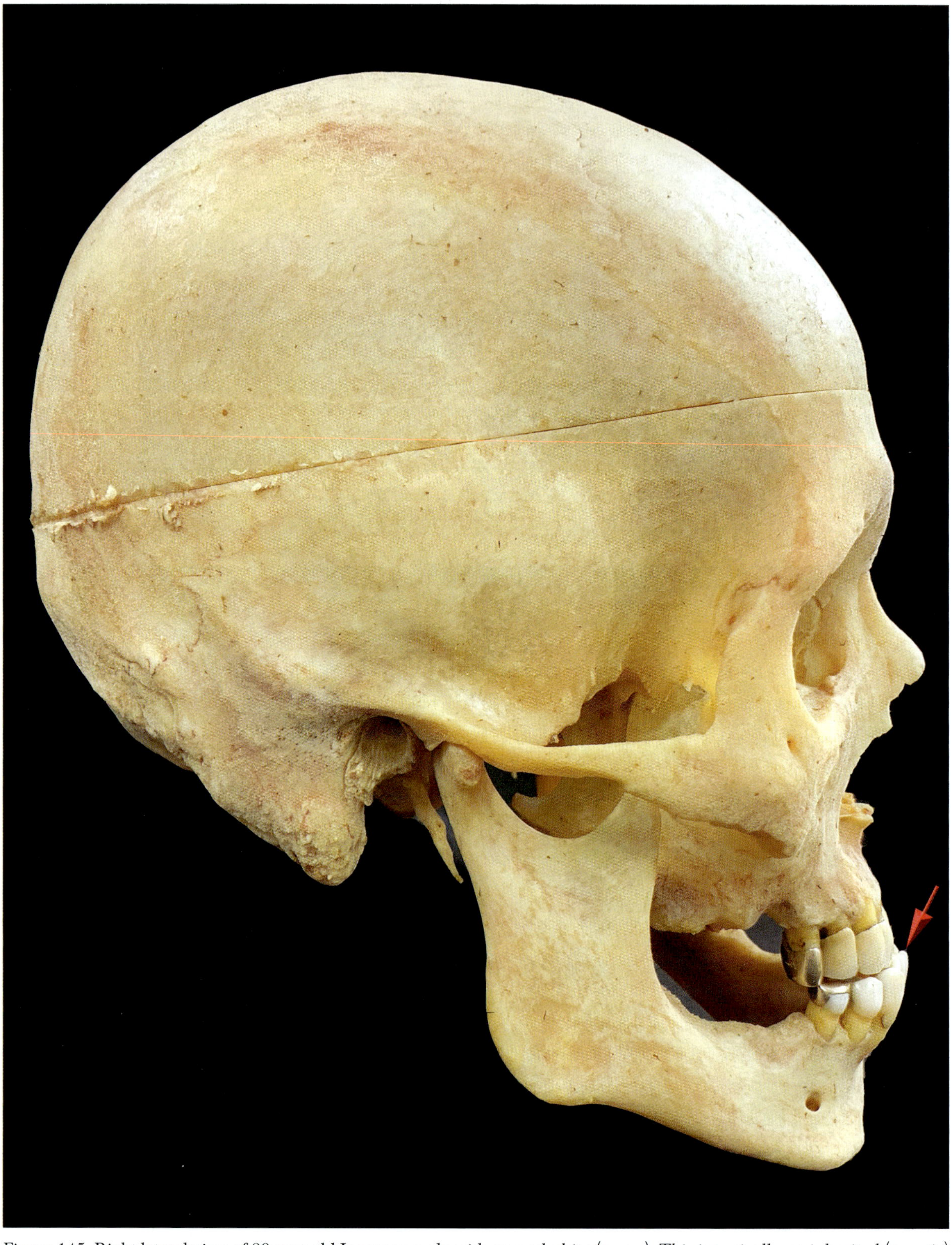

Figure 145. Right lateral view of 89-year-old Japanese male with an underbite (arrow). This is typically an inherited (genetic) trait that often runs in families.

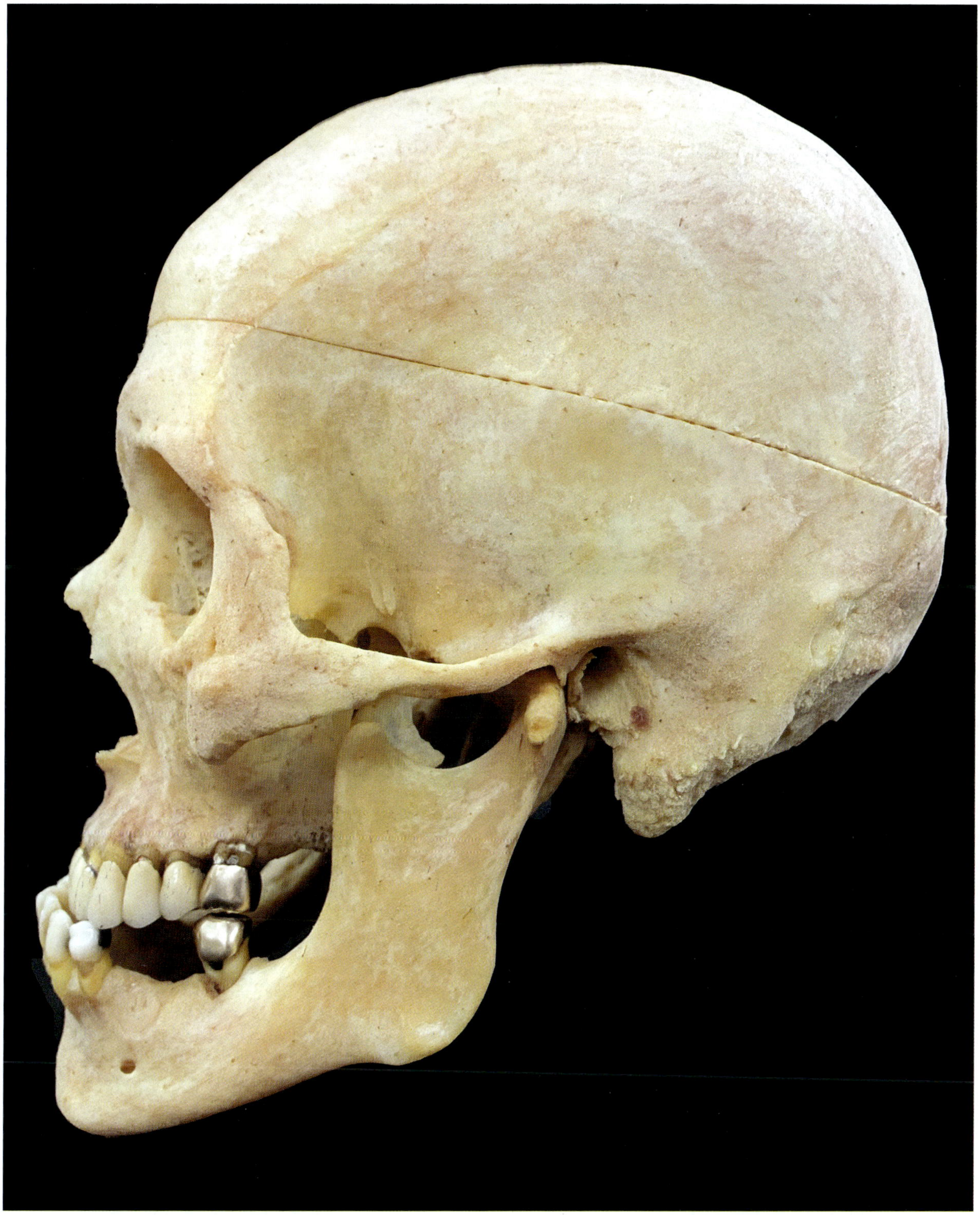

Figure 146. Left lateral view of 89-year-old Japanese male with an underbite. The temporomandibular joint is in approximate anatomical position.

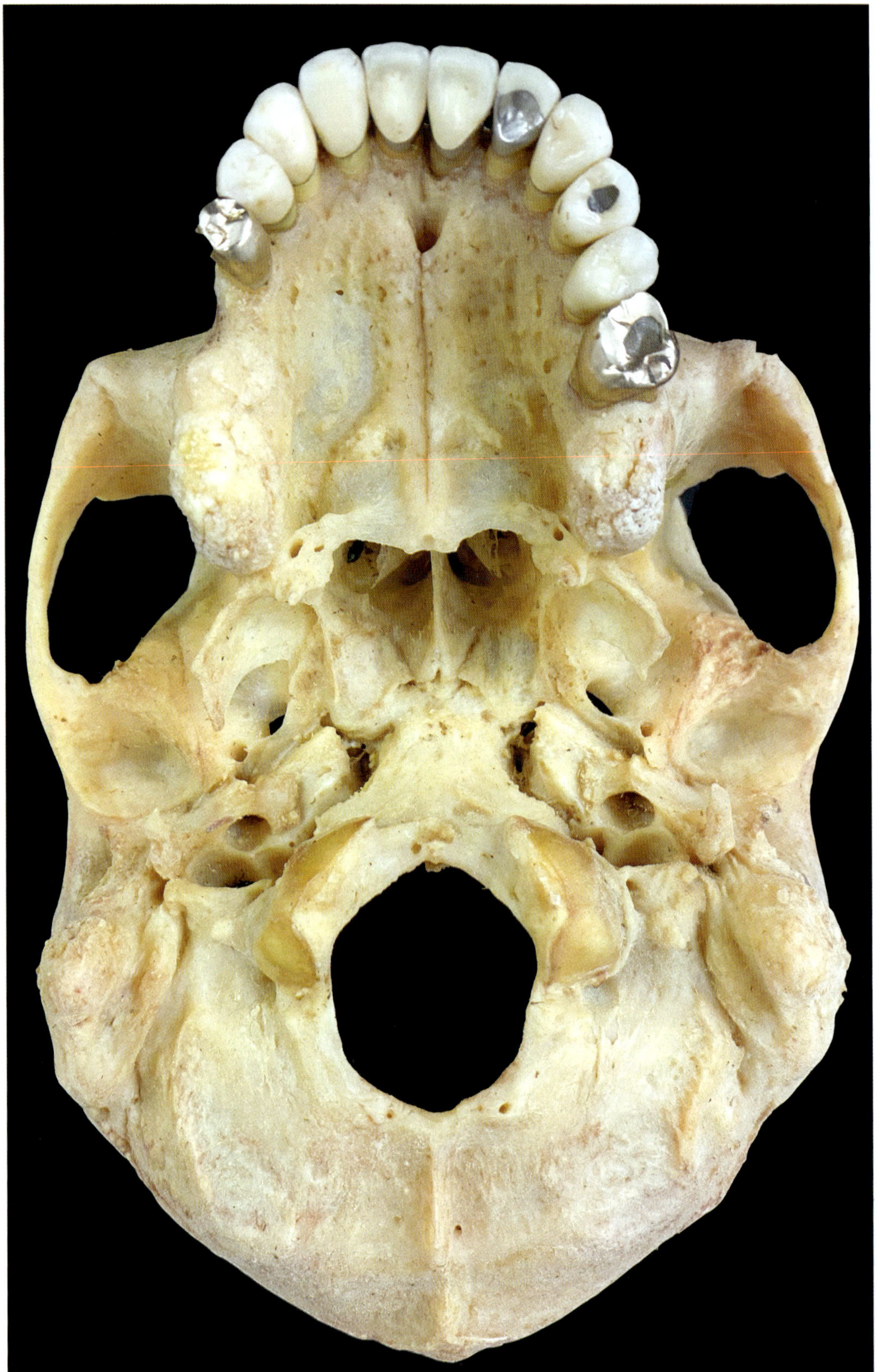

Figure 147. Inferior view of 89-year-old Japanese male.

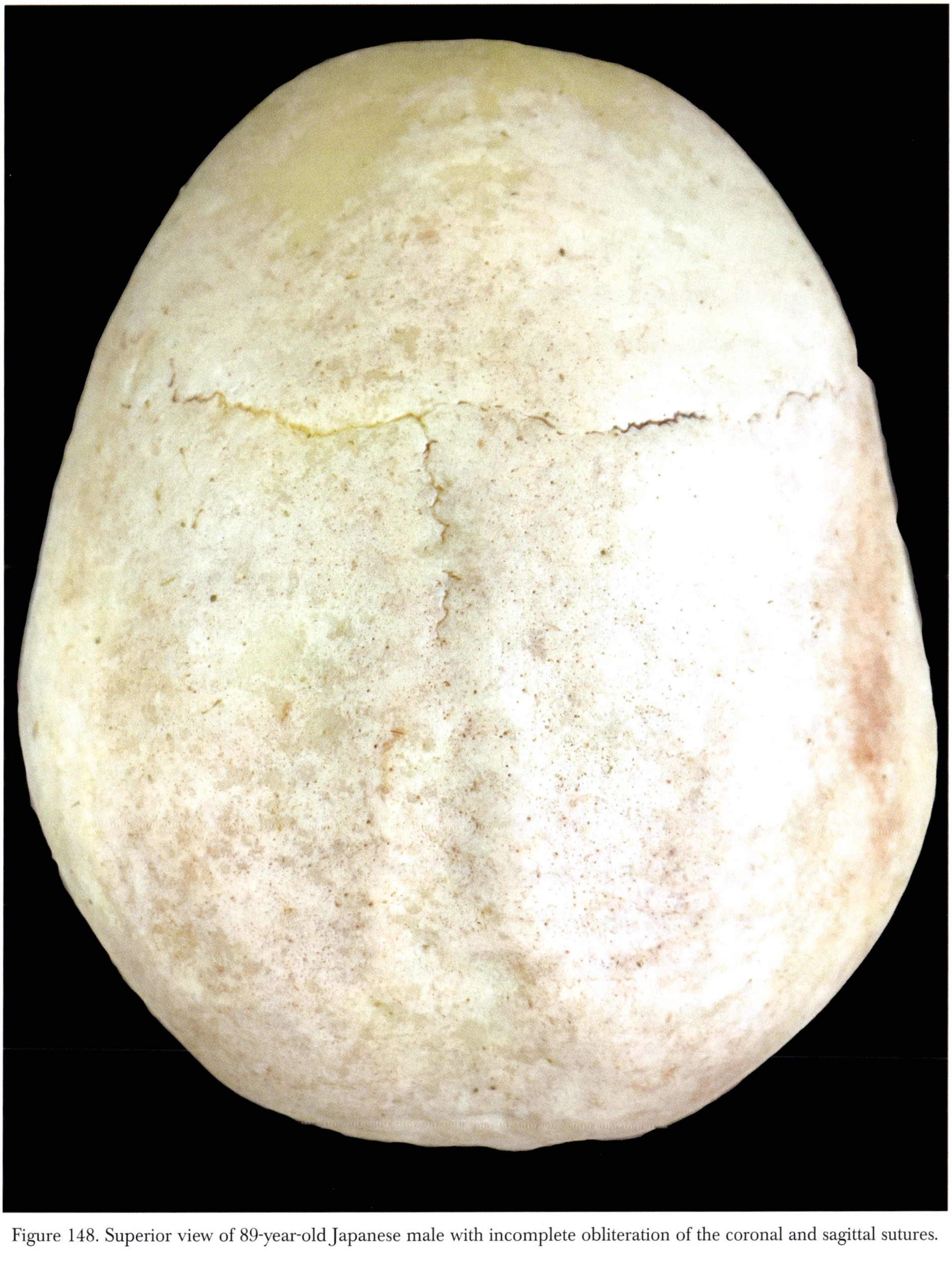

Figure 148. Superior view of 89-year-old Japanese male with incomplete obliteration of the coronal and sagittal sutures.

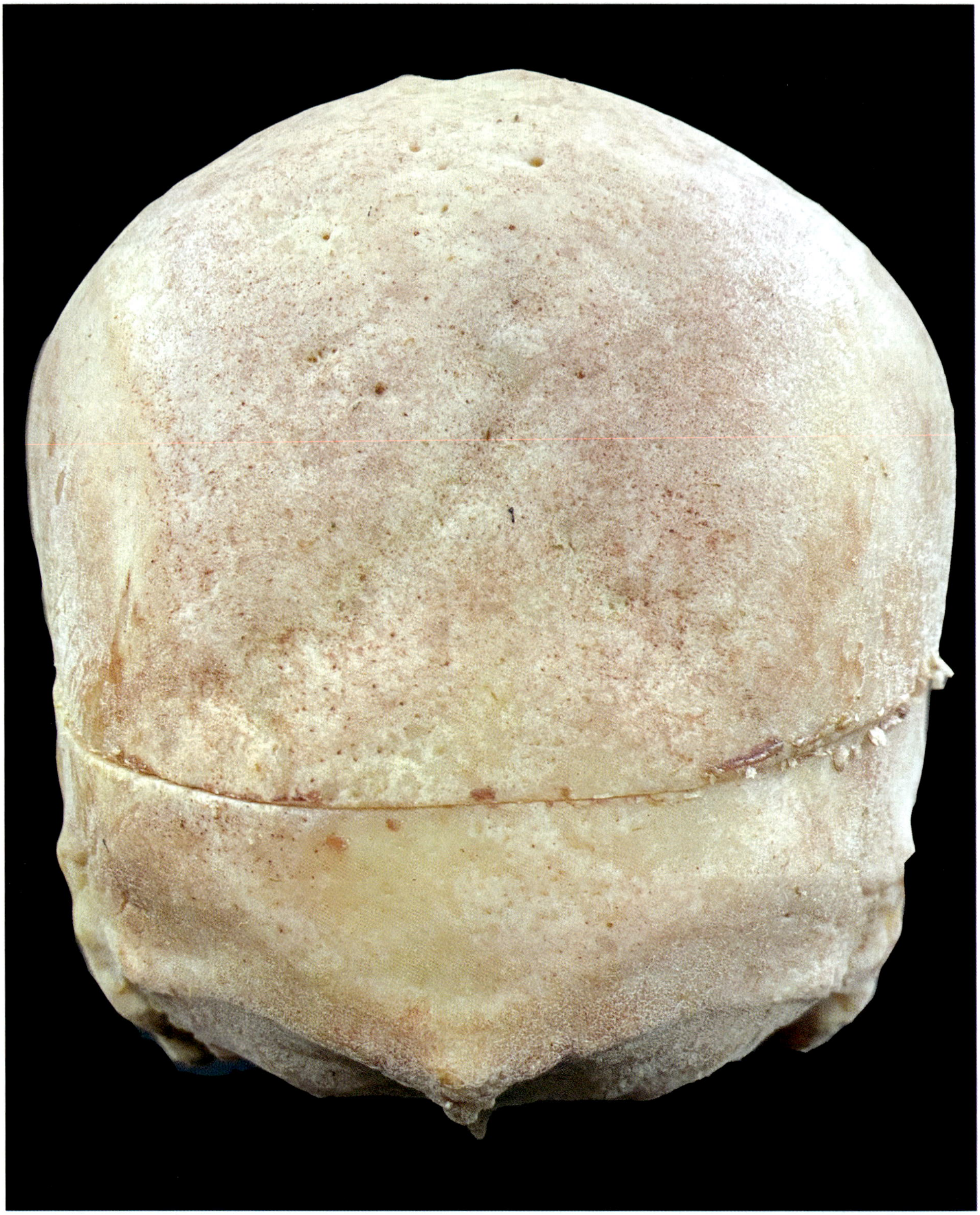

Figure 149. Posterior view of 89-year-old Japanese male with a pronounced nuchal crest and inion hook.

10. JAPANESE EUROPEAN MALE

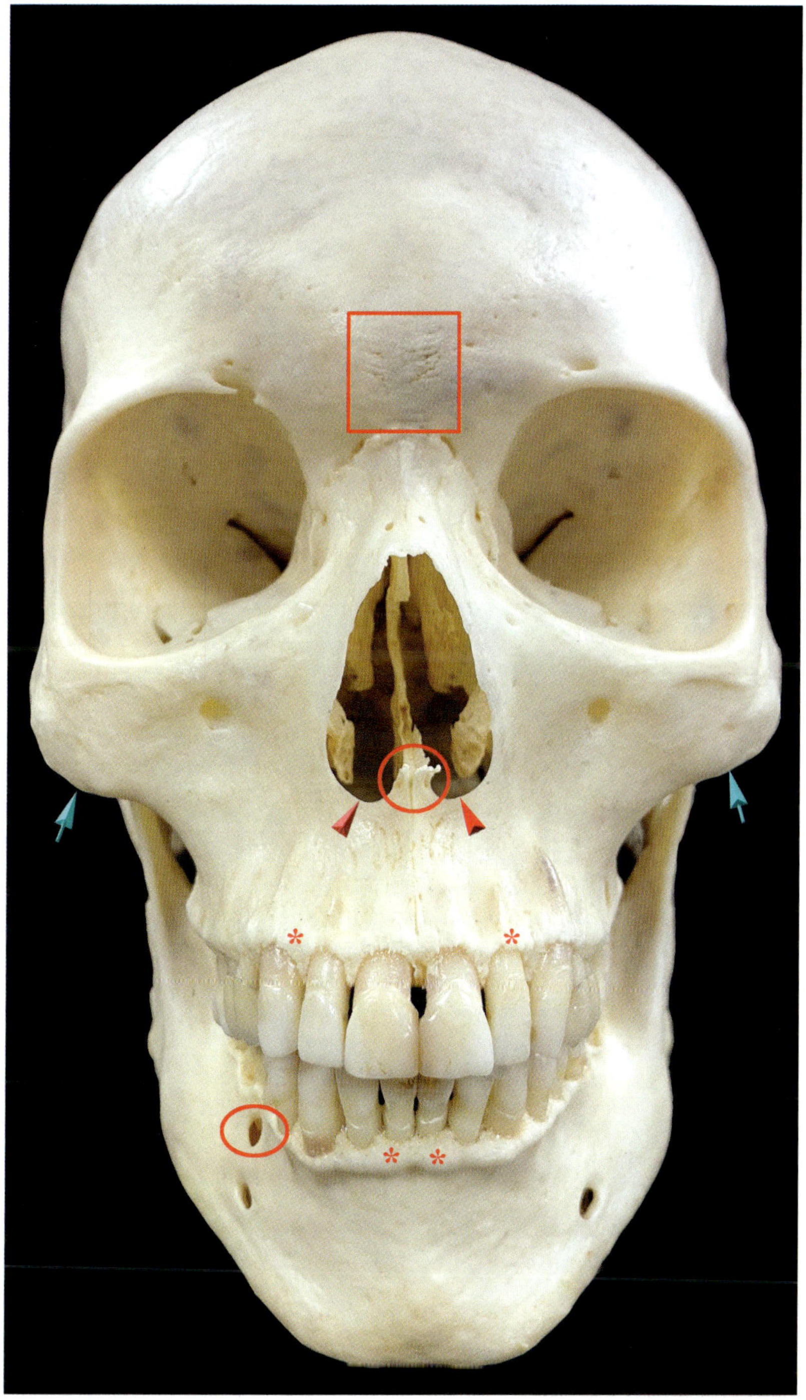

Figure 150. Anterior view of 68-year-old Japanese European male with a trace of the supranasal suture (square), sharp inferior nasal aperture/nasal sill (red arrowhead), and a long and sharp anterior nasal spine (circle). Note the absent malar tubercle (blue arrow), periapical lesion/abscess (red oval), and numerous buccal exostoses (*). Fordisc 3.0 (Jantz and Ousley 2005) using 16 variables classified this individual as White (European) female. Classification using Fordisc is dependent upon the database having a representative sample of the ancestral group in question.

Figure 151. Nasal bone contour in a 68-year-old Japanese European male.

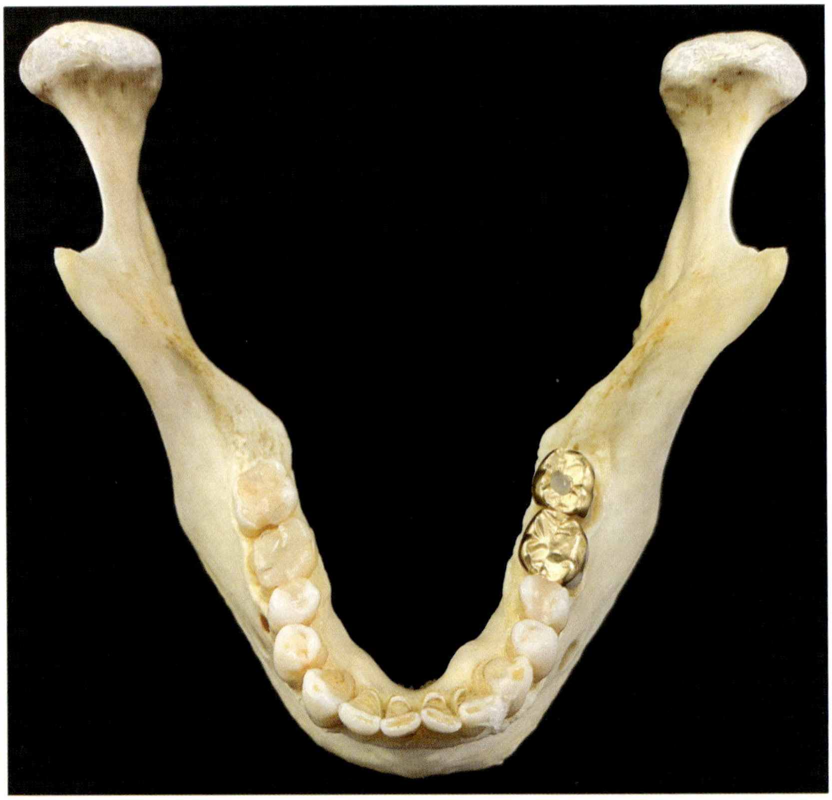

Figure 152. Mandible of 68-year-old Japanese European male with a wide, rounded chin.

TABLE 16.
Cranial Measurements in a 68-year-old Japanese European Male

Measurement	(mm)	Right	Left
Maximum Cranial Length (g-op, GOL)	189		
Maximum Cranial Breadth (eu-eu, XCB)	142		
Bizygomatic Breadth (zy-zy, ZYB)	120		
Basion-Bregma Height (ba-b, BBH)	149		
Cranial Base Length (ba-n, BNL)	106		
Basion-Prosthion (ba-pr, BPL)	100		
Maxillo-Alveolar Breadth (ecm-ecm, MAB)	62		
Maxillo-Alveolar Length (pr-alv, MAL)	55		
Biauricular Breadth (au-au, AUB)	127		
Upper Facial Height (n-pr)	68		
Minimum Frontal Breadth (ft-ft, WFB)	98		
Upper Facial Breadth (fmt-fmt)	109		
Nasal Height (n-ns, NLH)	45		
Nasal Breadth (al-al, NLB)	24		
Orbital Breadth (d-ec, OBB)	43		L
Orbital Height (OBH)	35		L
Biorbital Breadth (ec-ec, EKB)	107		
Interorbital Breadth (d-d, DKB)	27		
Frontal Chord (n-b, FRC)	120		
Parietal Chord (b-l, PAC)	124		
Occipital Chord (l-o, OCC)	101		
Foramen Magnum Length (ba-o, FOL)	35		
Foramen Magnum Breadth (FOB)	28		
Mastoid Length (MDH)		32	31

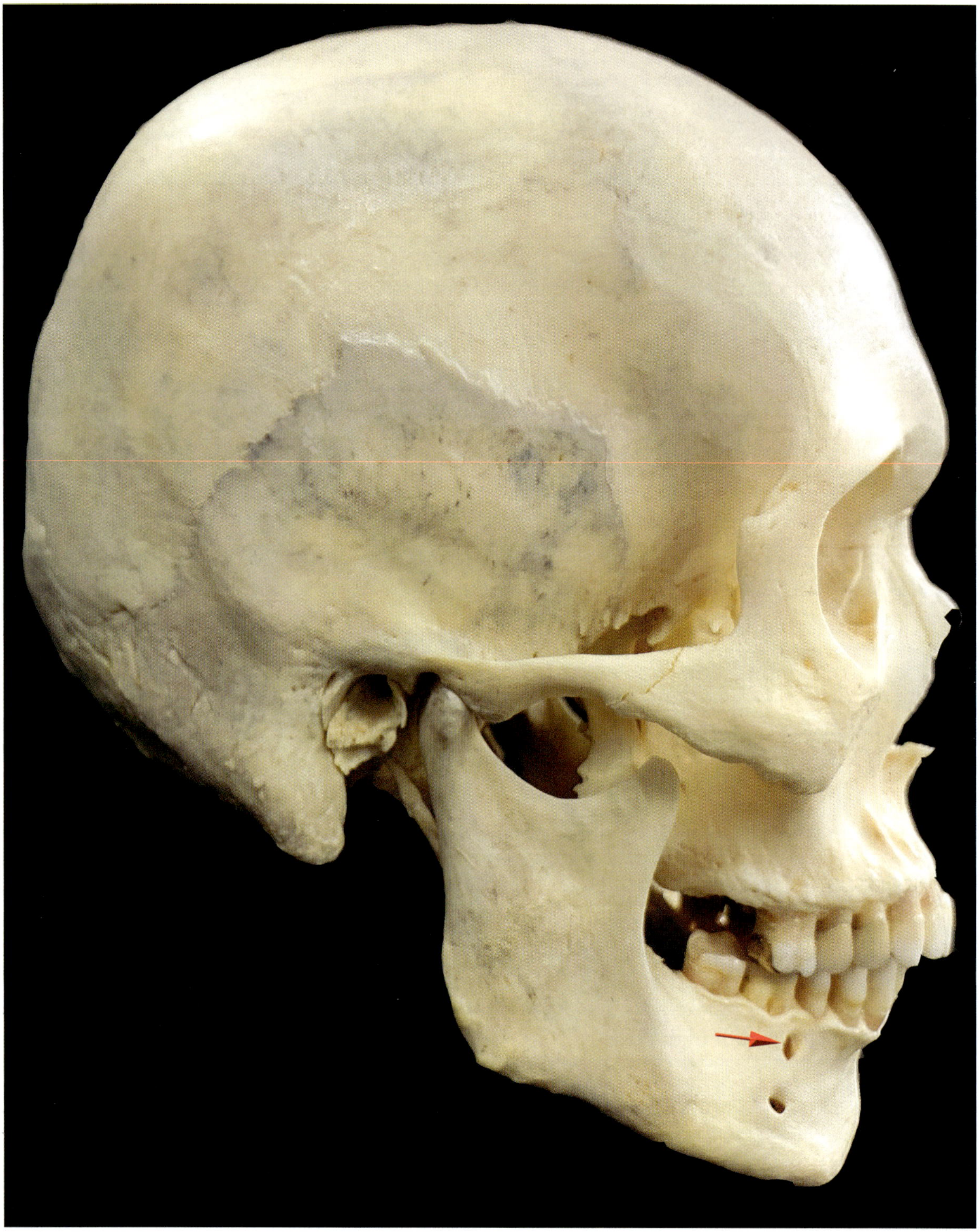

Figure 153. Right lateral view of 68-year-old Japanese European male. Note the developed brow ridge and sloping frontal bone, the small nasal bones, the projecting anterior nasal spine, and the absence of nasal overgrowth. Also note the robust mandible in comparison to overall size of the cranium, the developed posterior zygomatic tubercle, the supramastoid crest (zygomatic extension), and, superior to the right mental foramen, a periapical lesion (arrow) that could be mistaken for an accessory foramen.

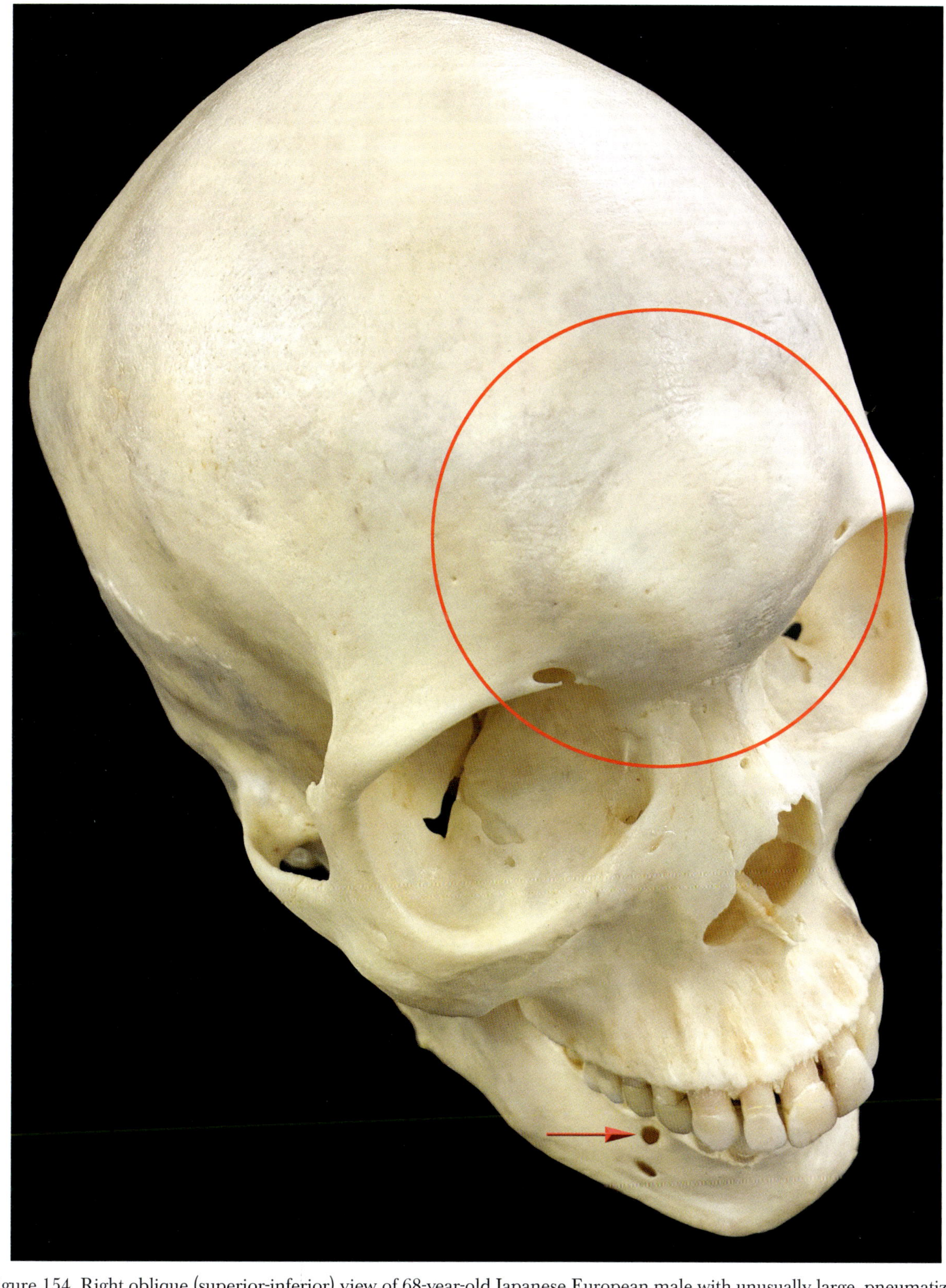

Figure 154. Right oblique (superior-inferior) view of 68-year-old Japanese European male with unusually large, pneumatized frontal sinuses (circle) and a periapical lesion (arrow) in the right half of the mandible that could be mistaken for an accessory mental foramen.

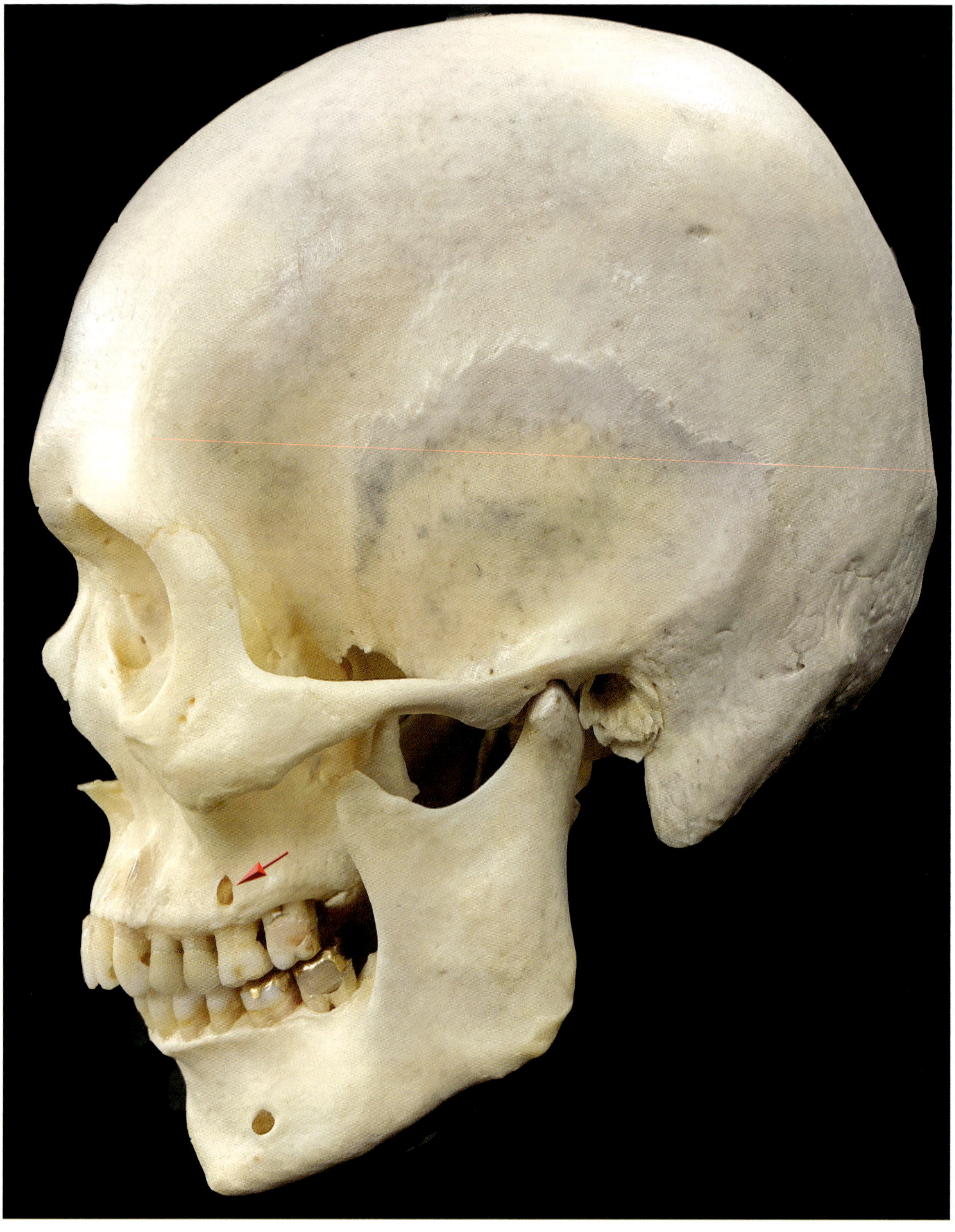

Figure 155. Left lateral view of 68-year-old Japanese European male with nasion depression, small nasal bones, and a periapical lesion (arrow) that could be mistaken for a fenestration of the maxillary left first molar.

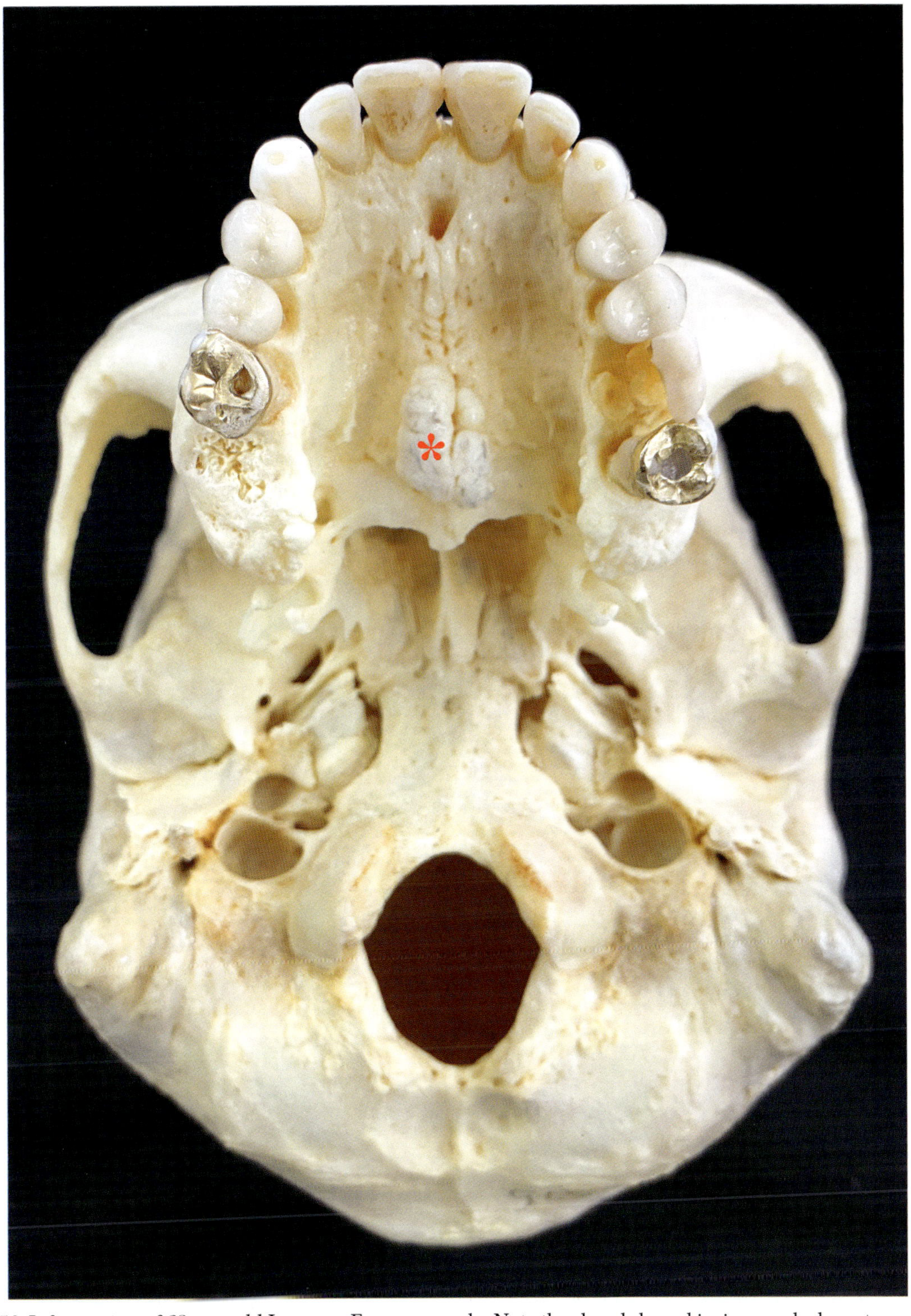

Figure 156. Inferior view of 68-year-old Japanese European male. Note the shovel-shaped incisors and a large torus palatinus (*). The torus palatinus, a raised plateau, is a trait more closely associated with African and African Americans and Europeans than with Asians and Native Americans according to Chohayeb and Volpe (2001). The author, however, has seen torus palatinus in many Southeast Asians.

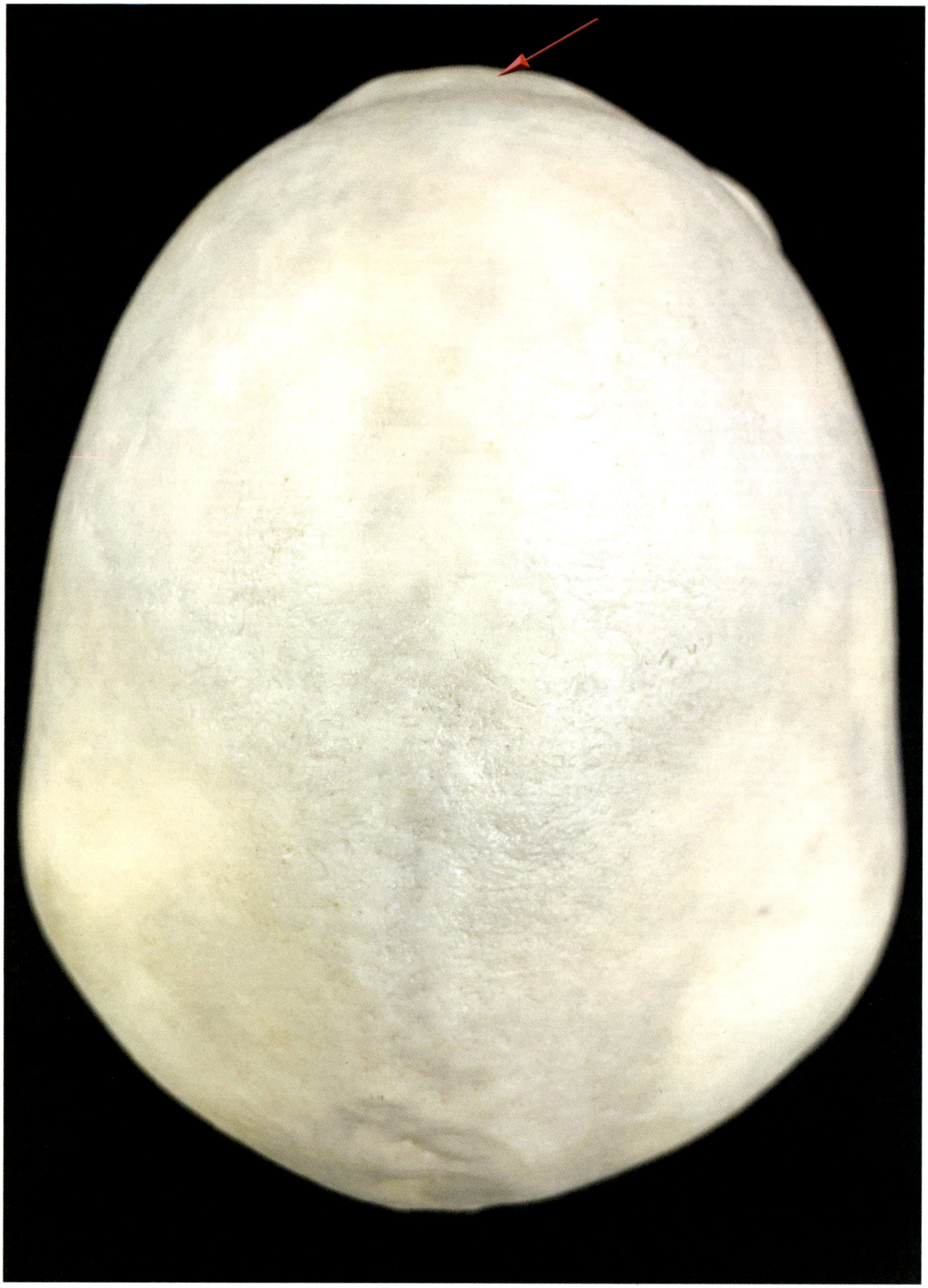

Figure 157. Superior view of 68-year-old Japanese European male showing an unusually large brow ridge (arrow) that houses the frontal sinuses that, in this individual, that resulted in a large bulge at the central portion of the frontal bone. The vault sutures have completely obliterated.

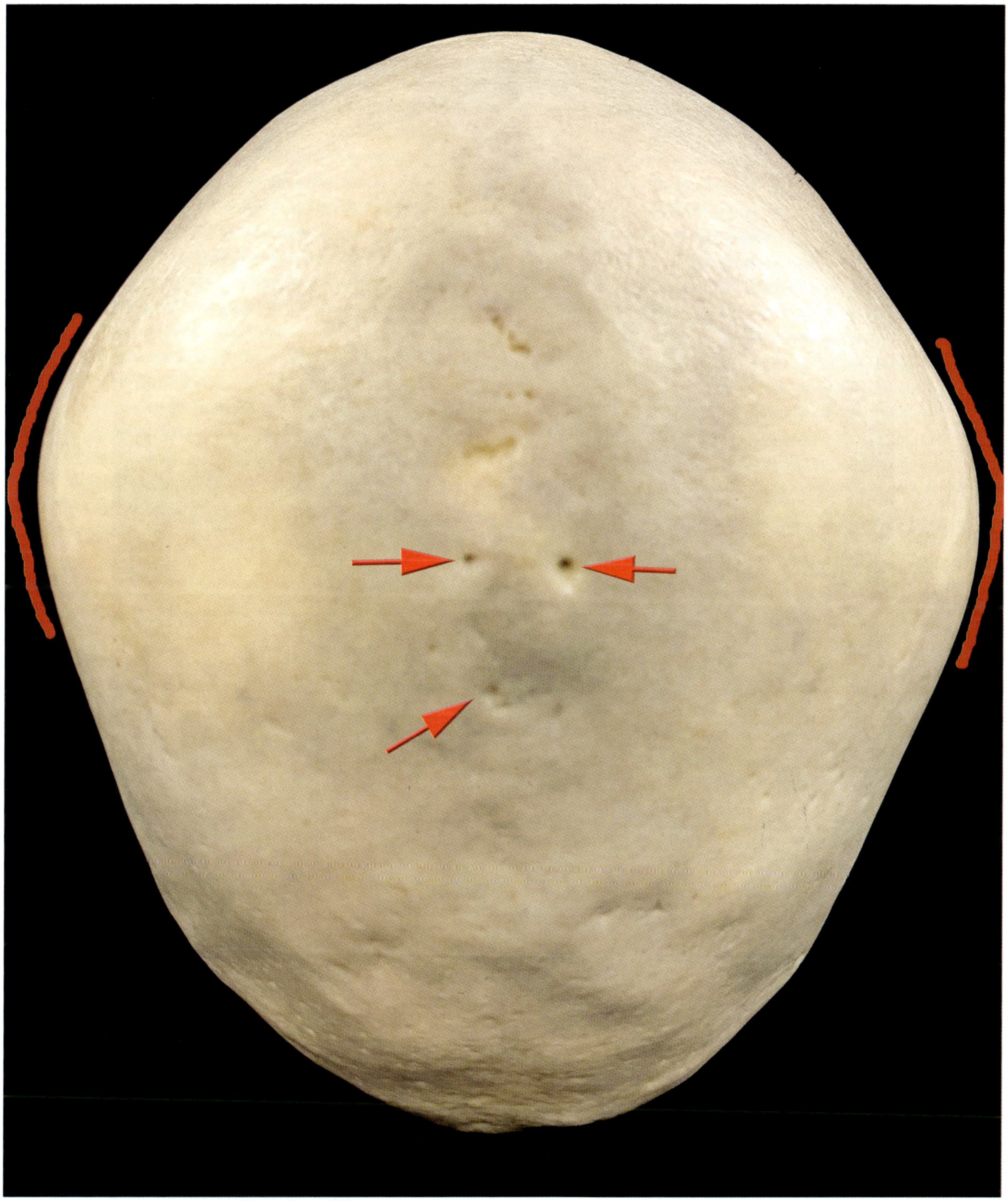

Figure 158. Posterior view of 68-year-old Japanese European male with three parietal foramina (arrow) (normal variant), parietal bossing of the eminences (curved red line), and from this perspective, a diamond-shaped cranial outline.

11. JAPANESE FEMALE

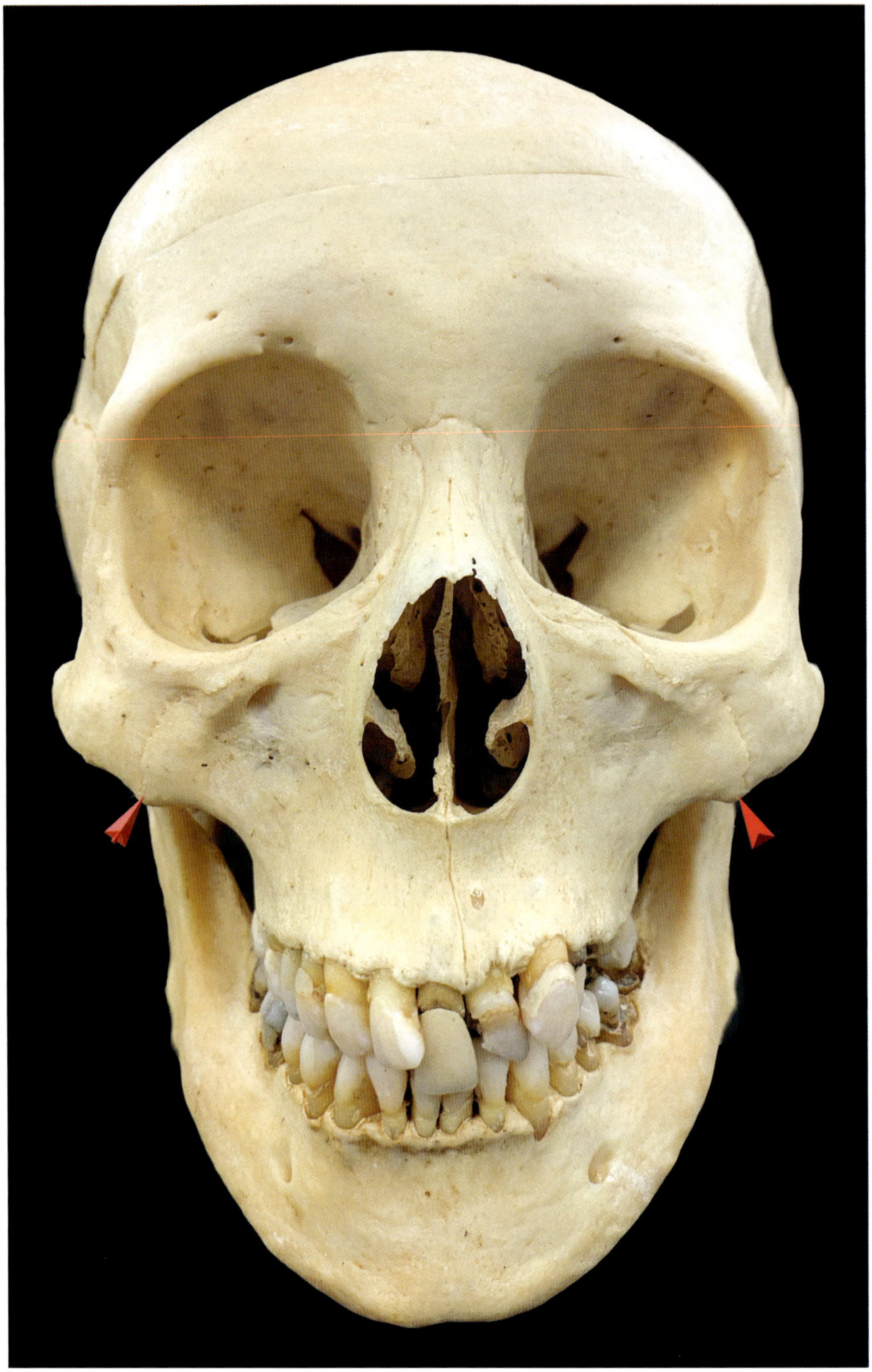

Figure 159. Anterior view 94-year-old Japanese female with a small nasal aperture, angular orbits, and small malar tubercles (arrowhead).

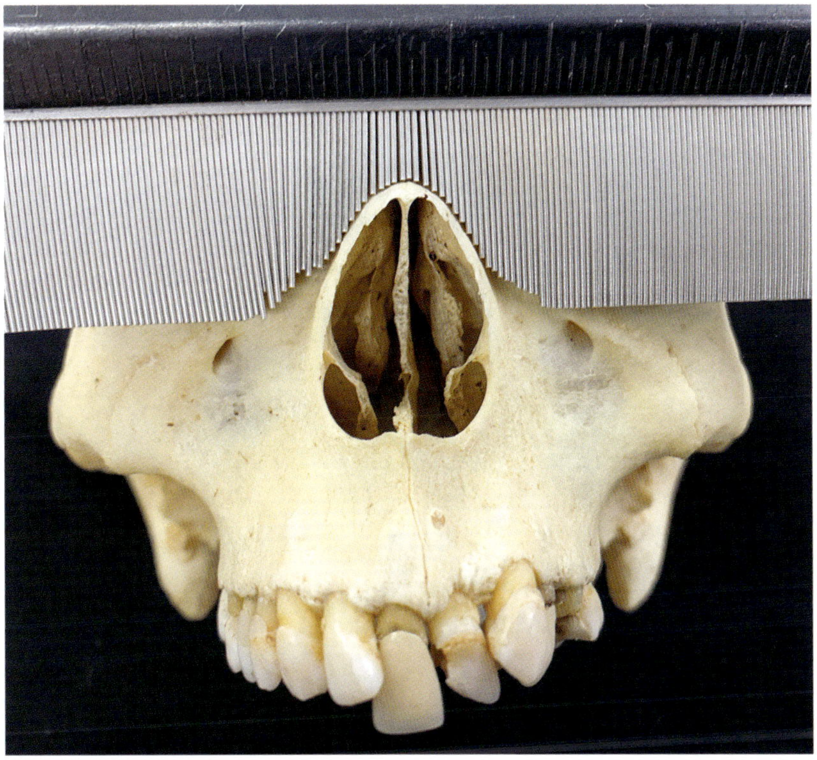

Figure 160. Nasal bone contour in a 94-year-old Japanese female.

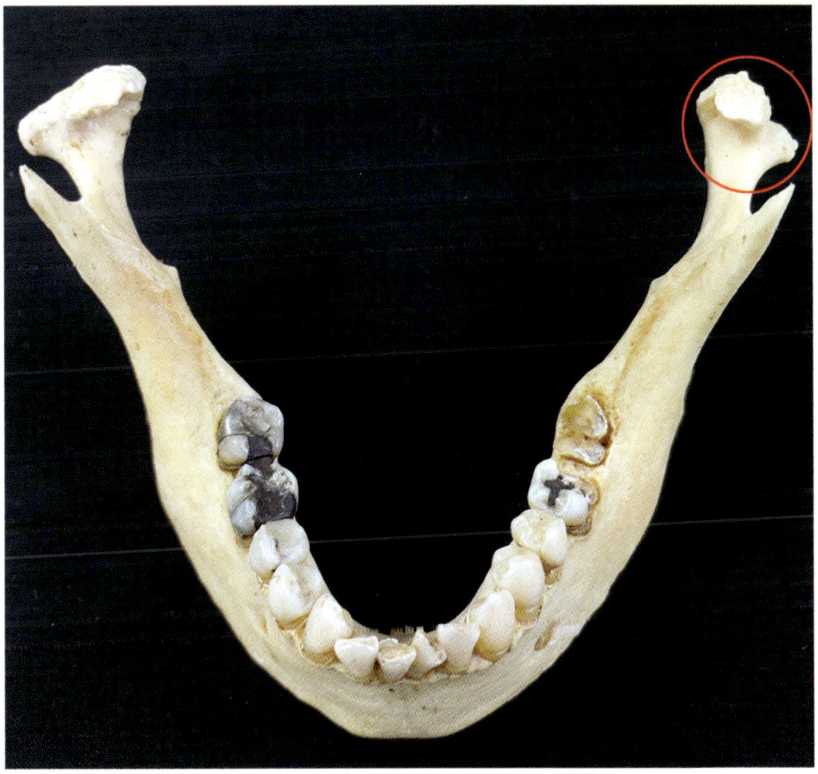

Figure 161. Mandible of 94-year-old Japanese female with degenerative joint disease (osteoarthritis) and what appears to be a bifurcated/divided left condyle (circle).

TABLE 17.
Cranial Measurements in a 94-year-old Japanese Female

Measurement	(mm)	Right	Left
Maximum Cranial Length (g-op, GOL)	172		
Maximum Cranial Breadth (eu-eu, XCB)	150		
Bizygomatic Breadth (zy-zy, ZYB)	129		
Basion-Bregma Height (ba-b, BBH)	123		
Cranial Base Length (ba-n, BNL)	90		
Basion-Prosthion (ba-pr, BPL)	83		
Maxillo-Alveolar Breadth (ecm-ecm, MAB)	57		
Maxillo-Alveolar Length (pr-alv, MAL)	47		
Biauricular Breadth (au-au, AUB)	129		
Upper Facial Height (n-pr)	70		
Minimum Frontal Breadth (ft-ft, WFB)	87		
Upper Facial Breadth (fmt-fmt)	98		
Nasal Height (n-ns, NLH)	50		
Nasal Breadth (al-al, NLB)	22		
Orbital Breadth (d-ec, OBB)	37		L
Orbital Height (OBH)	35		L
Biorbital Breadth (ec-ec, EKB)	88		
Interorbital Breadth (d-d, DKB)	18		
Frontal Chord (n-b, FRC)	108		
Parietal Chord (b-l, PAC)	108		
Occipital Chord (l-o, OCC)	96		
Foramen Magnum Length (ba-o, FOL)	34		
Foramen Magnum Breadth (FOB)	27		
Mastoid Length (MDH)		30	30

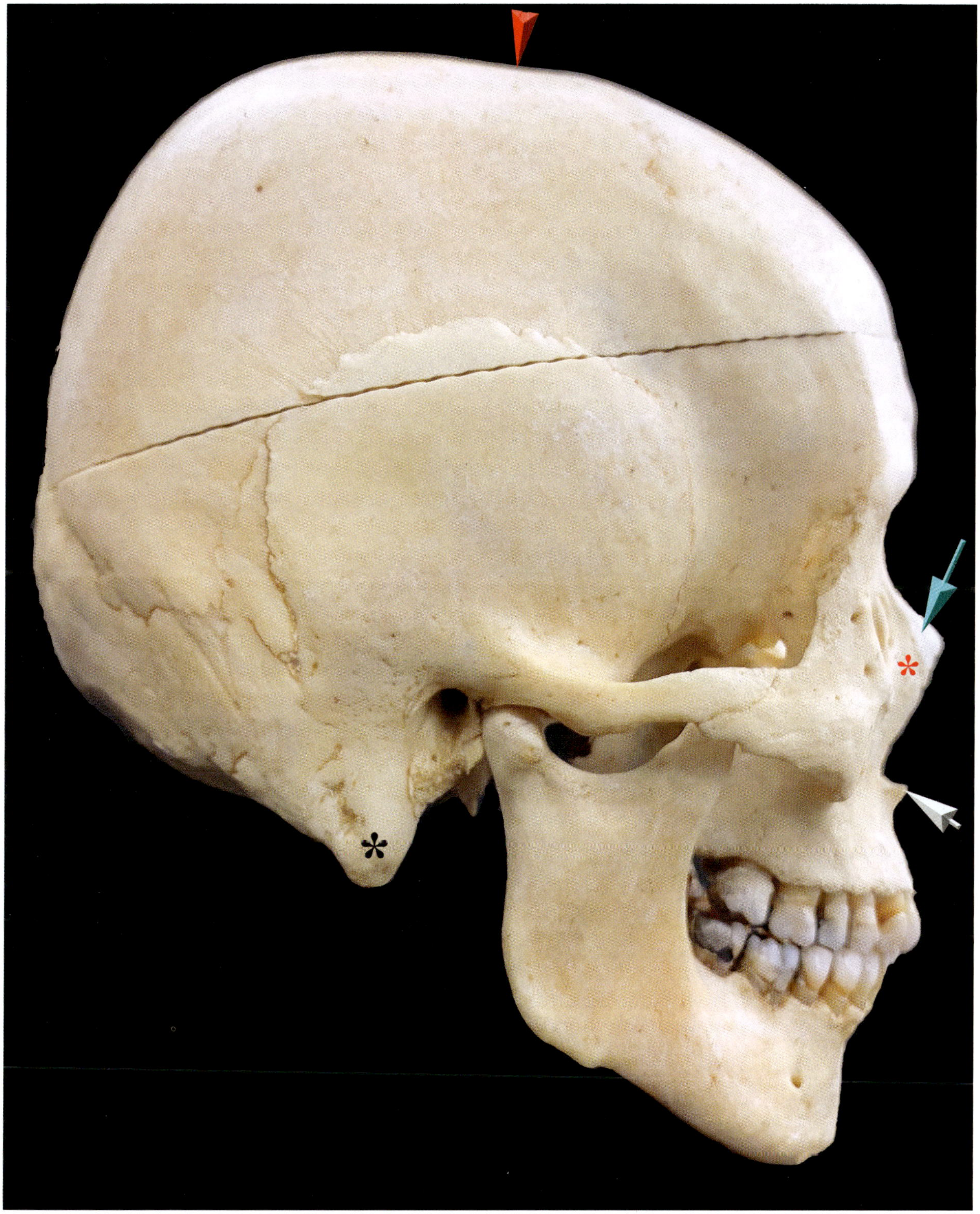

Figure 162. Right lateral view 94-year-old Japanese female with a postbregmatic depression (red arrowhead), small nasal bones (blue arrow), anterior nasal spine (white arrow), small mastoid process (black*), and small frontal process of the maxilla (red*).

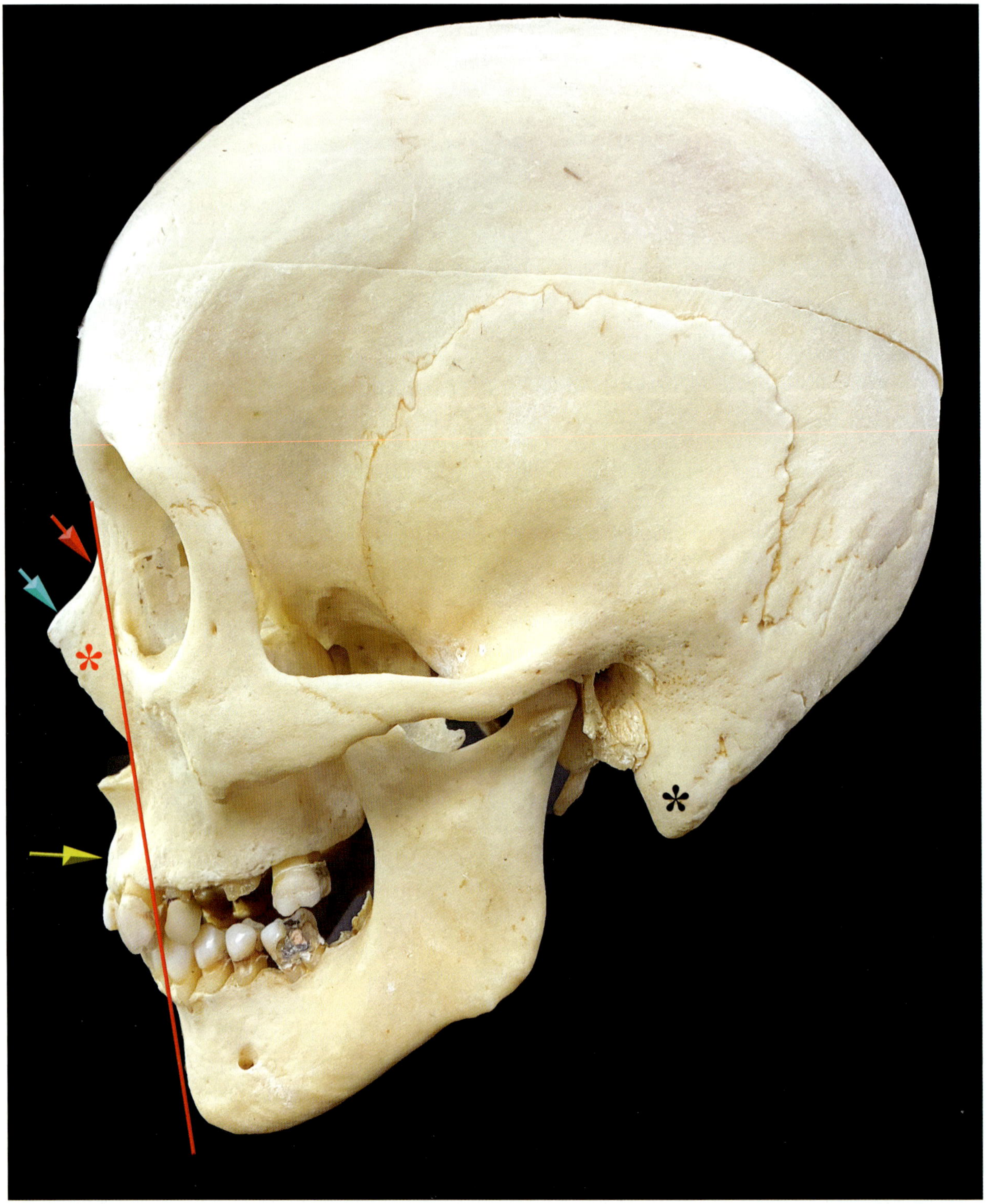

Figure 163. Left lateral view 94-year-old Japanese female with concave (red arrow) and convex (blue arrow) nasal bones, moderate frontal process of the maxilla (red*), mild alveolar prognathism of the maxilla (yellow arrow) extending beyond a line (red line) drawn from nasion to pogonian, and a small mastoid process (black*).

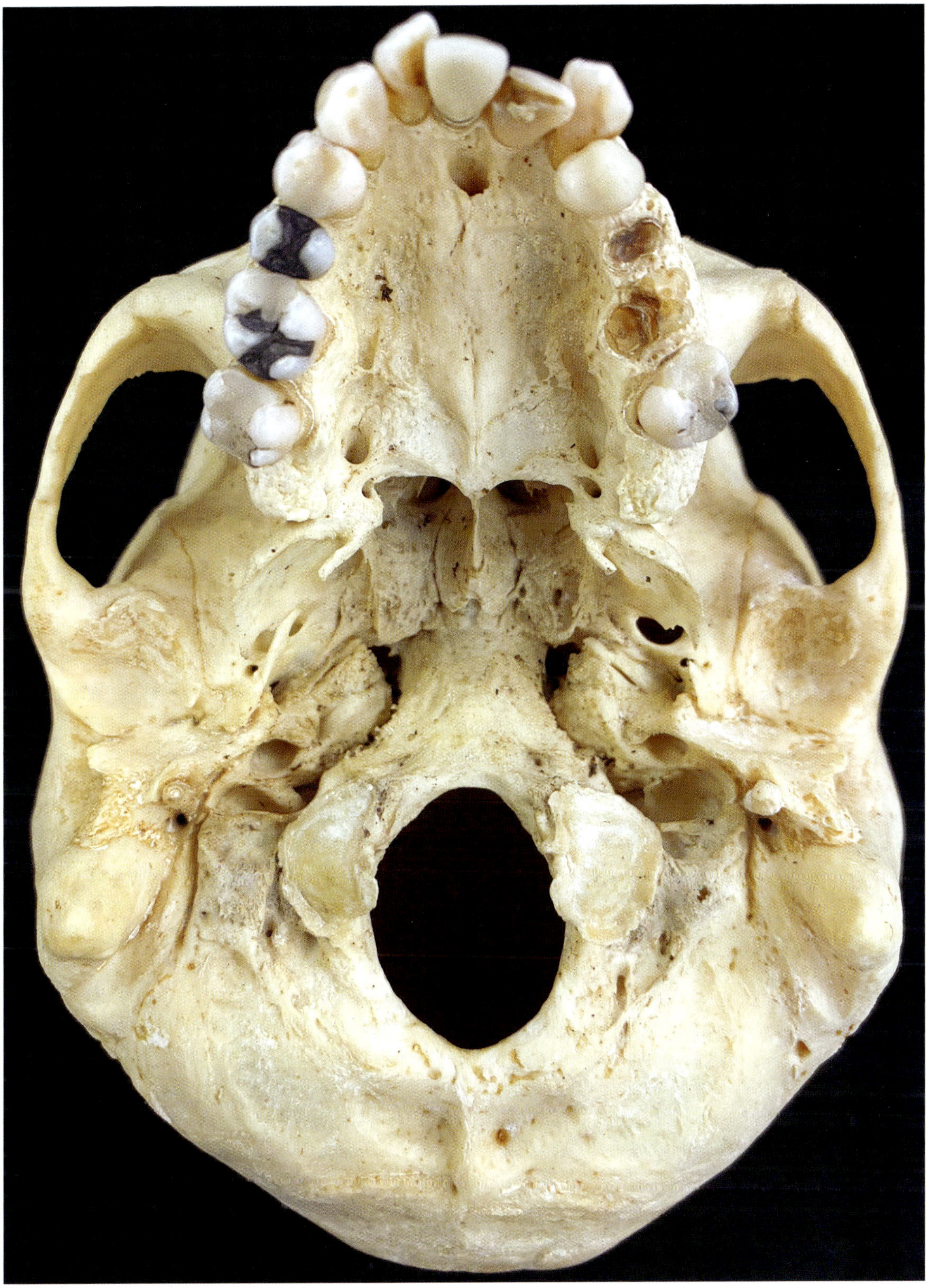

Figure 164. Inferior view 94-year-old Japanese female. Notice the crowding and rotation of the anterior teeth, a small torus palatinus, and osteoarthrithic destruction of the temporomandibular joints and occipital condyles.

Figure 165. Superior view 94-year-old Japanese female.

Figure 166. Posterior view 94-year-old Japanese female with parietal bossing and two parietal foramina.

12. JAPANESE FEMALE

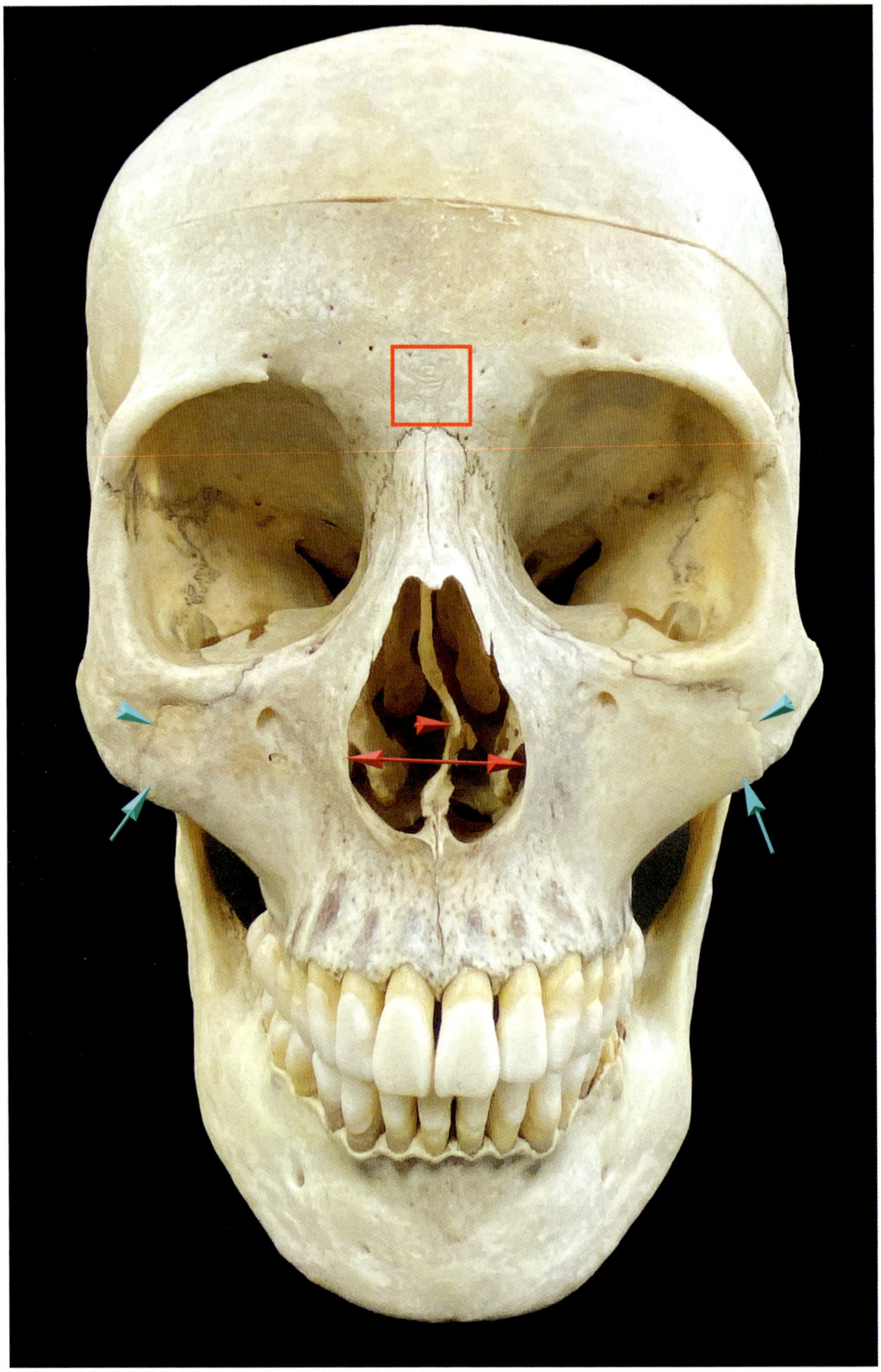

Figure 167. Anterior view of 64-year-old Japanese female. Notice the trace supranasal suture (square), the angular orbits, the blunt supraorbital margin, the projecting and rounded nasal bones, the deviated nasal septum red (arrowhead), the medium nasal breadth (double arrow), the absent malar tubercle (blue arrow), and the angled zygomaticomaxillary suture (blue arrowhead).

Figure 168. Nasal bone contour in a 64-year-old Japanese female.

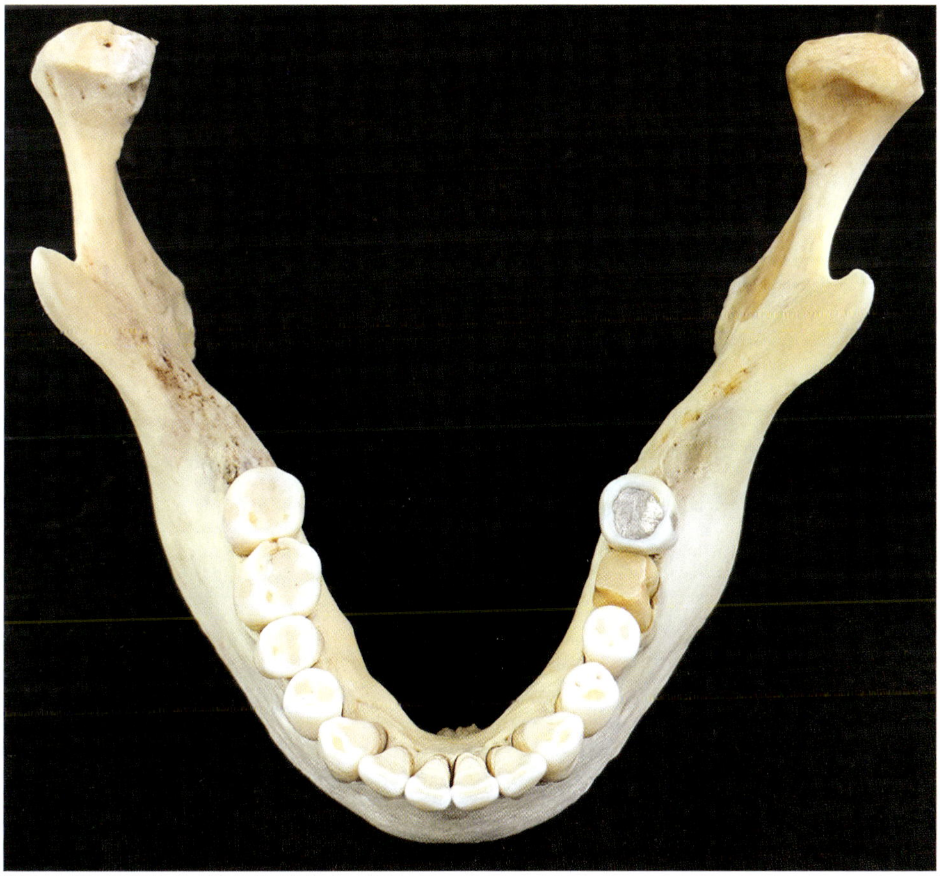

Figure 169. The mandible of 64-year-old Japanese female with a rounded chin.

TABLE 18.
Cranial Measurements in a 64-year-old Japanese Female

Measurement	(mm)	Right	Left
Maximum Cranial Length (g-op, GOL)	172		
Maximum Cranial Breadth (eu-eu, XCB)	145		
Bizygomatic Breadth (zy-zy, ZYB)	130		
Basion-Bregma Height (ba-b, BBH)	141		
Cranial Base Length (ba-n, BNL)	103		
Basion-Prosthion (ba-pr, BPL)	93		
Maxillo-Alveolar Breadth (ecm-ecm, MAB)	57		
Maxillo-Alveolar Length (pr-alv, MAL)	47		
Biauricular Breadth (au-au, AUB)	129		
Upper Facial Height (n-pr)	71		
Minimum Frontal Breadth (ft-ft, WFB)	85		
Upper Facial Breadth (fmt-fmt)	97		
Nasal Height (n-ns, NLH)	54		
Nasal Breadth (al-al, NLB)	24		
Orbital Breadth (d-ec, OBB)	40		L
Orbital Height (OBH)	36		L
Biorbital Breadth (ec-ec, EKB)	93		
Interorbital Breadth (d-d, DKB)	19		
Frontal Chord (n-b, FRC)	113		
Parietal Chord (b-l, PAC)	103		
Occipital Chord (l-o, OCC)	100		
Foramen Magnum Length (ba-o, FOL)	36		
Foramen Magnum Breadth (FOB)	33		
Mastoid Length (MDH)		22	23

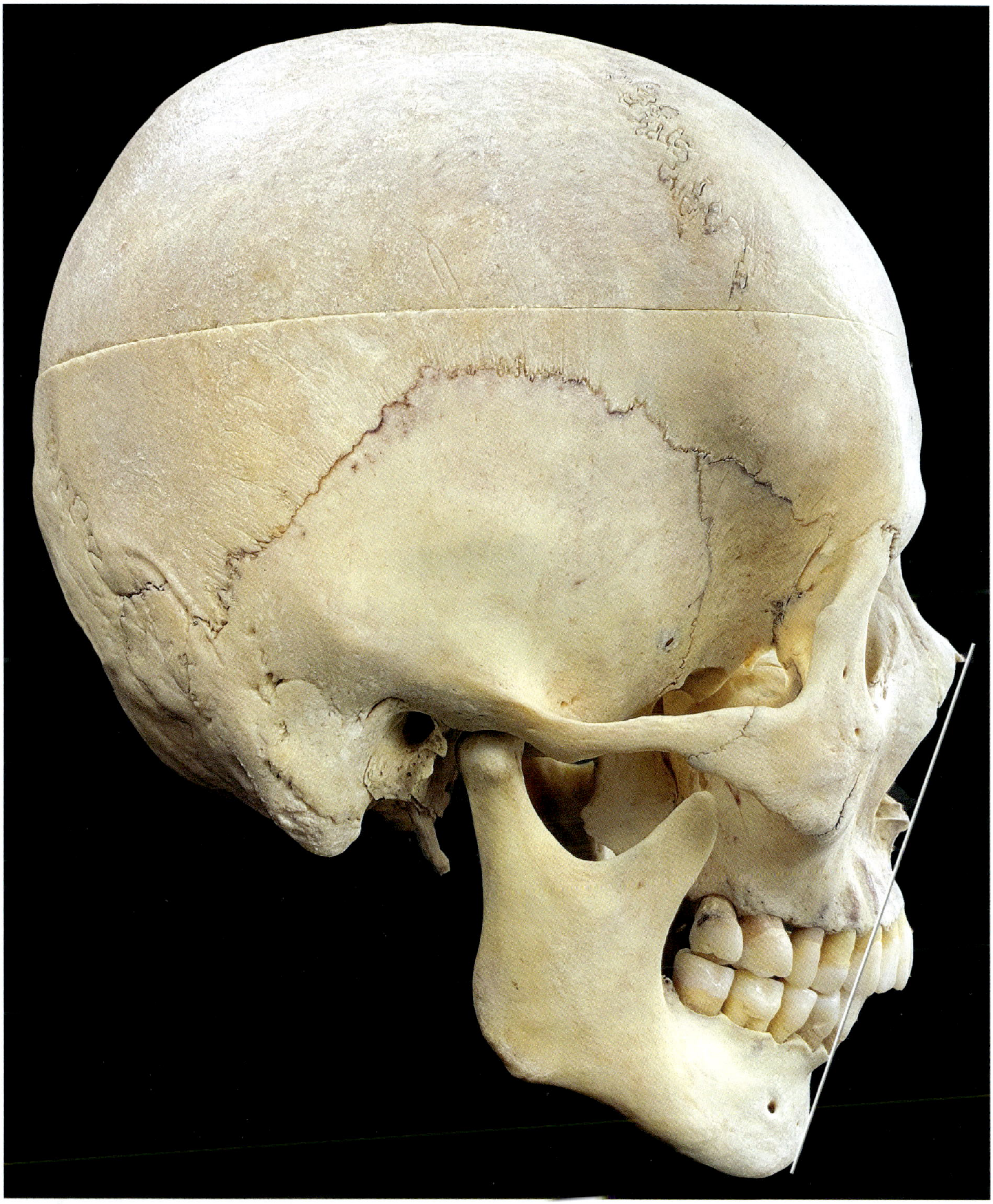

Figure 170. Right lateral view of 64-year-old Japanese female with medium size and blunt anterior nasal spine (red arrow). Notice the shallow antegonial notch (Singh et al. 2016) consisting of the curved area between the red line and inferior margin of the mandible, a small mastoid process, and a petrosquamousal suture (blue arrow) (Mann et al. 2016).

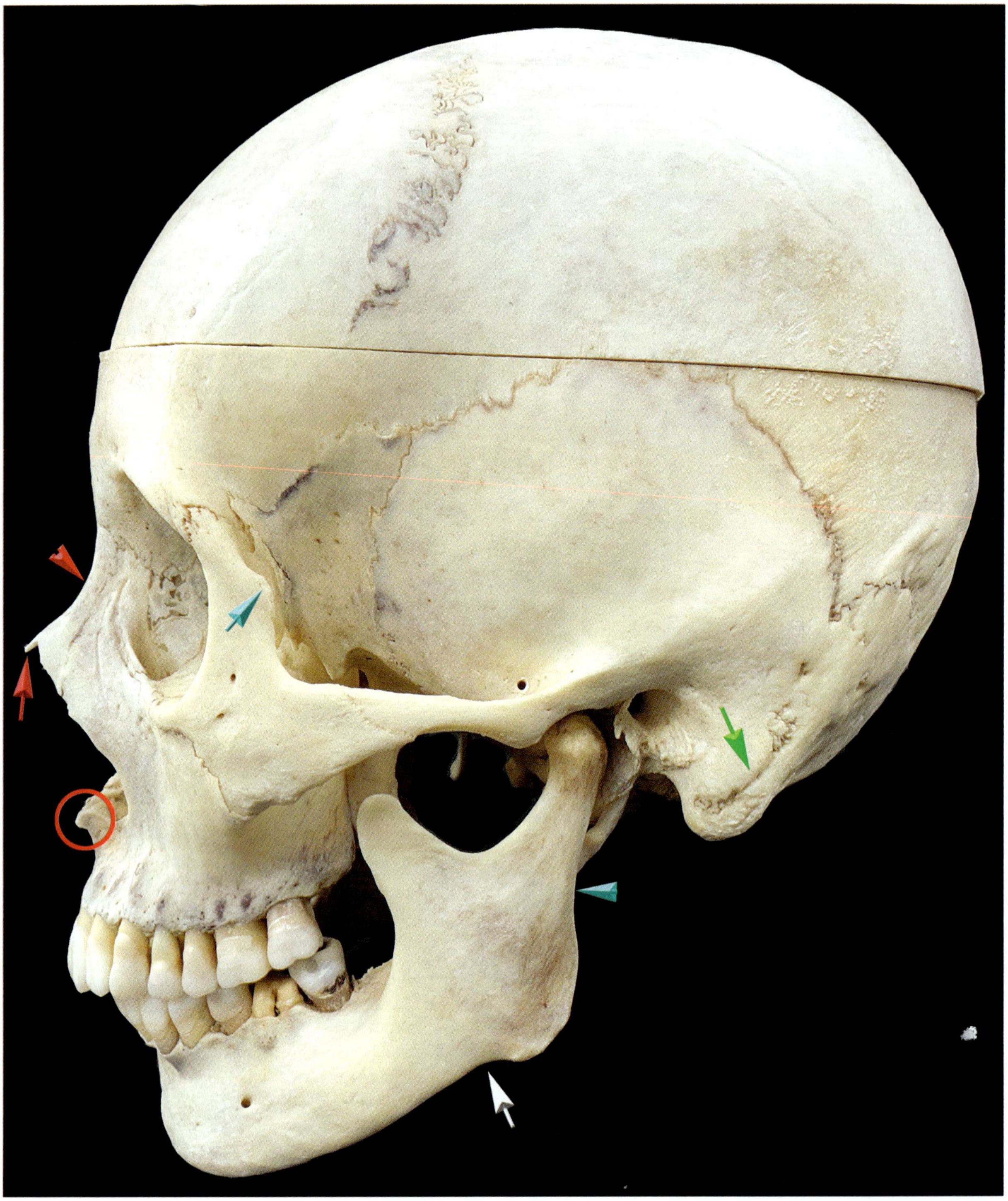

Figure 171. Left lateral view of 64-year-old Japanese female. Note the mild concave contour of the nasal bones (red arrowhead), the nasal bone overhang (red arrow), small anterior nasal spine (circle), a developed posterior zygomatic tubercle (blue arrow), antegonial notch (white arrow), concave posterior border of the mandible (blue arrowhead), and petrosquamousal suture (green arrow).

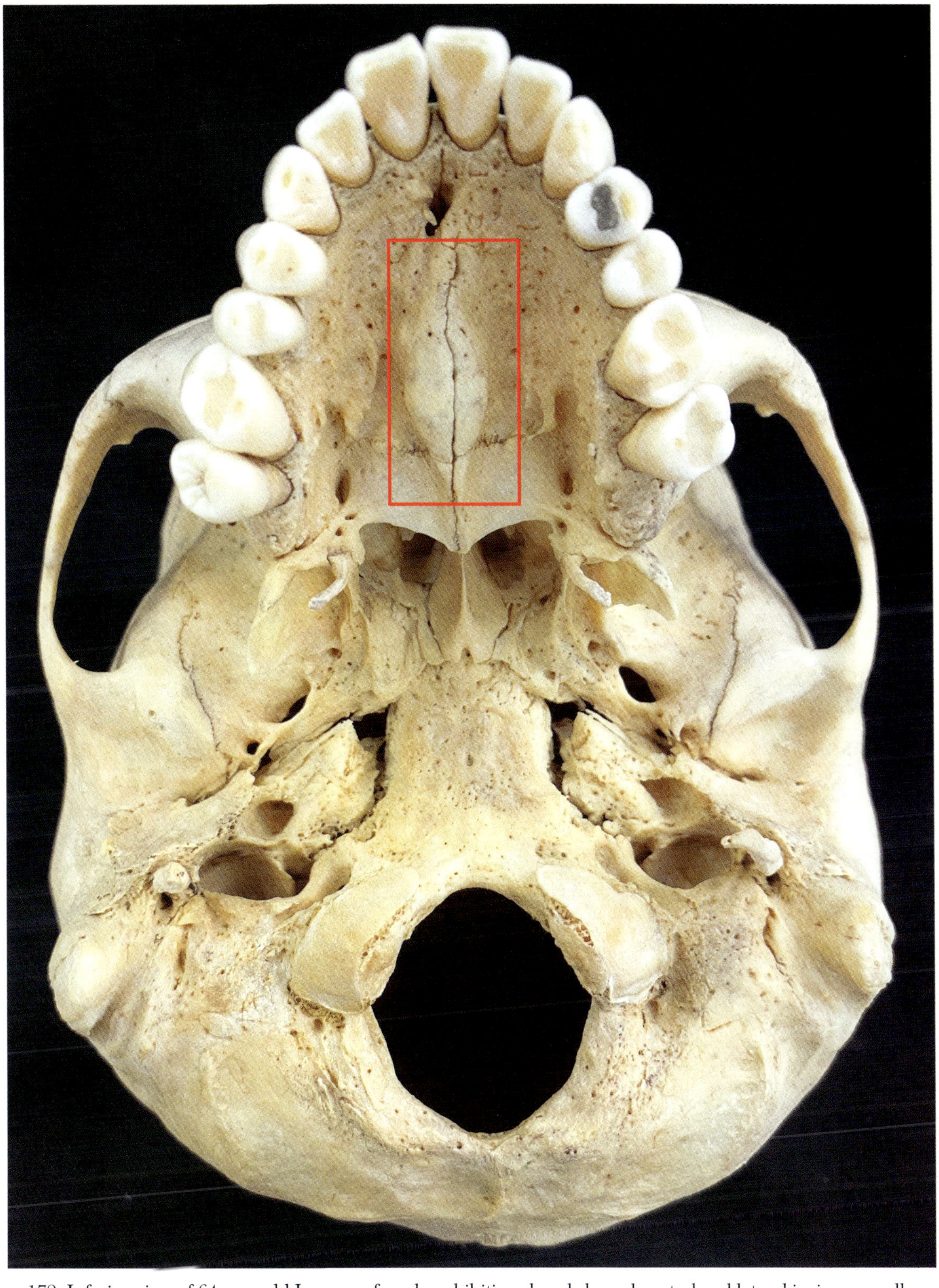

Figure 172. Inferior view of 64-year-old Japanese female exhibiting shovel-shaped central and lateral incisors, small mastoid processes, and torus palatinus (rectangle).

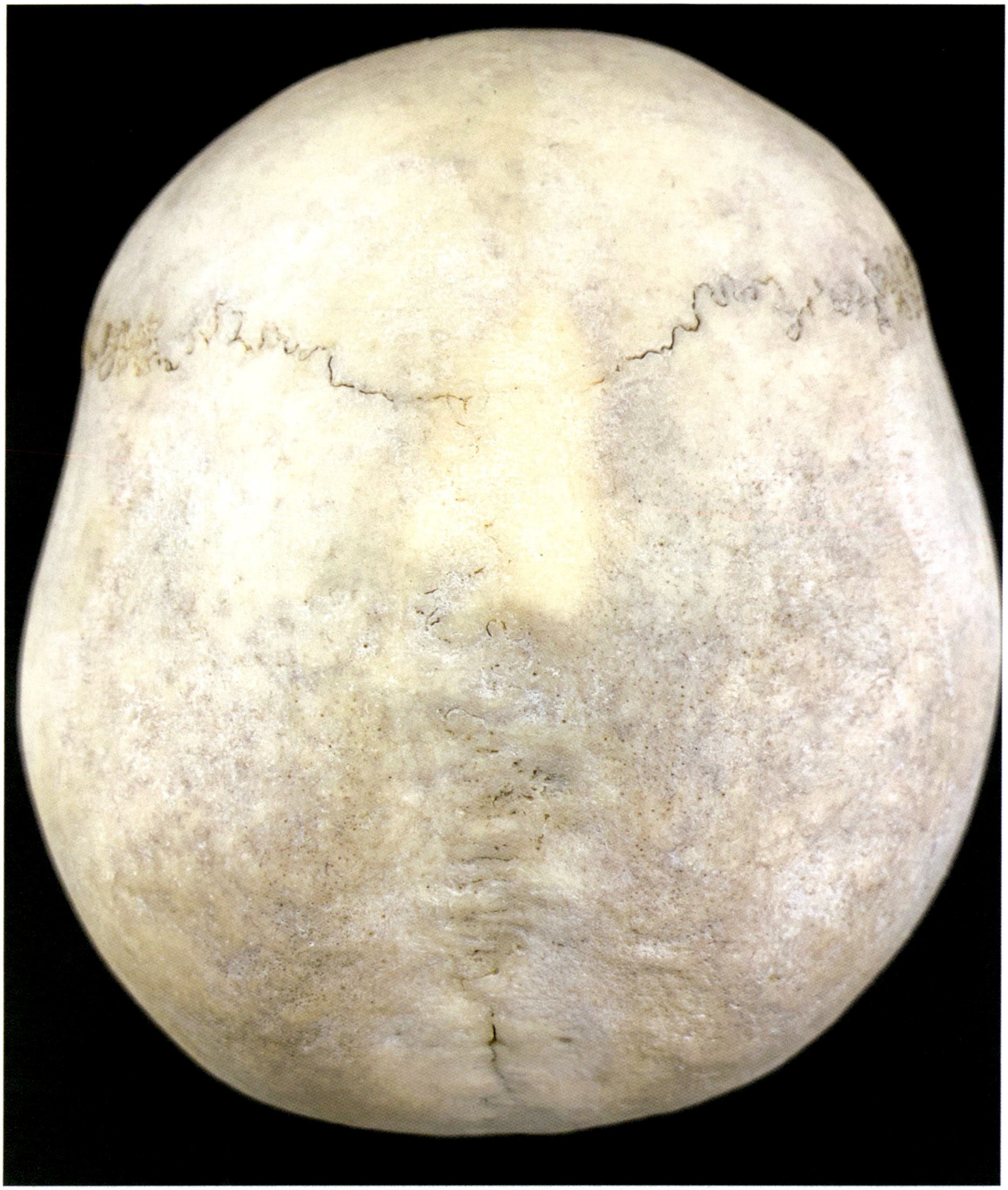

Figure 173. Superior view of 64-year-old Japanese female showing advanced obliteration of the coronal suture and partial obliteration of the sagittal suture. The light area near bregma is taphonomic.

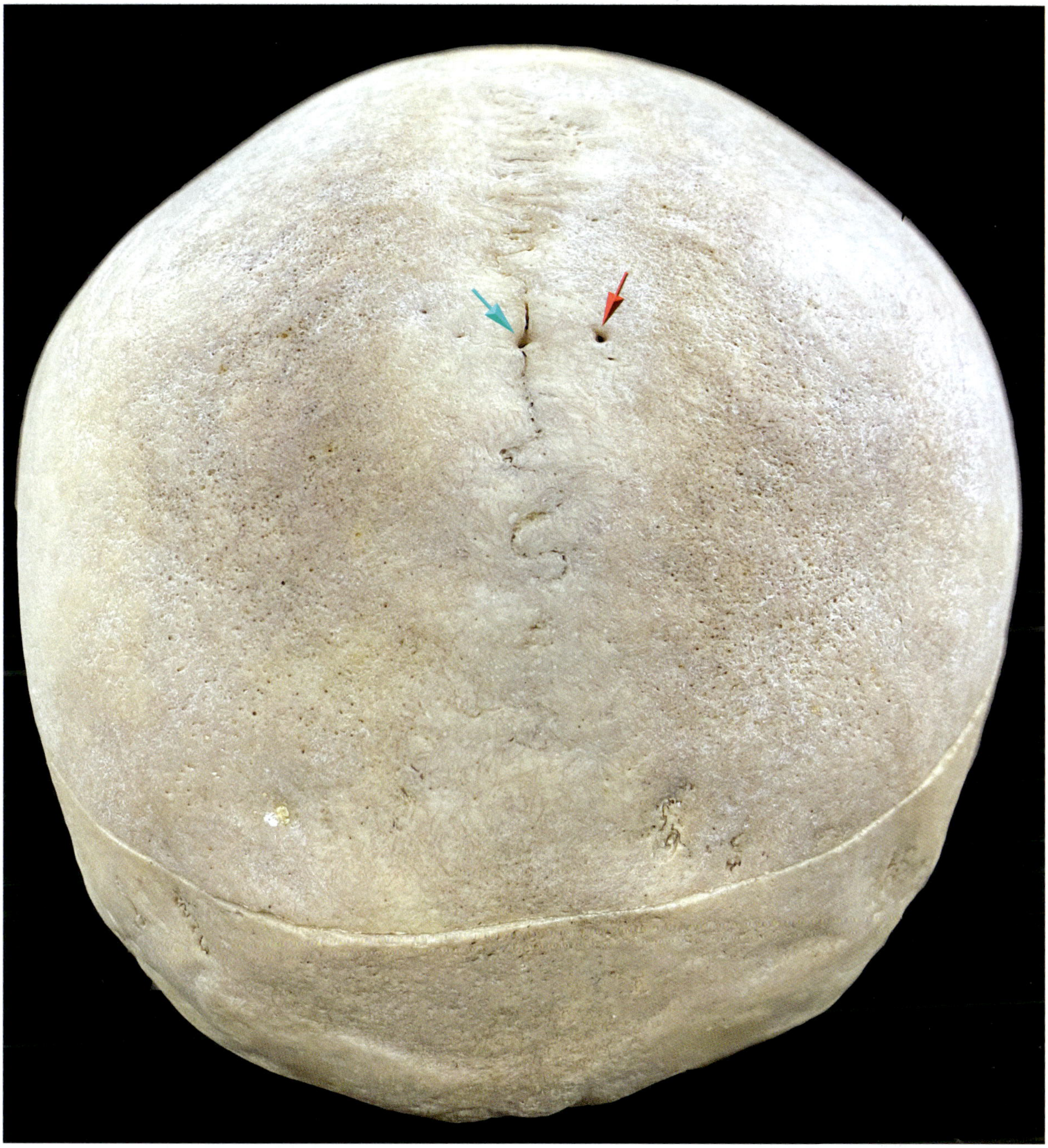

Figure 174. Posterior view of 64-year-old Japanese female with two parietal foramina, one along the sagittal suture (blue arrow) and another in the right parietal bone (red arrow). The lambdoidal suture has completely obliterated.

13. JAPANESE FEMALE

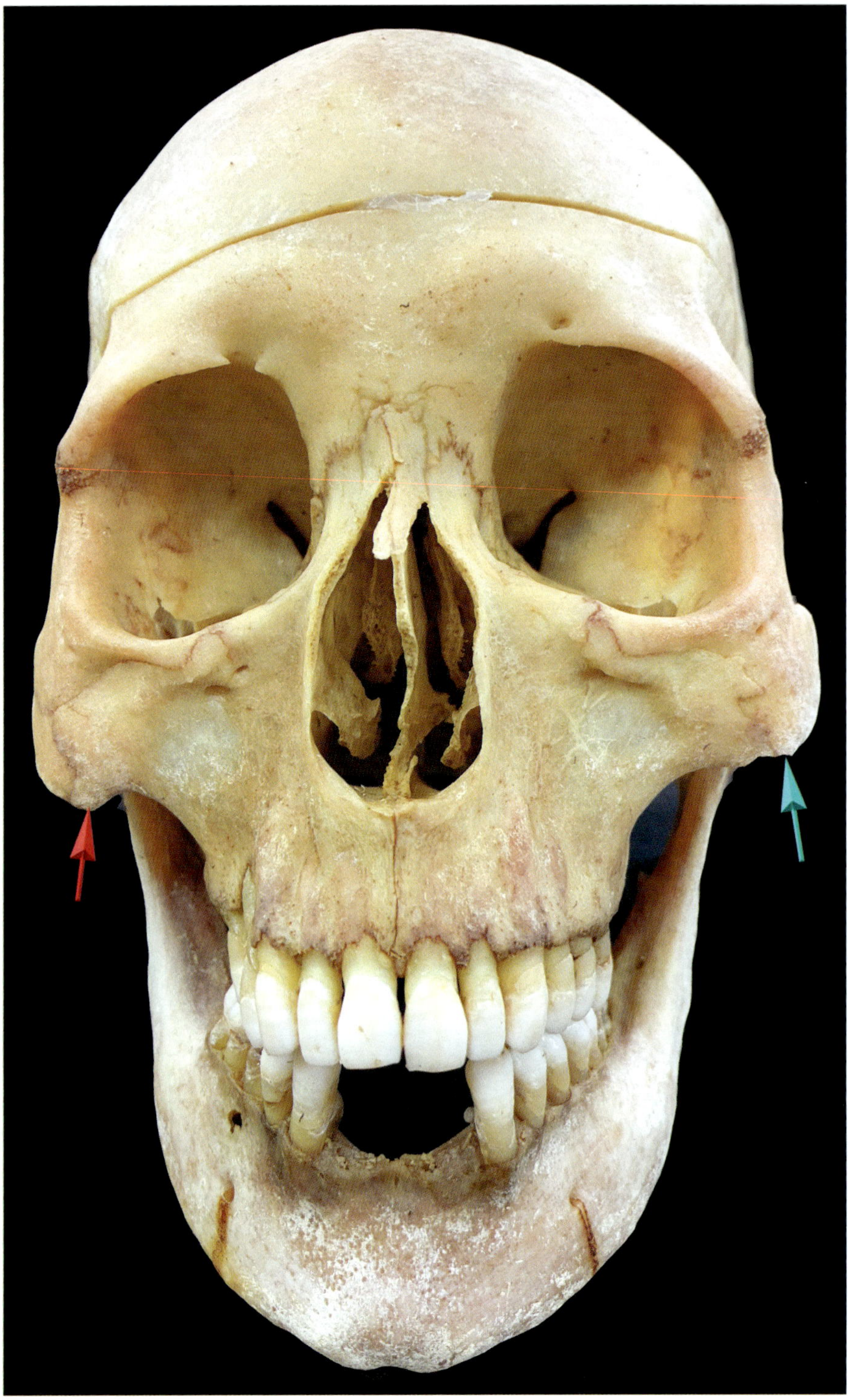

Figure 175. Anterior view 75-year-old Japanese female with what appears to be surgery (no evidence of healing) of the nasal bones, a right malar tubercle (red arrow), absent left tubercle (blue arrow), and a narrow chin with absence of several anterior teeth in the mandible. The dark "drainage" lines originating from the mental foramina are taphonomic.

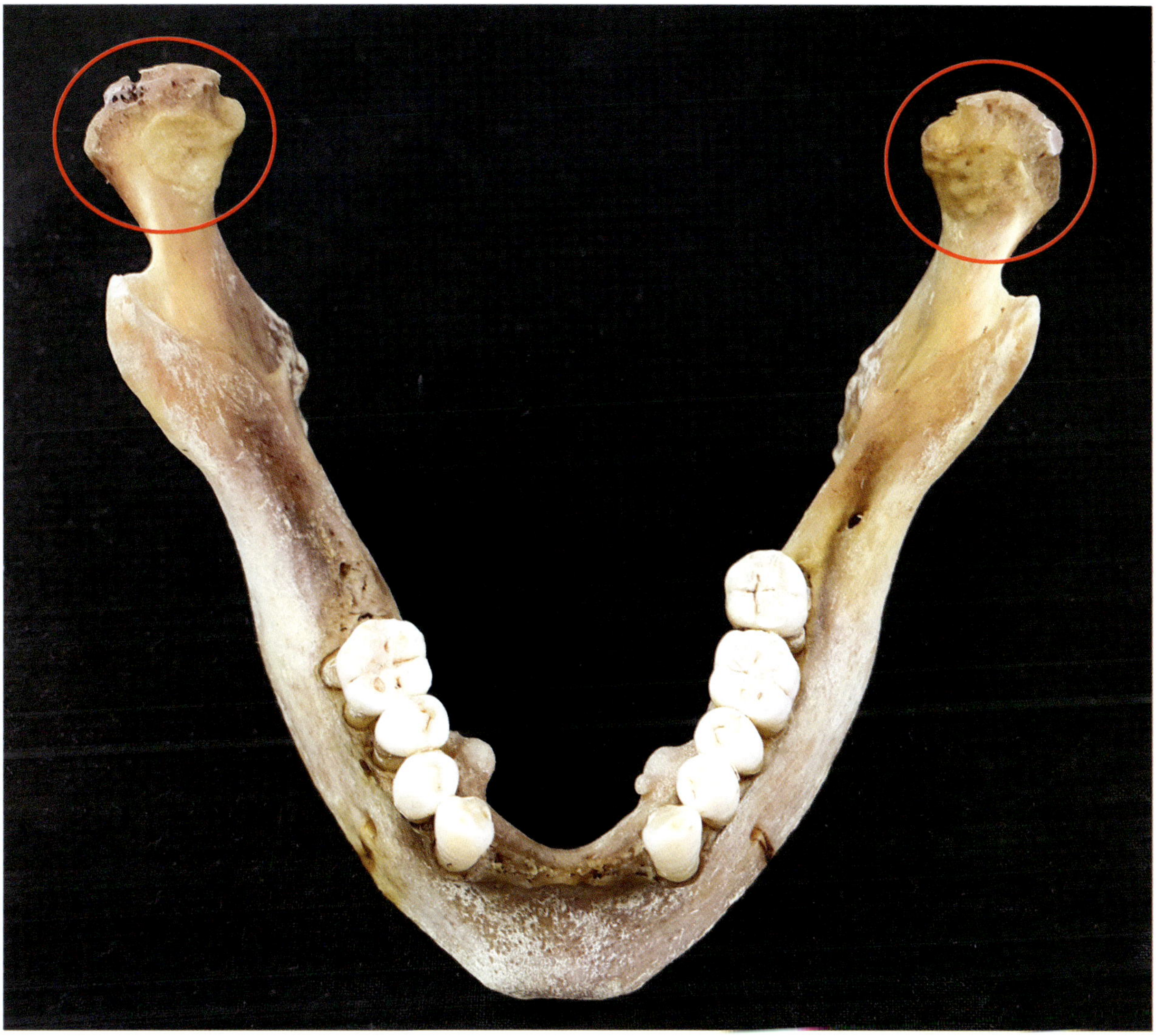

Figure 176 The mandible of 75-year-old Japanese female with degenerative joint disease (osteoarthritis) of the condyles (circle).

Figure 177. Right lateral view of 75-year-old Japanese female with os Japonicum (arrow), also known as os zygomaticum bipartitum, divided zygoma, or bipartite zygoma (Hanihara et al. 1998).

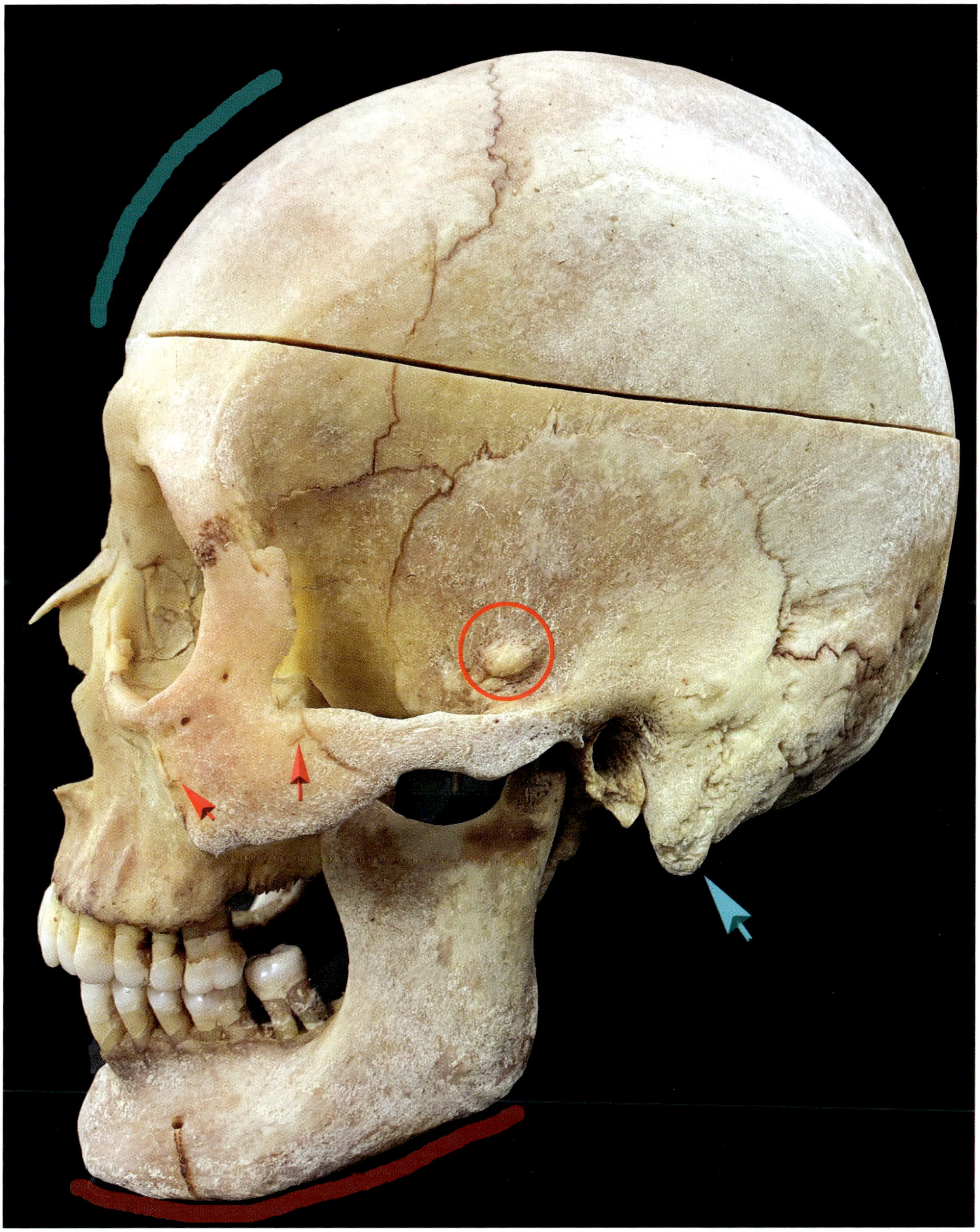

Figure 178. Left lateral view of 75-year-old Japanese female with a "rocker jaw" (curved red line) a feature often associated with Polynesian crania (Houghton 1977). Also note the sloping frontal bone (curved blue line), small mastoid process (blue arrow), a possible tumor (circle) in the temporal bone, and trace os Japonicum (os zygomaticum bipartitum, Hanihara et al. 1998) (red arrow). Although benign, a bony tumor, such as a button osteoma or button lesion, is relatively common (Chae et al. 2015; Eshed et al. 2002). An osteoma, or other type of lesion, in the temporal squamous is rare.

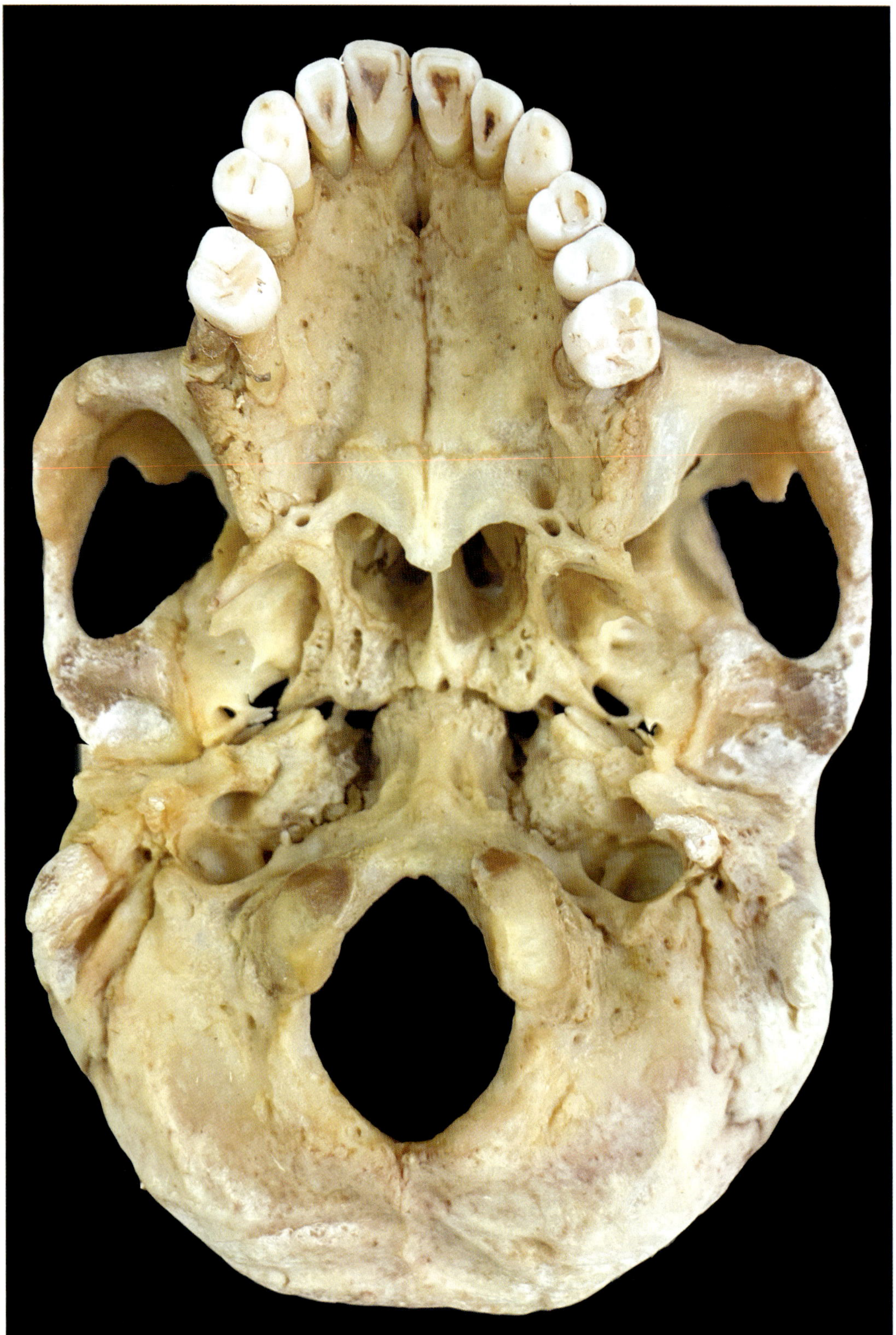

Figure 179. Inferior view of 75-year-old Japanese female exhibiting a shallow and wide palate, prominent shovel-shaped incisors, and small mastoid processes.

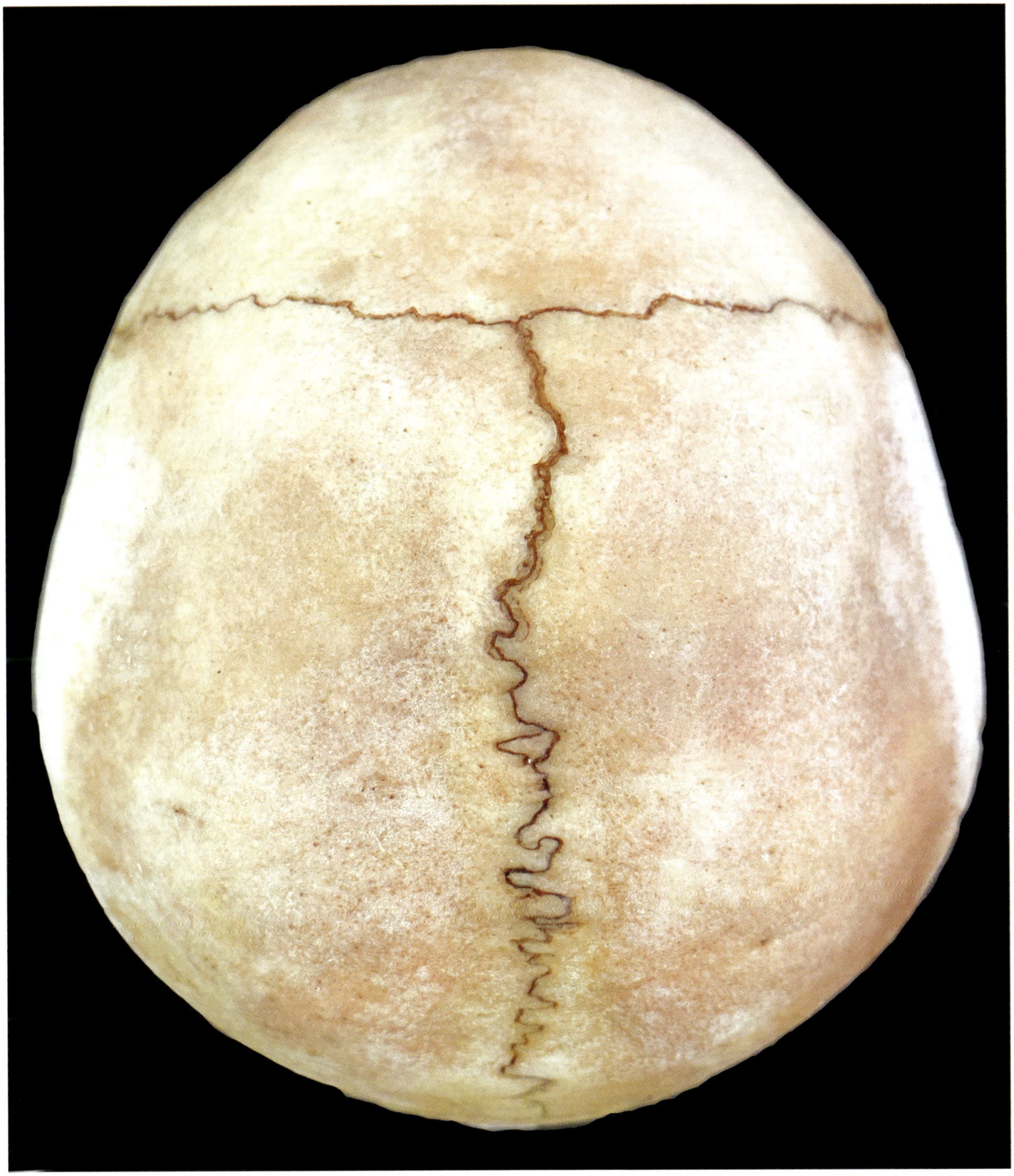

Figure 180. Superior view of 75-year-old Japanese female showing no obliteration of the coronal and sagittal sutures.

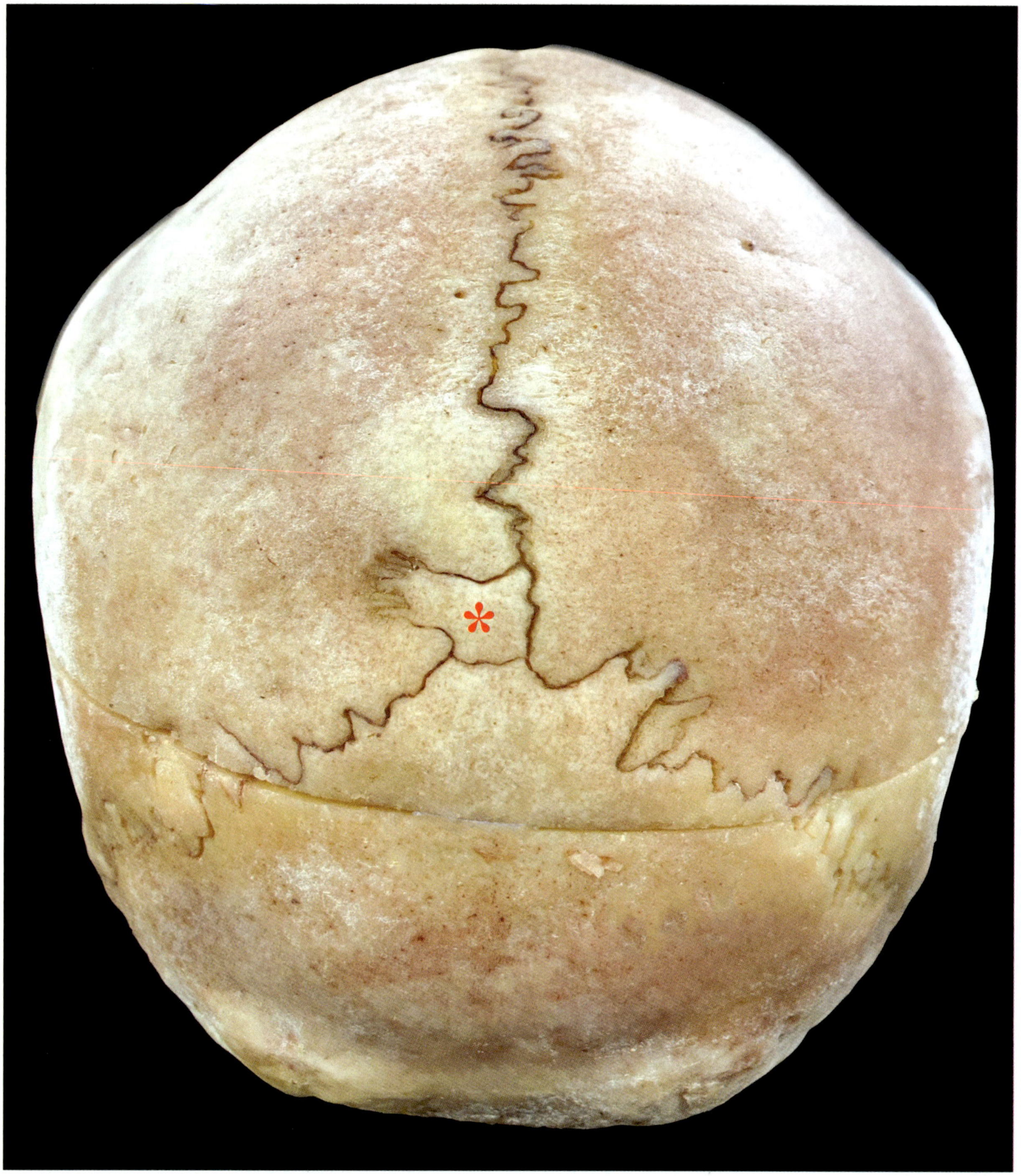

Figure 181. Posterior view of 75-year-old Japanese female with no obliteration of the sagittal and lambdoidal sutures. Note the ossicle (Wormian bone) at lambda (*).

14. JAPANESE FEMALE

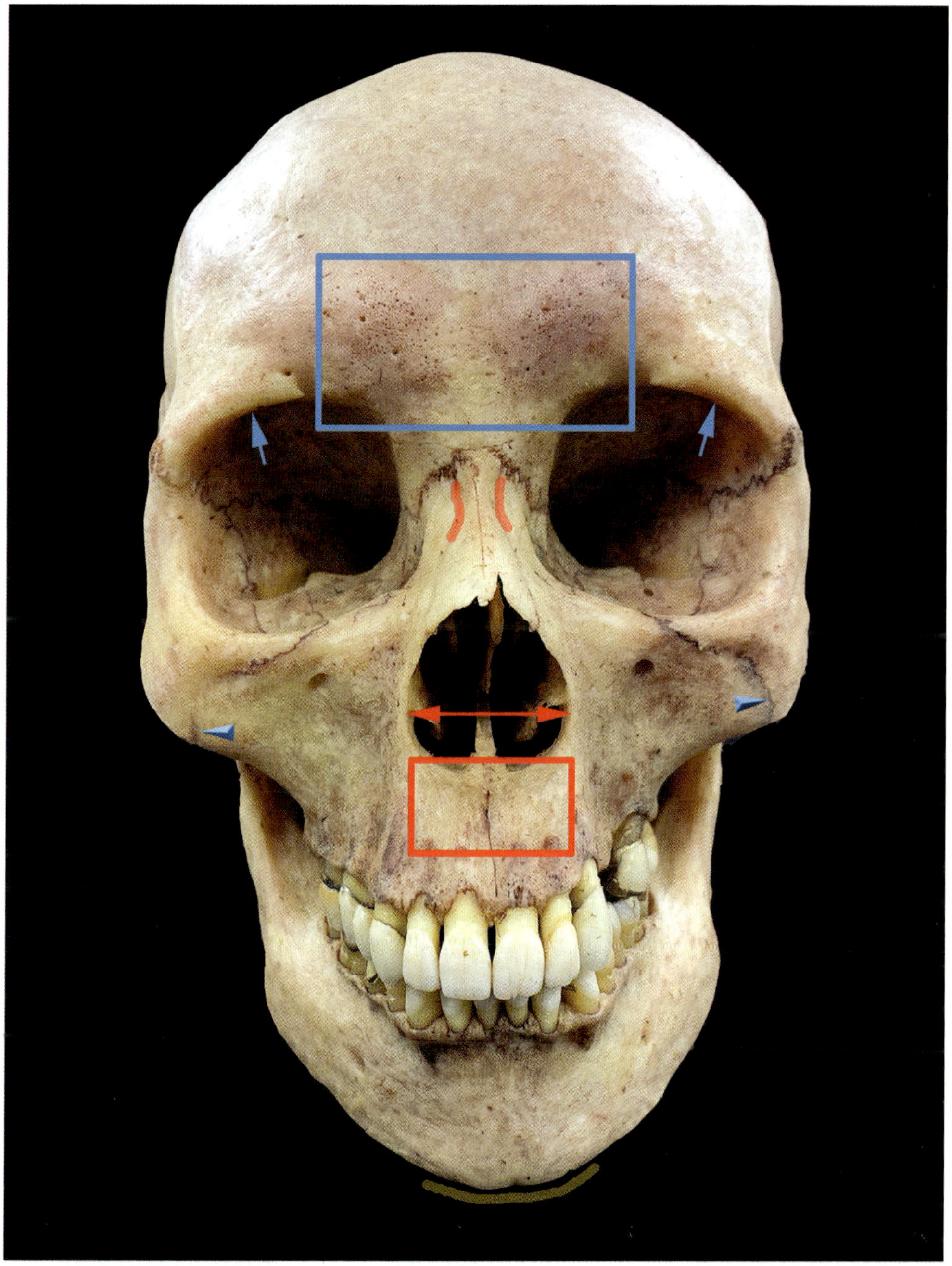

Figure 182. Anterior view 90-year-old Japanese female with blunt superior orbital margins (blue arrow). Note the pitting but absence of grooves/notches in the brow ridge (blue rectangle), the pinched nasal bones (curved red line), the wide and circular nasal aperture (double arrow), the smooth zygomaticomaxillary suture (blue arrowhead), the sloping subnasal area (red rectangle), and the narrow/rounded chin (curved yellow line).

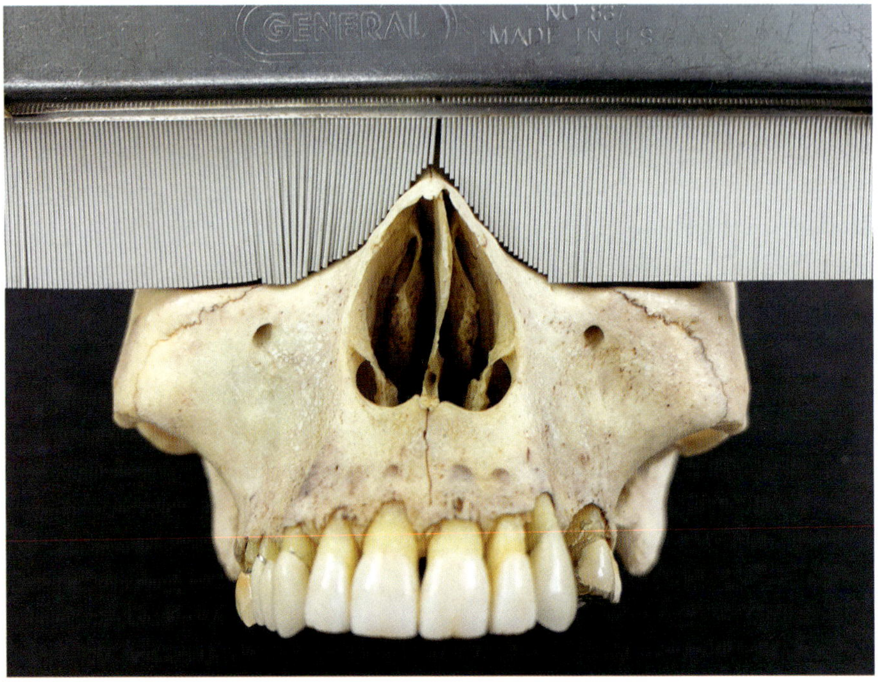

Figure 183. Nasal bone contour in a 90-year-old Japanese female.

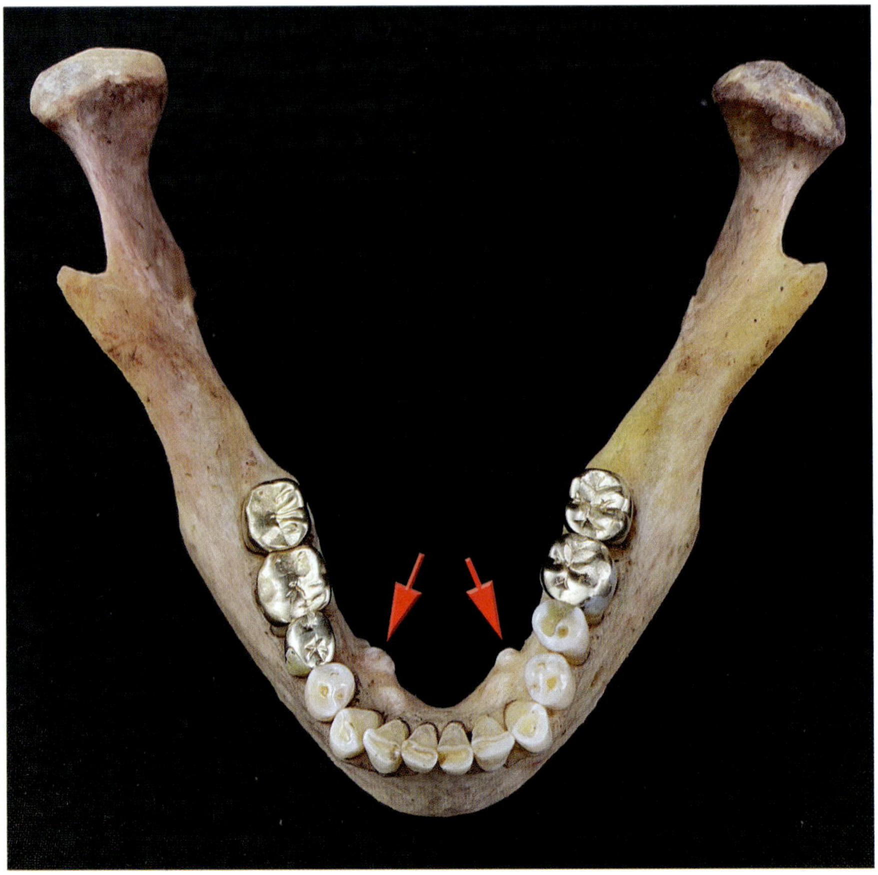

Figure 184. Mandible of 90-year-old Japanese female with symmetrical lingual exostoses (arrow).

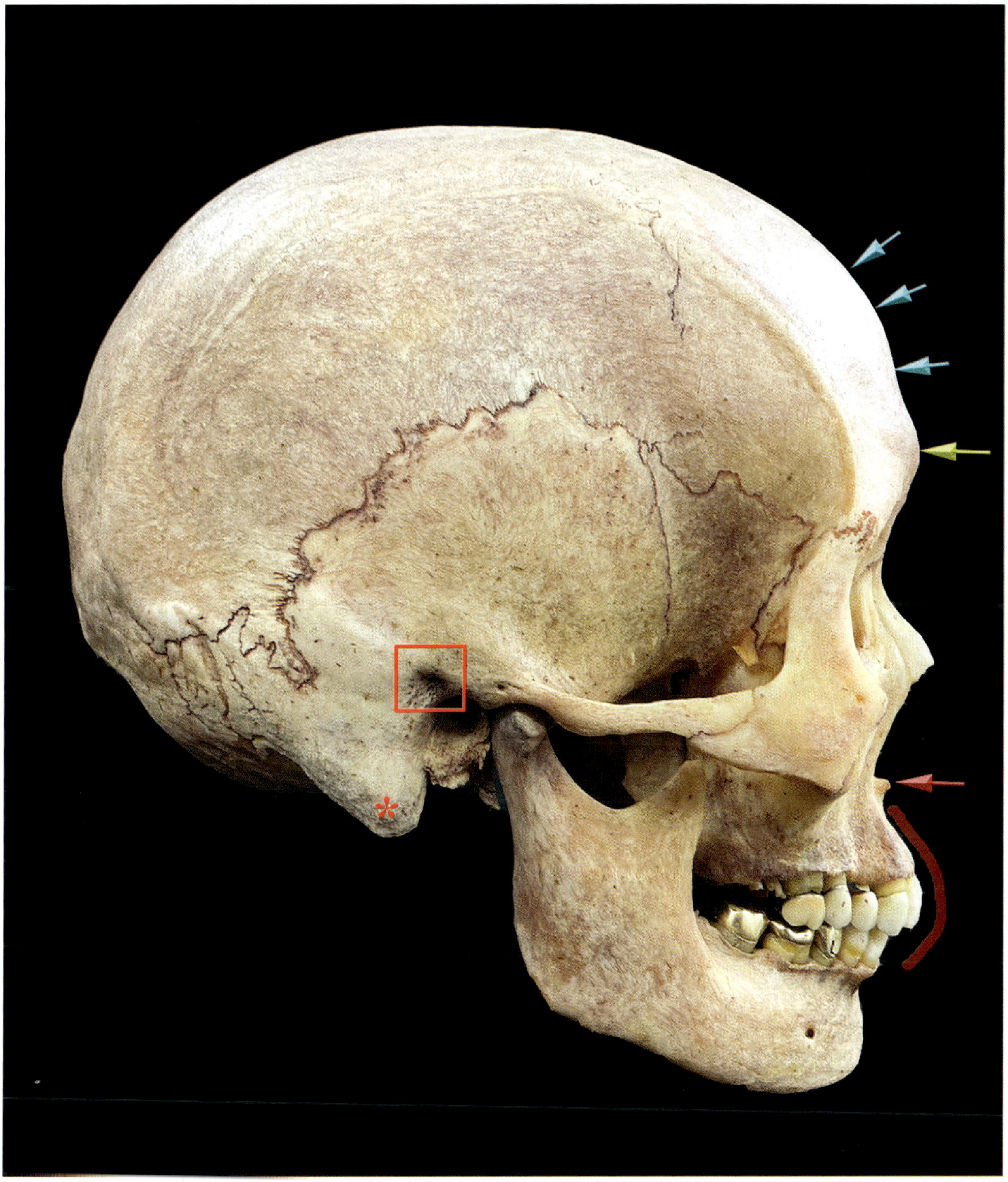

Figure 185. Right lateral view of 90-year-old Japanese female with a mildly sloping frontal bone (blue arrow), and a small brow ridge (yellow arrow). Note the small anterior nasal spine (red arrow) and mastoid process (asterisk), and the alveolar prognathism (red curved line). Observe the small perforated defect (square) of unknown etiology, possibly due to middle-ear infection (otitis media), superior to the acoustic meatus (see Figure 186).

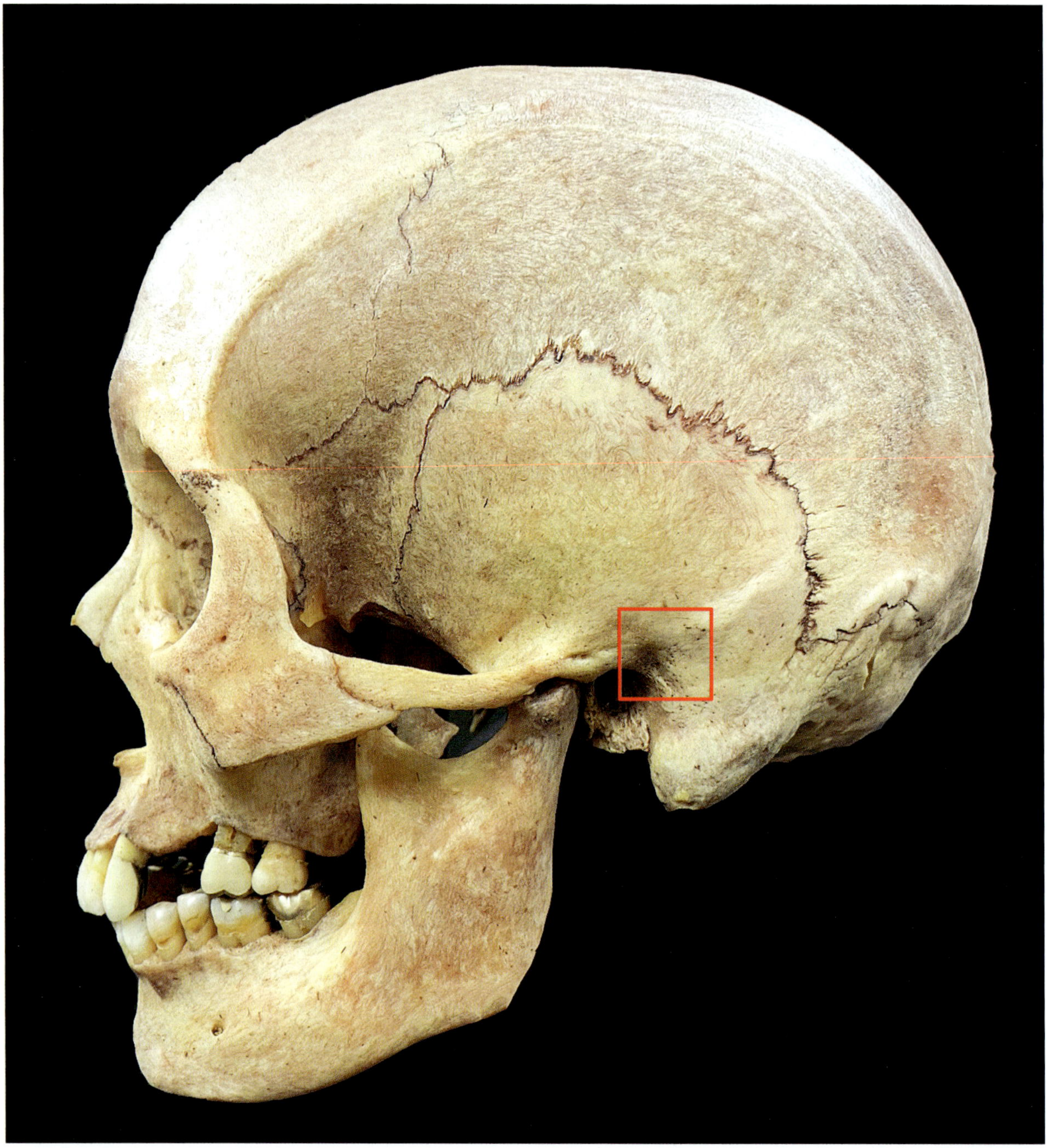

Figure 186. Left lateral view of 90-year-old Japanese female with a small cratered pit (square) superior-posterior to the auditory meatus. This bilateral pit is located in the same area as a suprameatal pit, a normal anatomical variant, but could also be the result of otitis media or middle ear infection superior to the external auditory meatus.

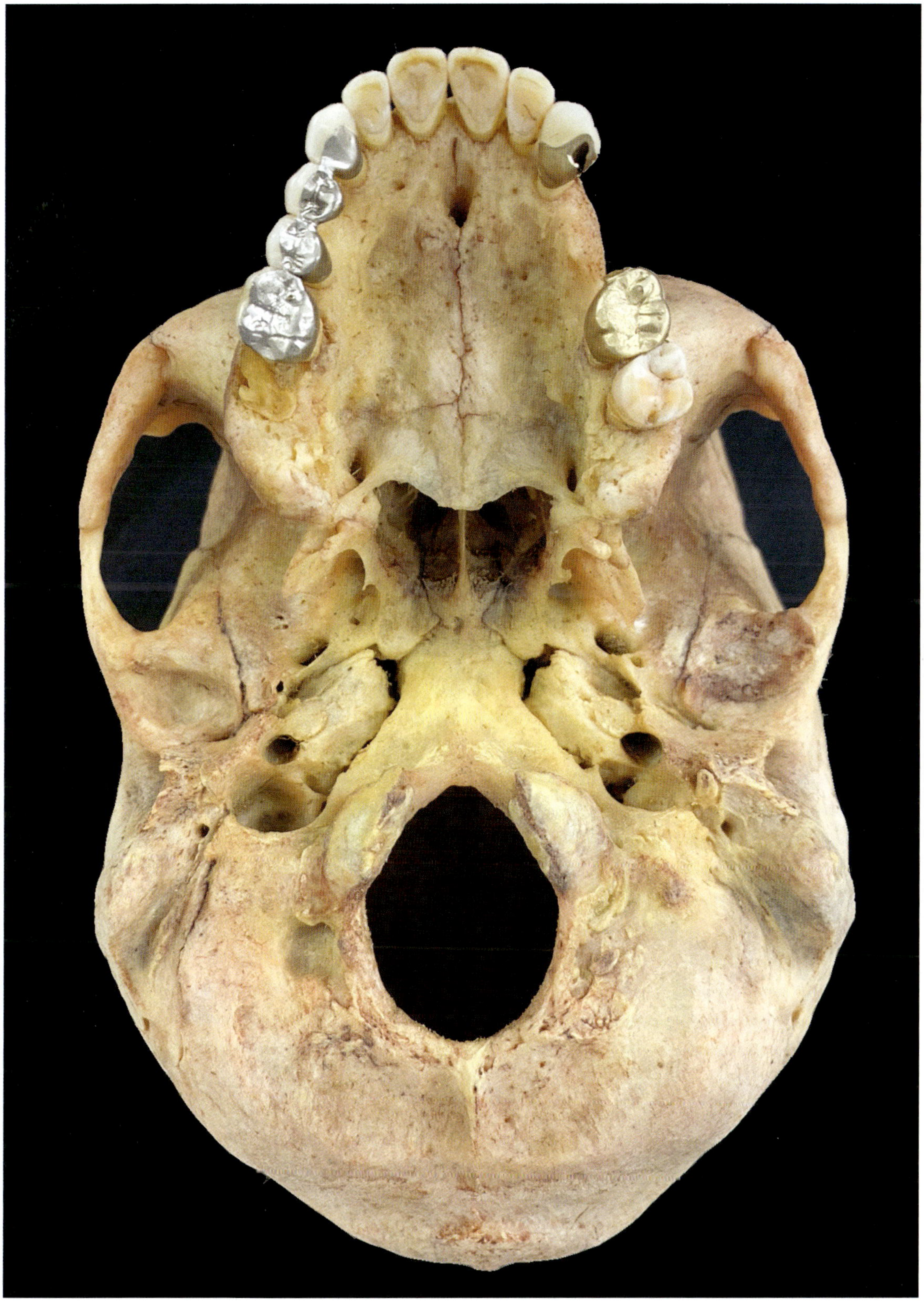

Figure 187. Inferior view of 90-year-old Japanese female with small mastoid processes and shovel-shaped central incisors.

Figure 188. Superior view of 90-year-old Japanese female with an oval-shaped vault and minimal obliteration of the sutures.

Figure 189. Posterior view of 90-year-old Asian Japanese female with minimal obliteration of the sutures, a small nuchal crest, and a squarish overall cranial shape. Note the vertical profile of the temporal bones.

15. KOREAN MALE

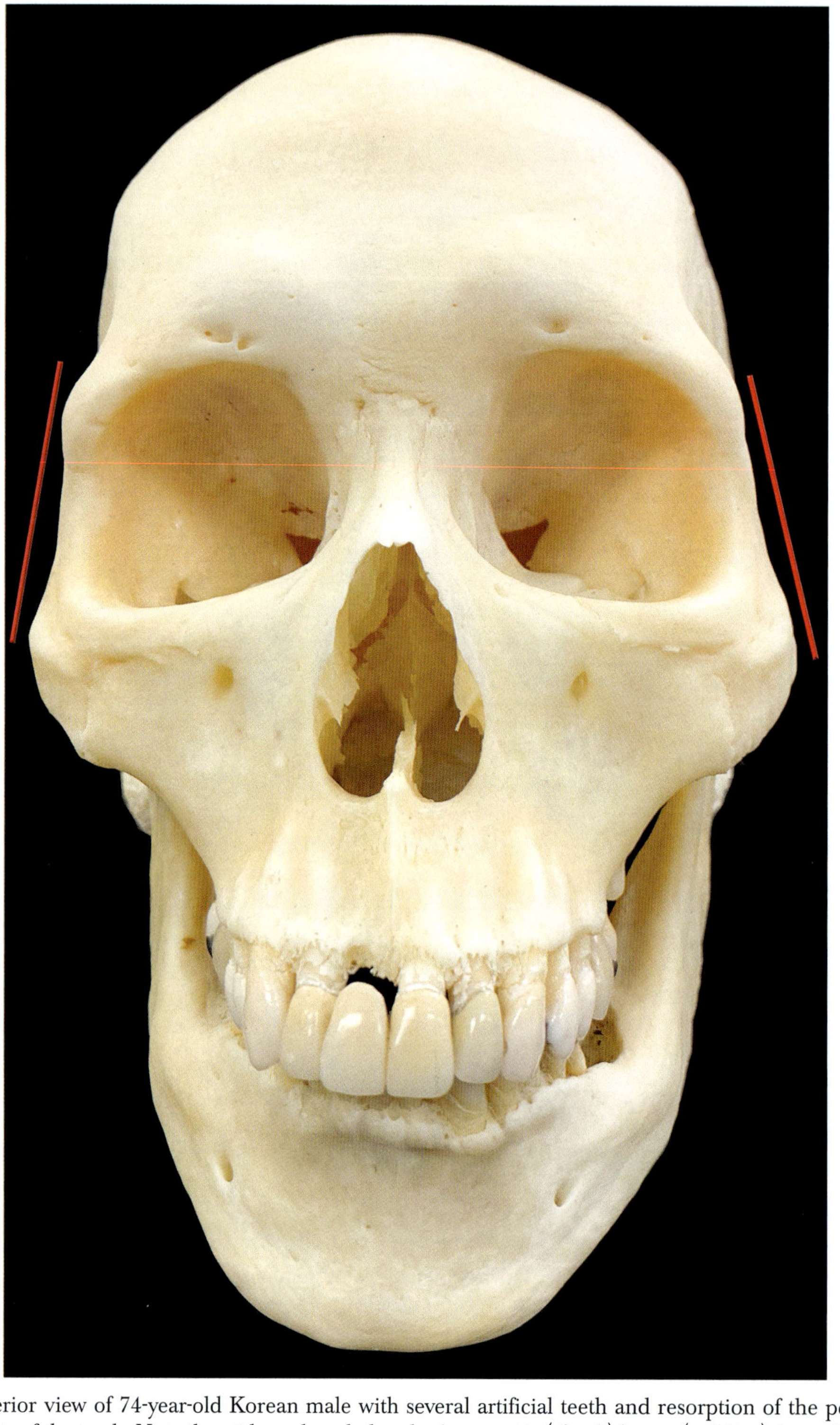

Figure 190. Anterior view of 74-year-old Korean male with several artificial teeth and resorption of the periodontal bone exposing the roots of the teeth. Note the wide and angled malar/zygomatic (cheek) bones (red line), buccal mandibular exostoses, and narrow, but contoured nasal bones (see Hwang et al. 2005).

Figure 191. Nasal bone contour in a 74-year-old Korean male.

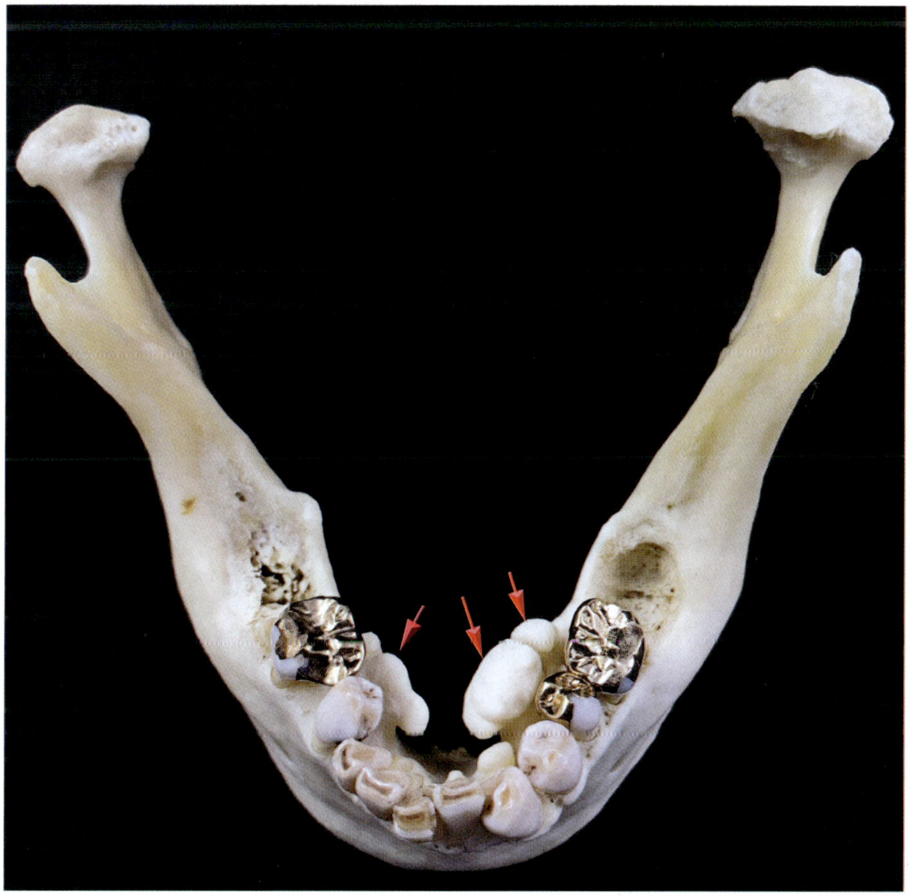

Figure 192. Mandible of 74-year-old Korean male with a wide and rounded chin. Note crowding of the anterior teeth and large lingual exostoses (arrow).

TABLE 19.
Cranial Measurements in a 74-year-old Korean Male

Measurement	(mm)	Right	Left
Maximum Cranial Length (g-op, GOL)	185		
Maximum Cranial Breadth (eu-eu, XCB)	136		
Bizygomatic Breadth (zy-zy, ZYB)	123		
Basion-Bregma Height (ba-b, BBH)	146		
Cranial Base Length (ba-n, BNL)	99		
Basion-Prosthion (ba-pr, BPL)	90		
Maxillo-Alveolar Breadth (ecm-ecm, MAB)	-		
Maxillo-Alveolar Length (pr-alv, MAL)	51		
Biauricular Breadth (au-au, AUB)	124		
Upper Facial Height (n-pr)	75		
Minimum Frontal Breadth (ft-ft, WFB)	87		
Upper Facial Breadth (fmt-fmt)	104		
Nasal Height (n-ns, NLH)	54		
Nasal Breadth (al-al, NLB)	23		
Orbital Breadth (d-ec, OBB)	39		L
Orbital Height (OBH)	34		L
Biorbital Breadth (ec-ec, EKB)	97		
Interorbital Breadth (d-d, DKB)	18		
Frontal Chord (n-b, FRC)	116		
Parietal Chord (b-l, PAC)	123		
Occipital Chord (l-o, OCC)	105		
Foramen Magnum Length (ba-o, FOL)	33		
Foramen Magnum Breadth (FOB)	31		
Mastoid Length (MDH)		26	23

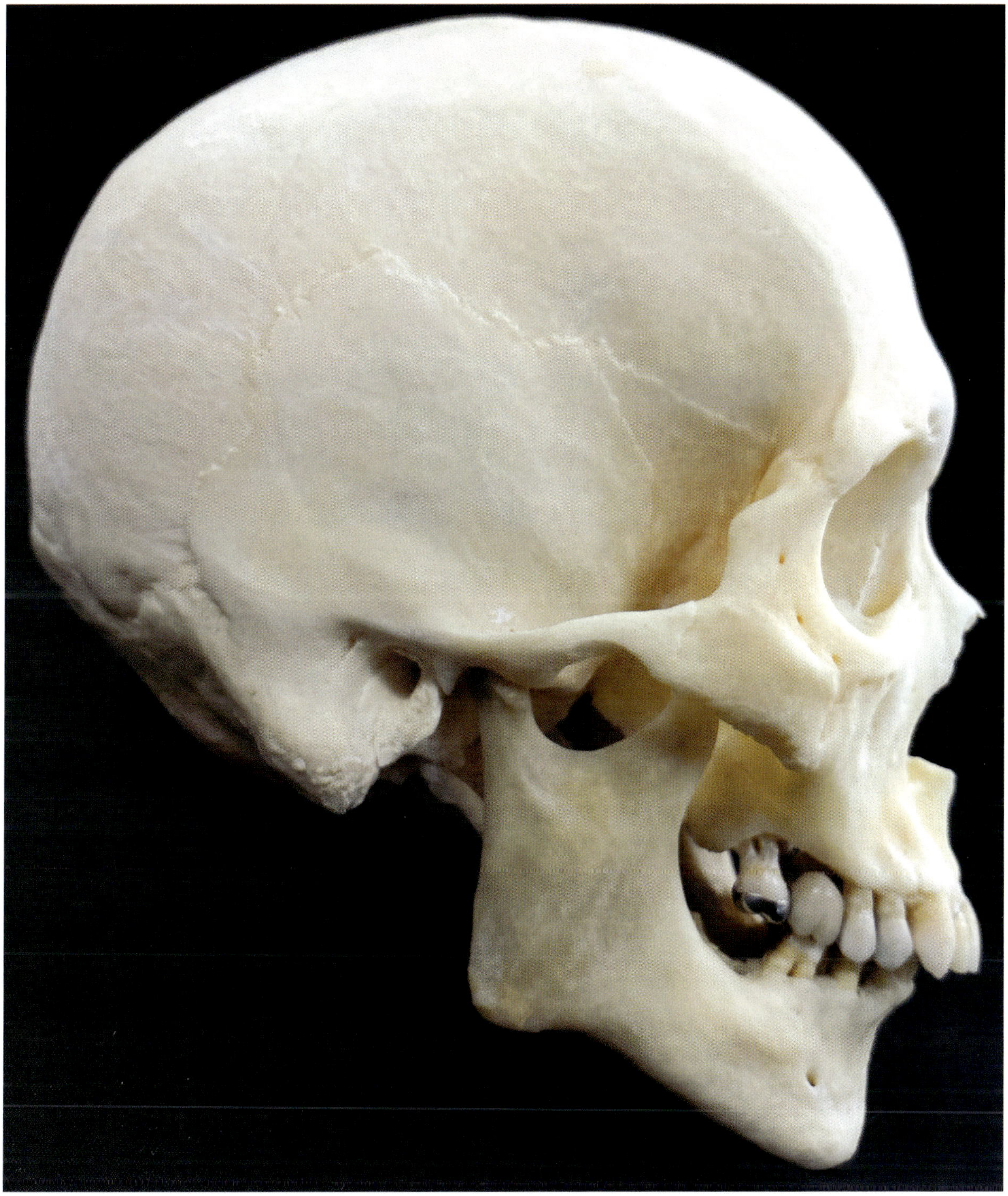

Figure 193. Right lateral view of 74-year-old Korean male with a moderate overbite of the teeth.

Figure 194. Left lateral view of 74-year-old Korean male.

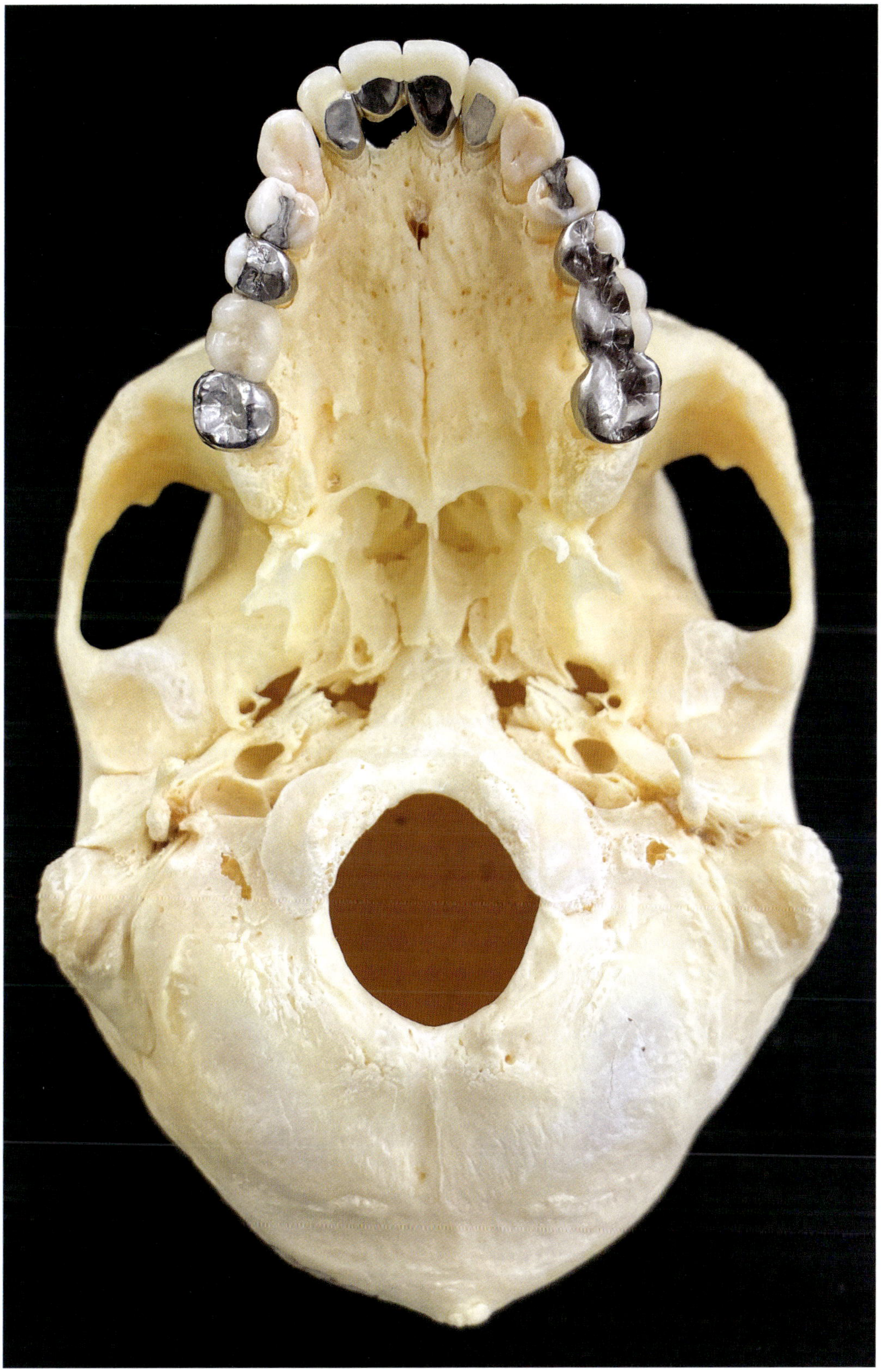

Figure 195. Inferior view of 74-year-old Korean male.

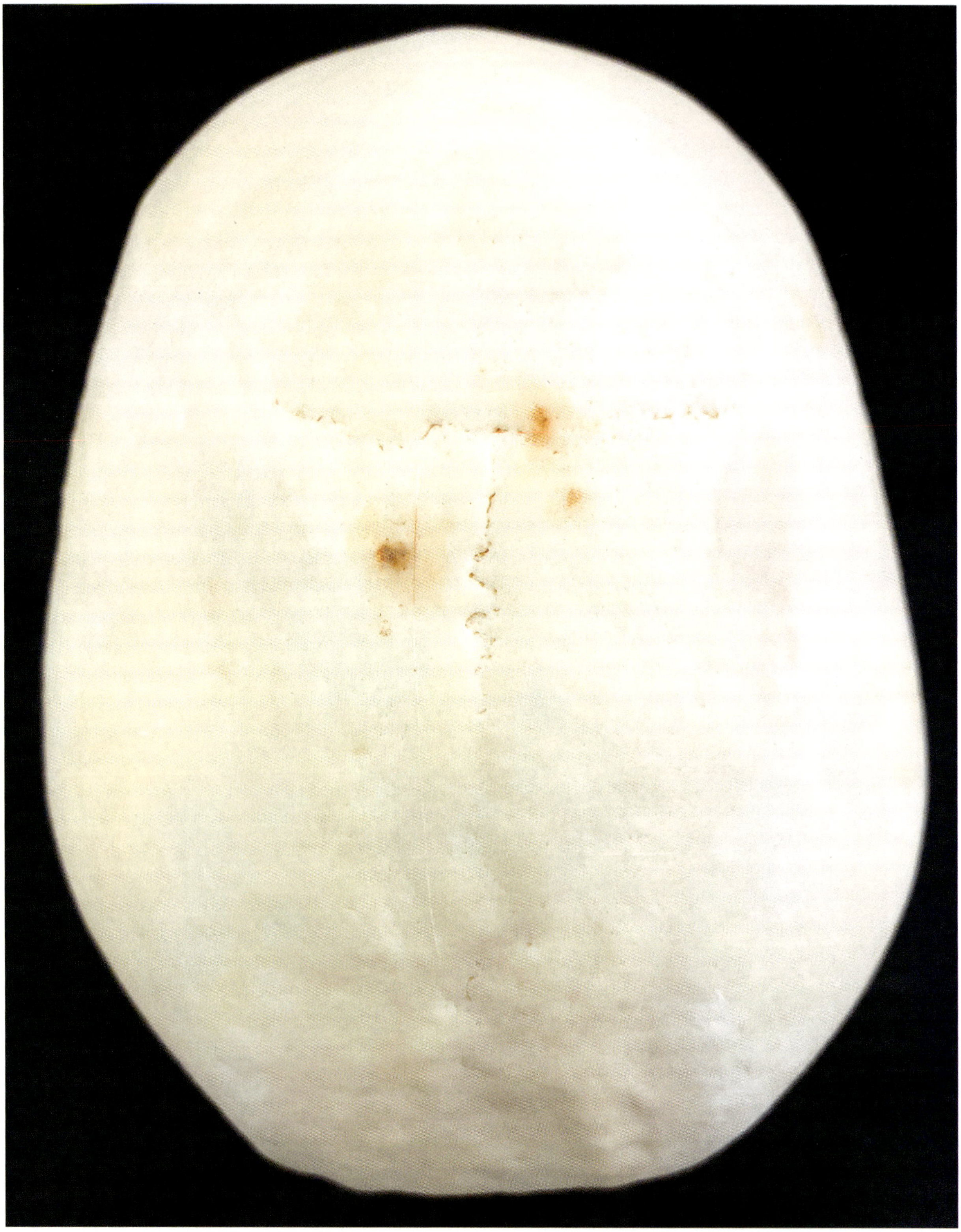

Figure 196. Superior view of 74-year-old Korean male showing nearly complete obliteration of the sutures. The brown stains are taphonomic as a result of the skull resting on its vertex while drying.

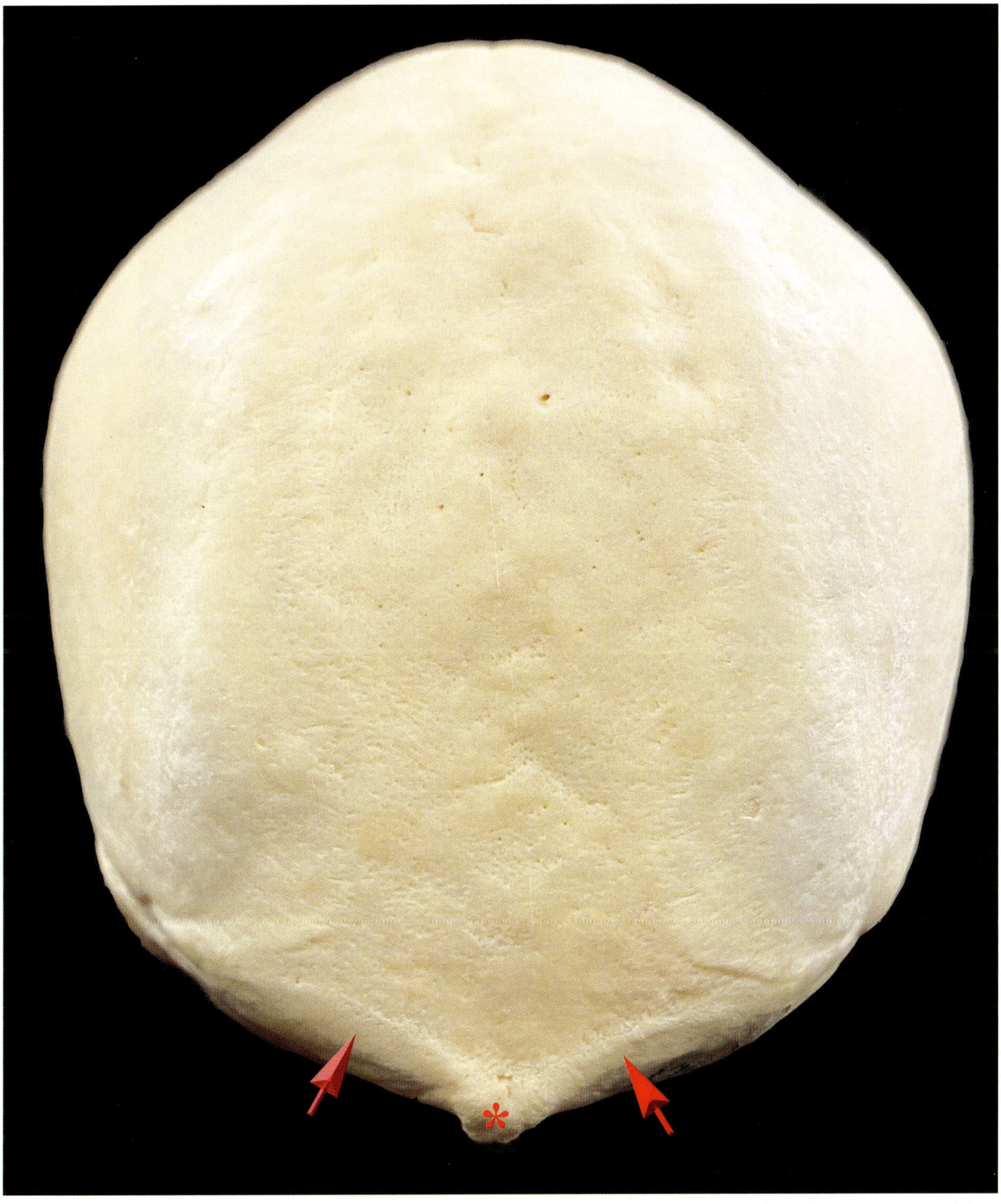

Figure 197. Posterior view of 74-year-old Korean male with a well-developed nuchal crest (arrow) and inion hook (*).

16. KOREAN MALE

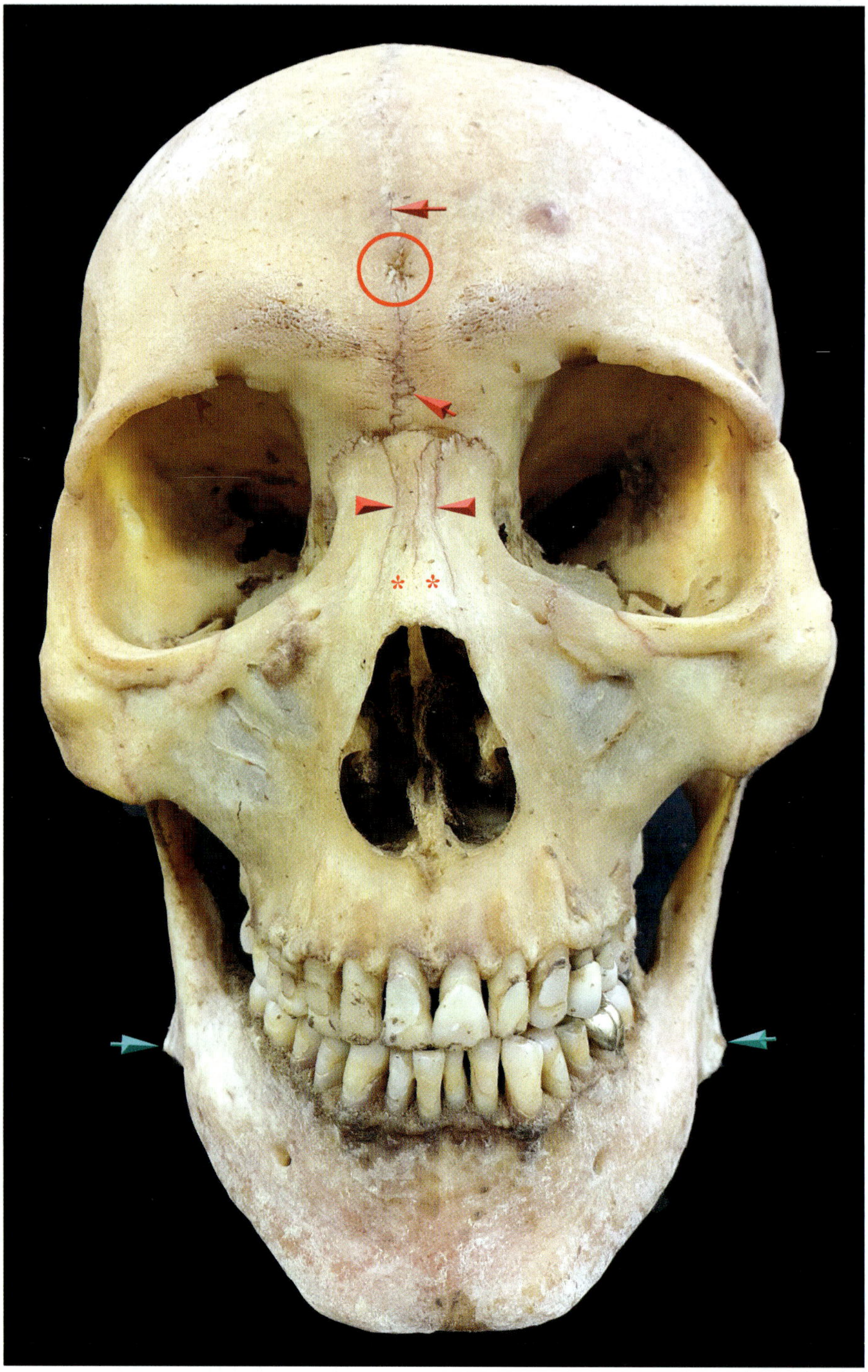

Figure 198. Anterior view 92-year-old Korean male. Note the trace metopic suture (red arrow), the metopic fissure (circle), and the flat (*), pinched/waisted nasal bones (arrowhead). Observe the wide nasal aperture, the wide interorbital breadth, and the numerous pits and grooves along the brow ridge. These brow ridge pits are sometimes referred to as the adult "vermiculate pattern" (Tappen 1983) and are often, but not exclusively, present in males. Note the trace malar tubercle, the gonial eversion (blue arrow), and the broad and rounded chin.

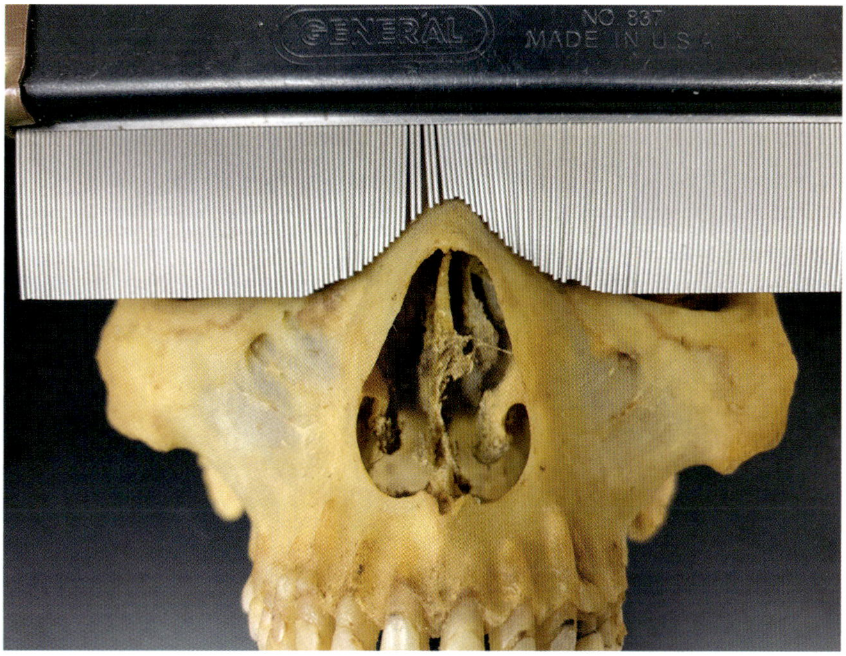

Figure 199. Contour of the nasal bones in a 92-year-old Korean male.

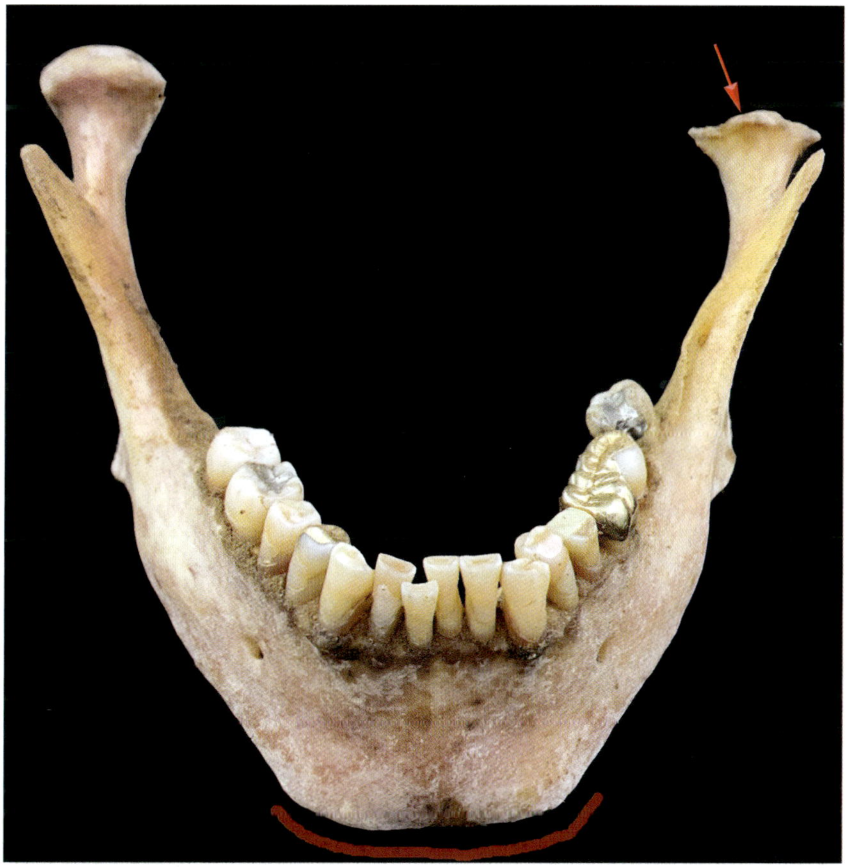

Figure 200. Mandible 92-year-old Korean male with a wide and rounded chin (curved red line) and degenerative joint disease (arrow) of the left condyle.

TABLE 20.
Cranial Measurements in a 92-year-old Korean Male

Measurement	(mm)	Right	Left
Maximum Cranial Length (g-op, GOL)	171		
Maximum Cranial Breadth (eu-eu, XCB)	151		
Bizygomatic Breadth (zy-zy, ZYB)	146		
Basion-Bregma Height (ba-b, BBH)	141		
Cranial Base Length (ba-n, BNL)	103		
Basion-Prosthion (ba-pr, BPL)	95		
Maxillo-Alveolar Breadth (ecm-ecm, MAB)	68		
Maxillo-Alveolar Length (pr-alv, MAL)	53		
Biauricular Breadth (au-au, AUB)	134		
Upper Facial Height (n-pr)	76		
Minimum Frontal Breadth (ft-ft, WFB)	101		
Upper Facial Breadth (fmt-fmt)	114		
Nasal Height (n-ns, NLH)	60		
Nasal Breadth (al-al, NLB)	26		
Orbital Breadth (d-ec, OBB)	41		L
Orbital Height (OBH)	34		L
Biorbital Breadth (ec-ec, EKB)	106		
Interorbital Breadth (d-d, DKB)	32		
Frontal Chord (n-b, FRC)	103		
Parietal Chord (b-l, PAC)	108		
Occipital Chord (l-o, OCC)	99		
Foramen Magnum Length (ba-o, FOL)	36		
Foramen Magnum Breadth (FOB)	33		
Mastoid Length (MDH)			29

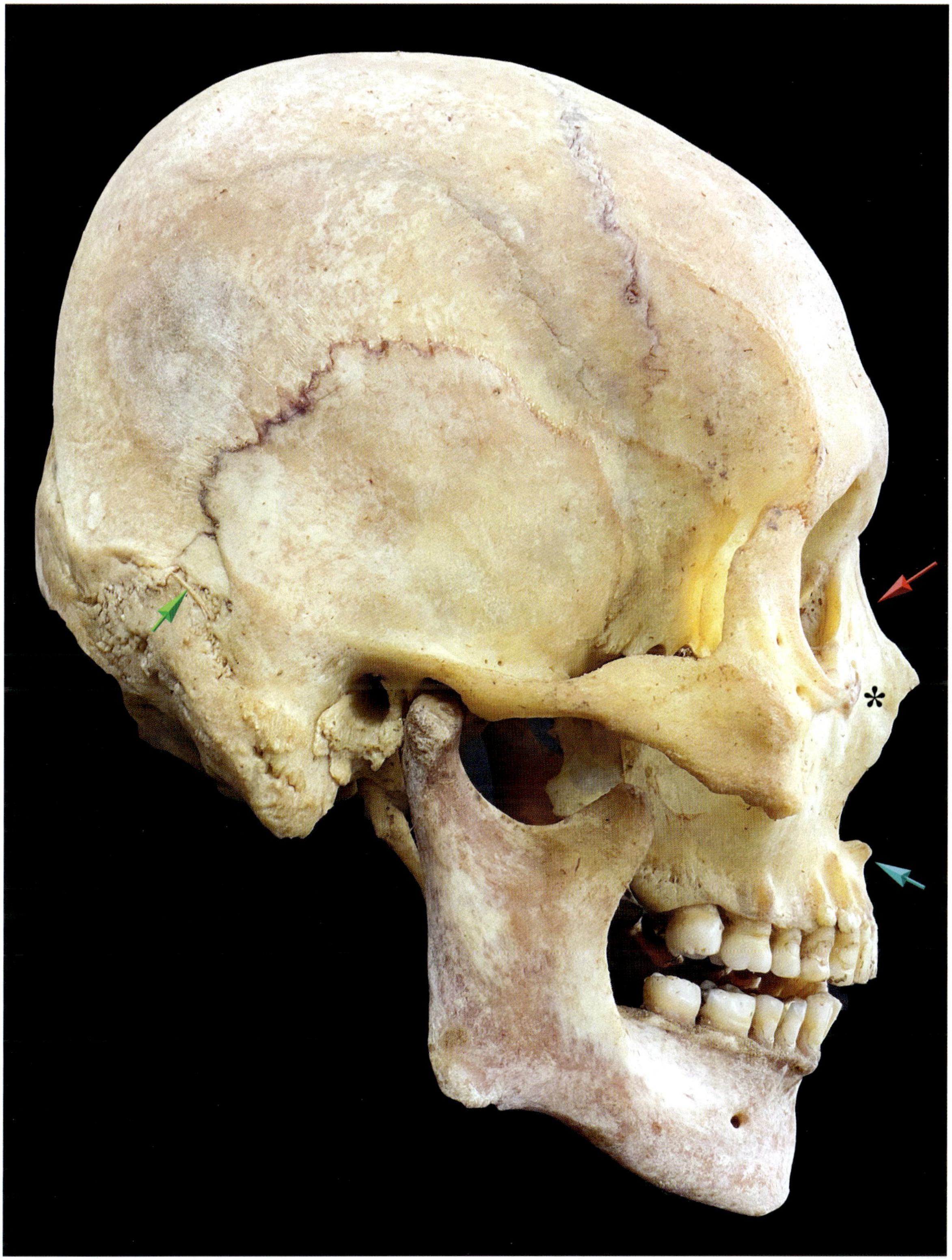

Figure 201. Right lateral view of 92-year-old Korean male with concave proximal (red arrow) and convex distal nasal bones, a small anterior nasal spine (blue arrow), and moderate-size frontal process of the maxilla (*). Note the presence of the loosely attached calcification (green arrow) often present along muscle attachment sites in elderly individuals.

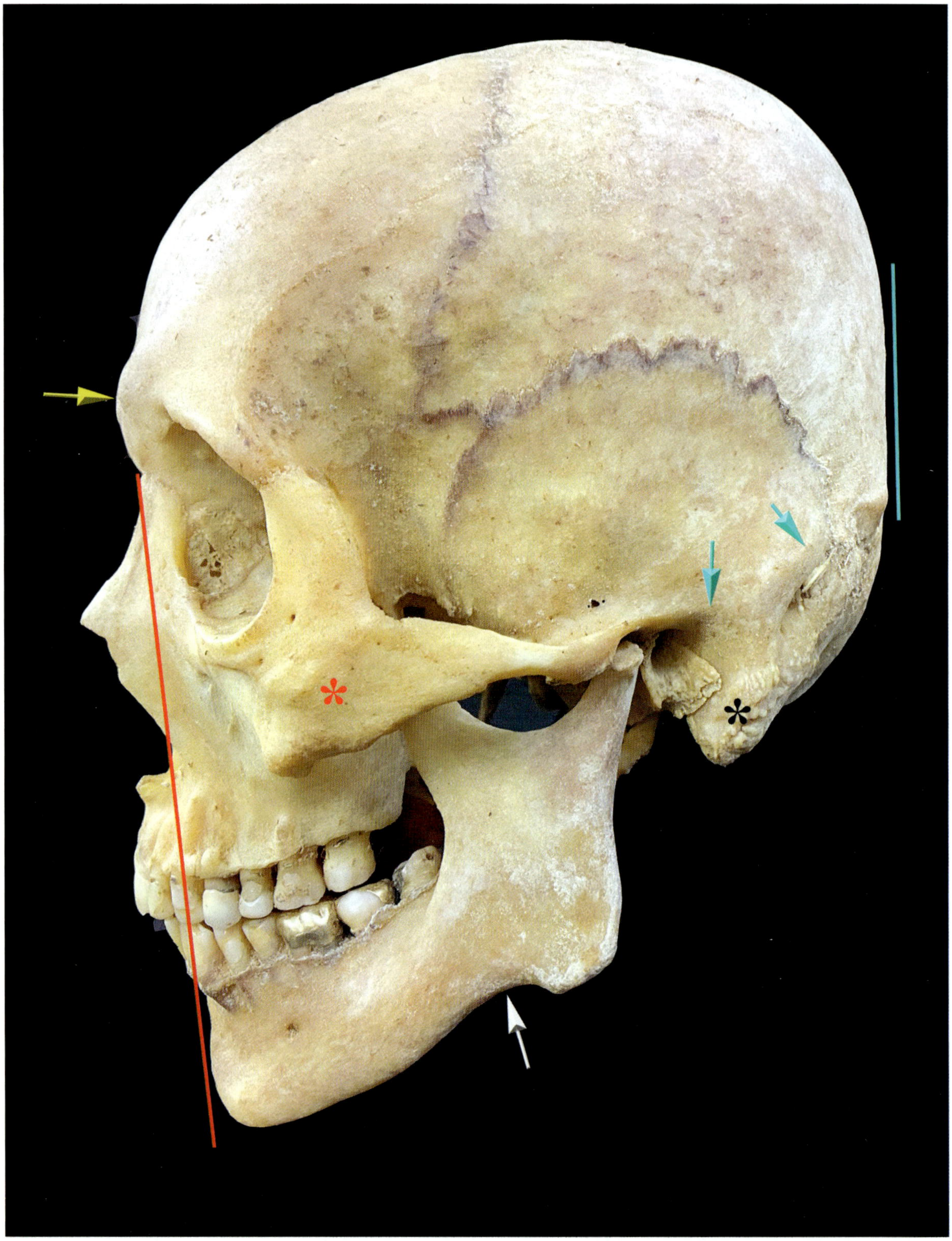

Figure 202. Left lateral view of 92-year-old Korean male with a small brow ridge (yellow arrow), mild alveolar prognathism (red line), large zygoma (red*), a deep antegonial notch (white arrow), developed supramastoid crest (blue arrow), small mastoid crest (black*), and a flat occipital region (blue line).

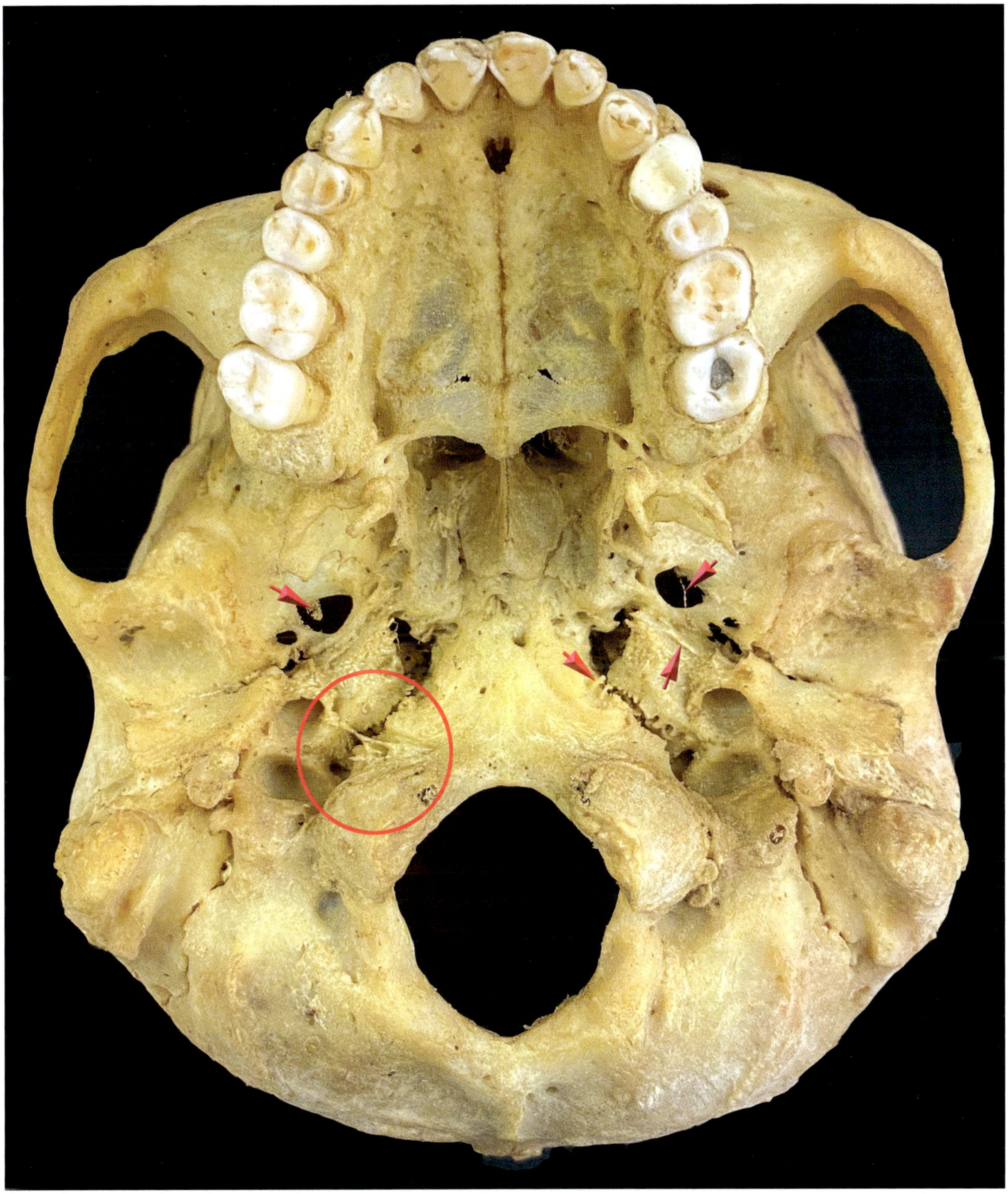

Figure 203. Inferior view of 92-year-old Korean male with osteoporosis, most noticeably in the hard palate that is wide and shallow. Note the numerous calcifications (circle and arrow), often present in elderly individuals.

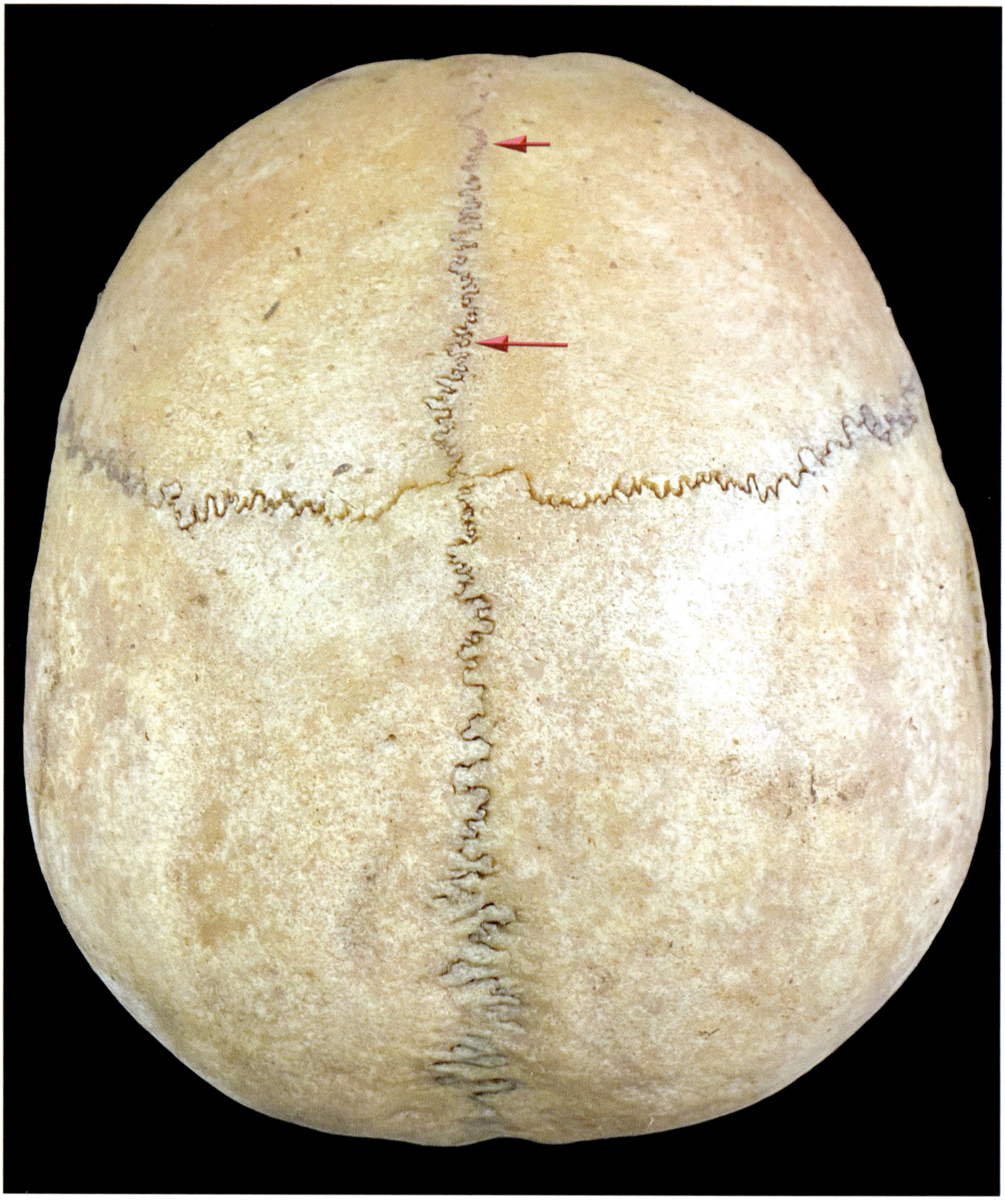

Figure 204. Superior view of 92-year-old Korean male with metopism (persistence of the interfrontal suture; arrow) and minimal obliteration of all sutures even at this age.

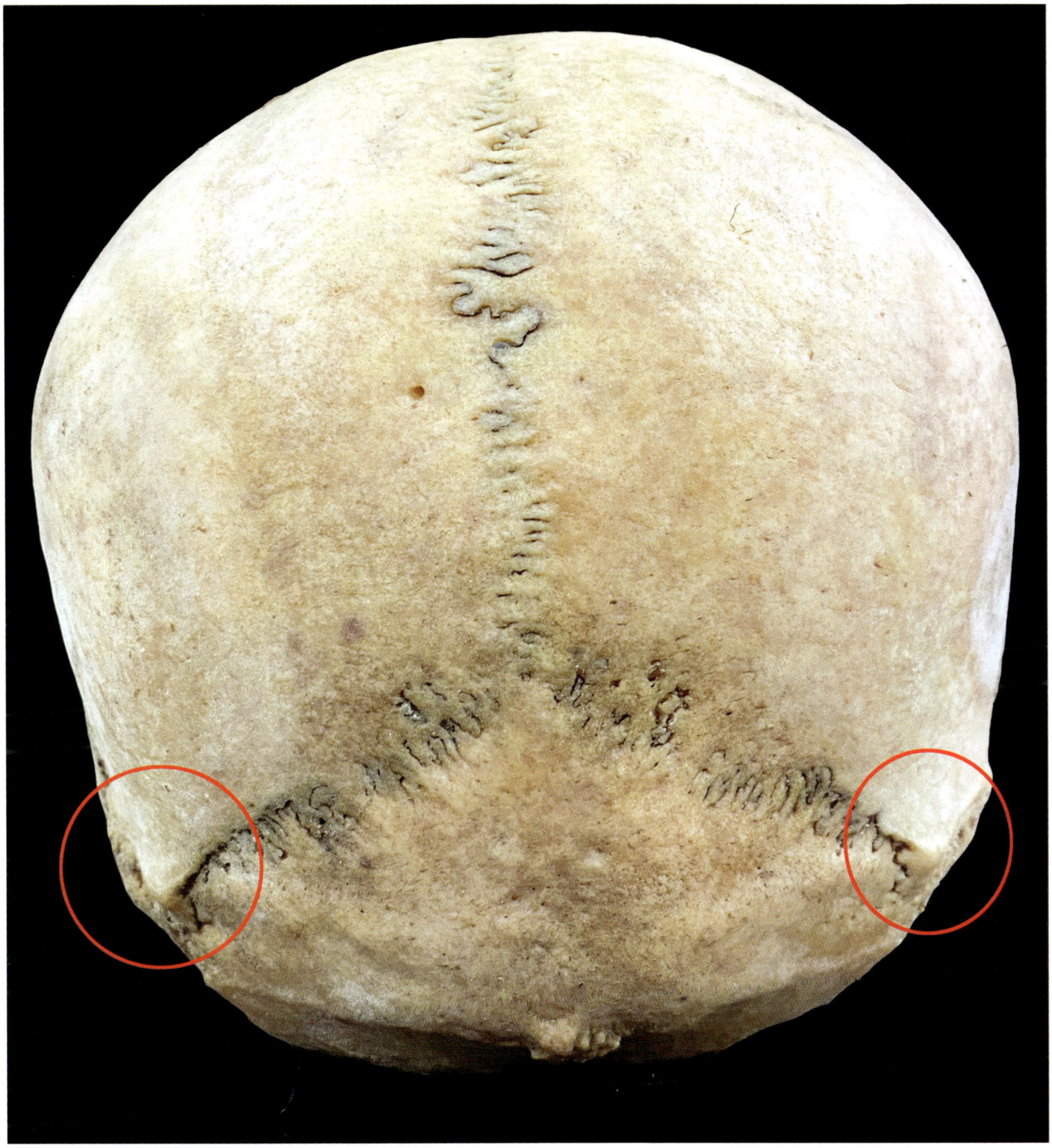

Figure 205. Posterior view of 92-year-old Korean male. Note the small occipital tori tubercles (circle).

17. OKINAWAN FEMALE

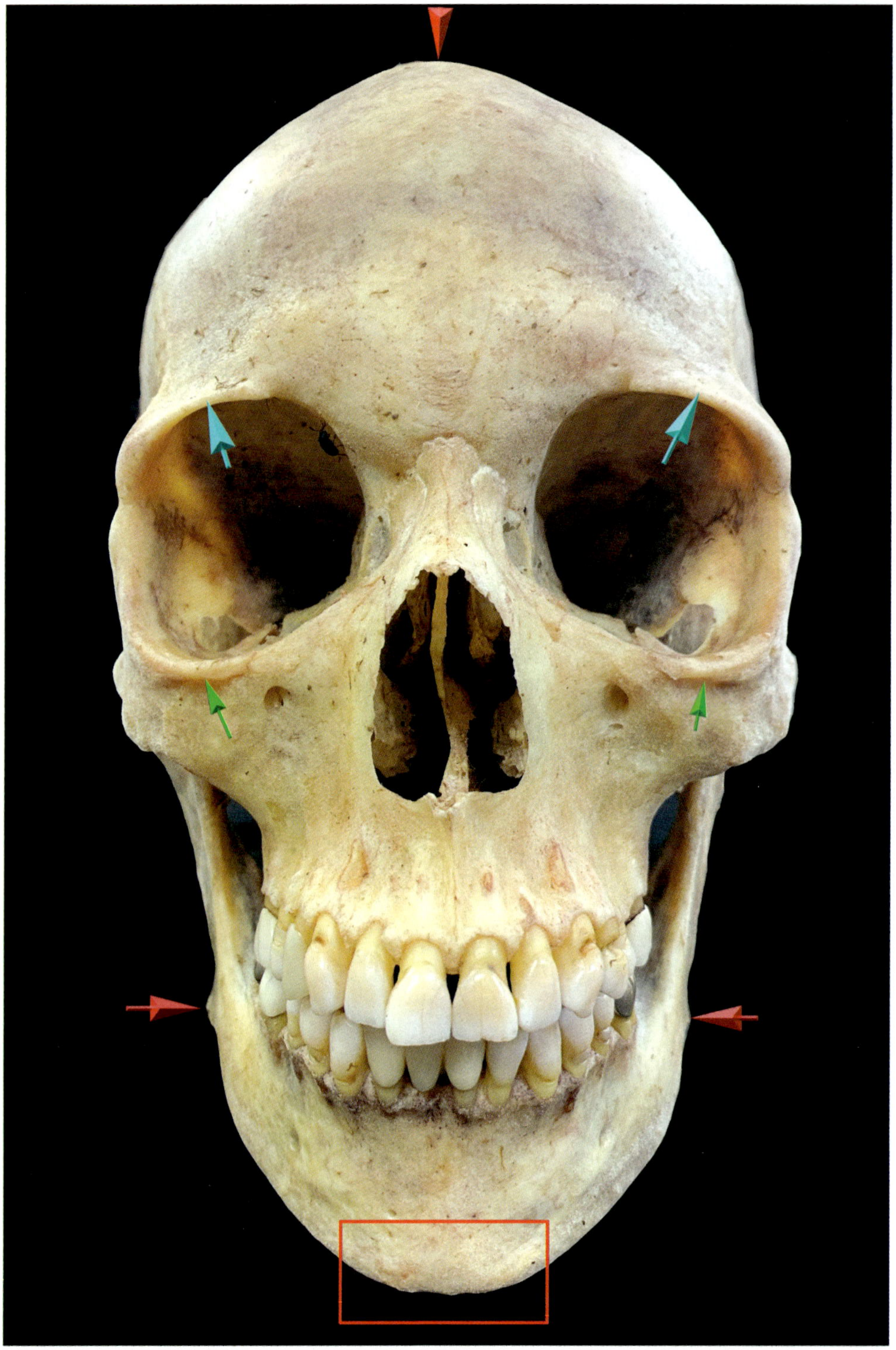

Figure 206. Anterior view of 70-year-old Okinawan female with pointed vertex (red arrowhead), sharp superior margin of the orbits (blue arrow), "undercut" inferior orbital margins (green arrow; see Figure 207), small/gracile zygomatic bones, mild gonial eversion (red arrow), and a rounded chin (rectangle).

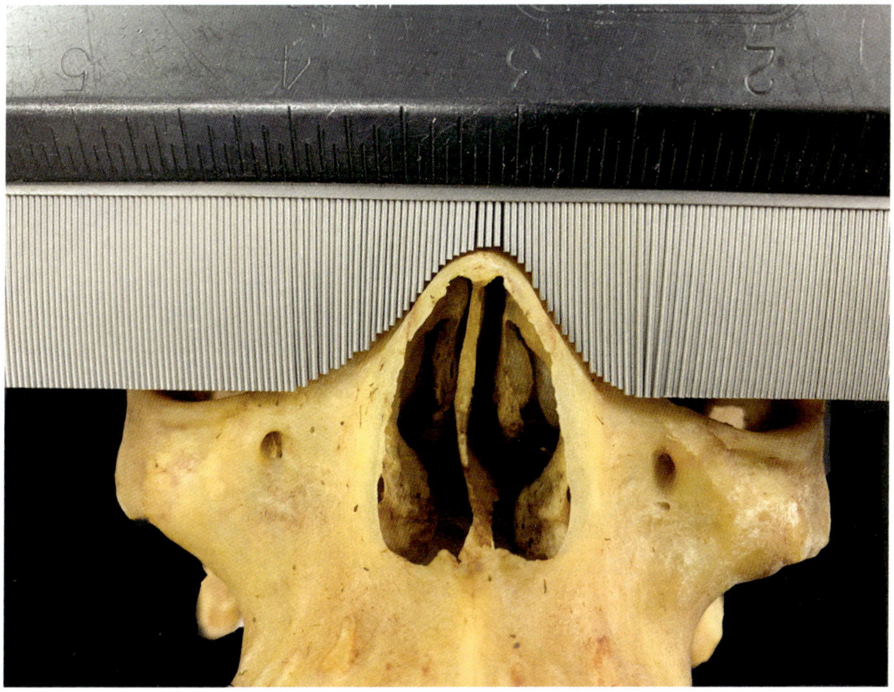

Figure 207. Nasal bone contour in a 70-year-old Okinawan female.

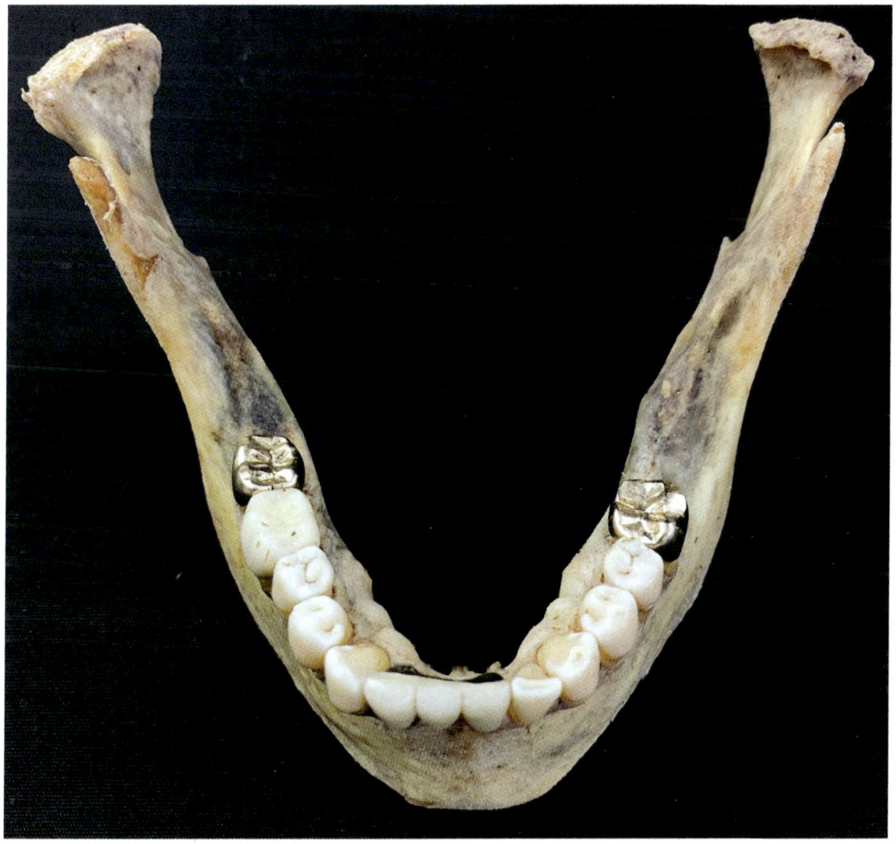

Figure 208. Mandible of 70-year-old Okinawan female with a small and rounded chin.

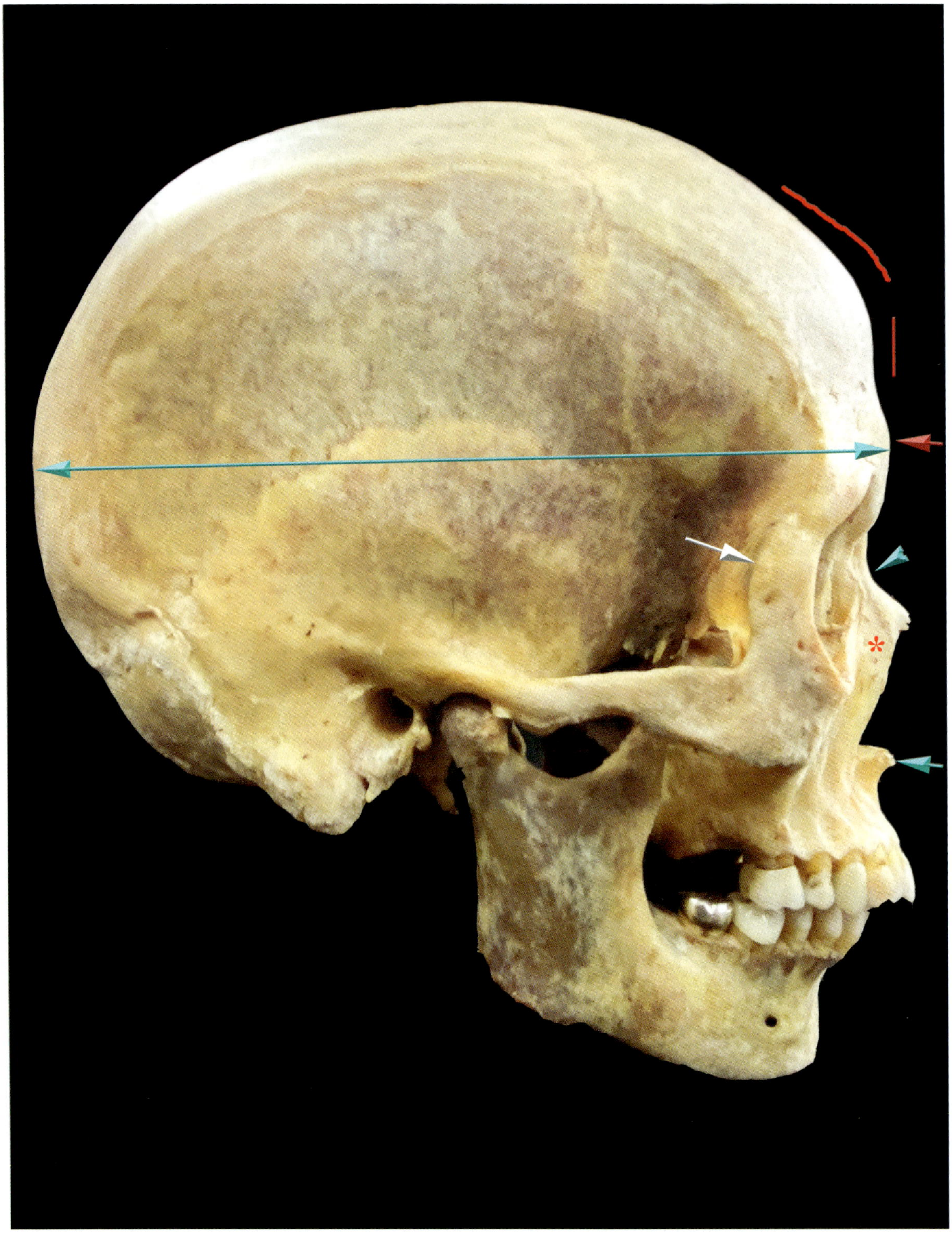

Figure 209. Right lateral view of 70-year-old Okinawan female with a sloping frontal bone (curved red line), vertical frontal bone superior to a small glabellar region (red arrow), concave nasal bone (blue arrowhead), moderate frontal process of the maxilla (*), anterior nasal spine (blue arrow), absent posterior zygomatic tubercle (white arrow), and a long vault (double blue arrow).

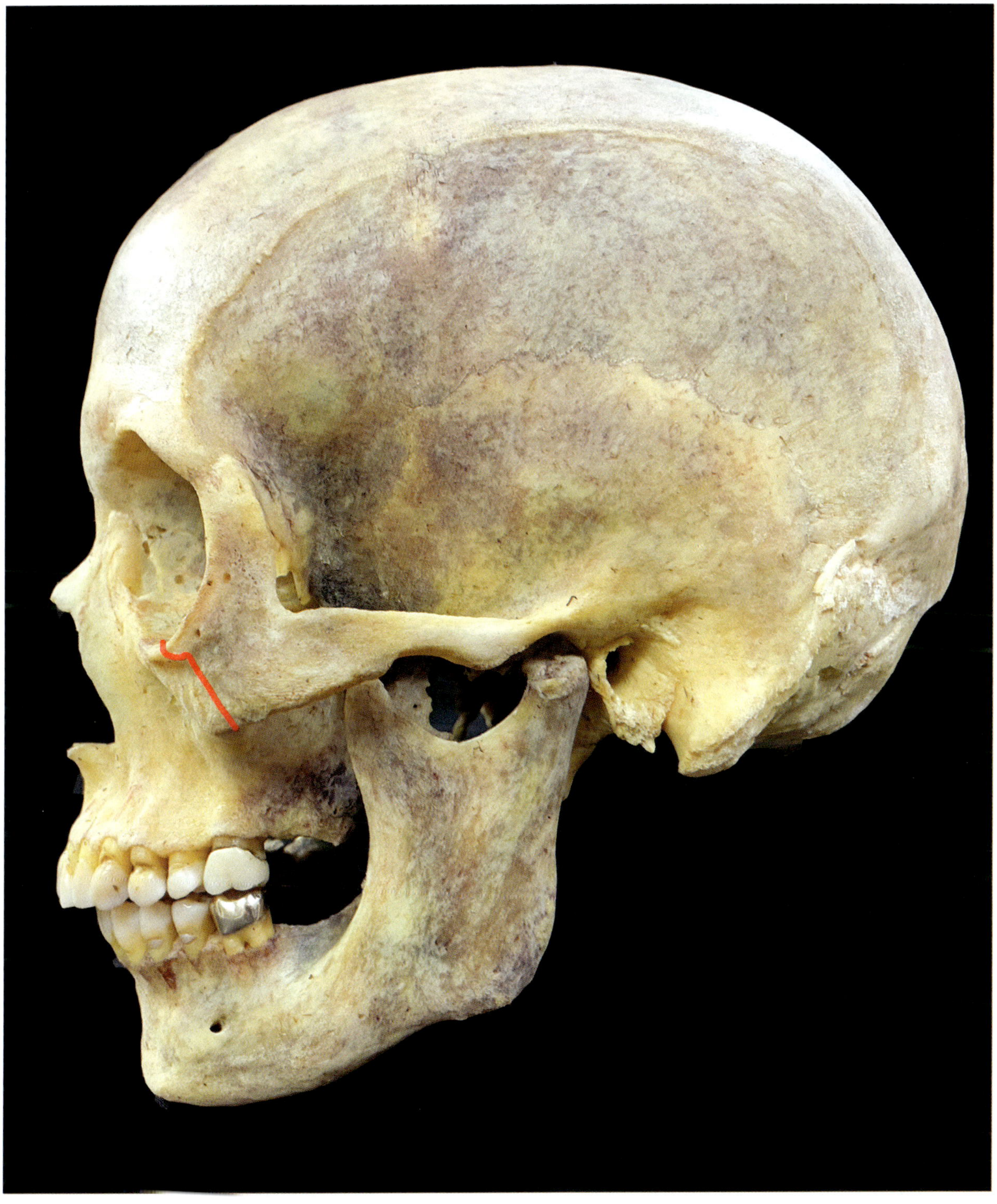

Figure 210. Left lateral view of 70-year-old Okinawan female with an overhanging inferior margin of the orbit and an "undercut" zygomatic bone (red line). While an "undercut" zygoma is a prominent feature in this individual, the author has seen this feature in males and females of all ancestries.

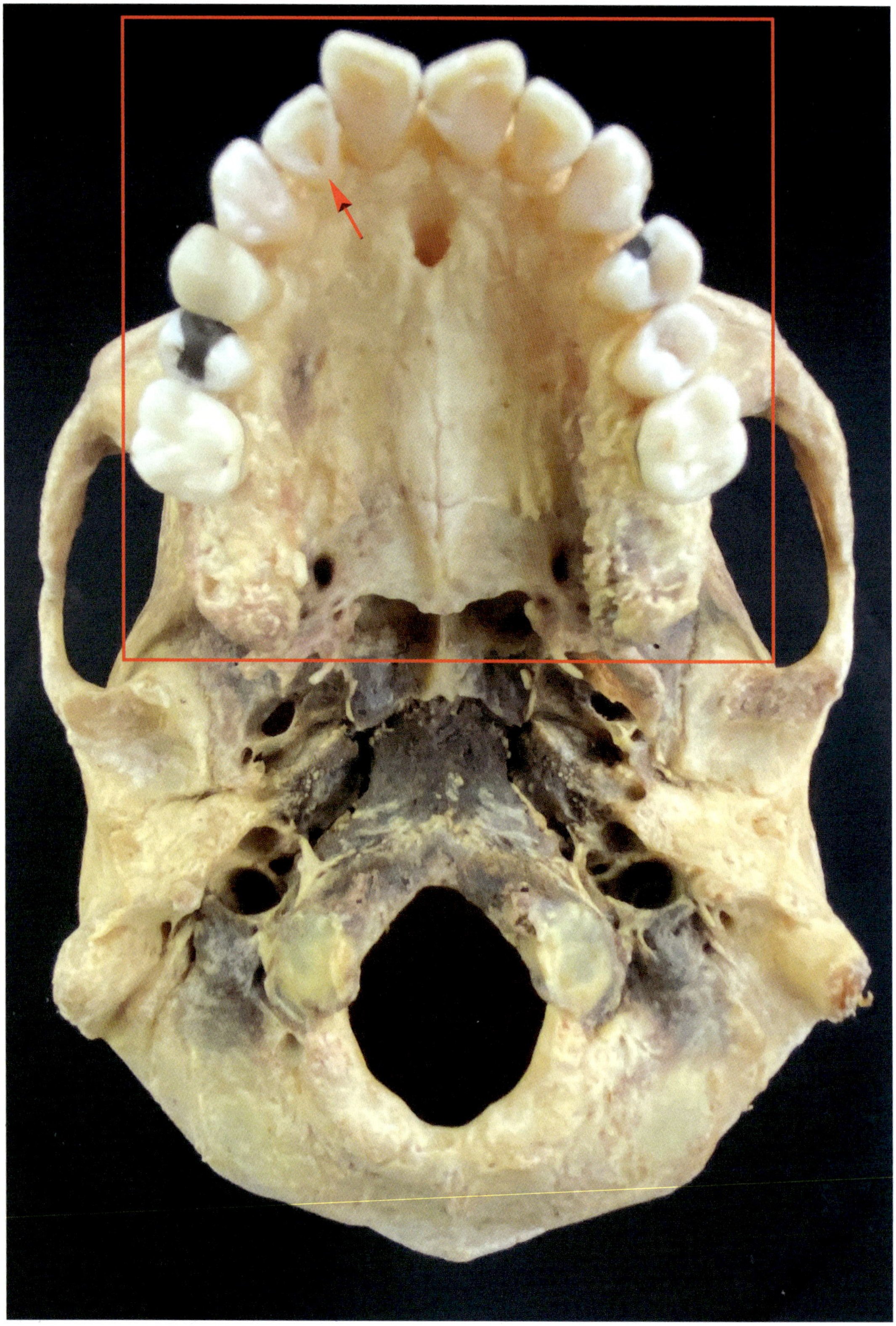

Figure 211. Inferior view of 70-year-old Okinawan female with a large palate (square) and shovel-shaped lateral incisors (arrow).

Figure 212. Superior view 70-year-old Okinawan female with an oval-shaped cranium and complete obliteration of the sutures.

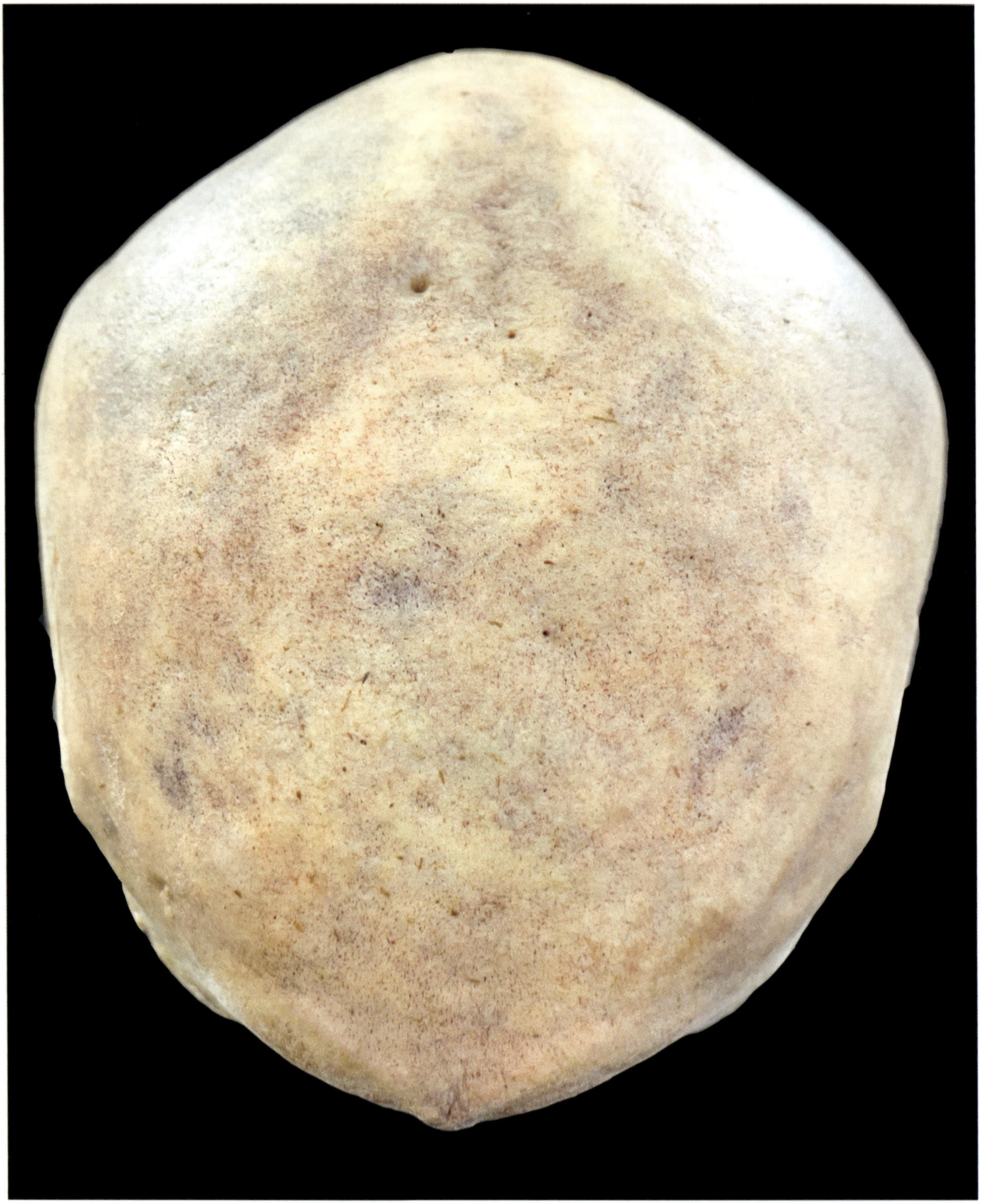

Figure 213. Posterior view of 70-year-old Okinawan female with parietal bossing (squaring) and complete obliteration of the sutures.

18. OKINAWAN MALE

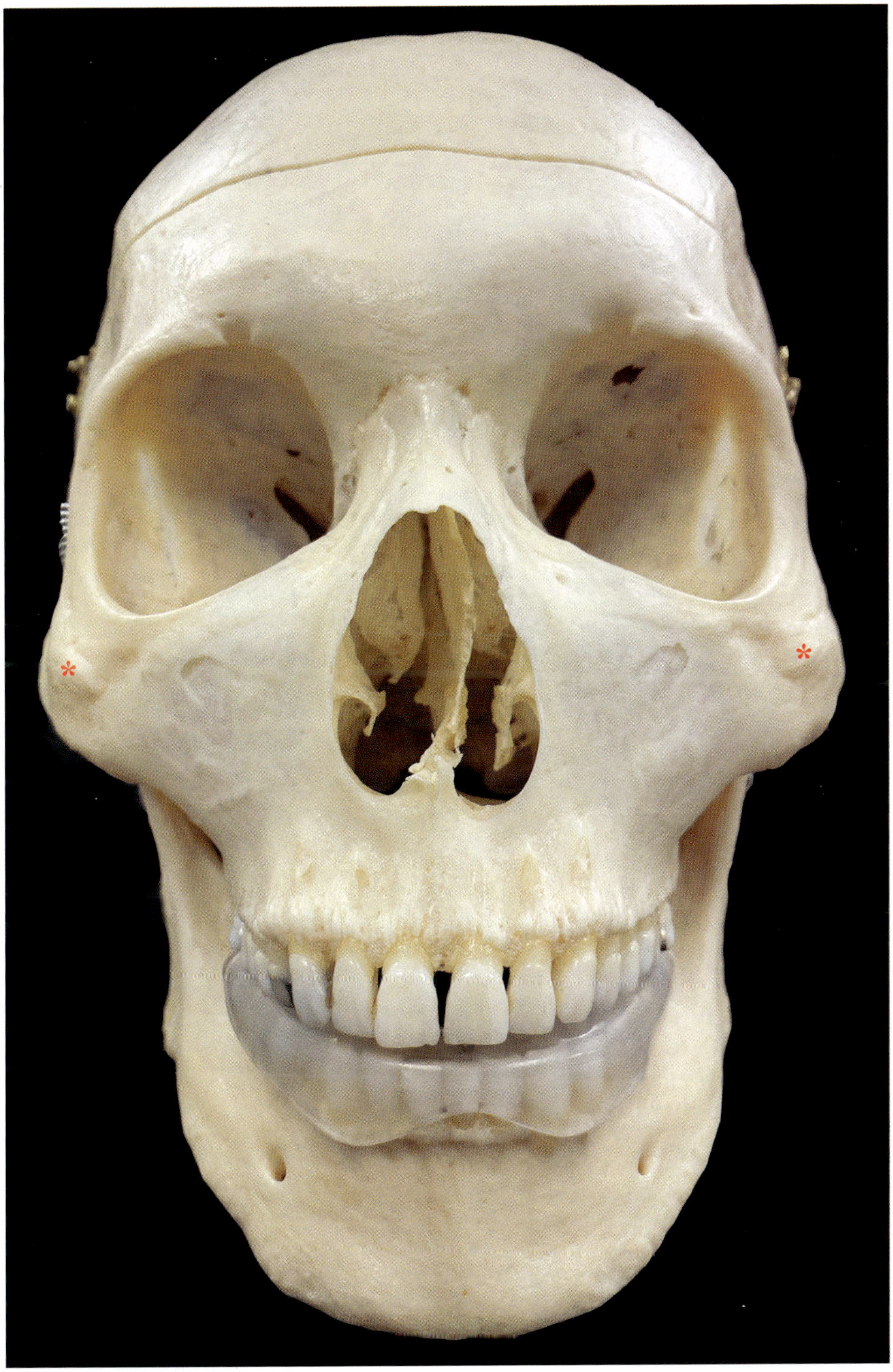

Figure 214. Anterior view of 71-year-old Okinawan male with developed muscle attachments (*) for the zygomaticus major muscle. The clear plastic athletic tooth guard is used to cushion the upper and lower teeth to prevent or reduce breakage in the laboratory. Note the broad and rounded chin.

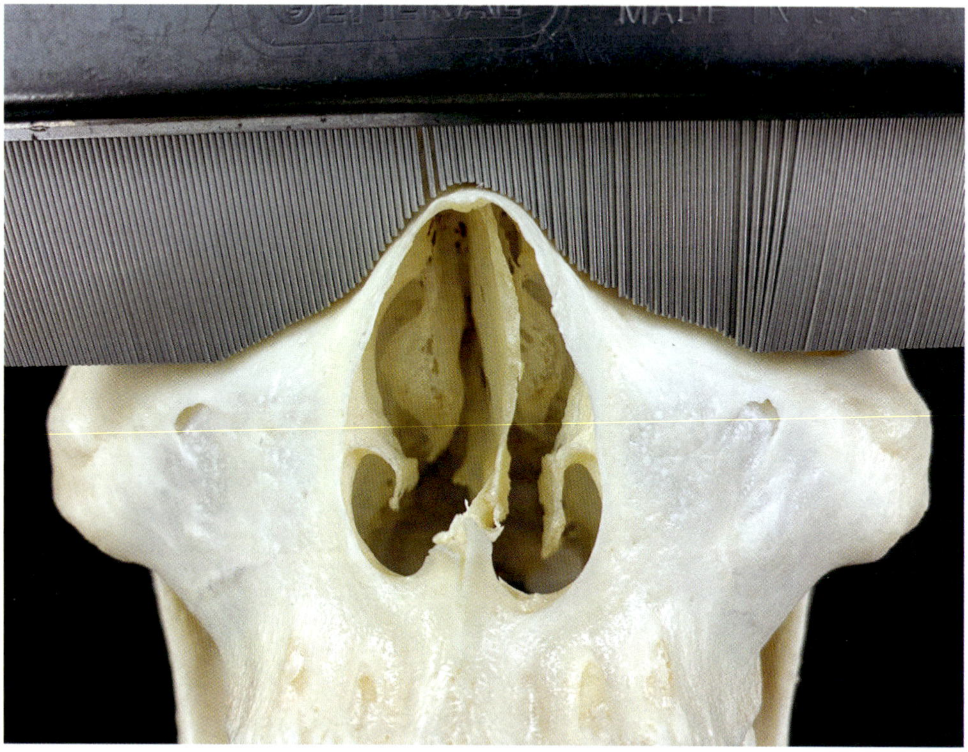

Figure 215. Nasal bone contour in a 71-year-old Okinawan male.

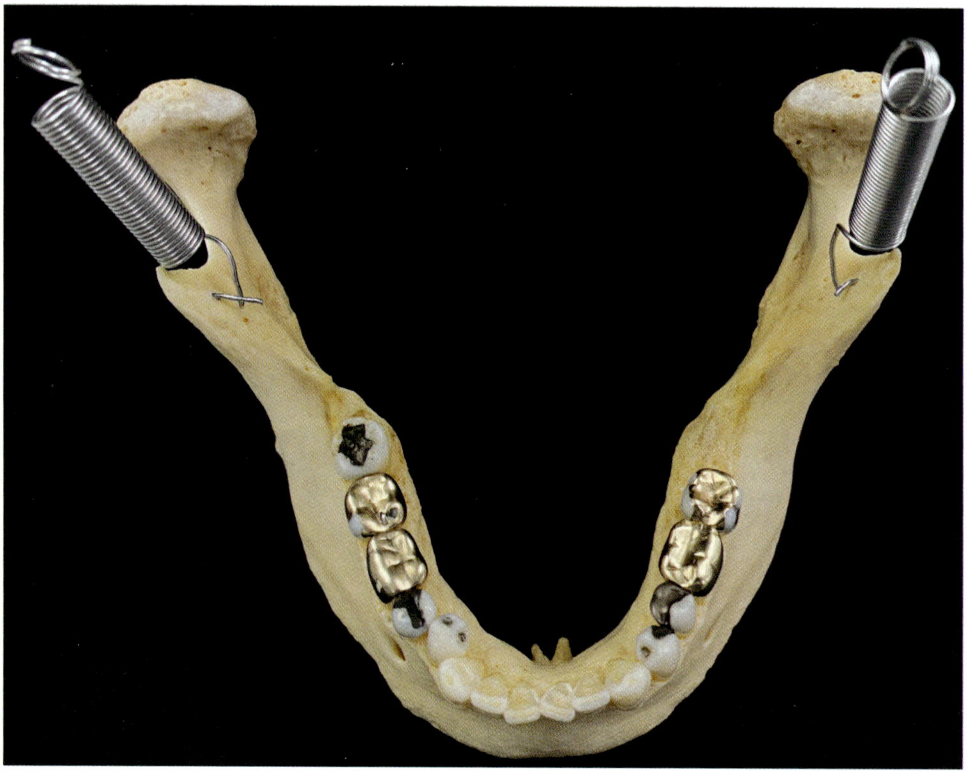

Figure 216. Mandible of 71-year-old Okinawan male with a broadly curved chin. The springs are used to articulate the mandible and skull.

TABLE 21.
Cranial Measurements in a 71-year-old Okinawan Male

Measurement	(mm)	Right	Left
Maximum Cranial Length (g-op, GOL)	180		
Maximum Cranial Breadth (eu-eu, XCB)	156		
Bizygomatic Breadth (zy-zy, ZYB)	148		
Basion-Bregma Height (ba-b, BBH)	143		
Cranial Base Length (ba-n, BNL)	107		
Basion-Prosthion (ba-pr, BPL)	92		
Maxillo-Alveolar Breadth (ecm-ecm, MAB)	73		
Maxillo-Alveolar Length (pr-alv, MAL)	55		
Biauricular Breadth (au-au, AUB)	143		
Upper Facial Height (n-pr)	75		
Minimum Frontal Breadth (ft-ft, WFB)	98		
Upper Facial Breadth (fmt-fmt)	112		
Nasal Height (n-ns, NLH)	60		
Nasal Breadth (al-al, NLB)	29		
Orbital Breadth (d-ec, OBB)	42		L
Orbital Height (OBH)	34		L
Biorbital Breadth (ec-ec, EKB)	107		
Interorbital Breadth (d-d, DKB)	28		
Frontal Chord (n-b, FRC)	111		
Parietal Chord (b-l, PAC)	109		
Occipital Chord (l-o, OCC)	102		
Foramen Magnum Length (ba-o, FOL)	37		
Foramen Magnum Breadth (FOB)	32		
Mastoid Length (MDH)		28	31

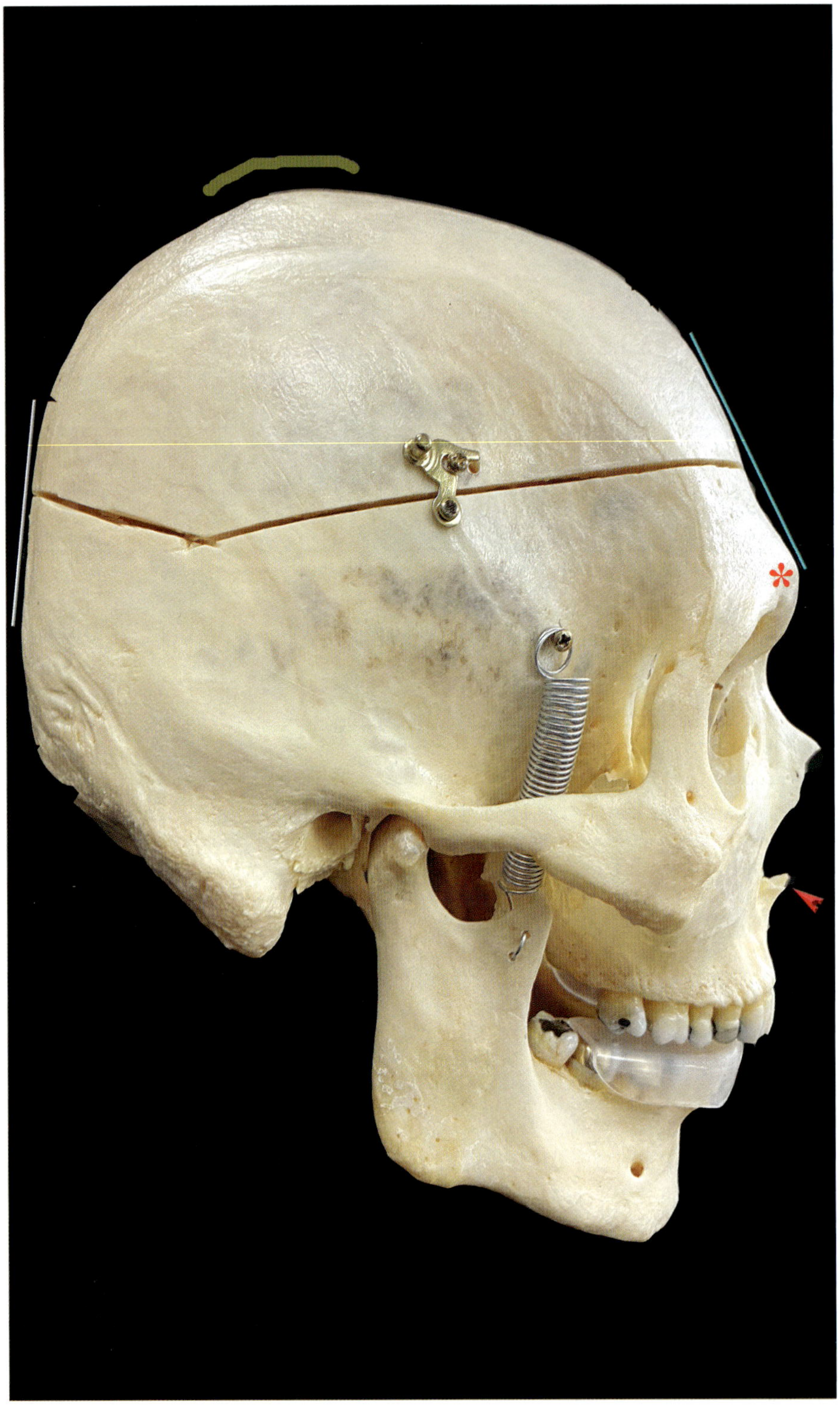

Figure 217a. Right lateral view of 71-year-old Okinawan male with a plastic tooth guard to protect the teeth from breakage. Note that the skull is short anterior-posterior (g-op), exhibits prominent curvature at the vertex (curved yellow line), a sloping frontal bone (blue line), prominent brow ridge (*), small anterior nasal spine (red arrowhead), and a flat occipital region (white line). See Figure 217b for a different perspective on the occipital flattening in this individual.

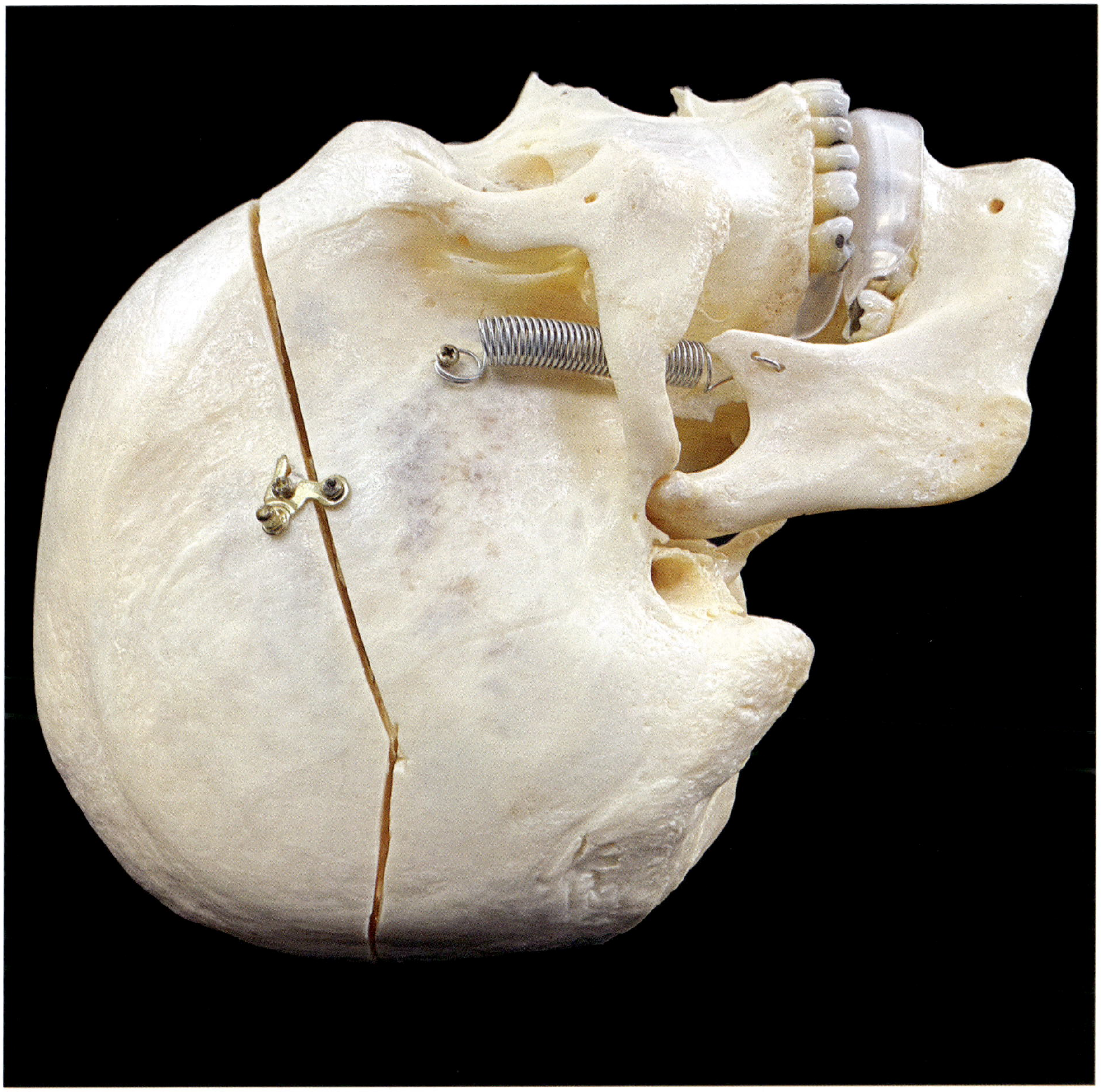

Figure 217b. Right lateral view of 71-year-old Okinawan male with occipital flattening such that the skull sits firmly and without any support on the occipital region. The calotte has been attached with hinges which allows access to the brain cavity and the mandible has been articula ted with springs to allow movement of the jaws (hardware attached by Steven Labrash).

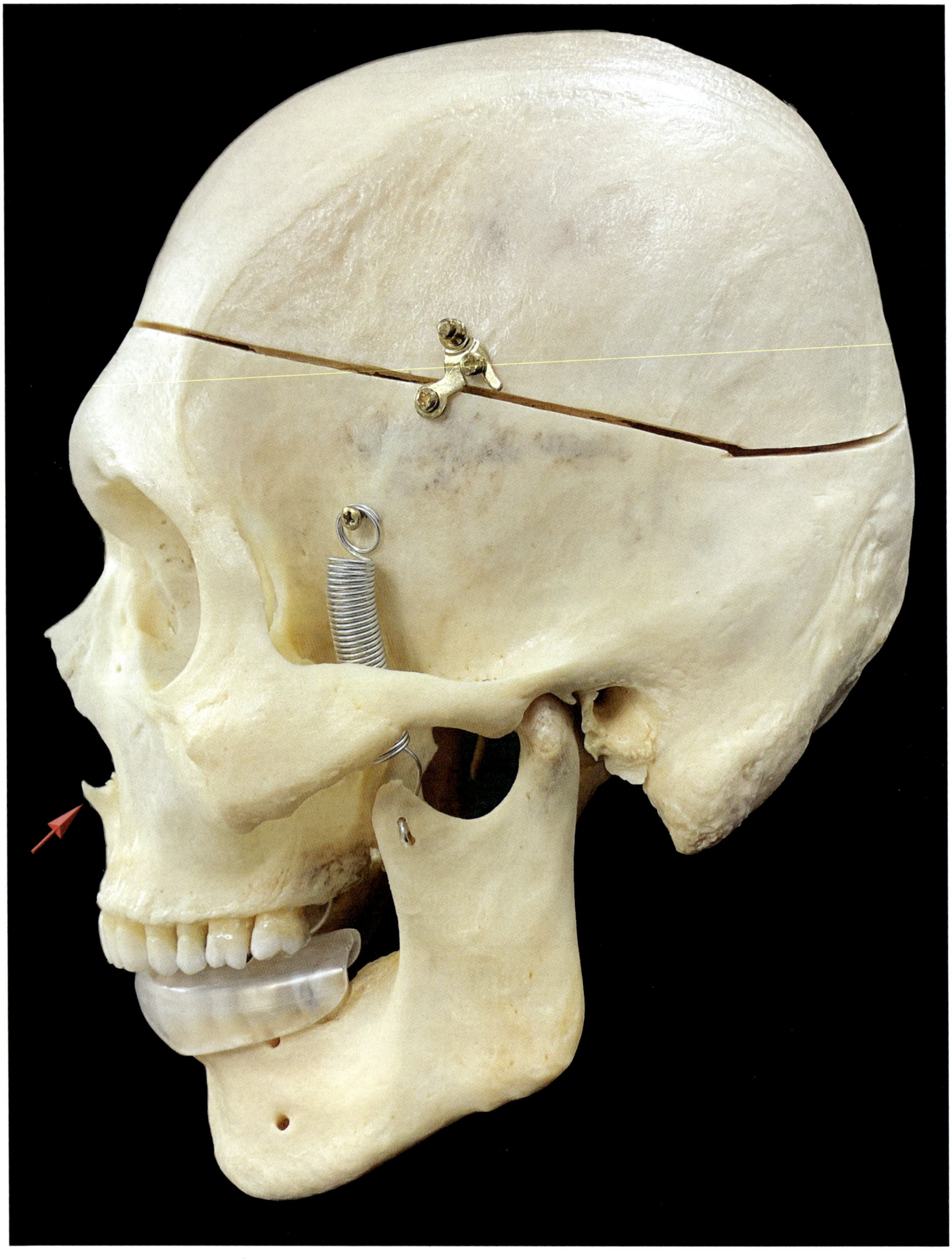

Figure 218. Left lateral view of 71-year-old Okinawan male. Note the small and upwardly curved anterior nasal spine (arrow), a normal anatomical variant, that may be present in any ancestral group.

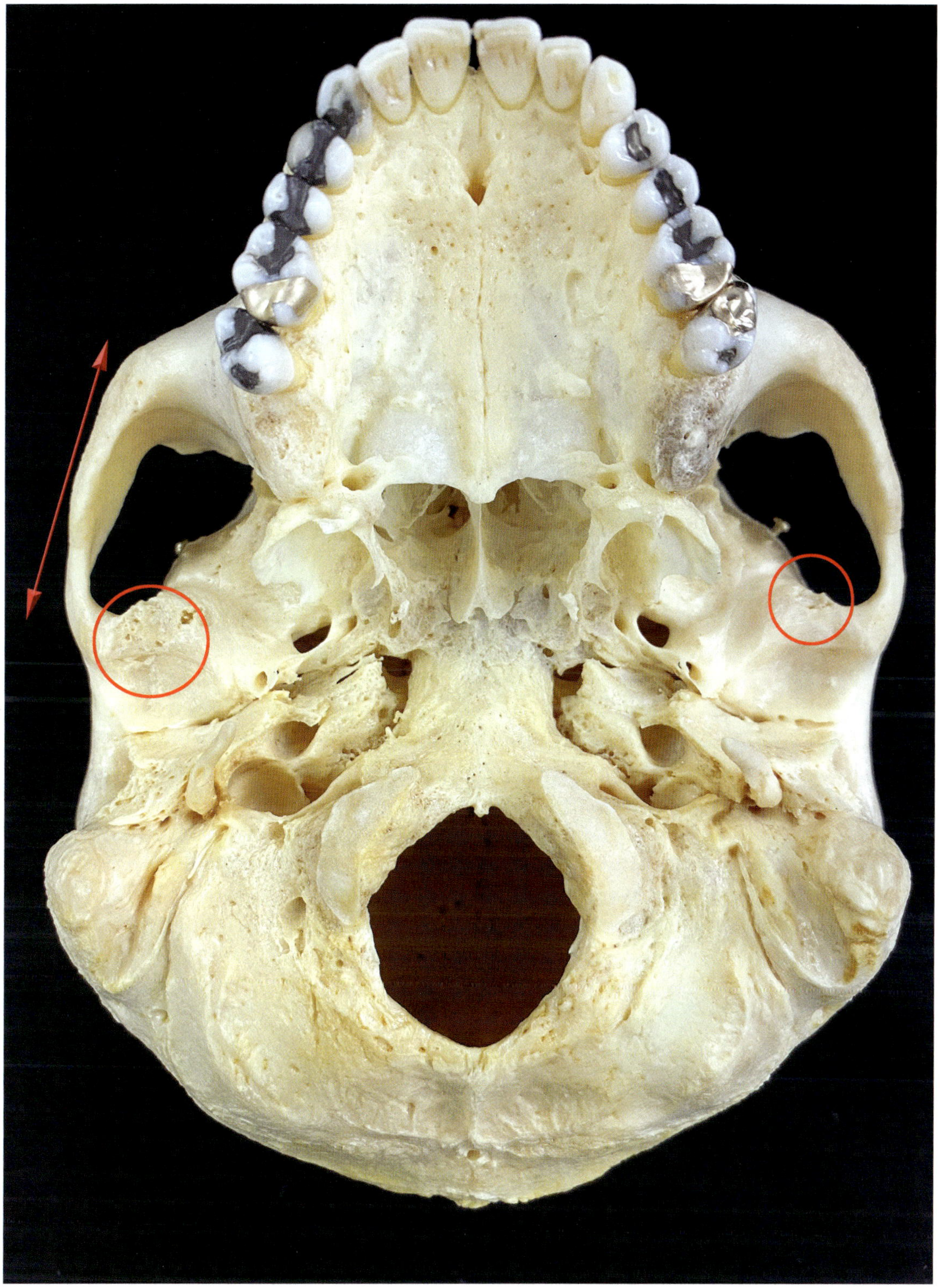

Figure 219. Inferior view of 71-year-old Okinawan male. The hard palate is very large, wide, and shallow. The zygomatic arches are short anterior-posterior (double arrow). The temporal articular eminences exhibit arthritic destruction (circle). The left digastric fossa (*) is wide and deep reflecting normal variation associated with movement of the mandible. The mastoid processes are medium size, and the nuchal crest is developed.

Figure 220. Superior view of 71-year-old Okinawan male exhibiting a nearly circular vault.

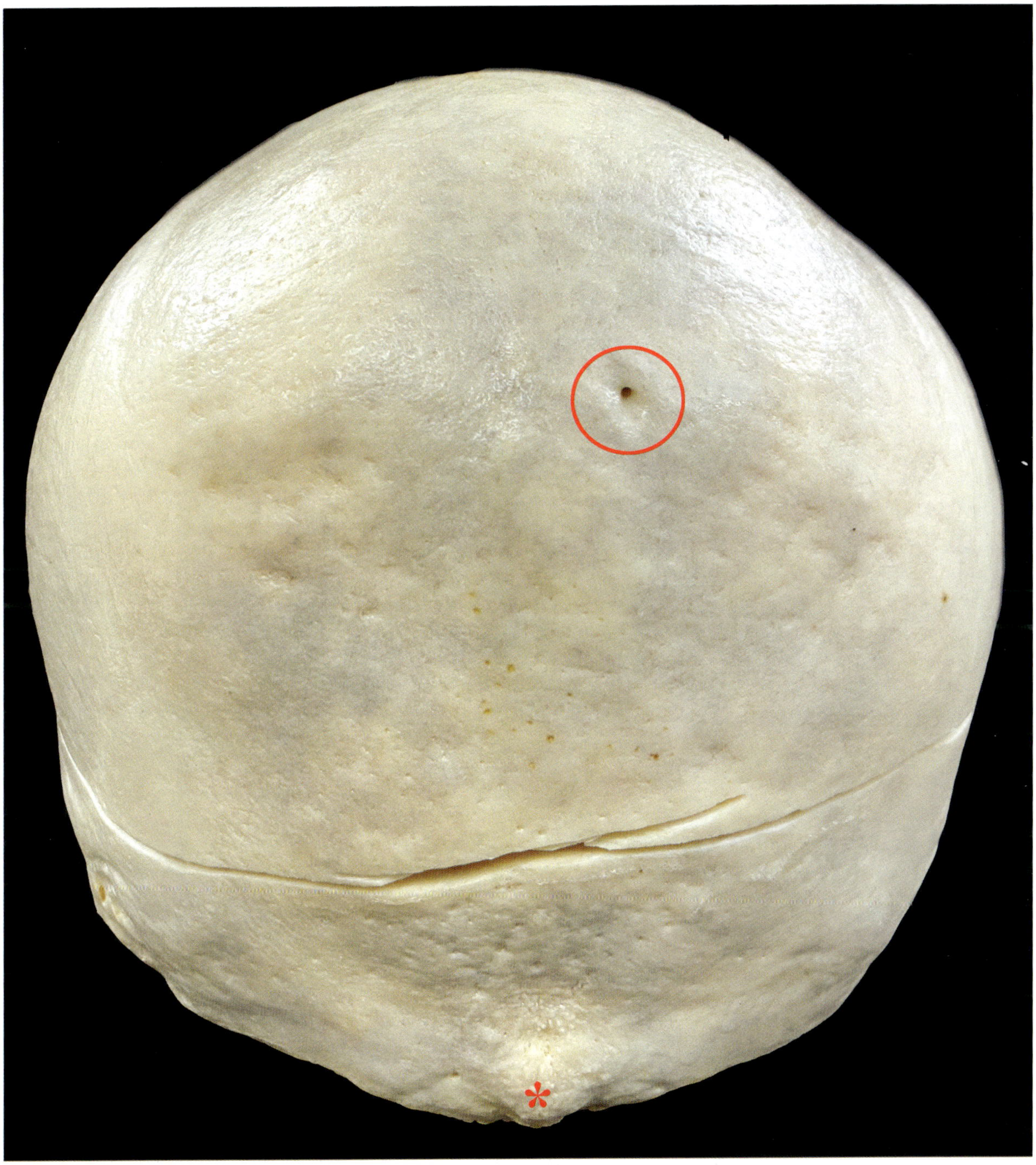

Figure 221. Posterior view of 71-year-old Okinawan male with complete obliteration of the sutures, a single right parietal foramen (circle), and a developed inion hook (*).

19. VIETNAMESE MALE

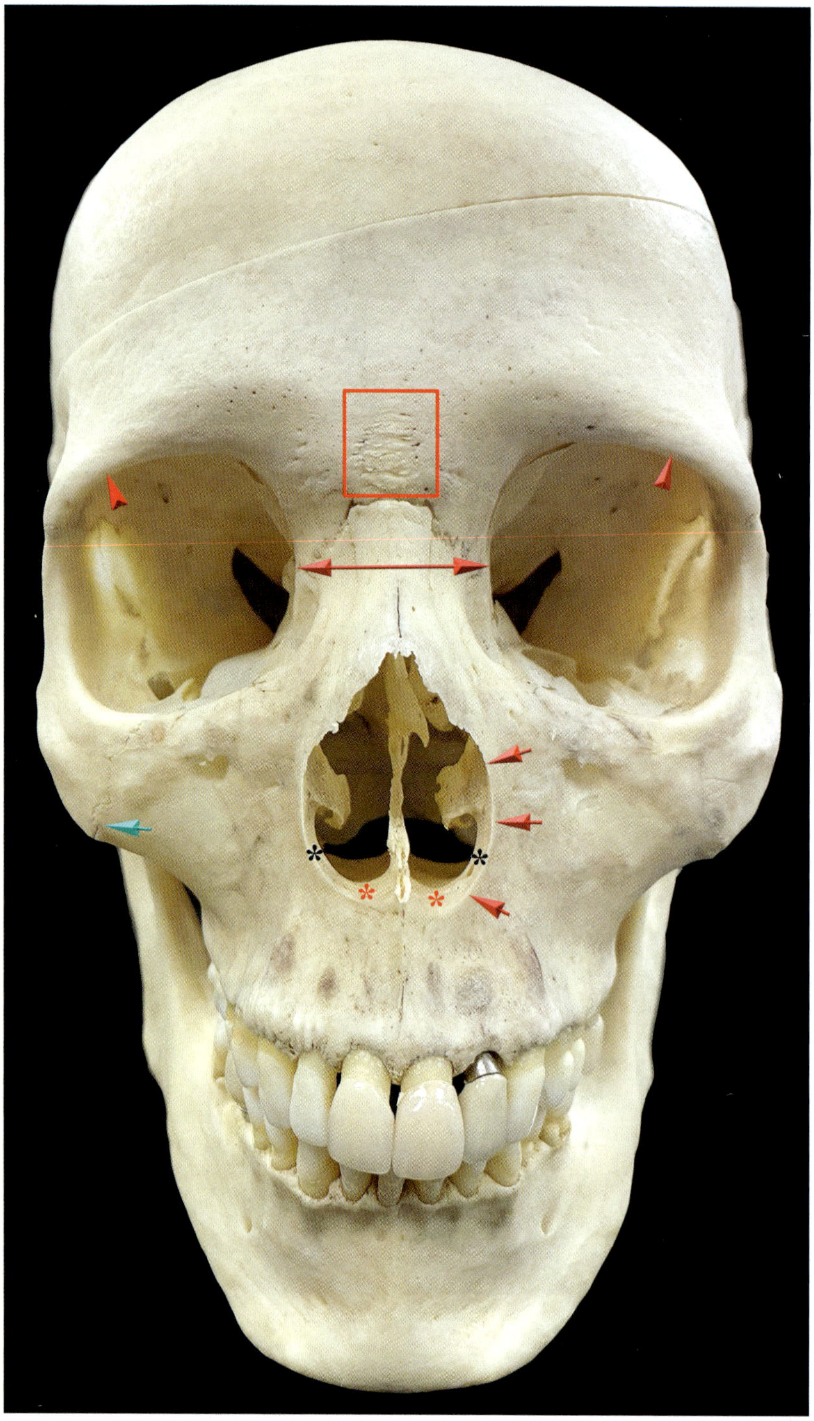

Figure 222. Anterior view 58-year-old Vietnamese male exhibiting a trace of the supranasal suture (rectangle), blunt superior margin of the orbits (arrowhead), wide interorbital breadth (double arrow), and a smooth zygomaticomaxillary suture (blue arrow). Note also the sharp nasal rim (red arrow) with mild nasal guttering (red *) that extends (black *) along most of the nasal aperture. The maxillary left lateral incisor consists of a metal implant and an artificial crown.

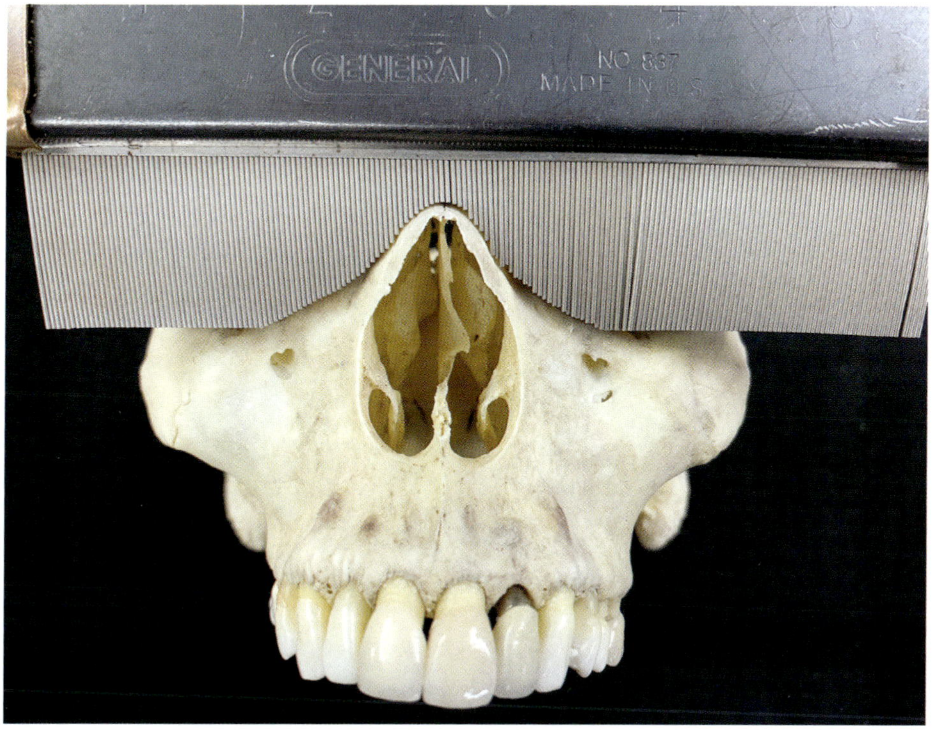

Figure 223. Nasal bone contour in a 58-year-old Vietnamese male.

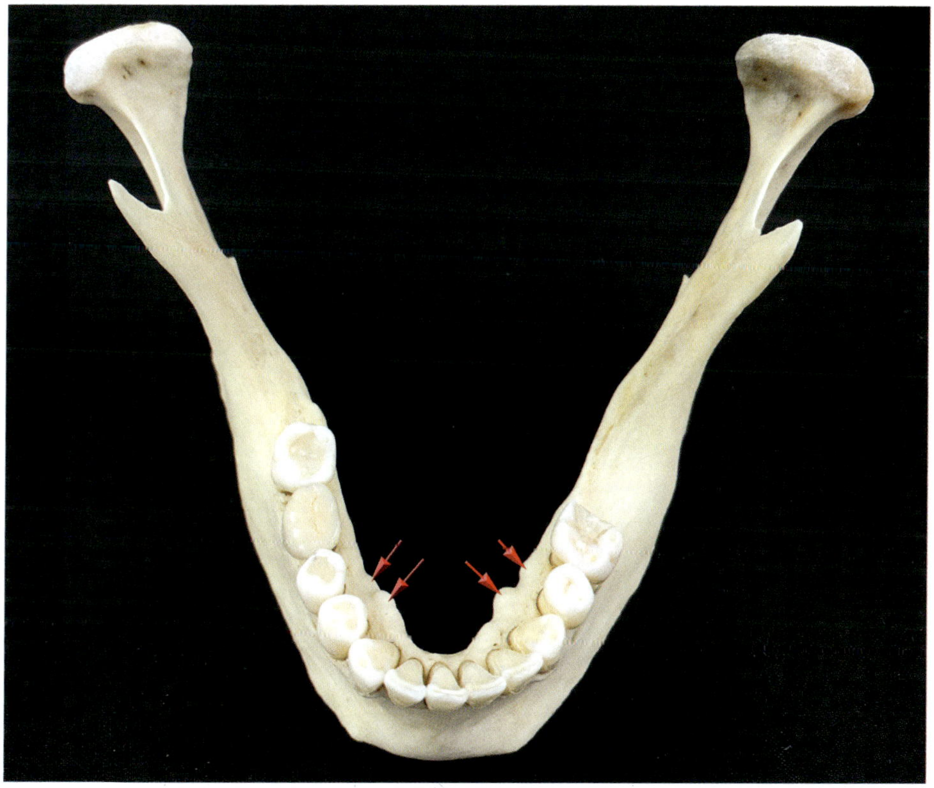

Figure 224. Mandible of 58-year-old Vietnamese male with a broad and square chin, and bilateral and symmetrical lingual exostoses (arrow).

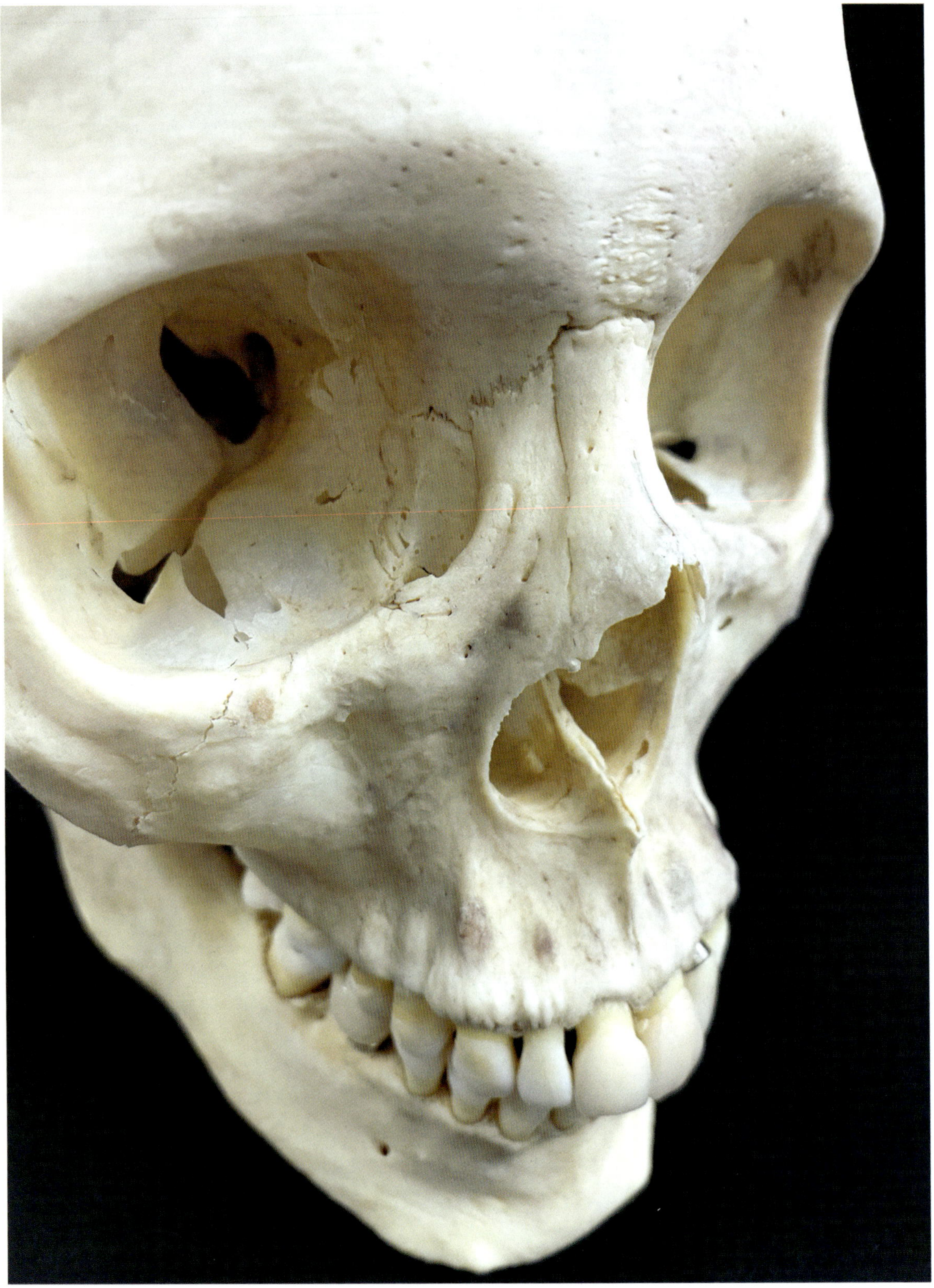

Figure 225. Right superior oblique view of 58-year-old Vietnamese male illustrating a different perspective of the facial morphology to include the frontal bone, superior margin of the orbits, nasal bones, nasal aperture, and chin.

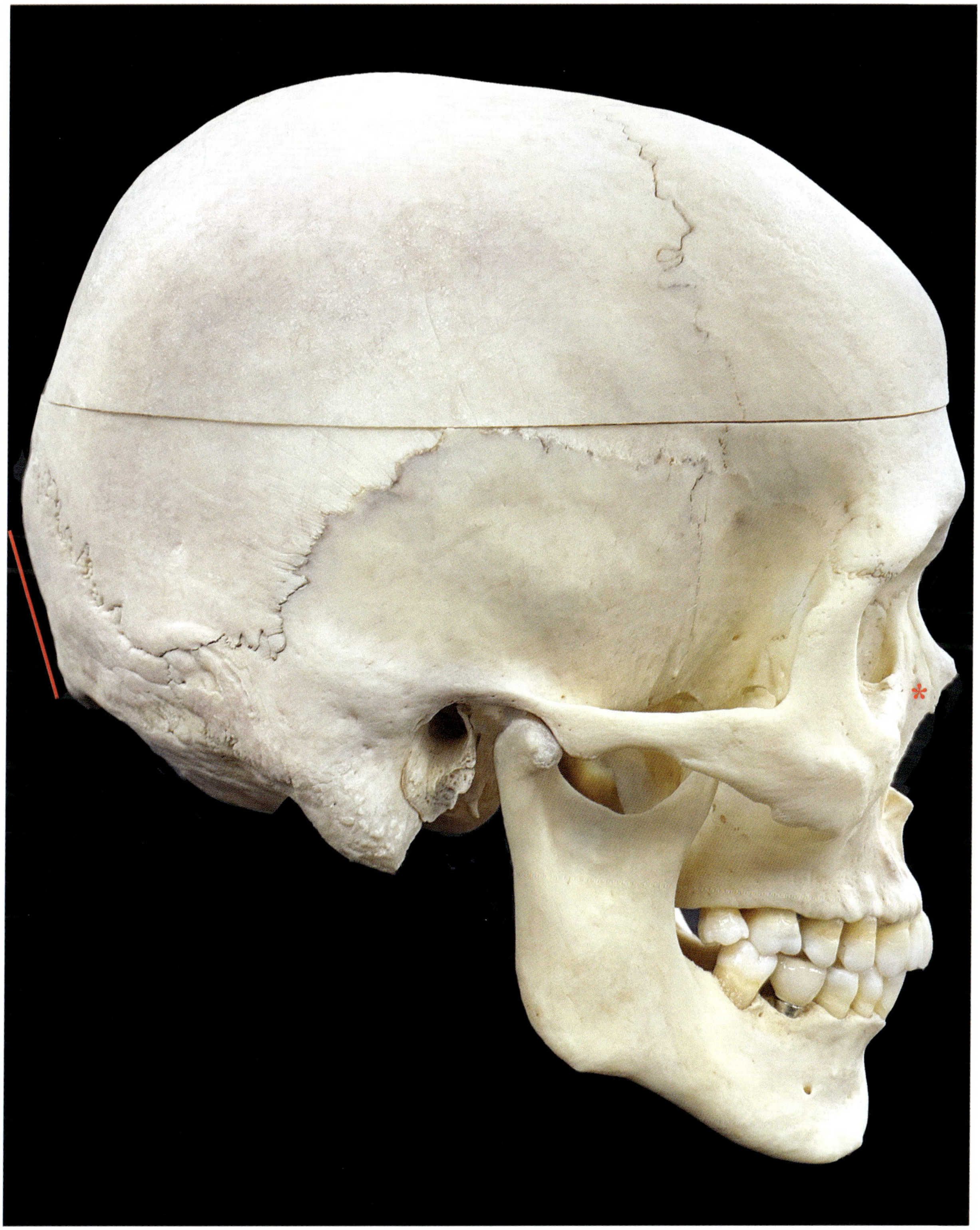

Figure 226. Right lateral view of 58-year-old Vietnamese male with mild alveolar prognathism, flat occipital region (red line), and a short frontal process of the maxilla (*).

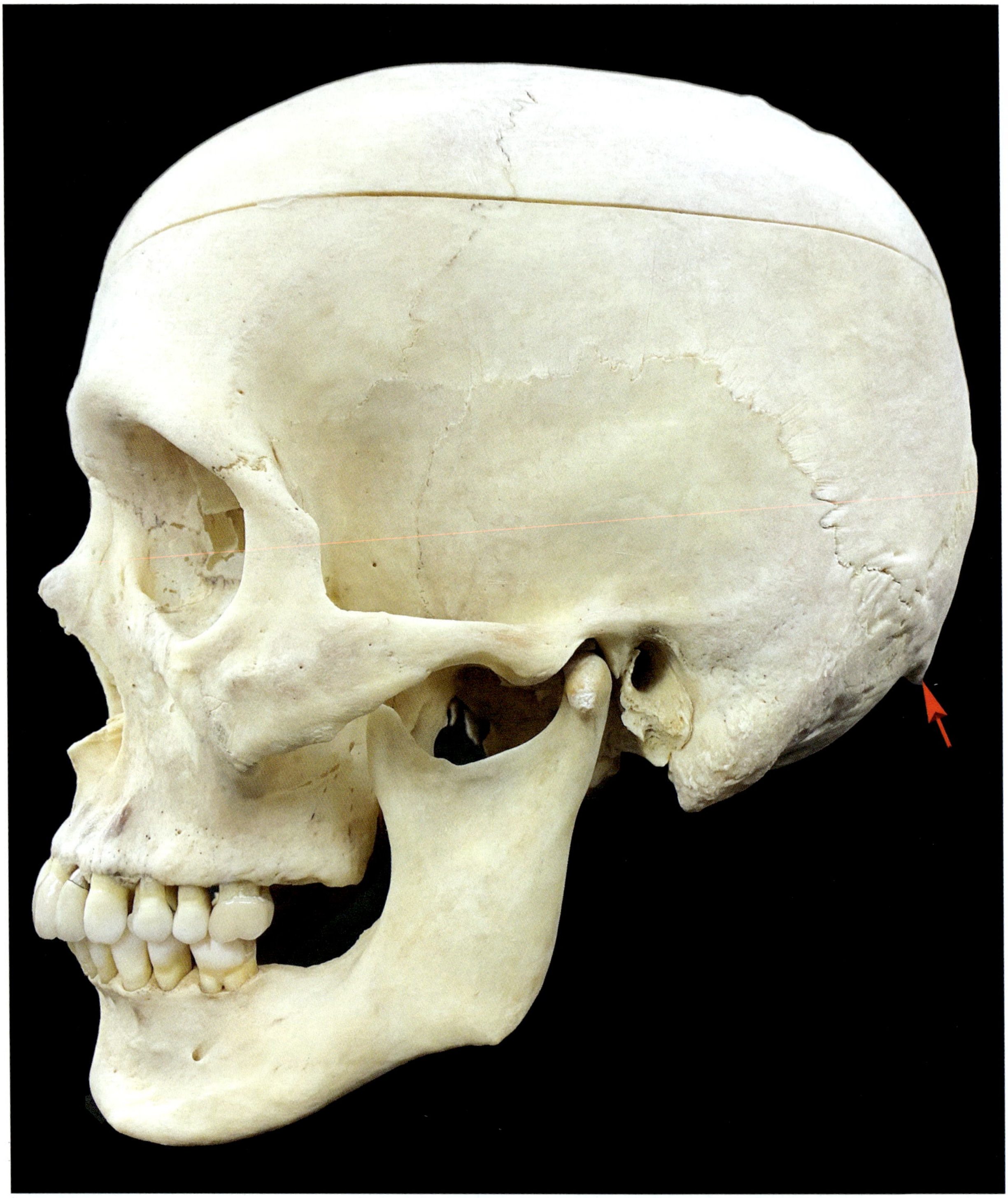

Figure 227. Left lateral view of 58-year-old Vietnamese male with mild alveolar prognathism. Notice the concave-convex nasal bone contour and the small anterior nasal spine. Also note the wide and shallow antegonial notch, small mastoid process, and the inion hook (arrow).

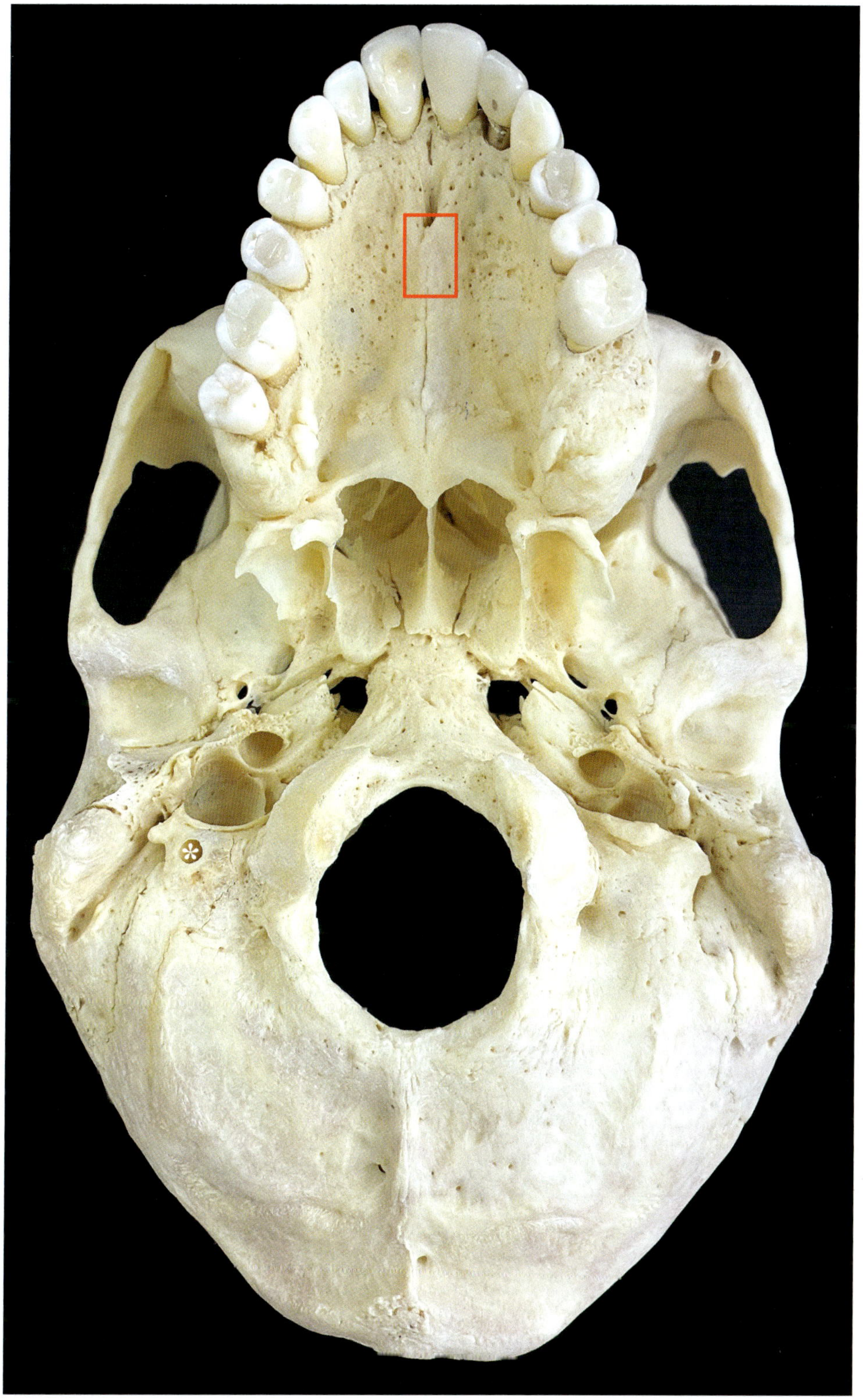

Figure 228. Inferior view of 58-year-old Vietnamese male. Observe the partial obliteration of the anterior median palatine suture (rectangle) consistent with an age of 50+ years (Mann et al. 1991). Note the perforation (blue*) in the cranial base lateral to the right occipital condyle, a rare feature that appears to be related to unusually thin bone in that area.

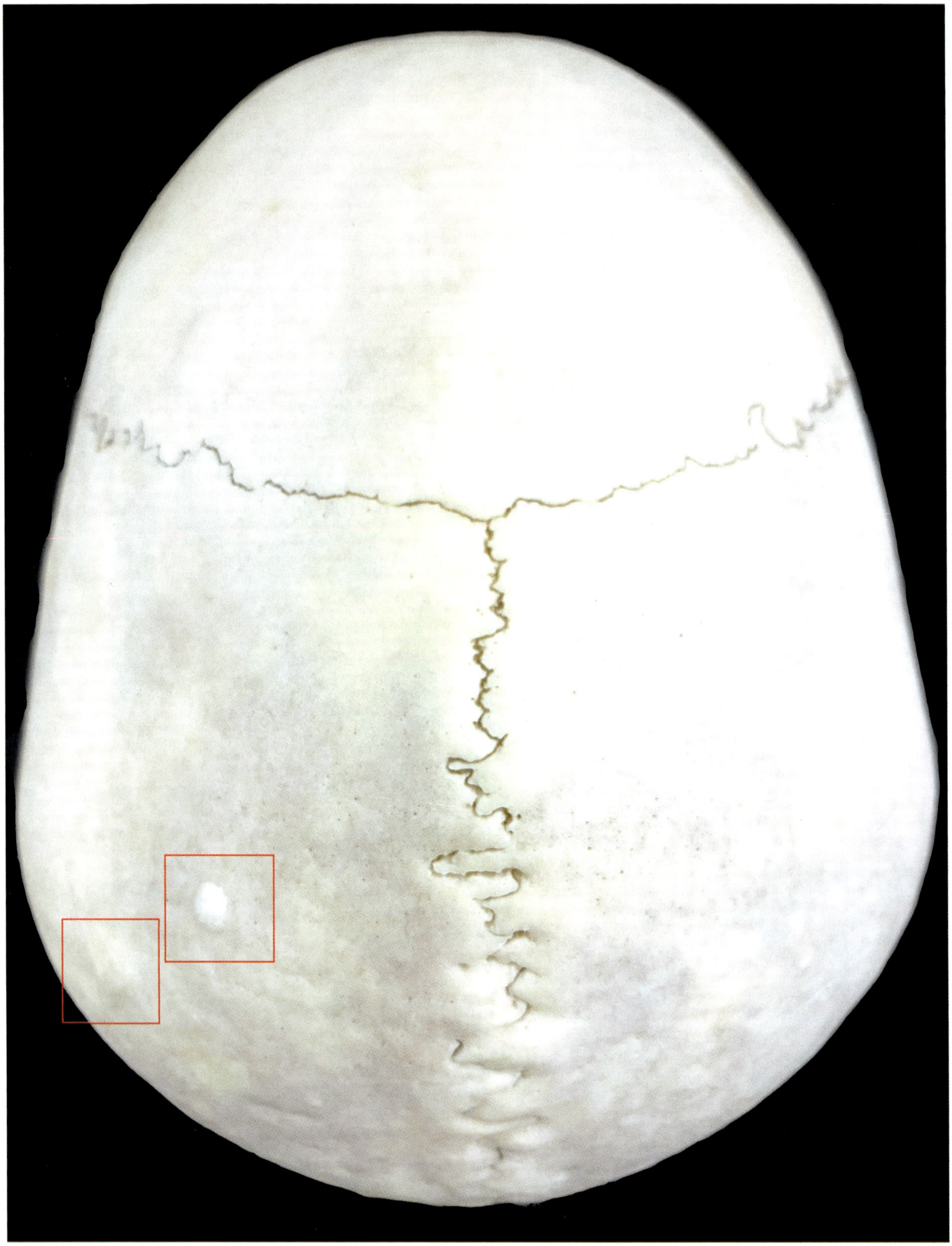

Figure 229. Superior view of 58-year-old Vietnamese male. Note the two small, benign button osteomas (square) in the left parietal bone.

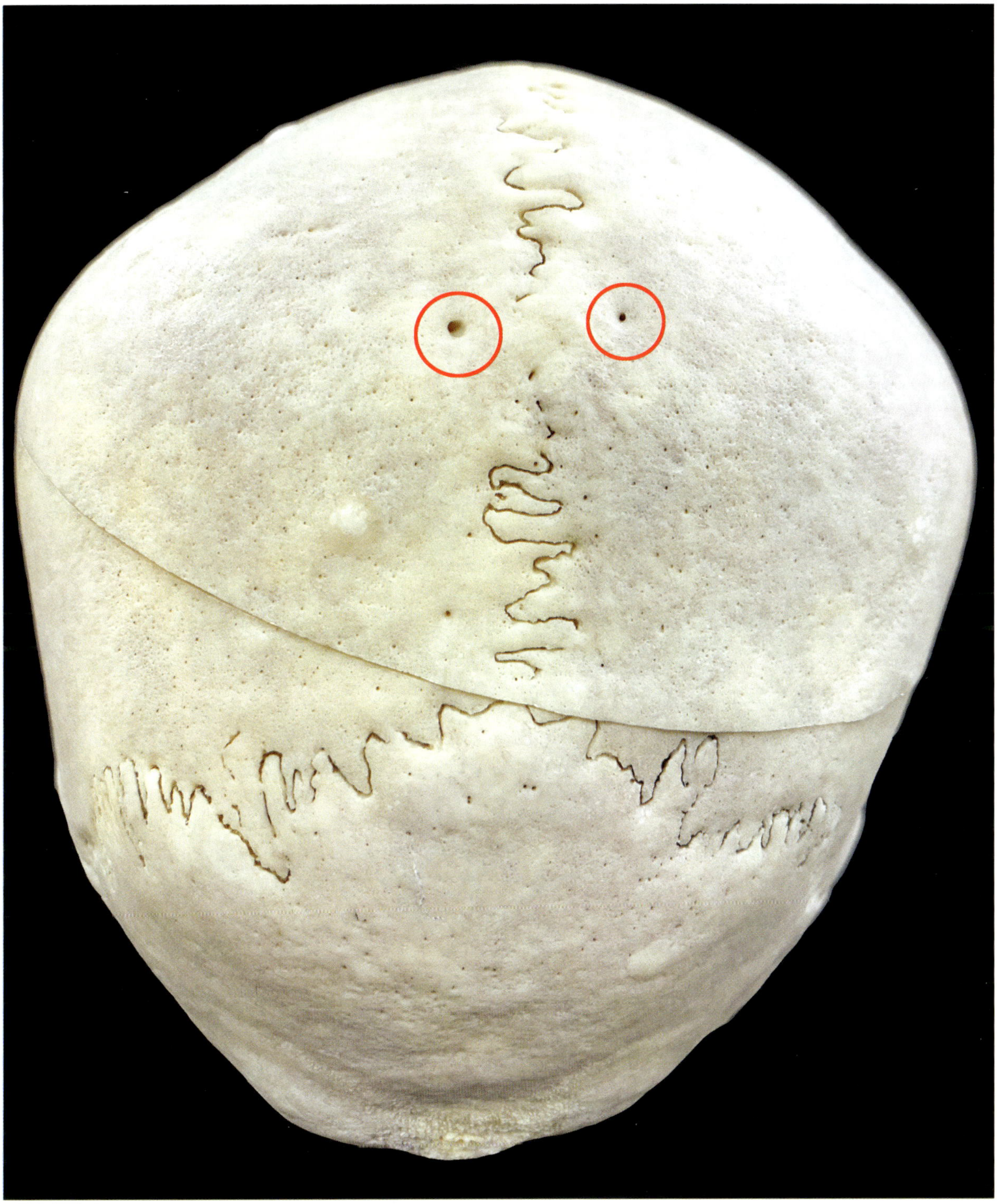

Figure 230. Posterior view of 58-year-old Vietnamese male with numerous vessel foramina, two parietal foramina (circle), and minimal obliteration of the sutures.

20. VIETNAMESE FEMALE

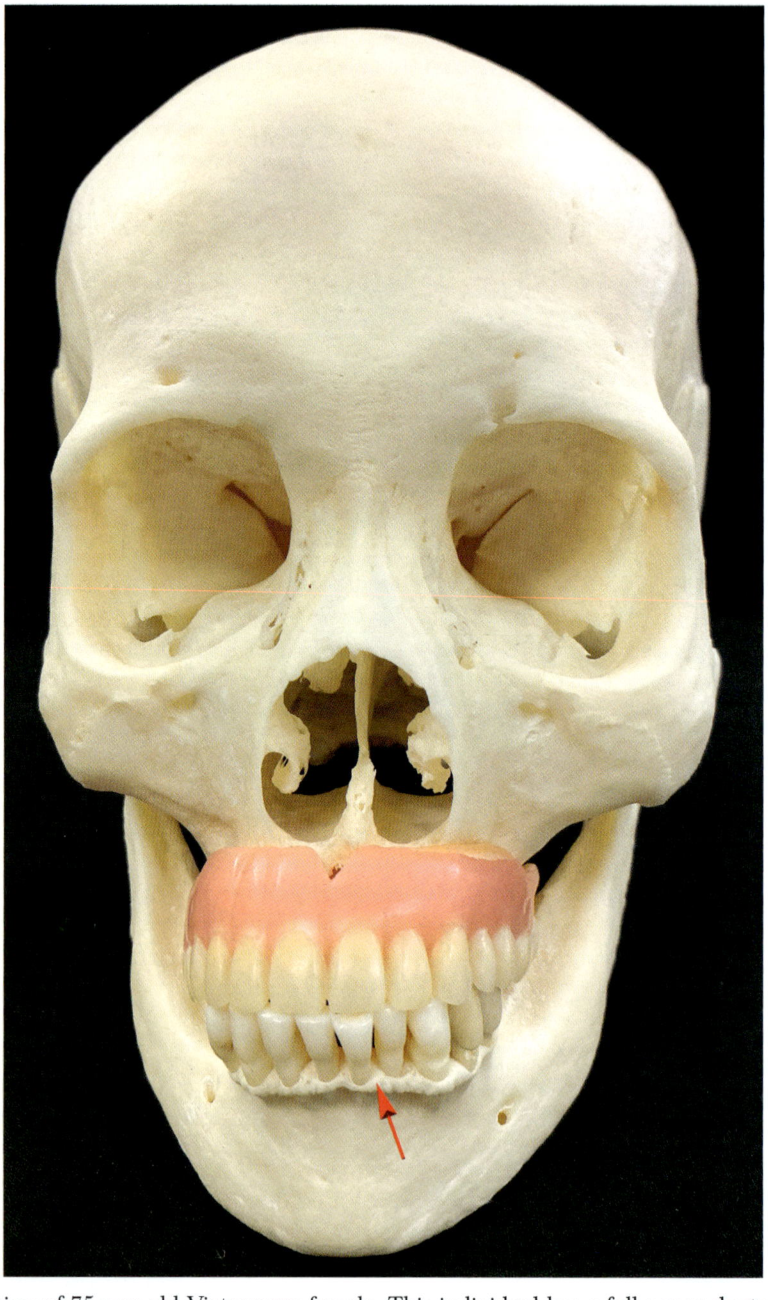

Figure 231. Anterior view of 75-year-old Vietnamese female. This individual has a full upper denture that has altered the morphology of the hard palate/maxilla. Also note the small nasal bones, and a very wide and round nasal aperture. The superior margins of the orbits are blunt, the brow ridge is small, and the chin is narrow and rounded. Note also the mandibular buccal exostoses that have formed a continuous ridge (arrow). Craniometric analysis using 20 variables classified this individual as Vietnamese male with a 0.962 posterior probability. Classification as male may be due to the number of Vietnamese individuals in Fordisc.

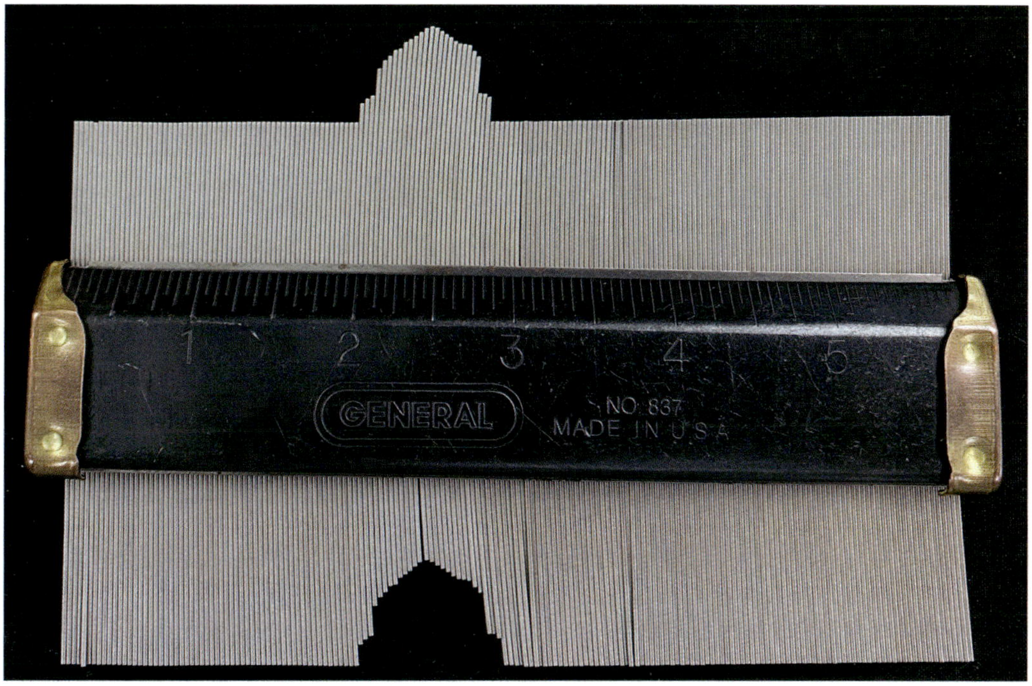

Figure 232. Nasal bone contour in a 75-year-old Vietnamese female.

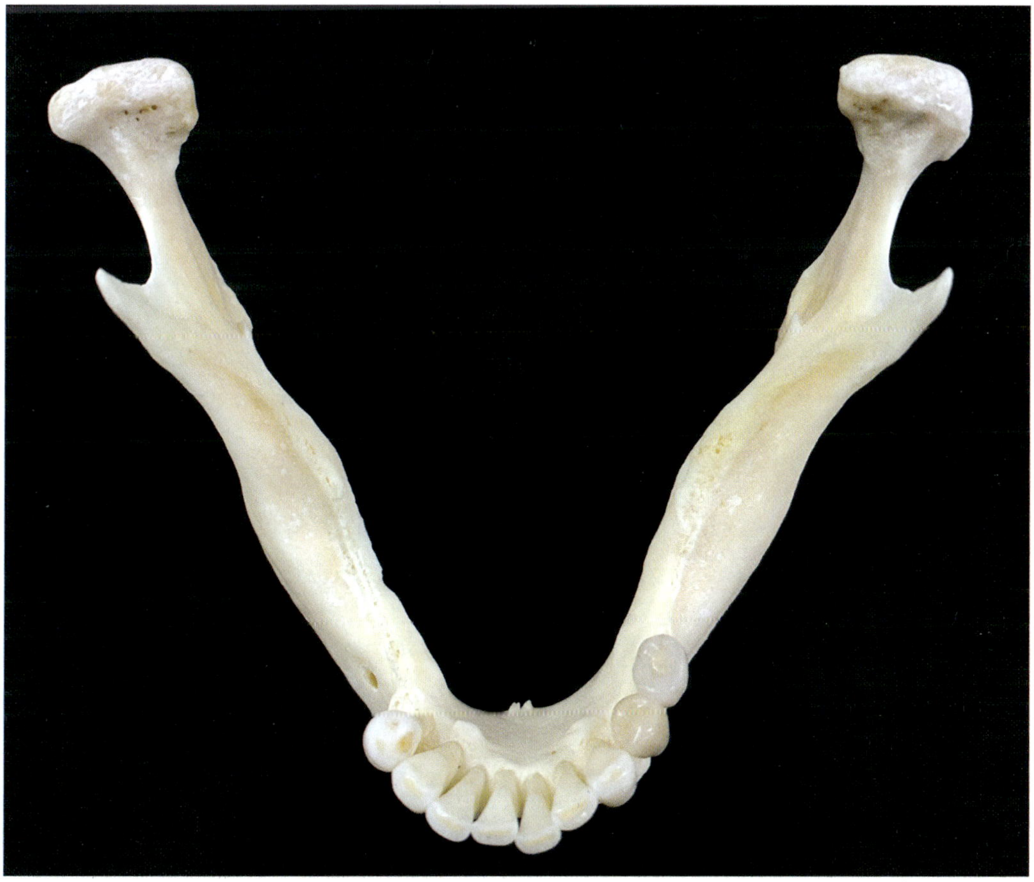

Figure 233. Mandible of 75-year-old Vietnamese female who wore a complete upper denture.

TABLE 22.
CRANIAL MEASUREMENTS IN A 75-YEAR-OLD VIETNAMESE FEMALE

MEASUREMENT	(MM)	RIGHT	LEFT
Maximum Cranial Length (g-op, GOL)	167		
Maximum Cranial Breadth (eu-eu, XCB)	141		
Bizygomatic Breadth (zy-zy, ZYB)	122		
Basion-Bregma Height (ba-b, BBH)	133		
Cranial Base Length (ba-n, BNL)	91		
Basion-Prosthion (ba-pr, BPL)	–		
Maxillo-Alveolar Breadth (ecm-ecm, MAB)	–		
Maxillo-Alveolar Length (pr-alv, MAL)	–		
Biauricular Breadth (au-au, AUB)	118		
Upper Facial Height (n-pr)	–		
Minimum Frontal Breadth (ft-ft, WFB)	86		
Upper Facial Breadth (fmt-fmt)	98		
Nasal Height (n-ns, NLH)	56		
Nasal Breadth (al-al, NLB)	28		
Orbital Breadth (d-ec, OBB)	39		L
Orbital Height (OBH)	34		L
Biorbital Breadth (ec-ec, EKB)	93		
Interorbital Breadth (d-d, DKB)	21		
Frontal Chord (n-b, FRC)	115		
Parietal Chord (b-l, PAC)	103		
Occipital Chord (l-o, OCC)	94		
Foramen Magnum Length (ba-o, FOL)	33		
Foramen Magnum Breadth (FOB)	28		
Mastoid Length (MDH)		25	26

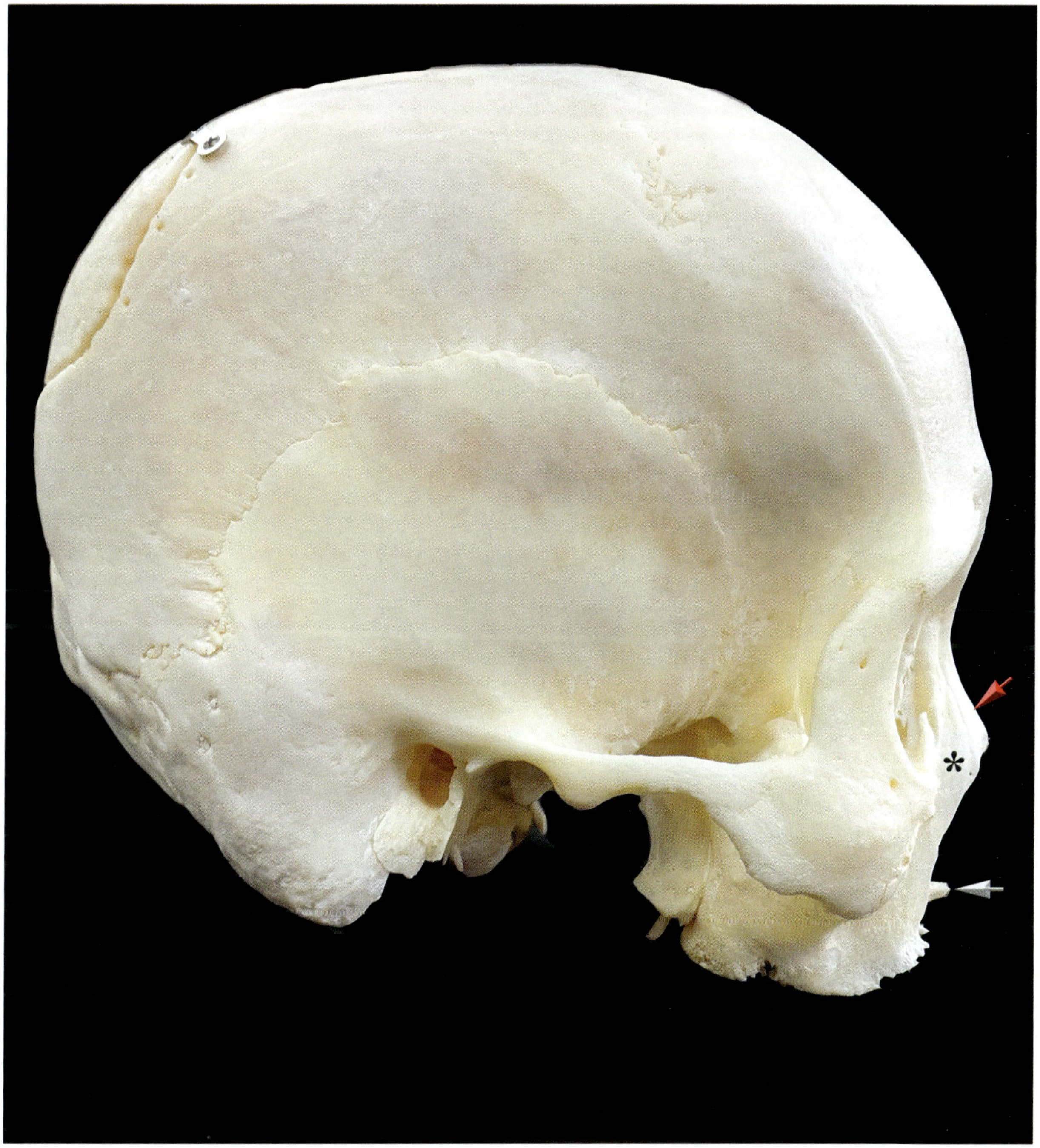

Figure 234. Right lateral view of 75-year-old Vietnamese female who wore a full upper denture resulting in severe resorption of the maxilla. Note the small nasal bones (red arrow) and anterior nasal spine (white arrow), and small, short frontal process of the maxilla (*).

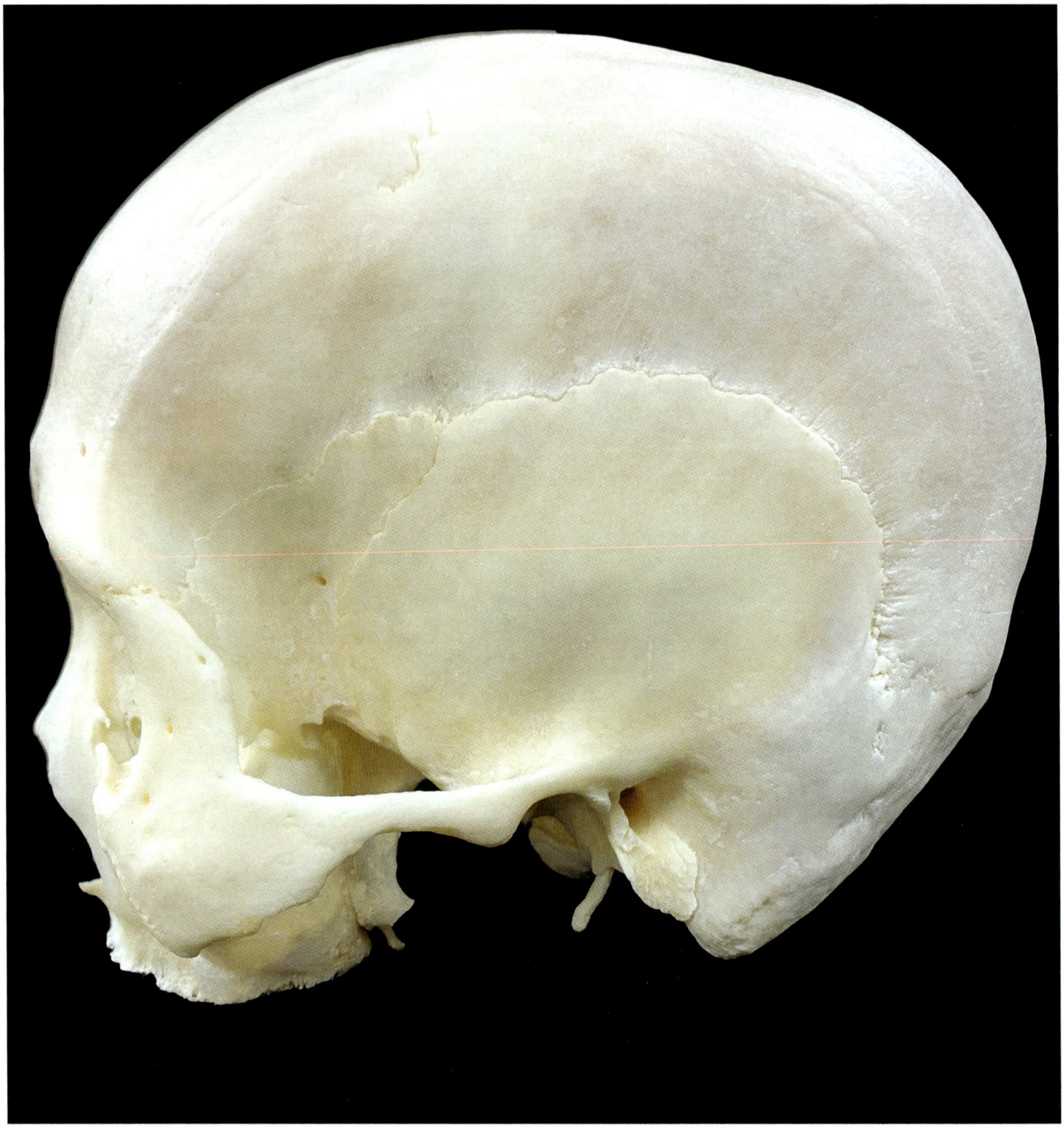

Figure 235. Left lateral view of 75-year-old Vietnamese female (without the denture) with severe resorption of the maxilla due to antemortem loss of the teeth. Note the very small and flat nasal bones.

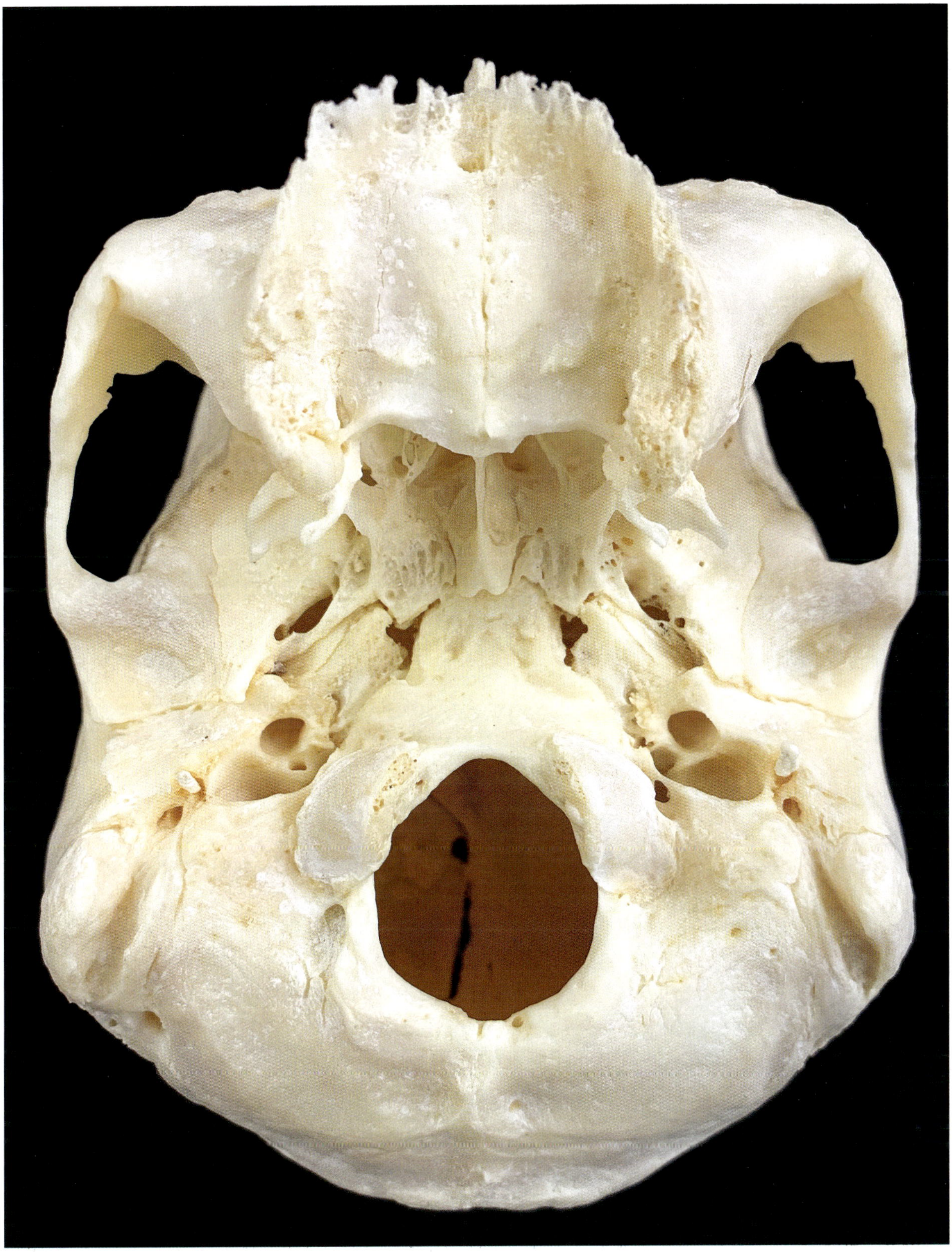

Figure 236. Inferior view of 75-year-old Vietnamese female who wore a full upper denture. Antemortem loss of the maxillary teeth and years of wearing a denture resulted in severe resorption of the maxilla.

Figure 237. Superior view 75-year-old Vietnamese female with healed cranioplasty (bone flap removed and refitted in place following surgery) and complete obliteration of the vault sutures.

Figure 238. Posterior view of 75-year-old Vietnamese female with healed cranioplasty in the right parietal bone, possibly to relieve intracranial bleeding or other surgical procedure. The surgical plates anchor the bone flap to the cranium.

V. EUROPEAN CRANIA

(8 Male and 4 Female)

1. EUROPEAN MALE

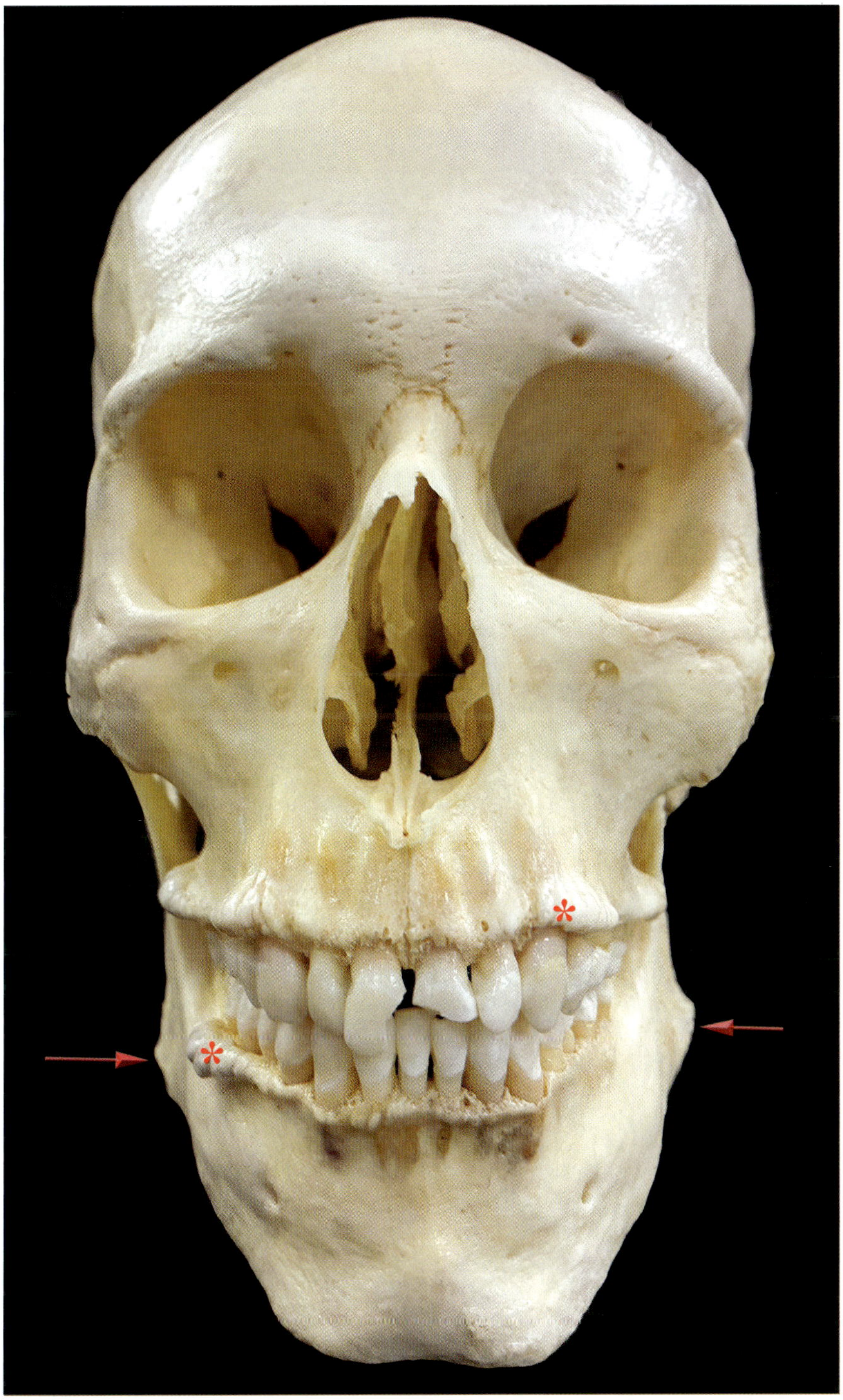

Figure 239. Anterior view of 59-year-old European male who died of facial cancer/carcinoma. Note the tall and narrow skull, the narrow nasal aperture with a sharp nasal sill (inferior nasal aperture), and the prominent nasal bones and nasal spine. Observe the narrow interorbital breadth and the blunt superior orbital margin, gonial eversion (arrow), and a bilobate (indented) chin. The maxillary and mandibular bones exhibit buccal exostoses (*).

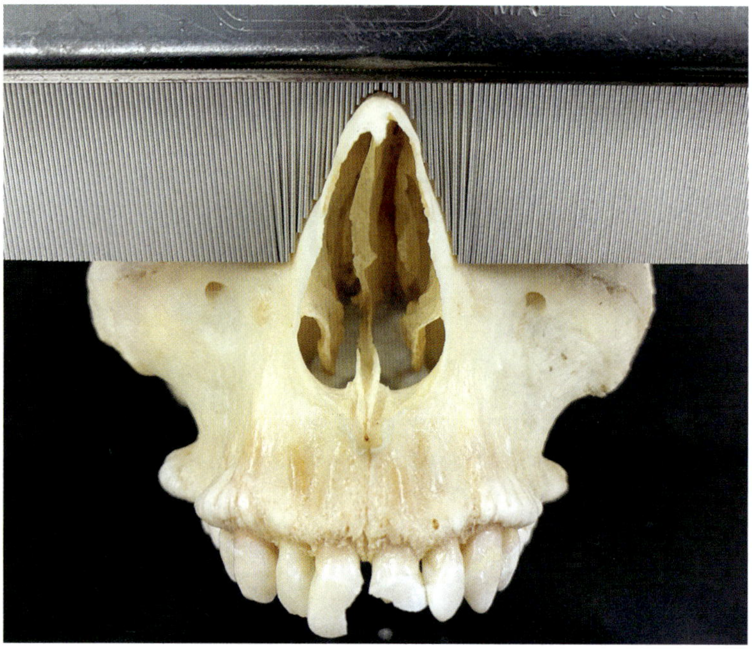

Figure 240. Nasal bone contour in a 59-year-old European male.

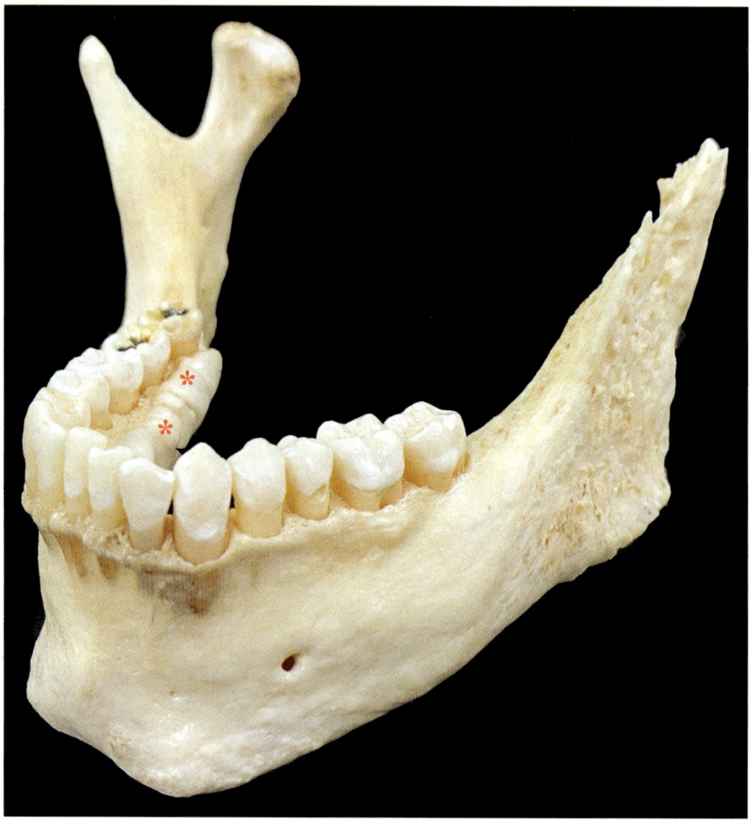

Figure 241. Mandible of 59-year-old European male. The left ramus and portions of the body of the mandible have been destroyed by carcinoma. Note the large lingual exostoses (*)

TABLE 23.
Cranial Measurements in a 59-year-old European Male

Measurement	(mm)	Right	Left
Maximum Cranial Length (g-op, GOL)	187		
Maximum Cranial Breadth (eu-eu, XCB)	140		
Bizygomatic Breadth (zy-zy, ZYB)	123		
Basion-Bregma Height (ba-b, BBH)	138		
Cranial Base Length (ba-n, BNL)	108		
Basion-Prosthion (ba-pr, BPL)	106		
Maxillo-Alveolar Breadth (ecm-ecm, MAB)	67		
Maxillo-Alveolar Length (pr-alv, MAL)	60		
Biauricular Breadth (au-au, AUB)	117		
Upper Facial Height (n-pr)	70		
Minimum Frontal Breadth (ft-ft, WFB)	87		
Upper Facial Breadth (fmt-fmt)	96		
Nasal Height (n-ns, NLH)	52		
Nasal Breadth (al-al, NLB)	23		
Orbital Breadth (d-ec, OBB)	40		L
Orbital Height (OBH)	30		L
Biorbital Breadth (ec-ec, EKB)	88		
Interorbital Breadth (d-d, DKB)	16		
Frontal Chord (n-b, FRC)	118		
Parietal Chord (b-l, PAC)	111		
Occipital Chord (l-o, OCC)	105		
Foramen Magnum Length (ba-o, FOL)	33		
Foramen Magnum Breadth (FOB)	30		
Mastoid Length (MDH)		29	30

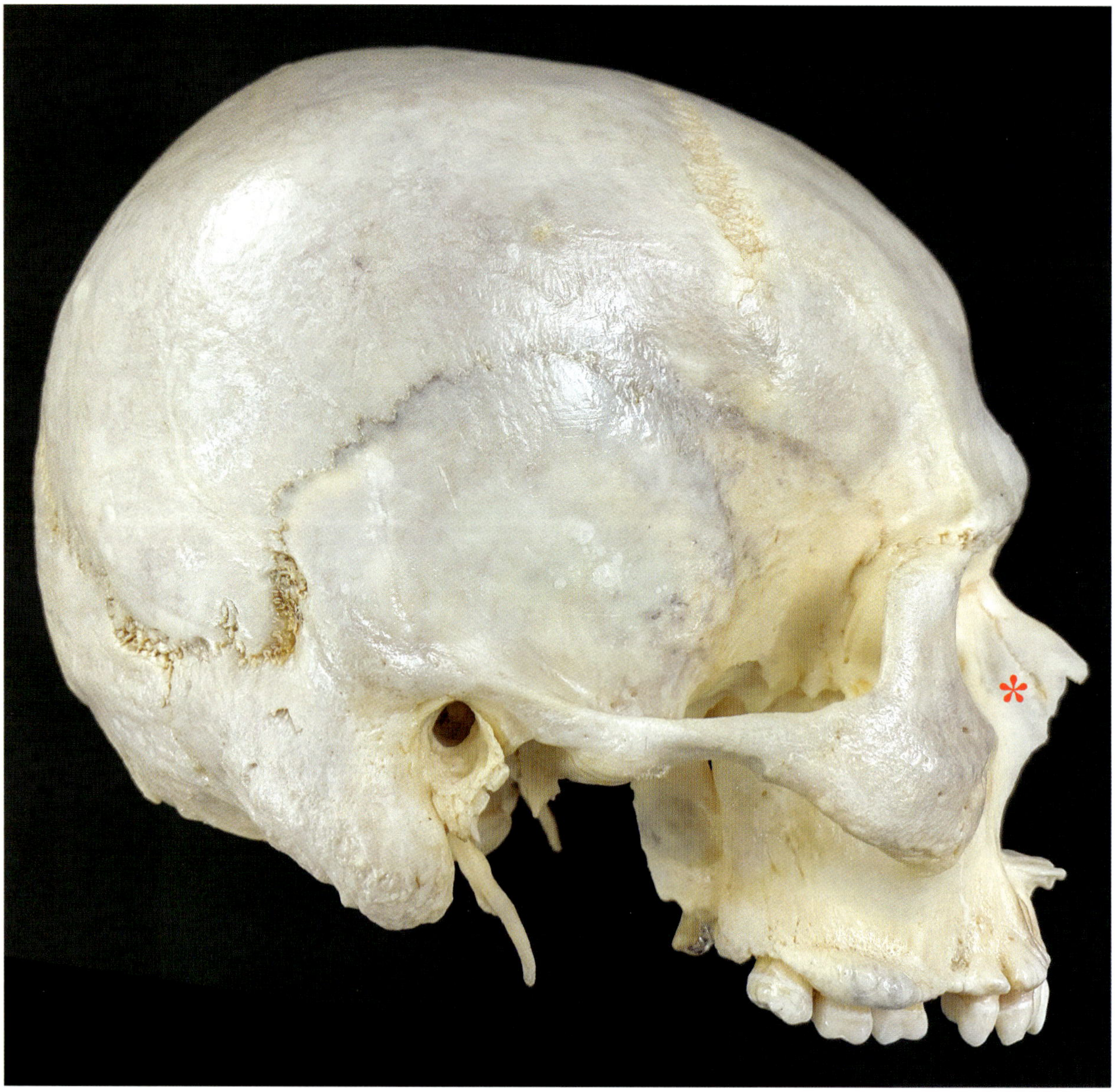

Figure 242. Right lateral view of 59-year-old European male with a high, rounded cranial vault, sloping frontal bone, and protruding brow ridge. Observe the overhanging and prominent nasal bones and frontal process of the maxilla (*), the long anterior nasal spine, and the absence of alveolar prognathism. Note the large mastoid processes and malars, and the supramastoid crest. Notice also the slight development of the superior line for attachment of the temporalis muscle.

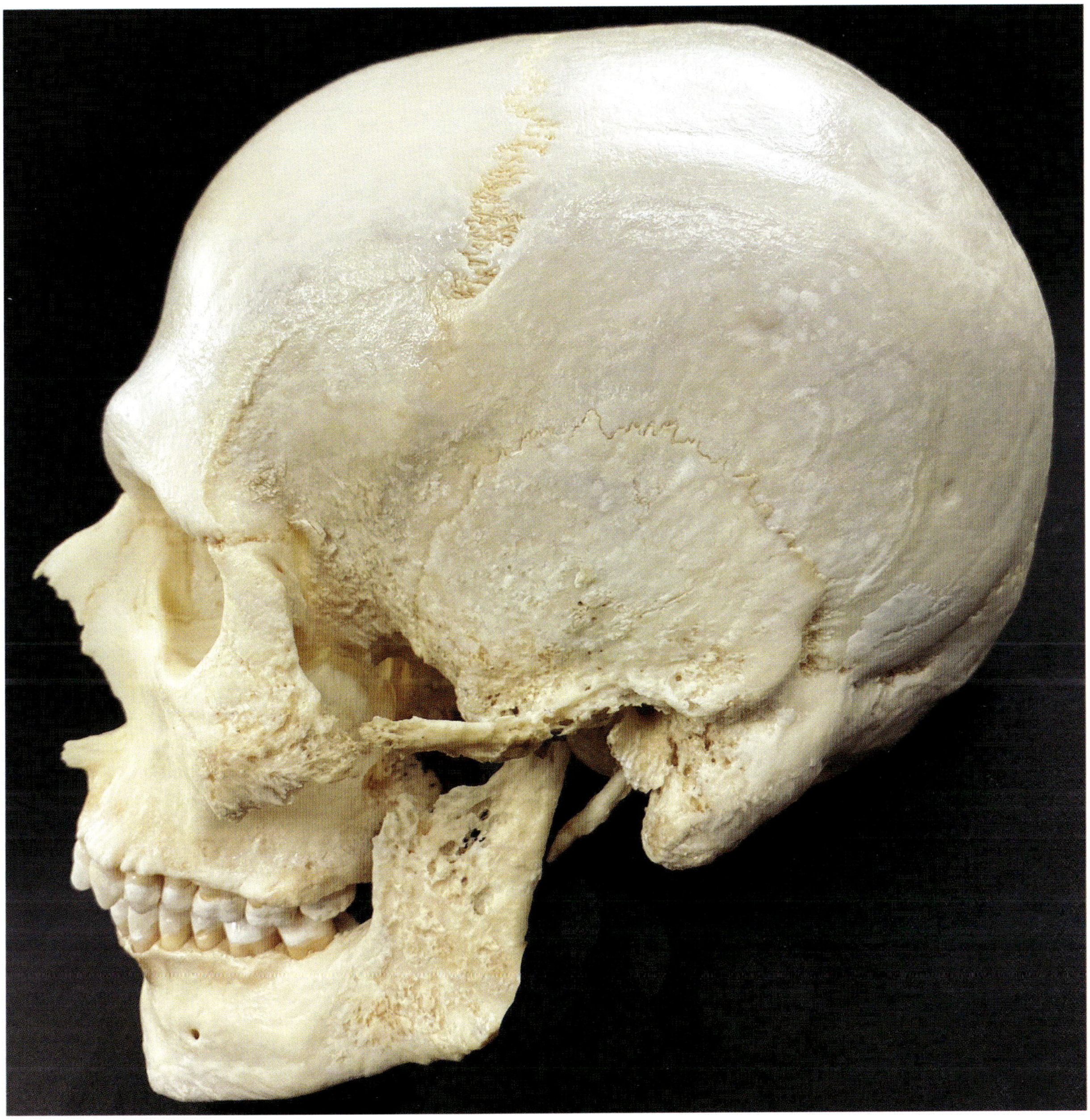

Figure 243. Left lateral view of 59-year-old European male. Note the cancerous destruction of the left mandibular ramus, the zygoma, the left sphenoid, and left temporal bones.

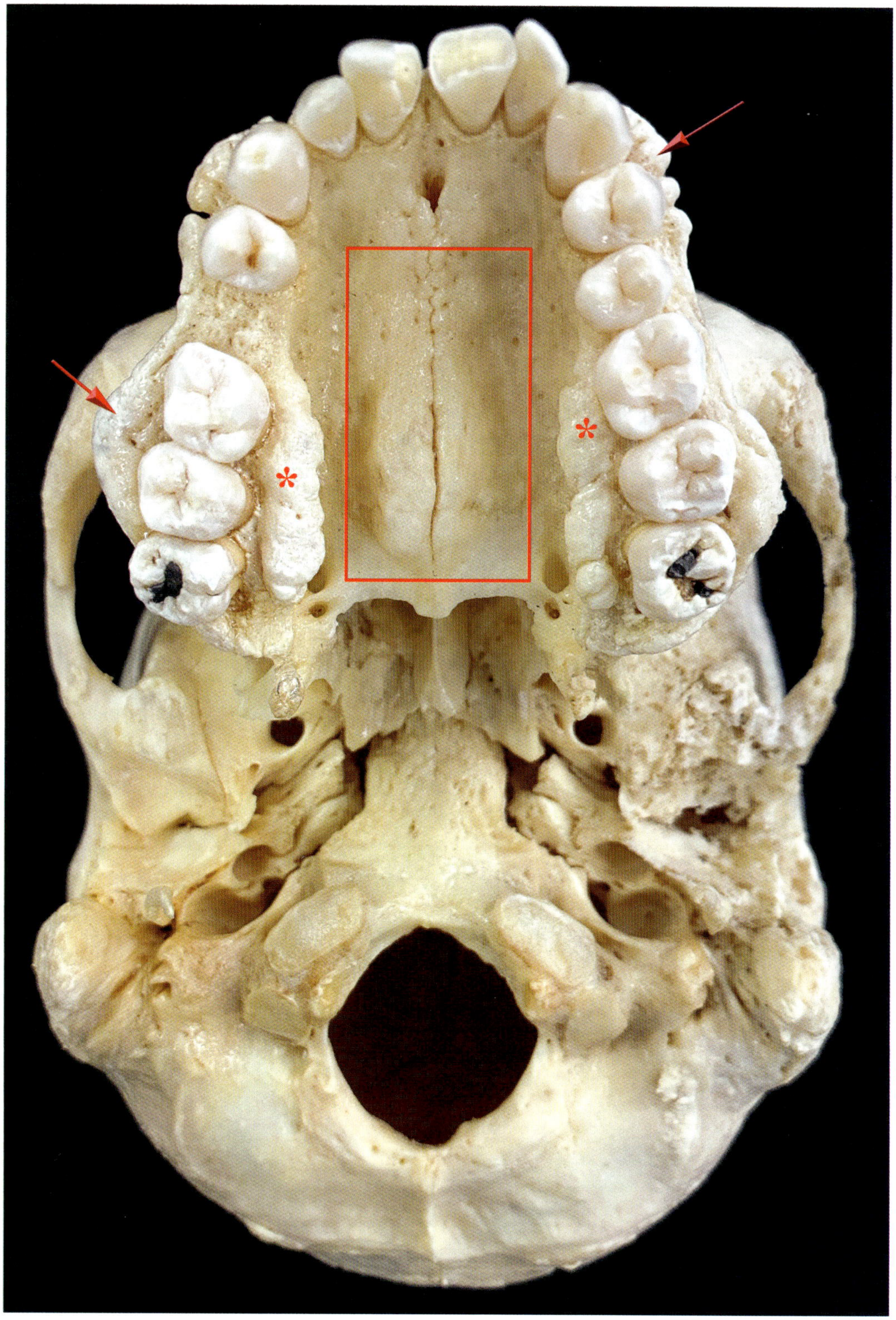

Figure 244. Inferior view of 59-year-old European male showing cancerous destruction of the left sphenoid, the zygoma, and the temporal bones. Note the large buccal (arrow) and lingual exostoses (*) and torus palatinus (rectangle). This male exhibits wide zygomatic arches, a large U-shaped hard palate, spatulate central incisors, and developed muscle attachments on the occipital bone.

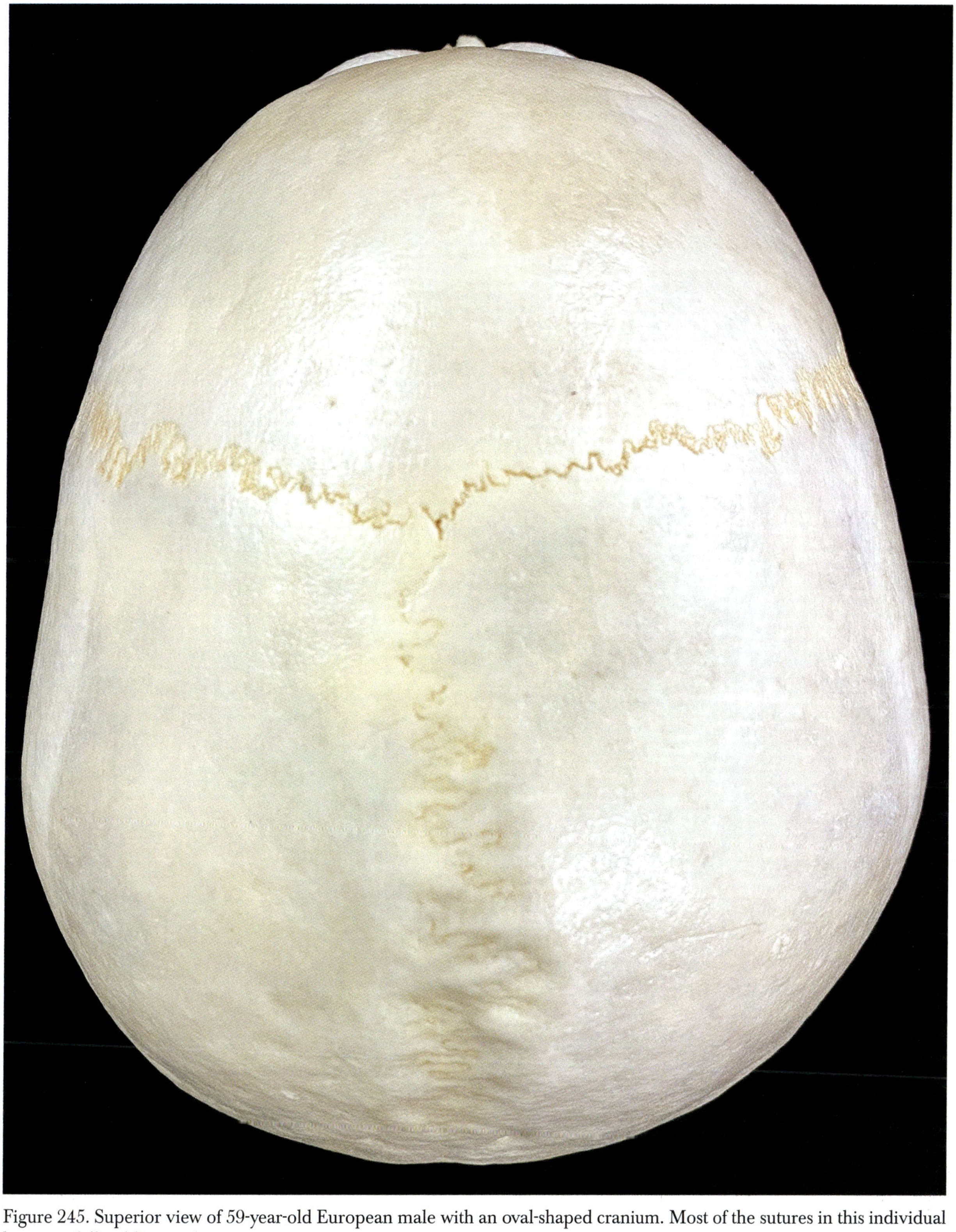

Figure 245. Superior view of 59-year-old European male with an oval-shaped cranium. Most of the sutures in this individual have not fully obliterated.

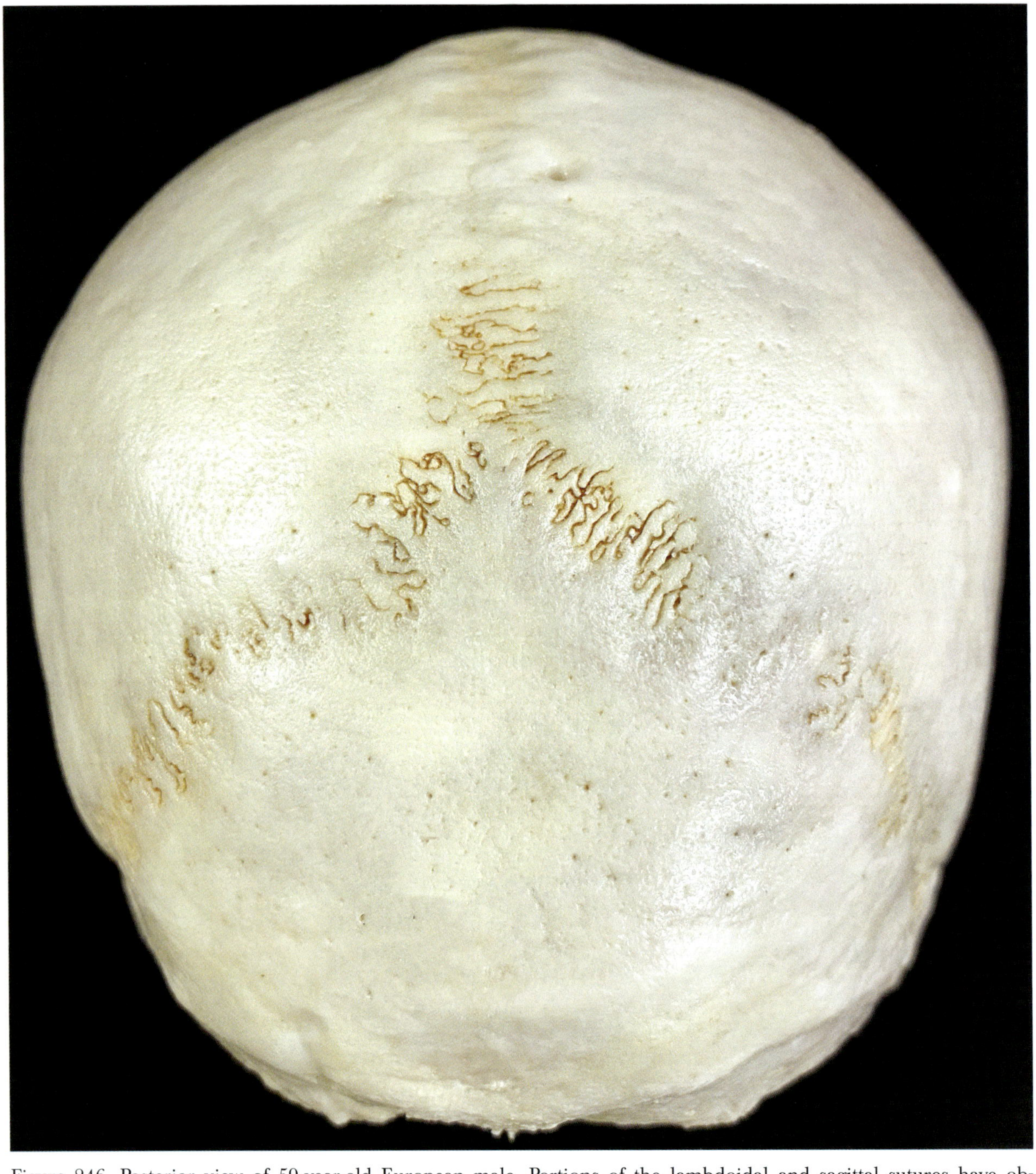

Figure 246. Posterior view of 59-year-old European male. Portions of the lambdoidal and sagittal sutures have obliterated/fused. Note the developed muscle attachment sites along the nuchal crest.

2. EUROPEAN MALE

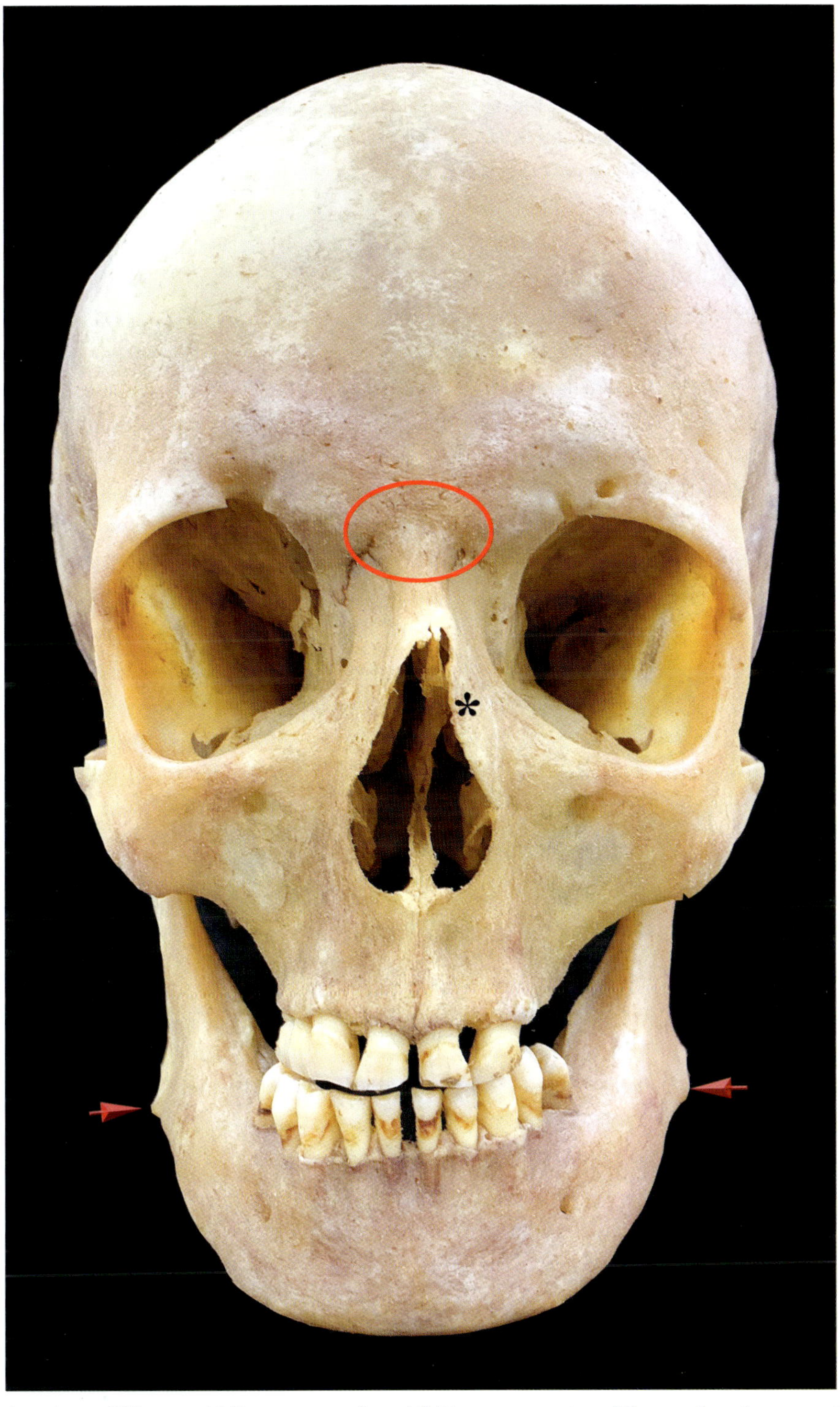

Figure 247. Anterior view of 66-year-old European male exhibiting asymmetry of the nasal region, prominent nasal bones and maxilla, obliteration of the frontonasal suture (oval), medially flattened left maxilla (*), and gonial eversion (arrow).

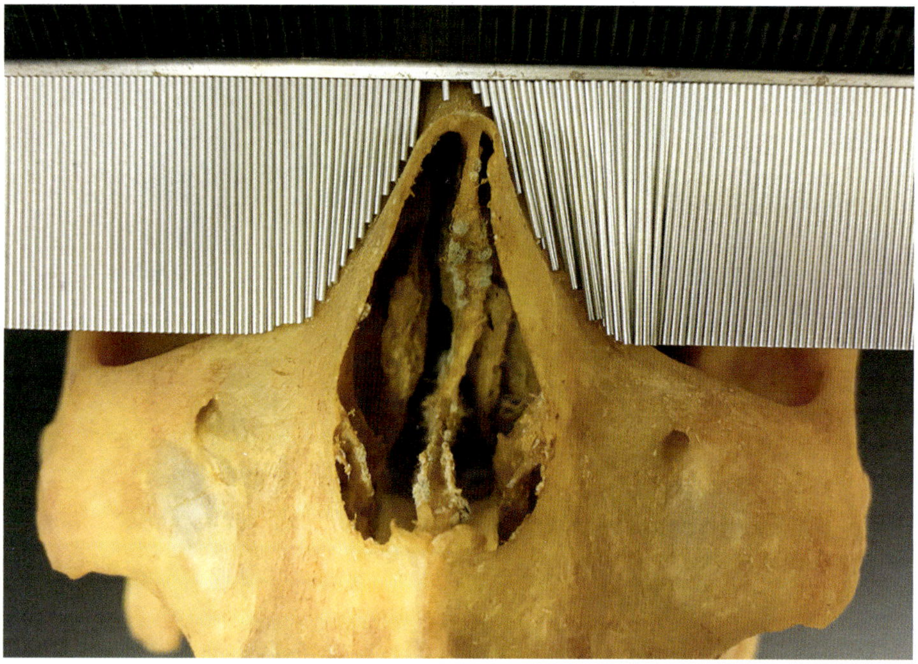

Figure 248. Nasal bone contour in a 66-year-old European male.

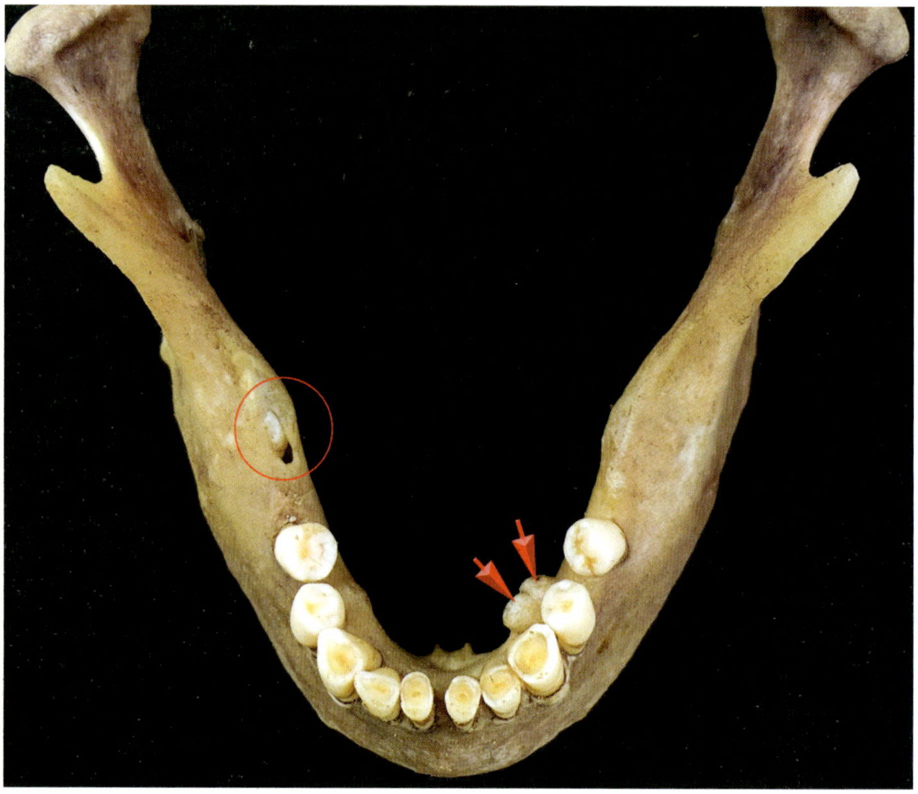

Figure 249. Mandible of 66-year-old European male with severe occlusal wear exposing the pulp in the anterior teeth. Also present are an unerupted/impacted right molar (circle) and asymmetrical lingual exostoses (arrow).

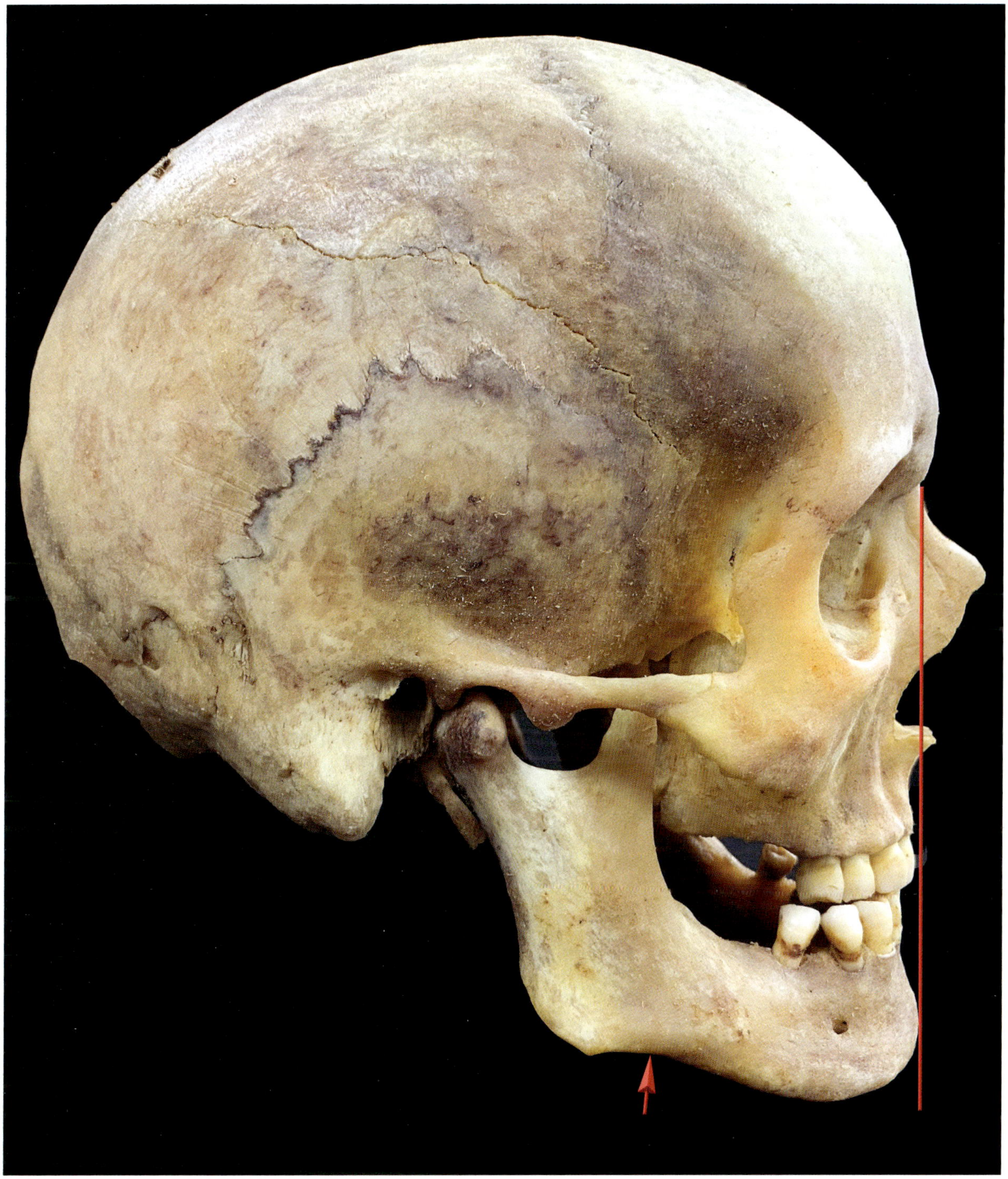

Figure 250. Right lateral view of 66-year-old European male presenting a straight profile and absence of alveolar prognathism as indicated by a vertical line from pogonion to nasion. Also note the antegonial notch (arrow). The fractured right parietal bone is taphonomic.

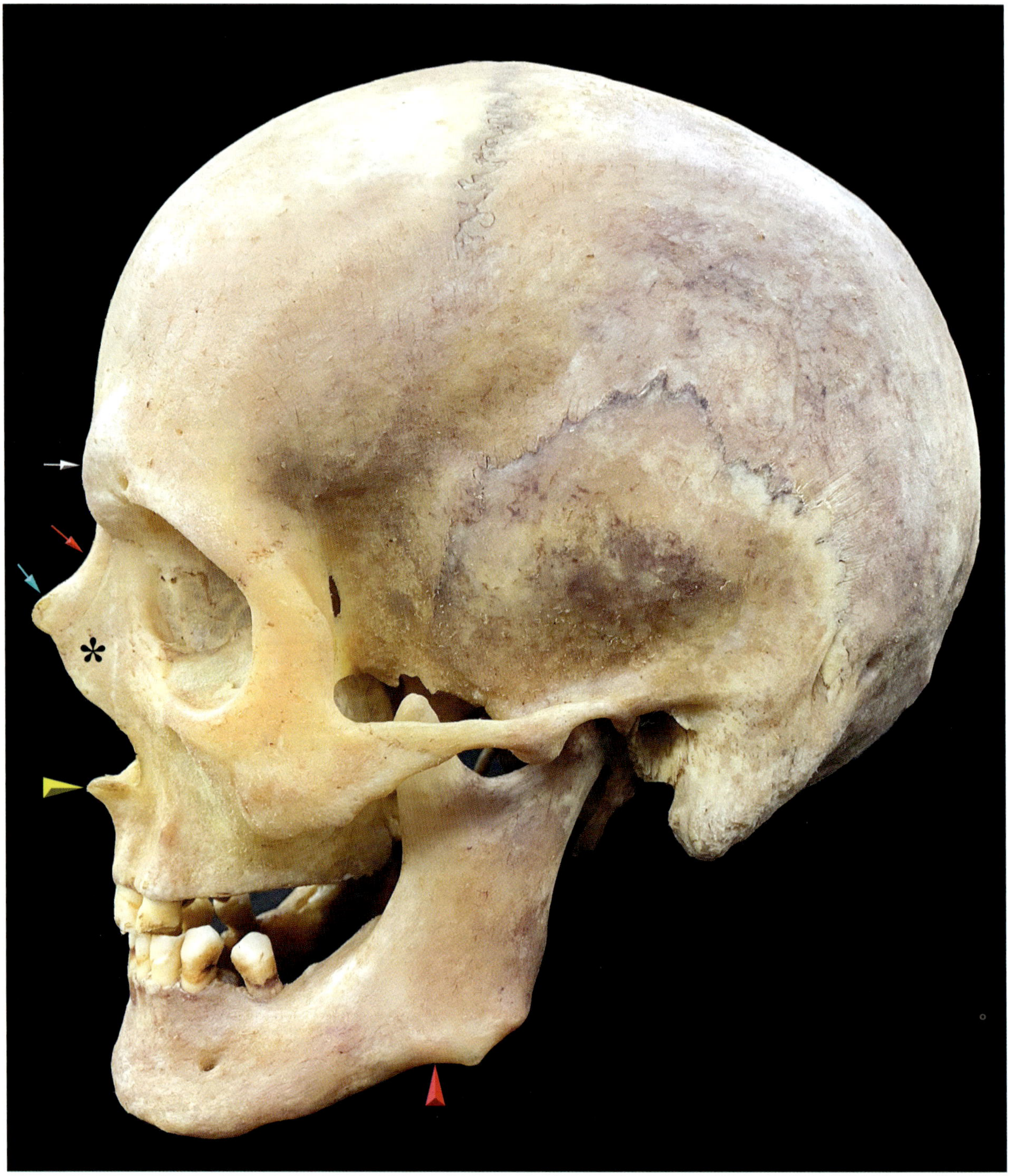

Figure 251. Left lateral view of 66-year-old European male with a sloping frontal bone, small brow ridge (white arrow), concave (red arrow) and convex (blue arrow) nasal bones, prominent frontal process of the maxilla (*), an anterior nasal spine (yellow arrowhead), and an antegonial notch (red arrowhead).

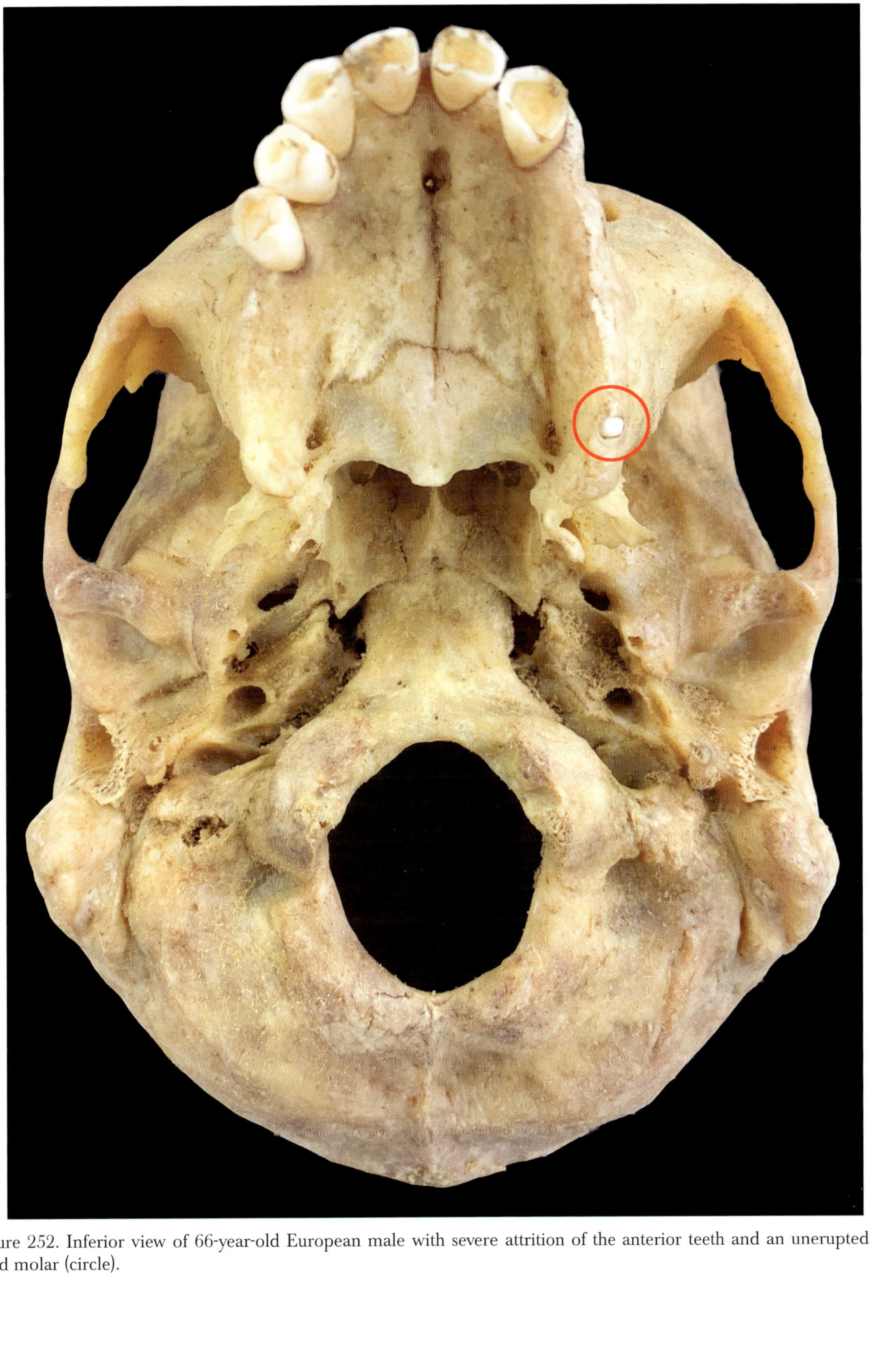

Figure 252. Inferior view of 66-year-old European male with severe attrition of the anterior teeth and an unerupted left third molar (circle).

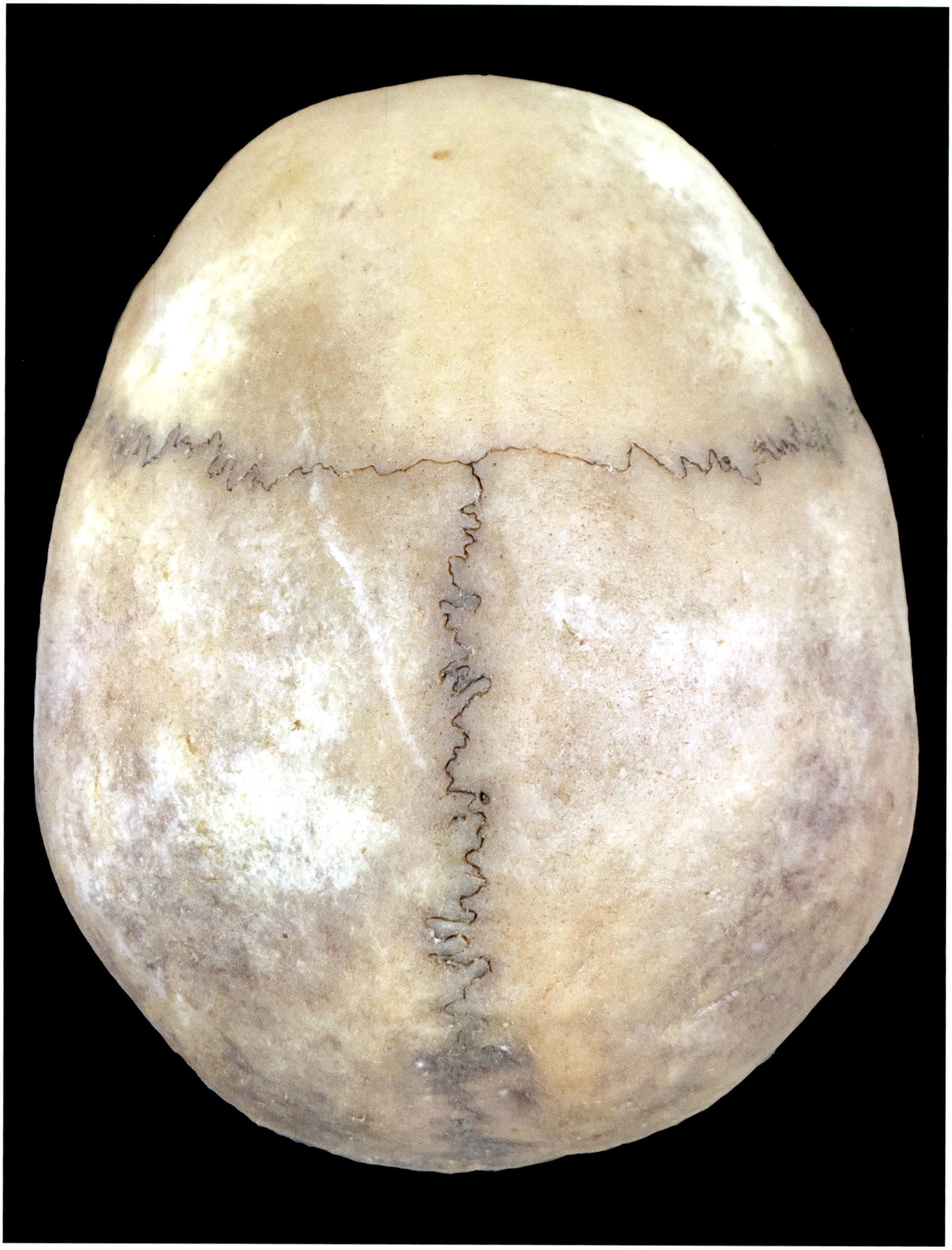

Figure 253. Superior view of 66-year-old European male with minimal obliteration of the sutures.

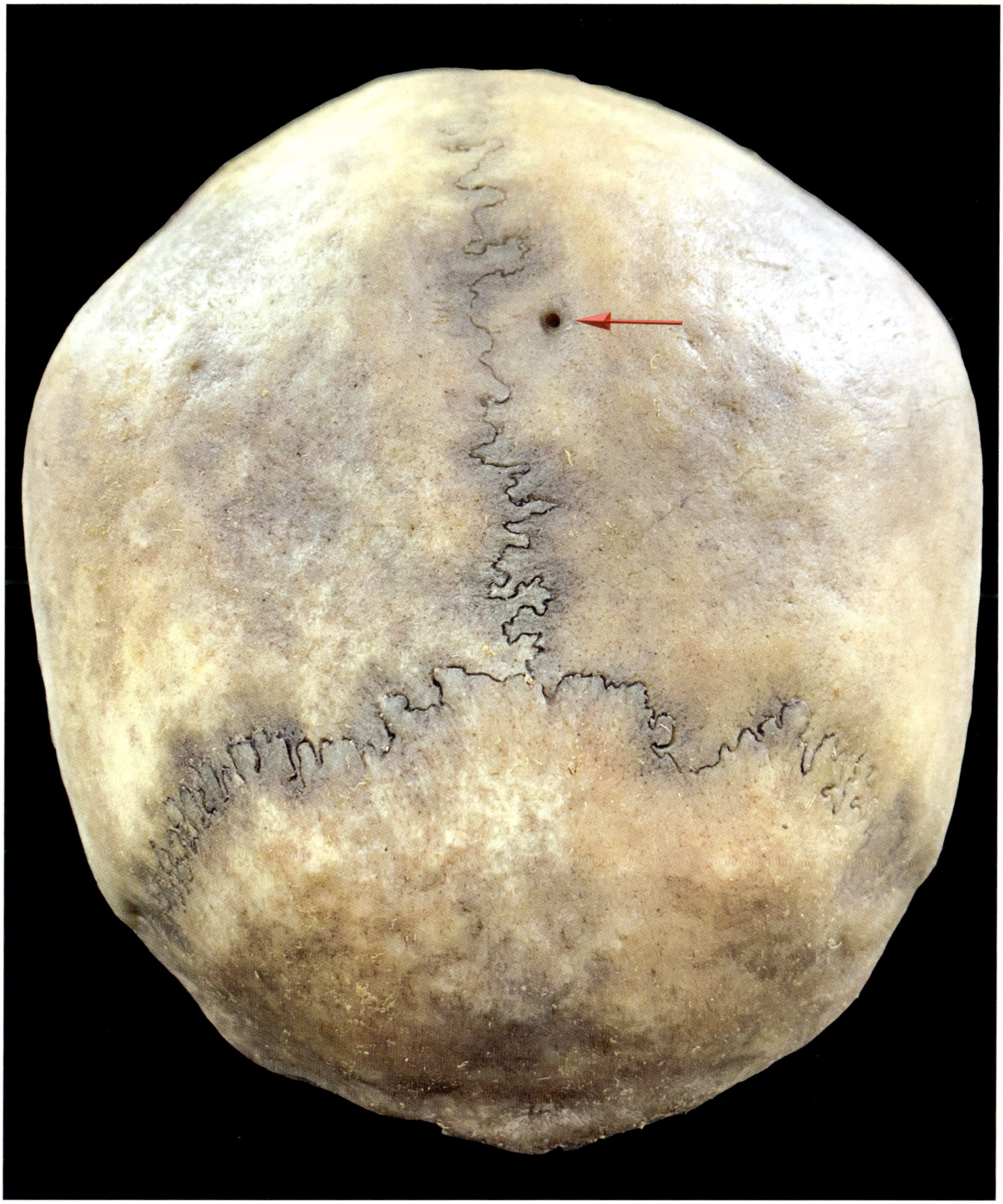

Figure 254. Posterior view of 66-year-old European male with a right parietal foramen (arrow) and minimal obliteration of the sagittal and lambdoidal sutures. Normal parietal foramen range in size and number from very small to large (5mm maximum diameter) and from none to several in one or both parietal bones.

3. EUROPEAN MALE

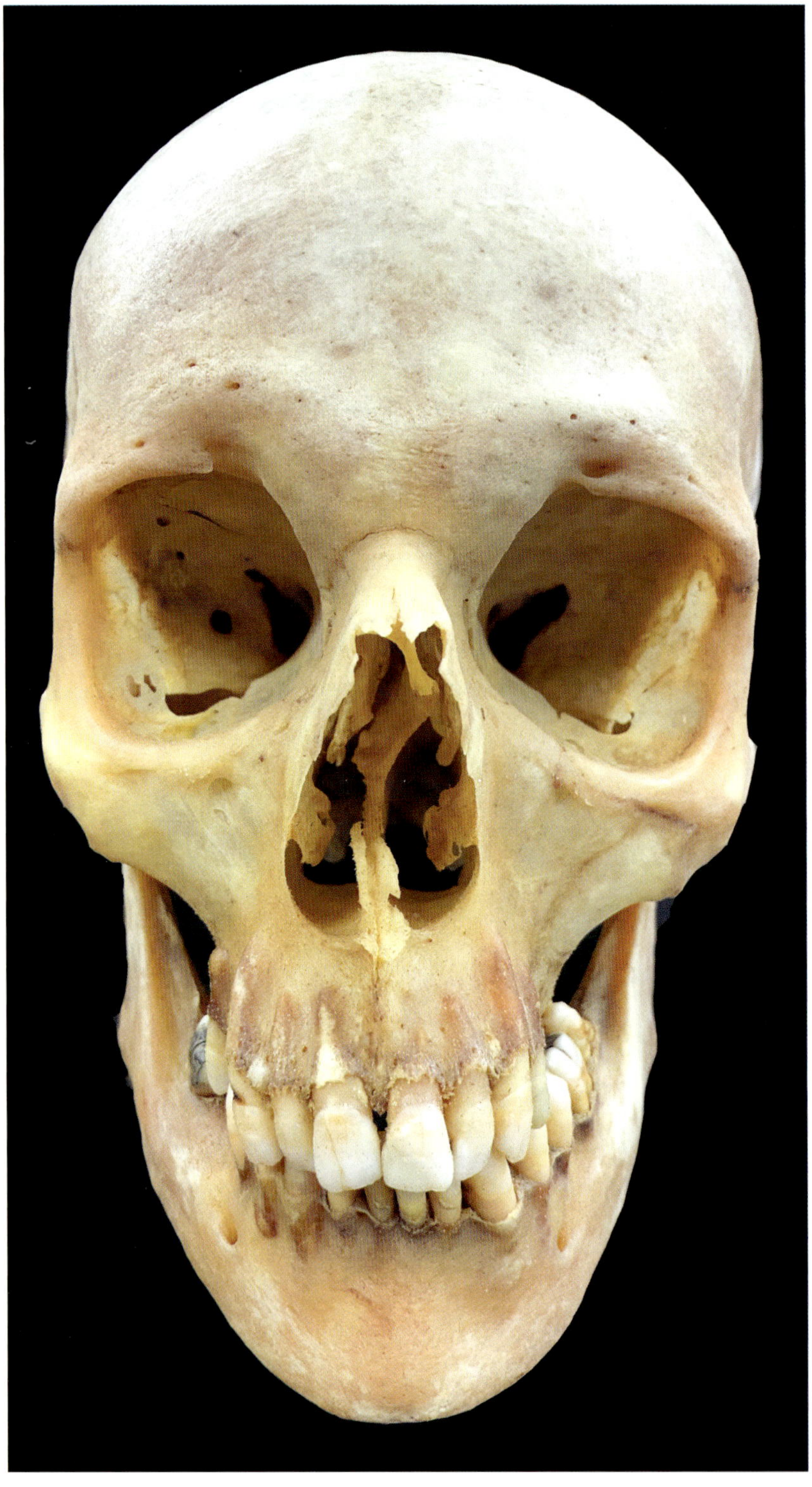

Figure 255. Anterior view of 77-year-old European male with prominent nasal bones, a narrow nasal aperture and interorbital breadth, rectangular orbits, small and narrow zygomatic bones, and a deviated nasal septum (*).

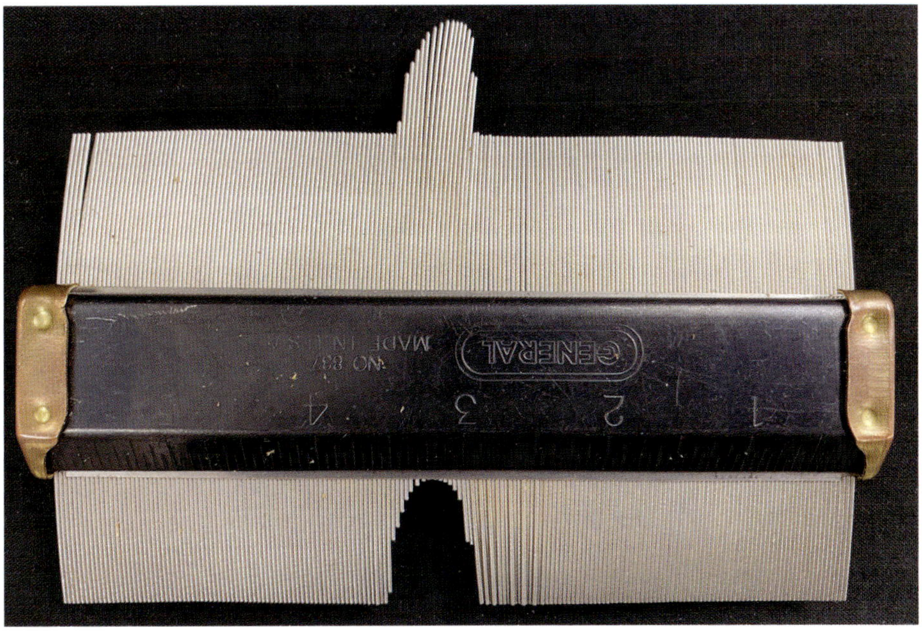

Figure 256. Nasal bone contour in a 77-year-old European male.

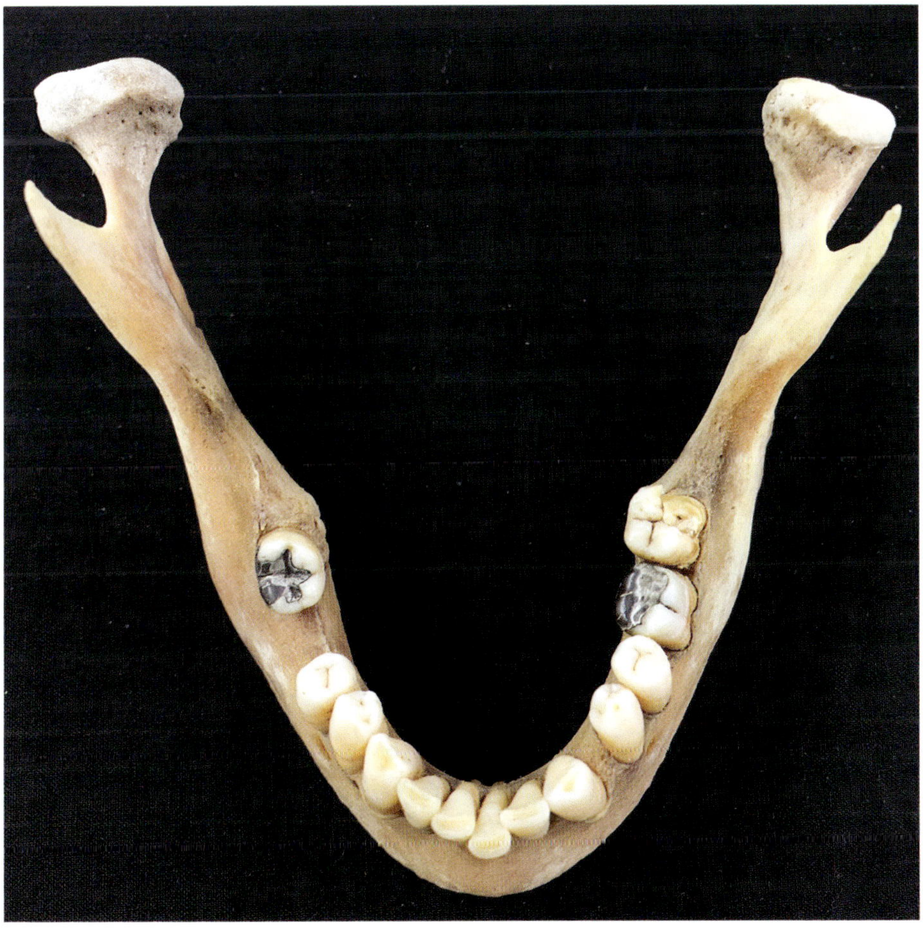

Figure 257. Mandible of 77-year-old European male.

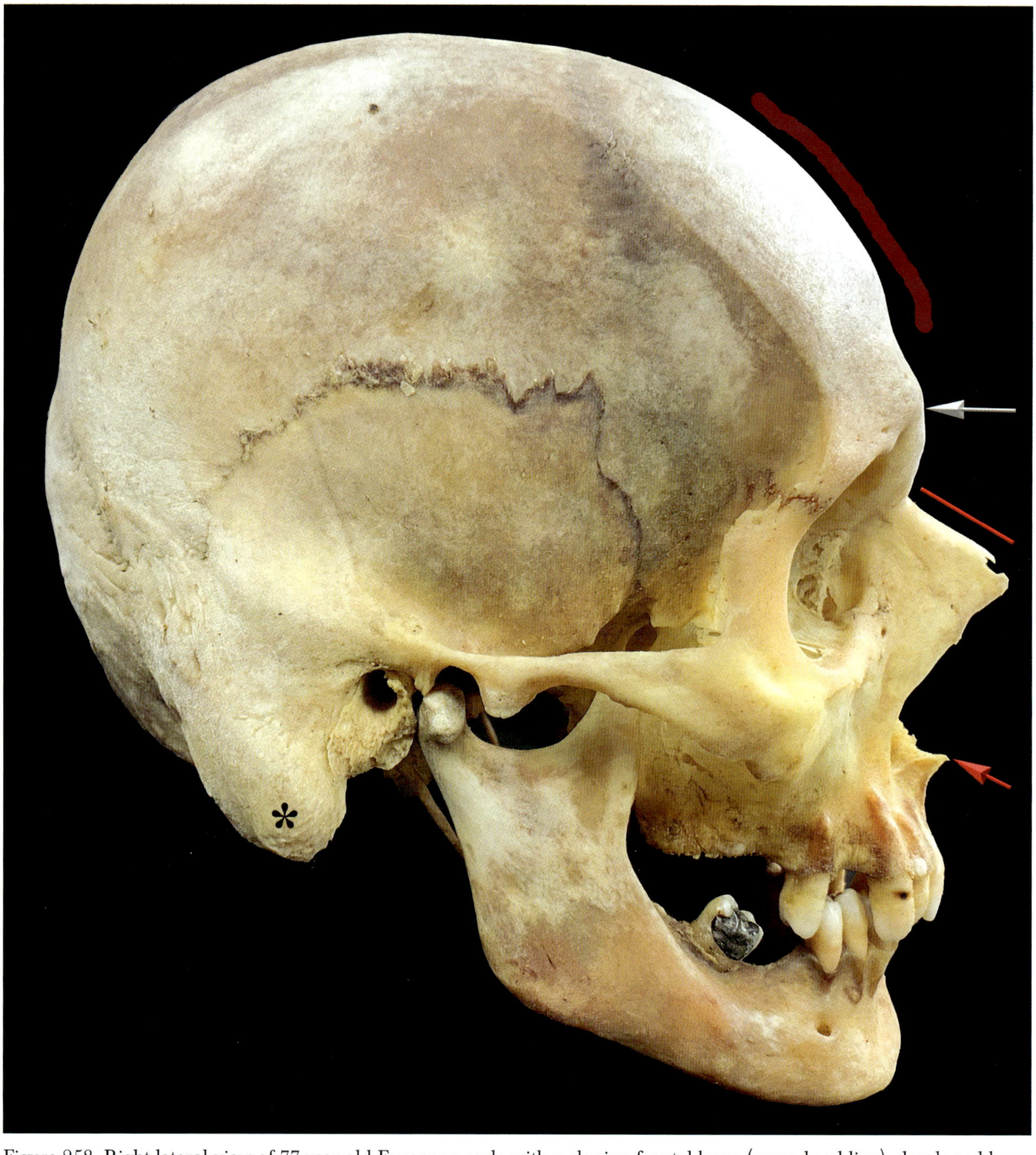

Figure 258. Right lateral view of 77-year-old European male with a sloping frontal bone (curved red line), developed brow ridge (white arrow), straight and prominent nasal bones (straight red line), a moderate-sized anterior nasal spine (red arrow), and a large mastoid process (*).

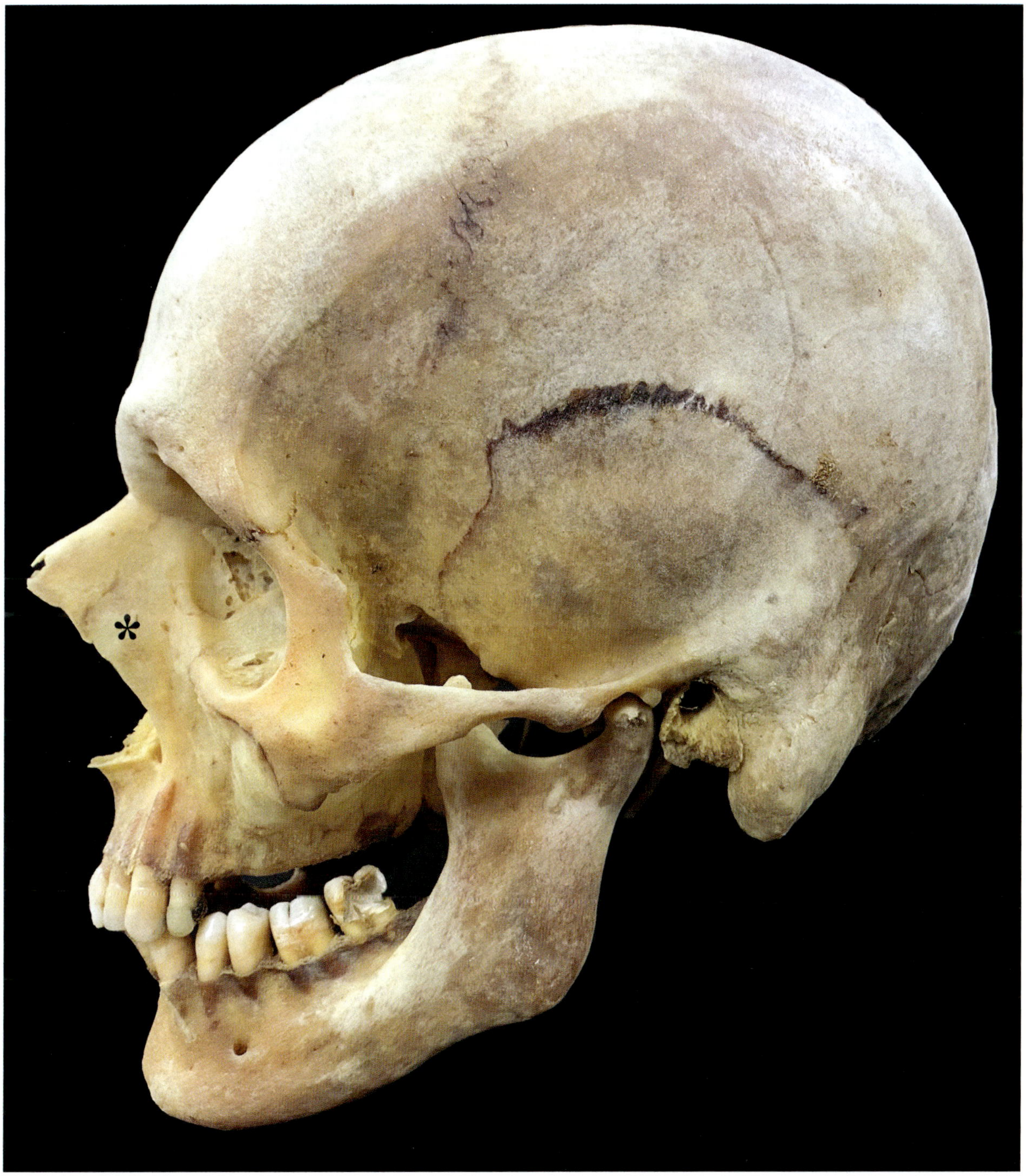

Figure 259. Left lateral view of 77-year-old European male with a rounded vault, prominent nasal bones, frontal process of the maxilla (*) and anterior nasal sill, and absence of alveolar prognathism.

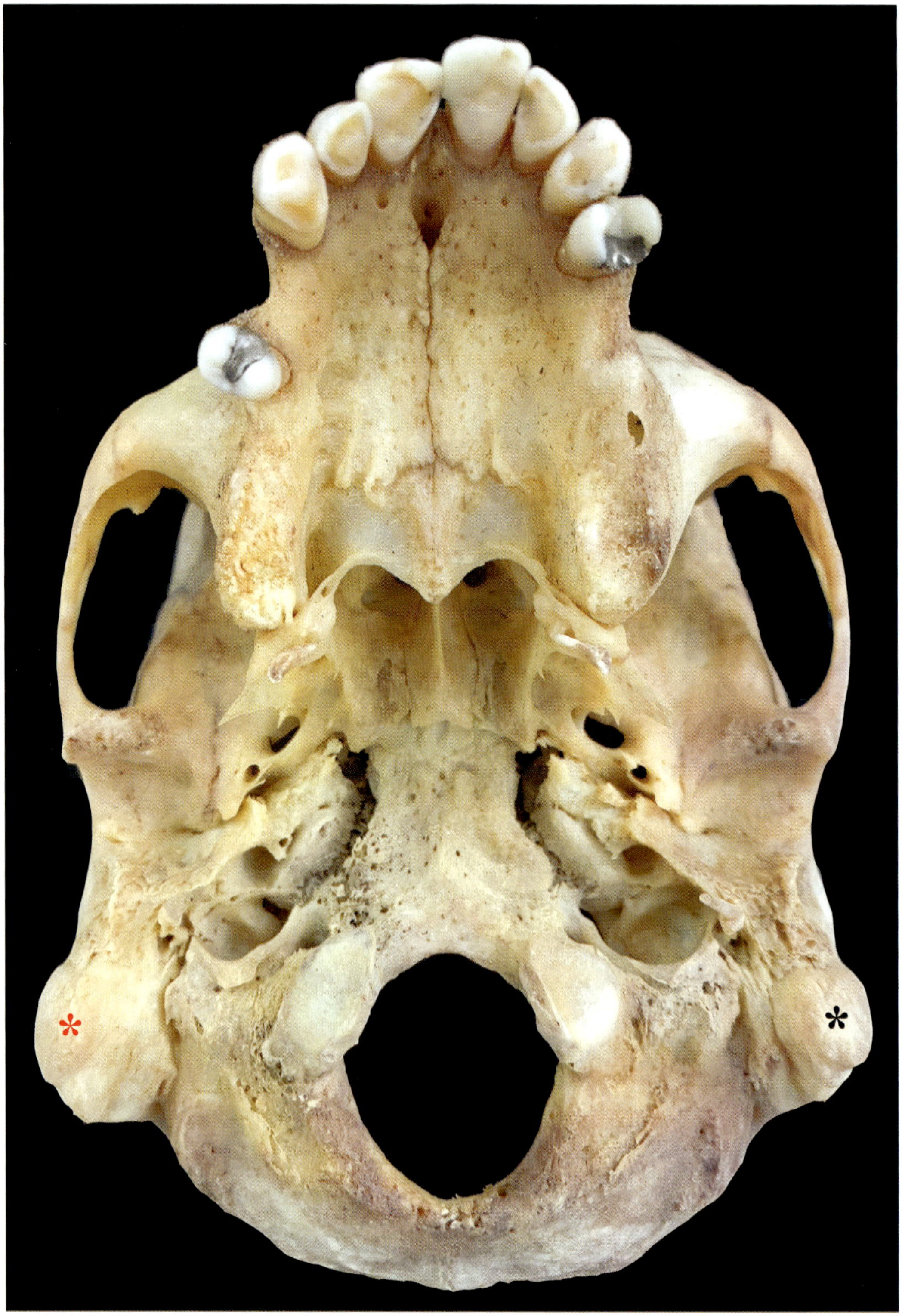

Figure 260. Inferior view of 77-year-old European male with asymmetrical mastoid processes. The right mastoid process (red*) is much larger than the left mastoid process (black*) likely reflecting pathological involvement of the right middle ear (Schumacher 1936) during early childhood (Tos and Strangerup 2010). At birth, the mastoid process is a small mound that gradually increases in size with age.

Figure 261. Superior view of 77-year-old European male showing obliteration of the coronal and sagittal sutures, but visible traces of their morphology.

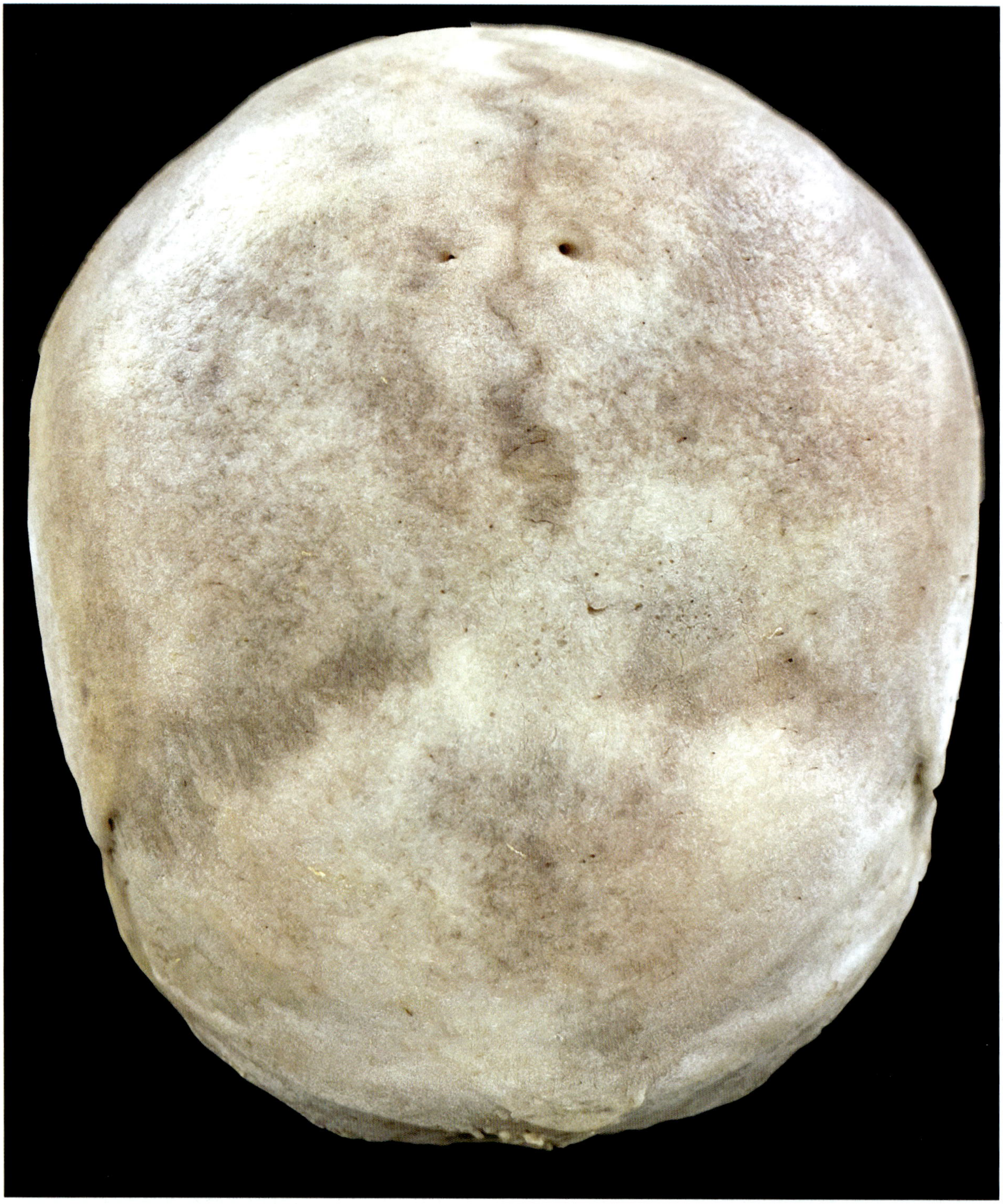

Figure 262. Posterior view of 77-year-old European male with bilateral parietal foramina.

4. EUROPEAN MALE

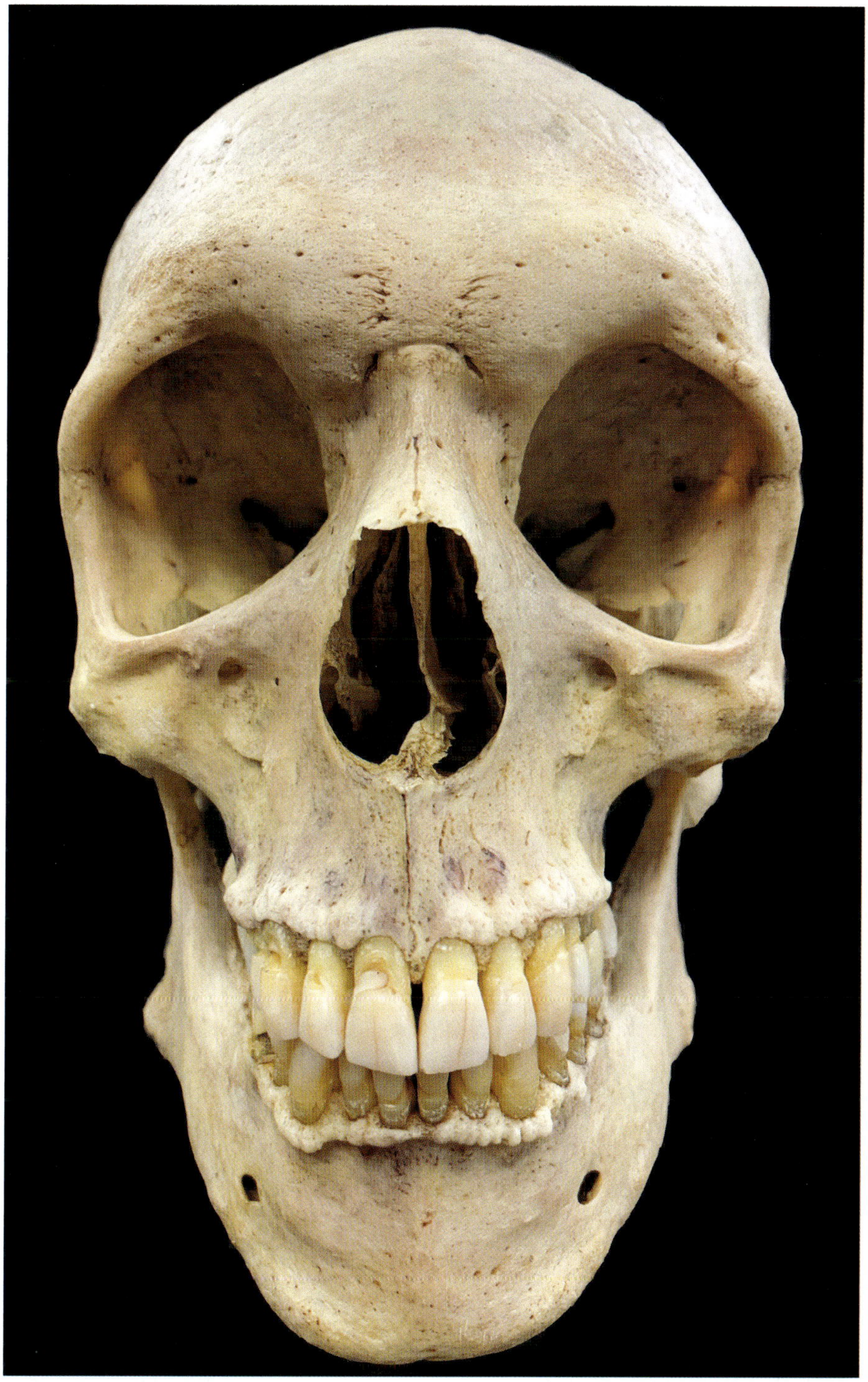

Figure 263. Anterior view 75-year-old European male. The cranium is narrow and tall with trace of the supranasal suture and mild deviation (to the left) of the nasal septum. Note the angular "aviator glasses" orbits, and blunt supraorbital margins. The mandible exhibits gonial eversion, and notice the maxillary and mandibular buccal exostoses. Fordisc (3.0) analysis using 24 variables classified this individual as White (European) male with a posterior probability of 0.999.

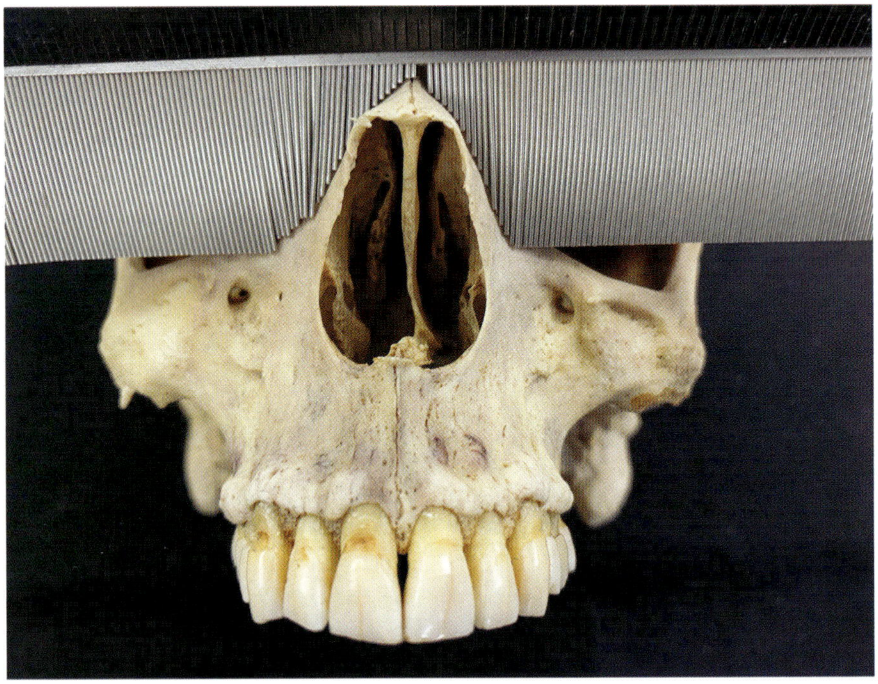

Figure 264. Nasal bone contour in a 75-year-old European male.

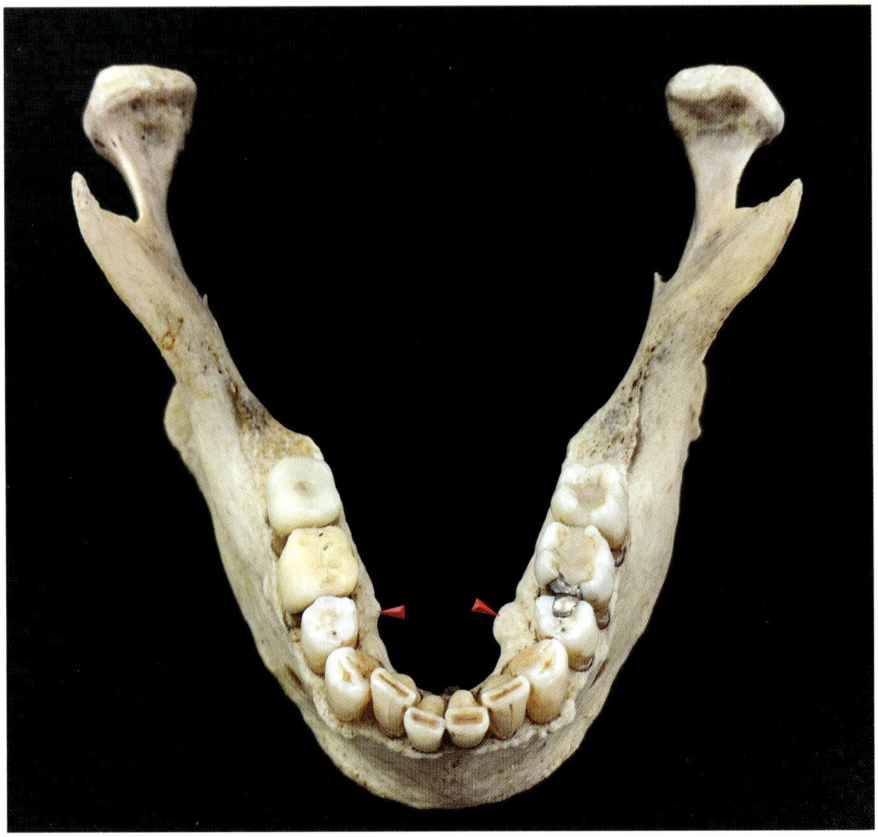

Figure 265. Mandible of 75-year-old European male with lingual exostoses (arrowhead).

TABLE 24.
Cranial Measurements in a 75-year-old European Male

Measurement	(mm)	Right	Left
Maximum Cranial Length (g-op, GOL)	183		
Maximum Cranial Breadth (eu-eu, XCB)	139		
Bizygomatic Breadth (zy-zy, ZYB)	123		
Basion-Bregma Height (ba-b, BBH)	138		
Cranial Base Length (ba-n, BNL)	100		
Basion-Prosthion (ba-pr, BPL)	93		
Maxillo-Alveolar Breadth (ecm-ecm, MAB)	55		
Maxillo-Alveolar Length (pr-alv, MAL)	55		
Biauricular Breadth (au-au, AUB)	122		
Upper Facial Height (n-pr)	66		
Minimum Frontal Breadth (ft-ft, WFB)	94		
Upper Facial Breadth (fmt-fmt)	101		
Nasal Height (n-ns, NLH)	49		
Nasal Breadth (al-al, NLB)	23		
Orbital Breadth (d-ec, OBB)	38		L
Orbital Height (OBH)	37		L
Biorbital Breadth (ec-ec, EKB)	93		
Interorbital Breadth (d-d, DKB)	20		
Frontal Chord (n-b, FRC)	111		
Parietal Chord (b-l, PAC)	111		
Occipital Chord (l-o, OCC)	105		
Foramen Magnum Length (ba-o, FOL)	35		
Foramen Magnum Breadth (FOB)	31		
Mastoid Length (MDH)		25	24

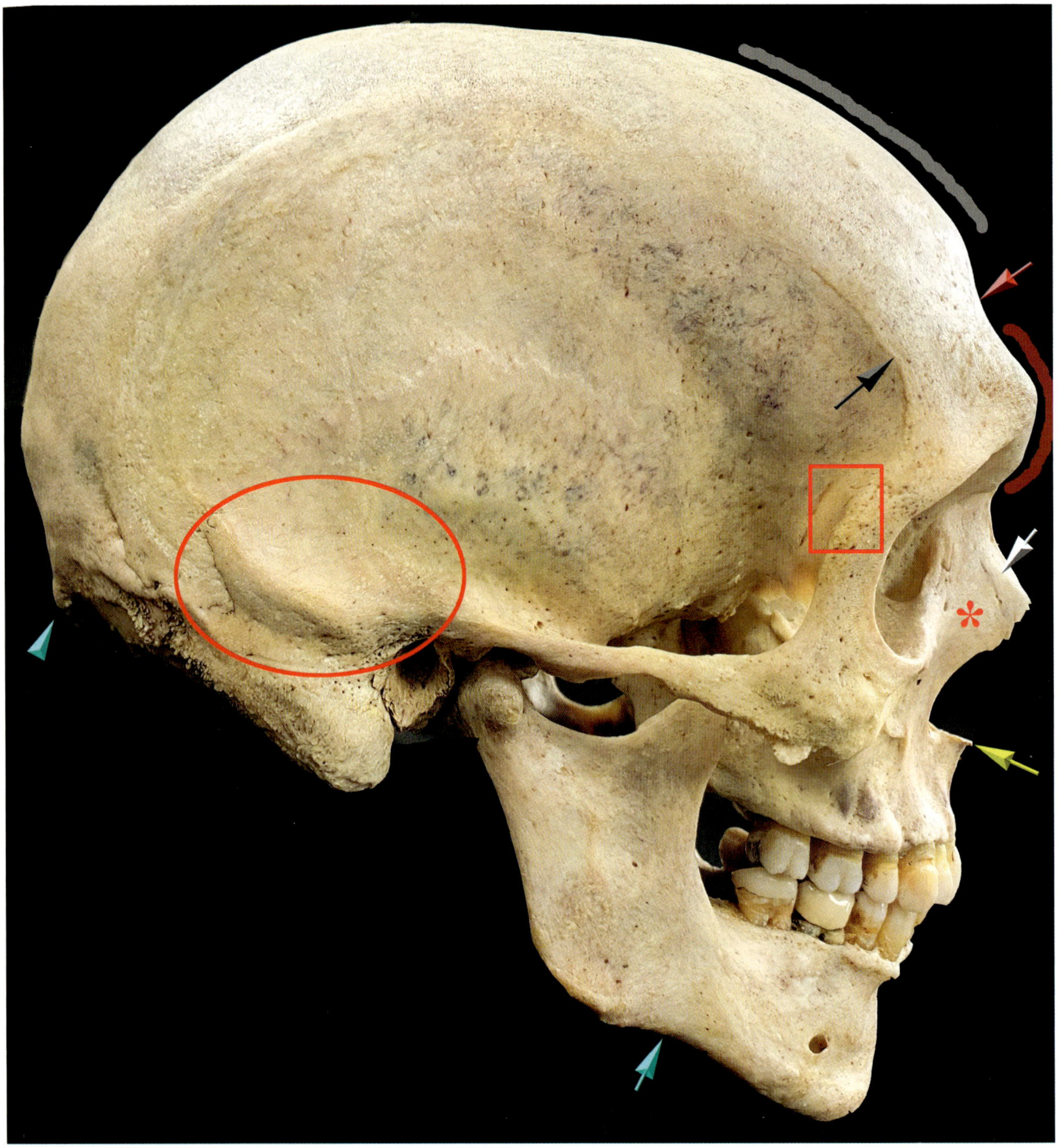

Figure 266. Right lateral view 75-year-old European male with a long (anterior-posterior), low cranial contour, a sloping frontal bone with trace frontal keeling (curved gray line), and small concavity (red arrow) superior to a prominent brow ridge (curved red line). Additionally, note the prominent and projecting nasal bones (white arrow), the anterior nasal spine (yellow arrow), prominent frontal process of the maxilla (*), and absence of alveolar prognathism and a posterior zygomatic tubercle (square). Observe the antegonial notch (blue arrow), the prominent supramastoid crest (oval), slight temporal line (black arrow), and the developed nuchal crest and inion hook (blue arrowhead).

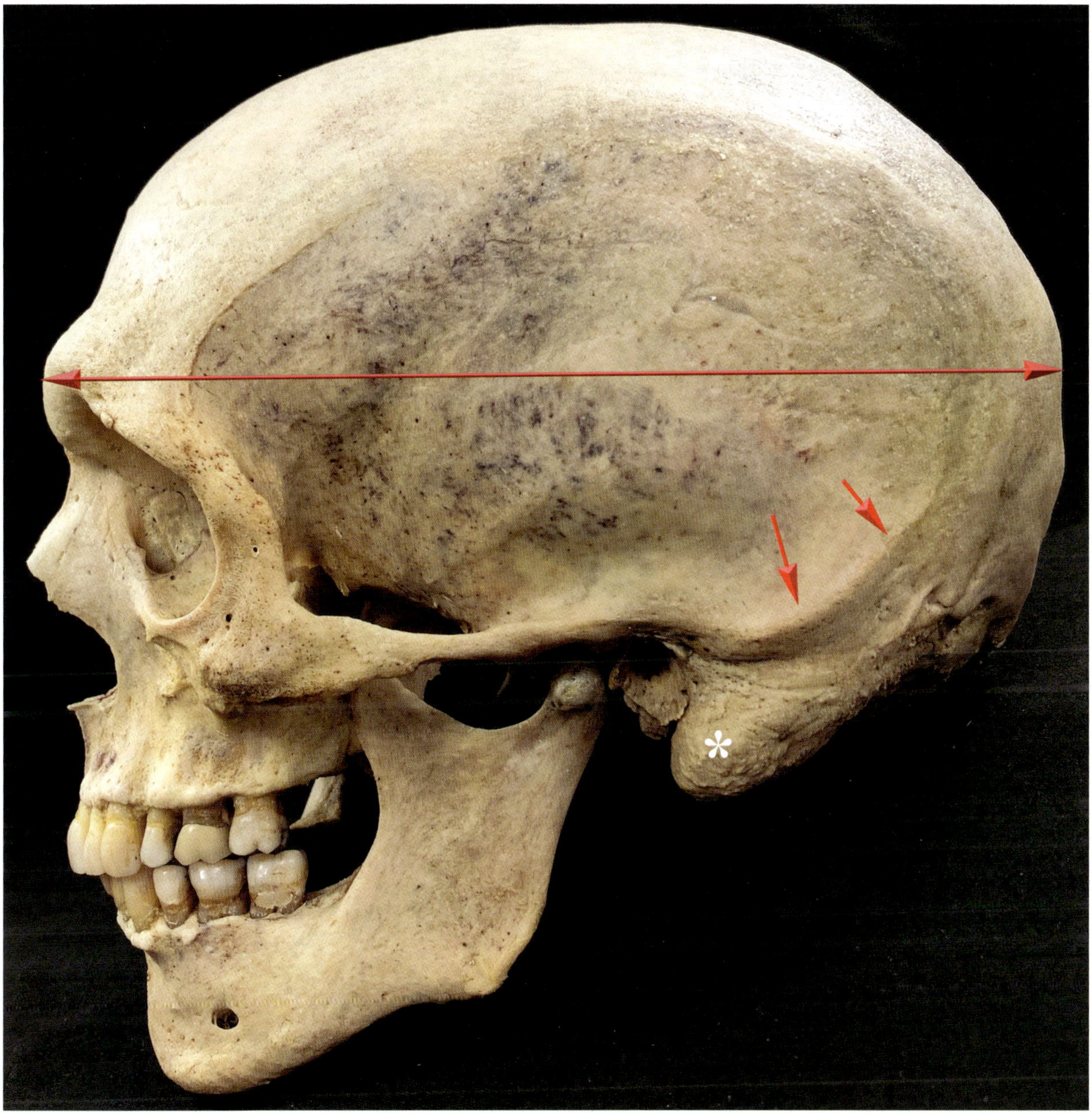

Figure 267. Left lateral view 75-year-old European male with an elongated vault (double arrow), a prominent supramastoid crest/zygomatic extension (arrow) and glabellar region, a large mastoid process (*), and a small/gracile zygoma and zygomatic arch.

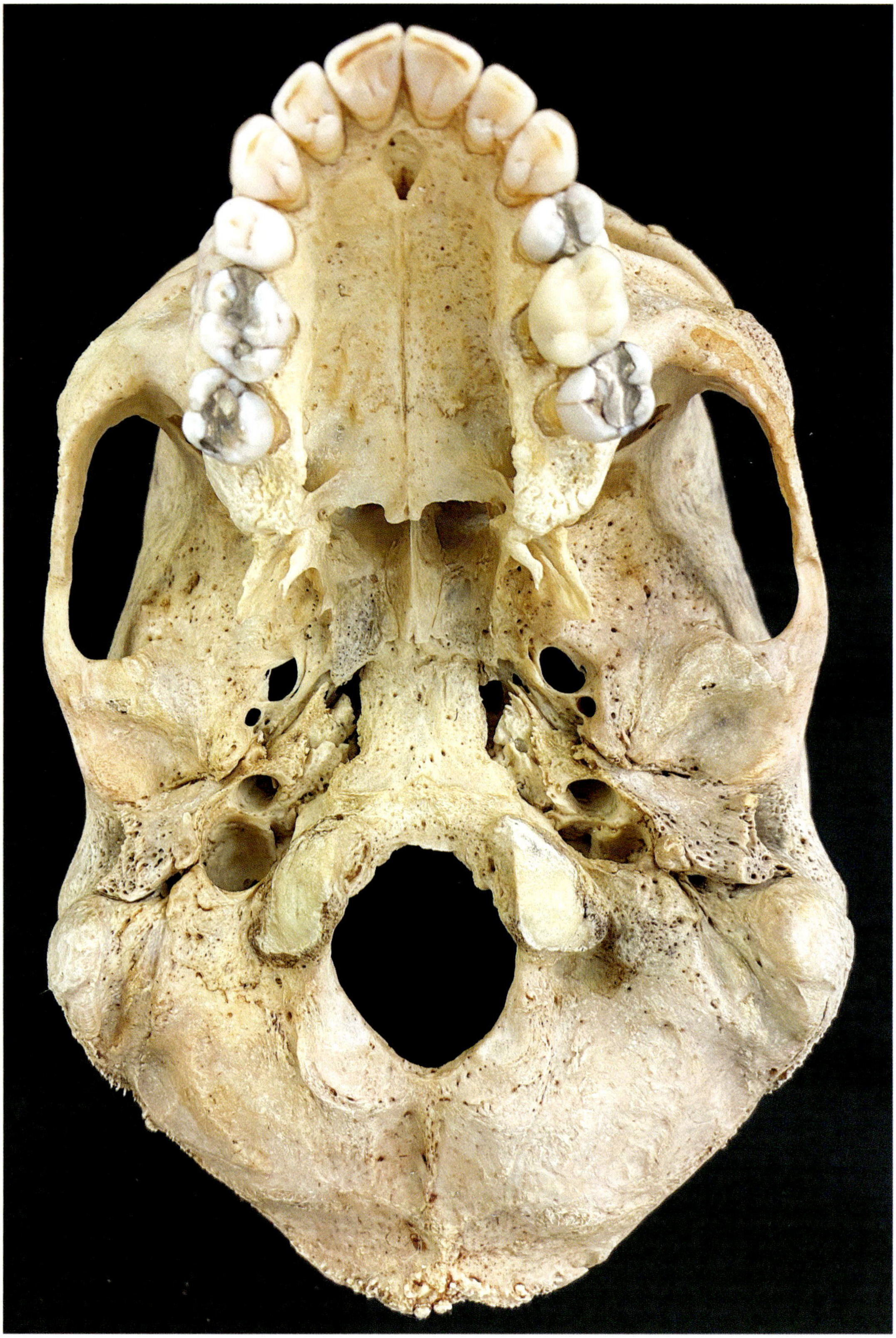

Figure 268. Inferior view 75-year-old European male showing narrow, deep and long palate, and mild shovel-shaping of some of the incisors. Note the triangular-shaped posterior cranial portion.

Figure 269. Superior view 75-year-old European male with a long and narrow cranium with an especially long frontal bone.

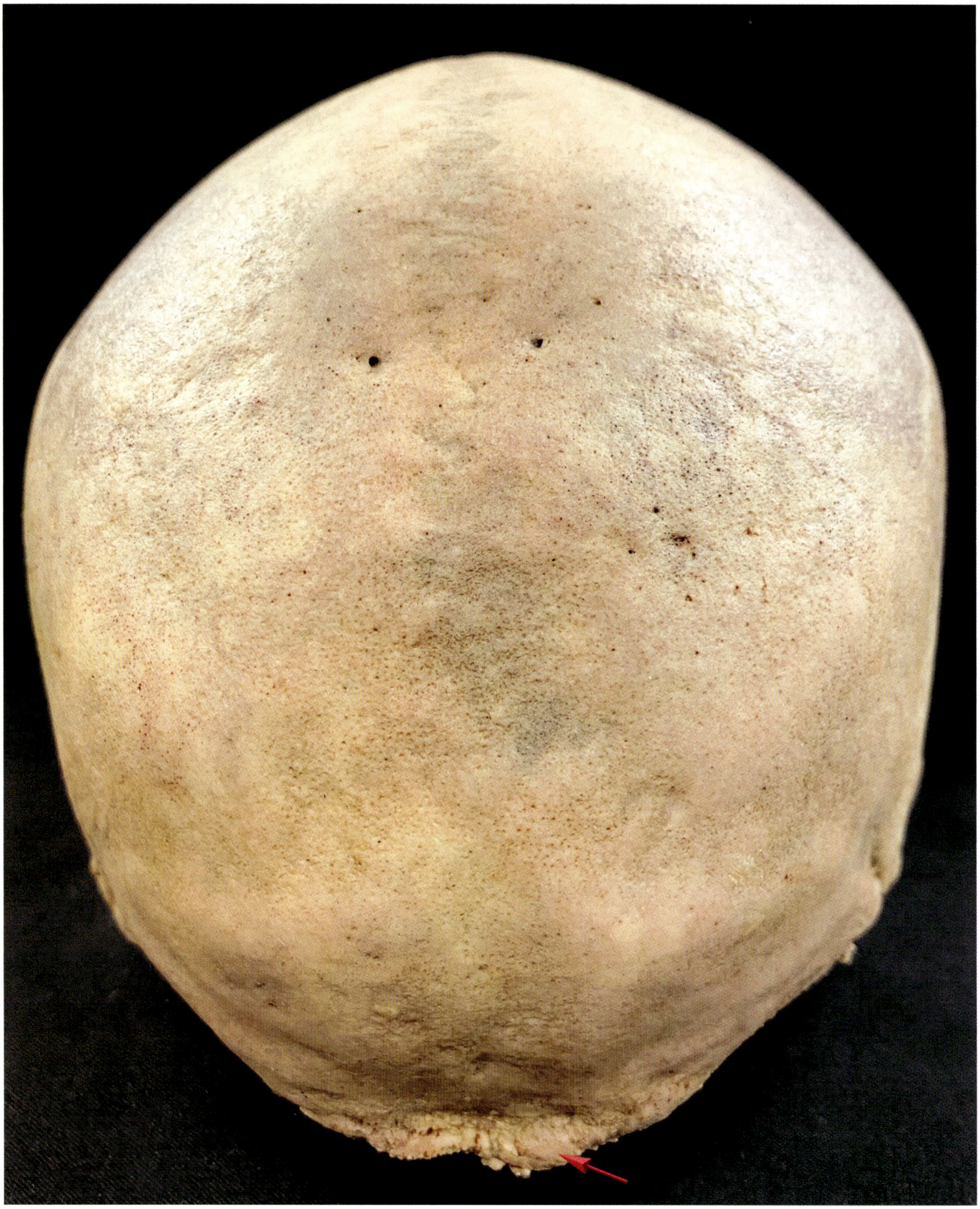

Figure 270. Posterior view 75-year-old European male with complete obliteration of the sutures, bilateral parietal foramina, and a developed nuchal crest (arrow).

5. EUROPEAN MALE

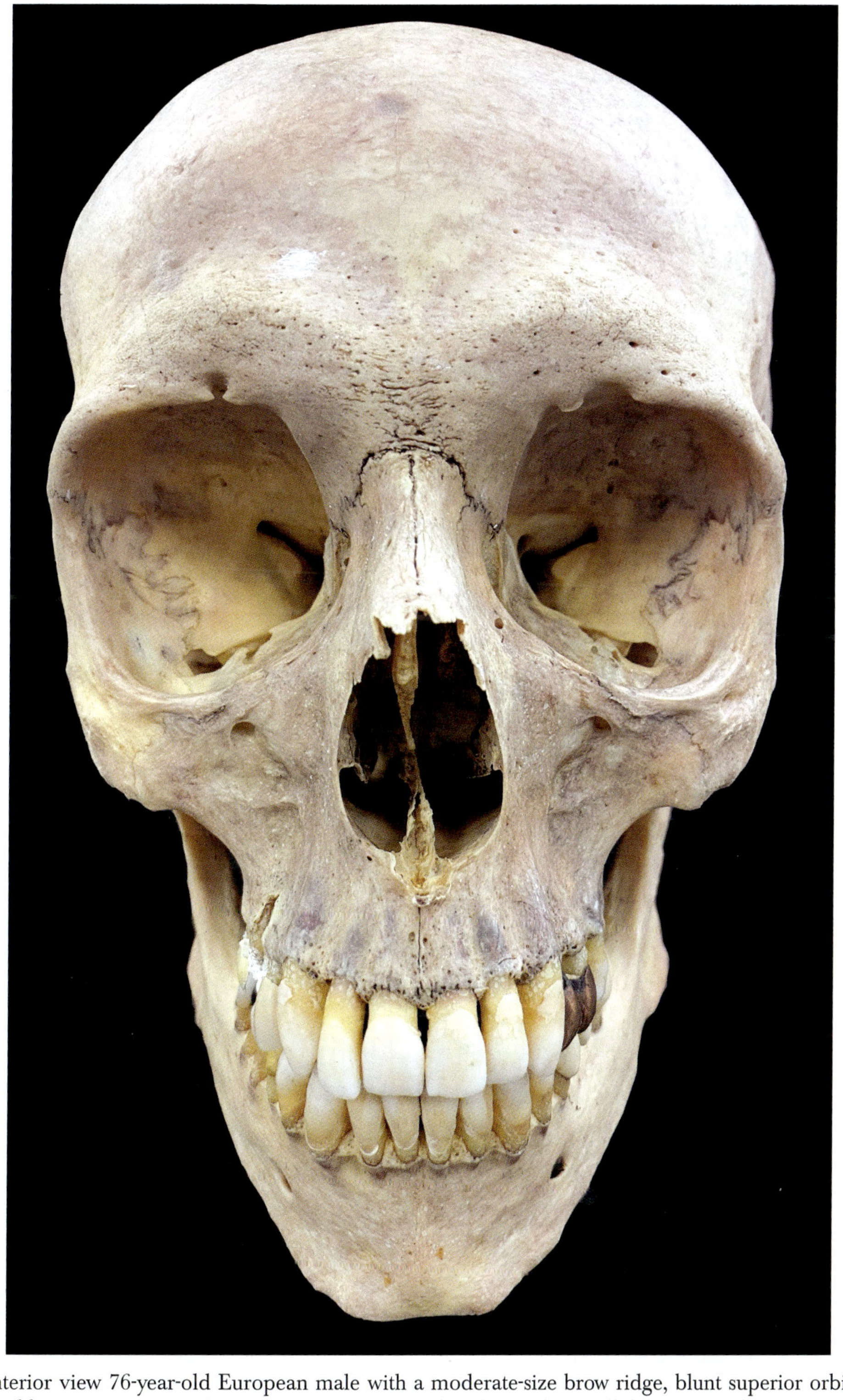

Figure 271. Anterior view 76-year-old European male with a moderate-size brow ridge, blunt superior orbital rims, large and straight nasal bones, narrow nasal aperture, small malar tubercles, and a small, but square chin.

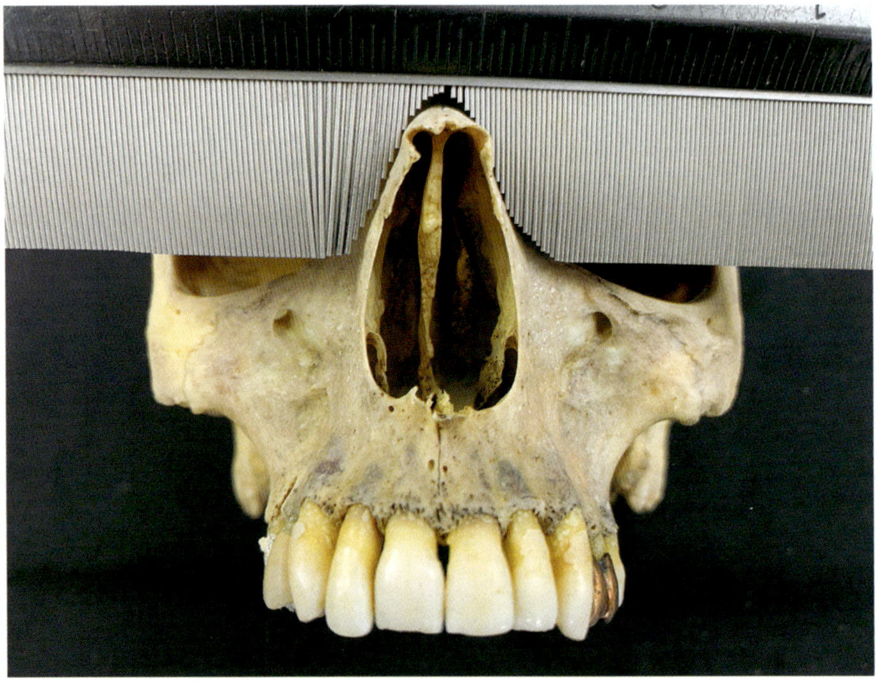

Figure 272. Nasal bone contour in a 76-year-old European male.

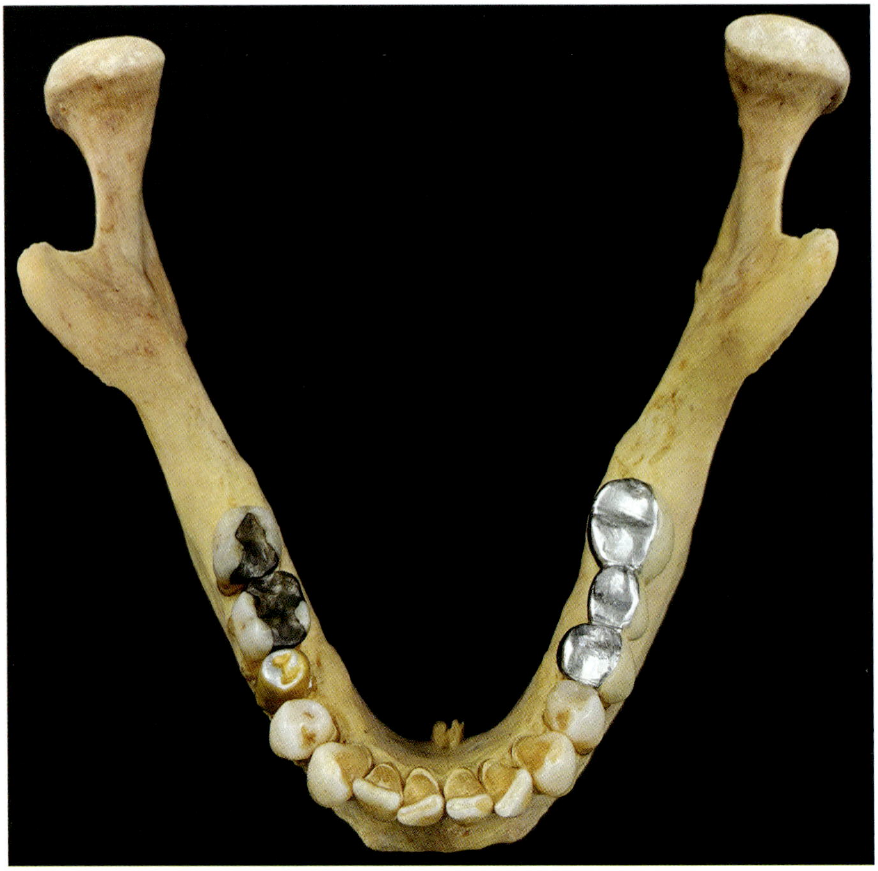

Figure 273. Mandible of 76-year-old European male.

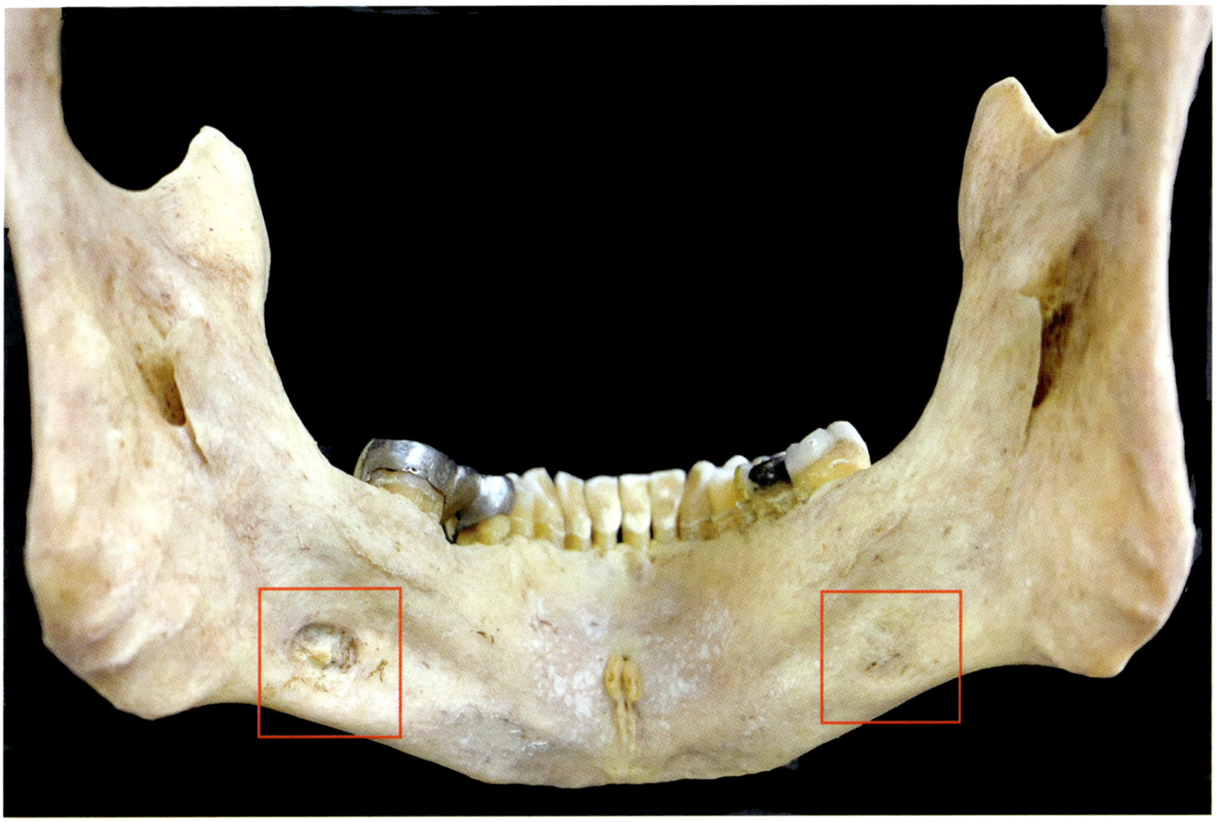

Figure 274. Lingual view of mandible of 76-year-old European male with bilateral and symmetrical Stafne's defects in the lingual plate (squares). These defects, also known as Stafne's idiopathic bone cavity and Stafne's static bone cavity, among others, result from contact with the submandibular salivary gland situated inferior to the molars and mandibular canal. The left and right defects in this male measure approximately 11 mm (anterior-posterior) by 7.5 mm (superior-inferior). Stafne's defects are normal anatomical variants found overwhelmingly in middle-age and older males (Mann 2001).

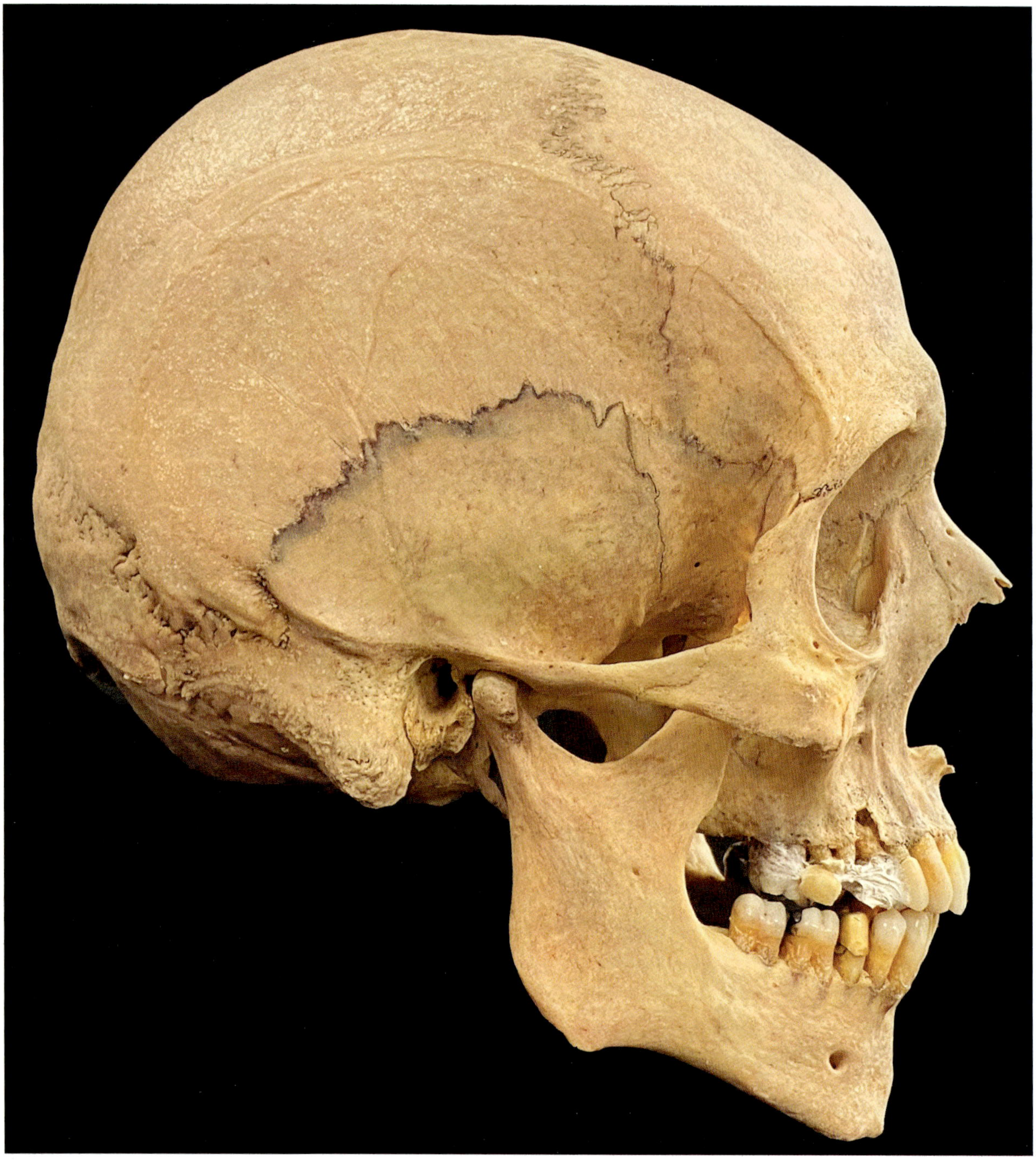

Figure 275. Right lateral view 76-year-old European male with a sloping frontal bone, brow ridge, overhanging nasal bones, small/gracile zygoma and zygomatic arch, and a developed anterior nasal spine. Note the small mastoid, prominent supramastoid crest, and antegonial notch.

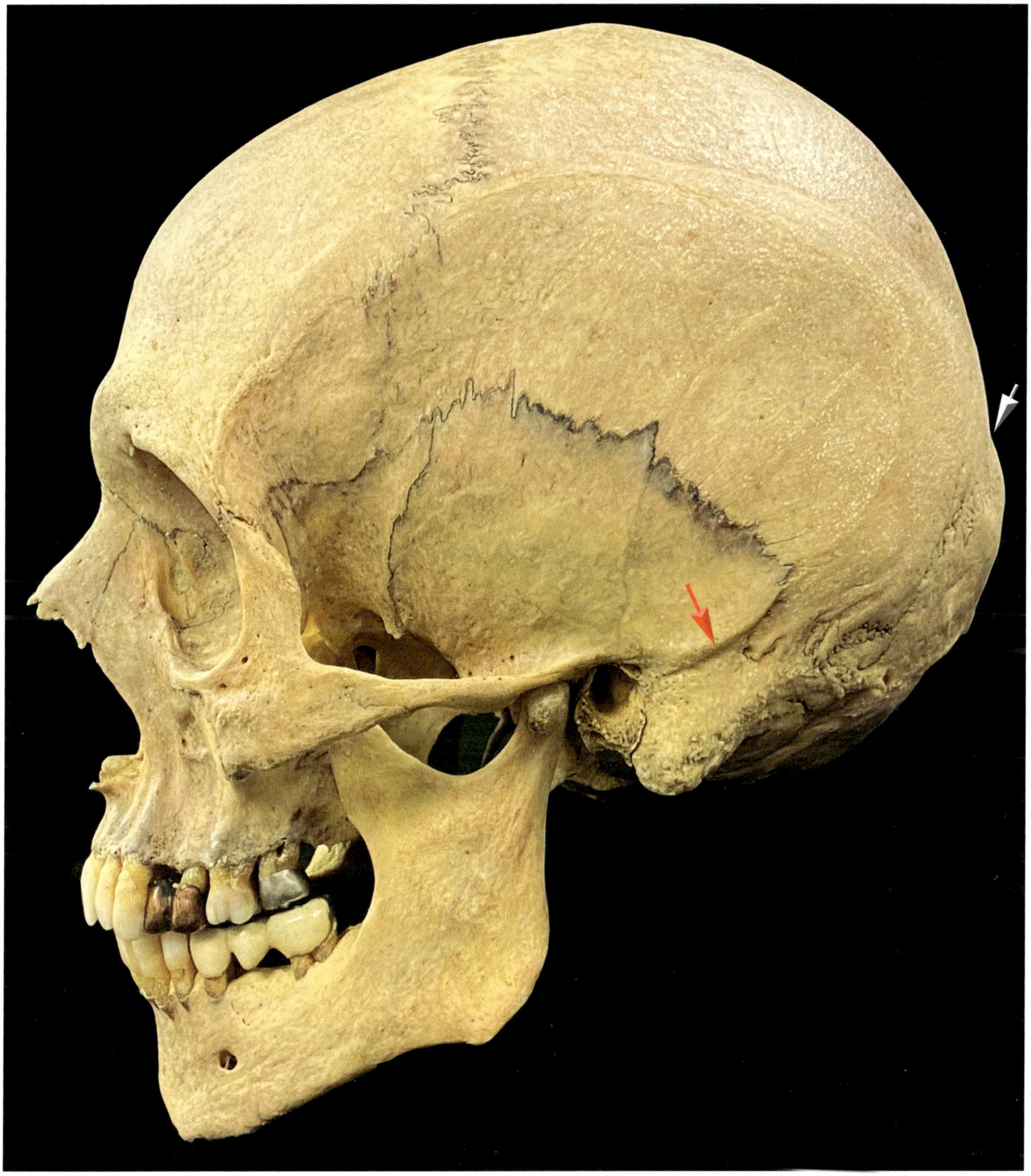

Figure 276. Left lateral view 76-year-old European male with a prominent supramastoid crest (red arrow) and indention (white arrow) superior to the lambdoidal suture resulting in protrusion of the occipital bone. This condition is sometimes referred to as bathrocephaly. The author has often seen bathrocepahly in ancient and modern European crania.

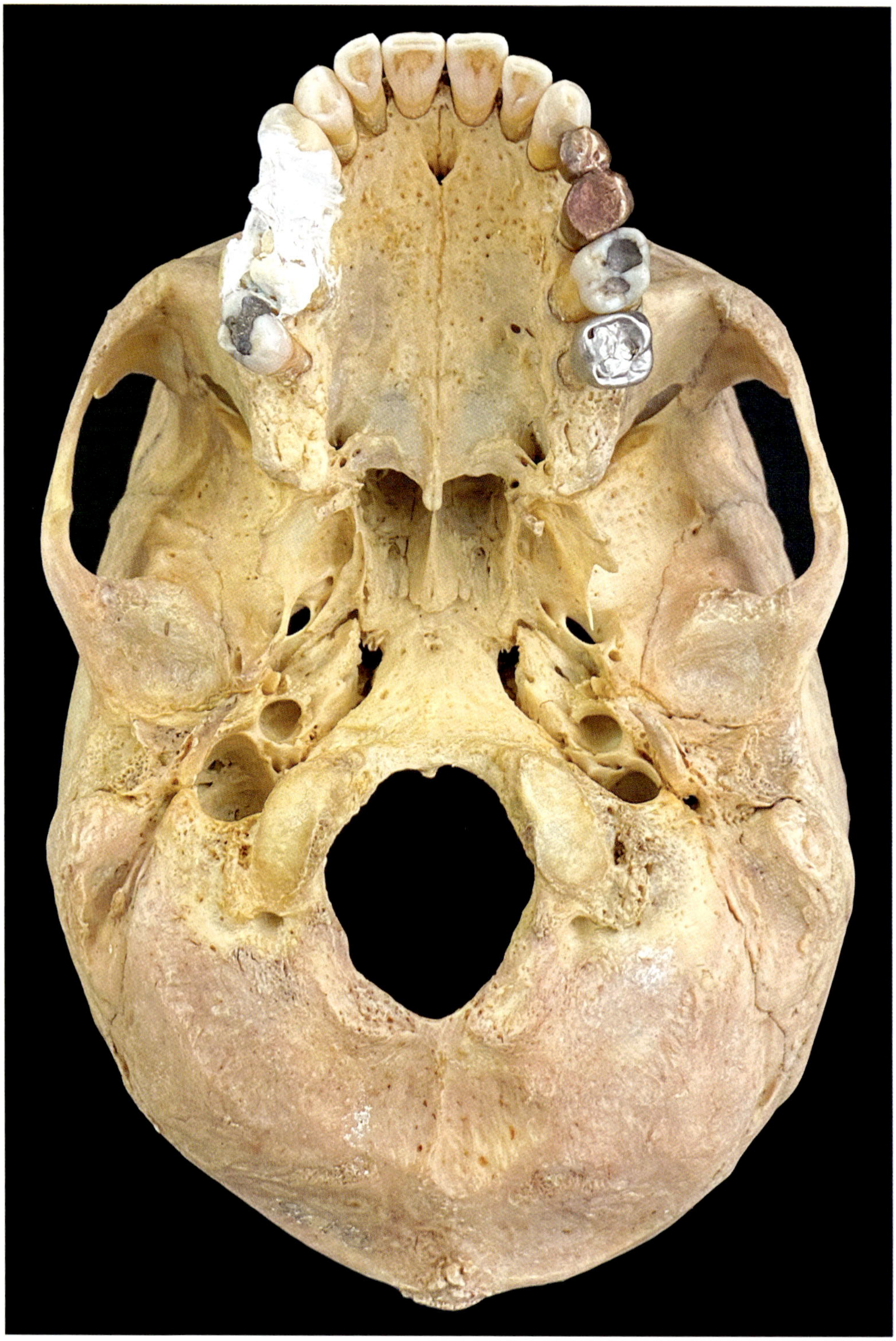

Figure 277. Inferior view 76-year-old European male with shovel-shaped incisors, rounded and shallow hard palate, and small mastoid processes that are more often present in Asians. Atypical combinations of skeletal traits and features are common among different ancestral groups and represent normal variation reflecting the combined effect of genetics and the environment. The whitish substance (hydrogel) surrounding the right teeth was used to promote healing of the tooth sockets while this individual was alive (Personal Communication Dr. J. DeMeo).

Figure 278. Superior view of 76-year-old European male with partial obliteration of the sutures.

Figure 279. Posterior view 76-year-old European male with a prominent inion hook.

6. EUROPEAN MALE

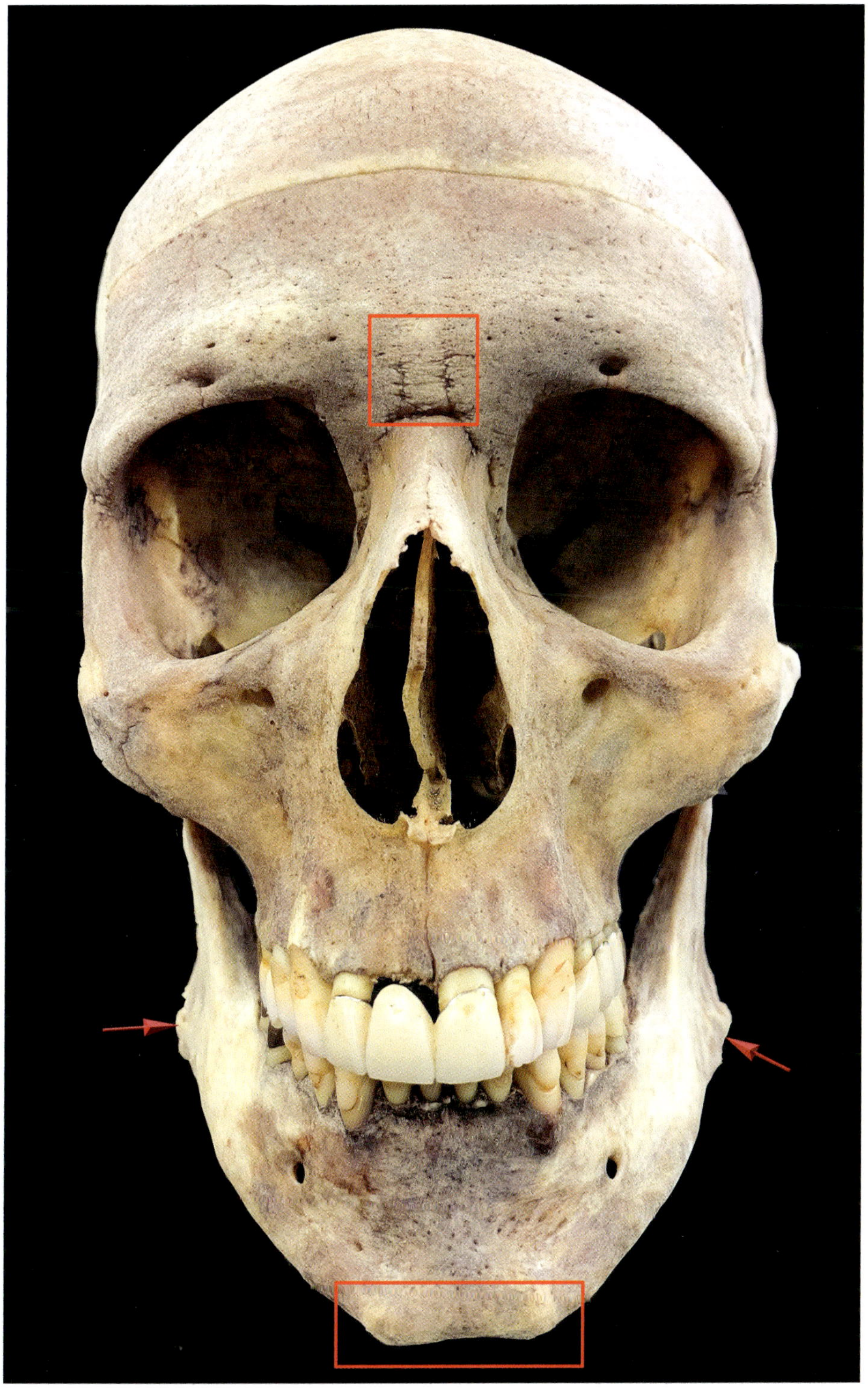

Figure 280. Anterior view 79-year-old European male with a trace of the supranasal suture (square), gonial eversion (arrow), and a mildly bilobate or square chin (rectangle).

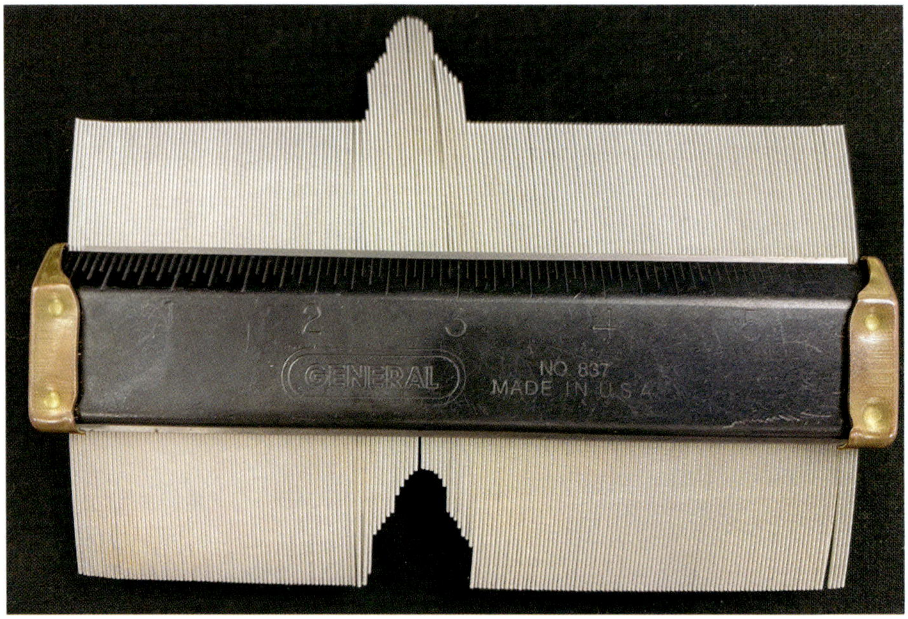

Figure 281. Nasal bone contour in a 79-year-old European male.

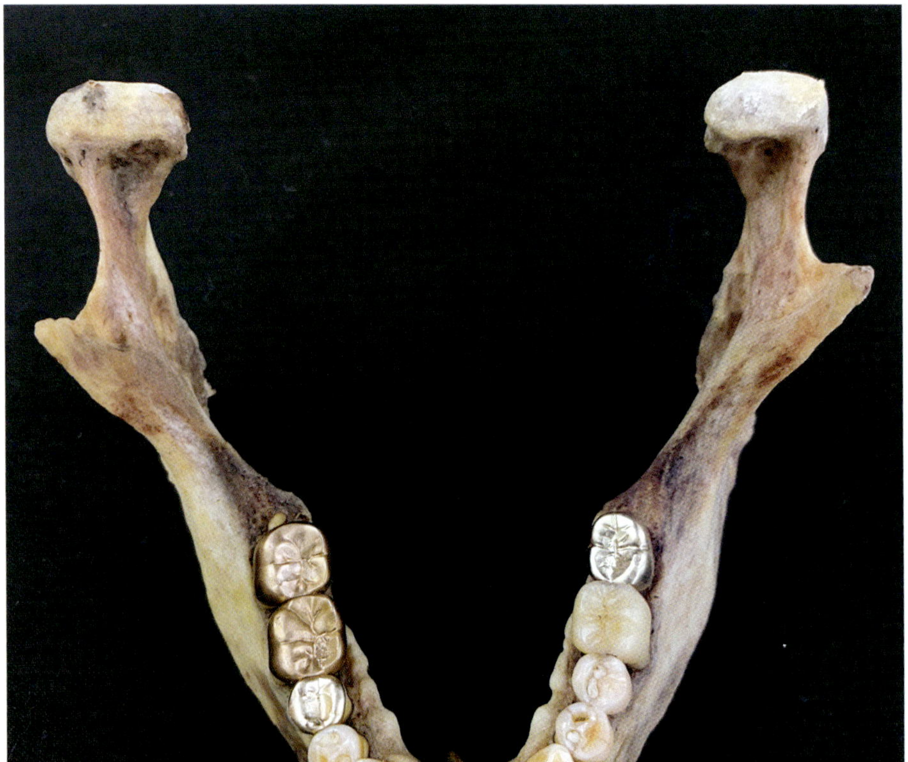

Figure 282. Mandible of 79-year-old European male with a square chin (rectangle).

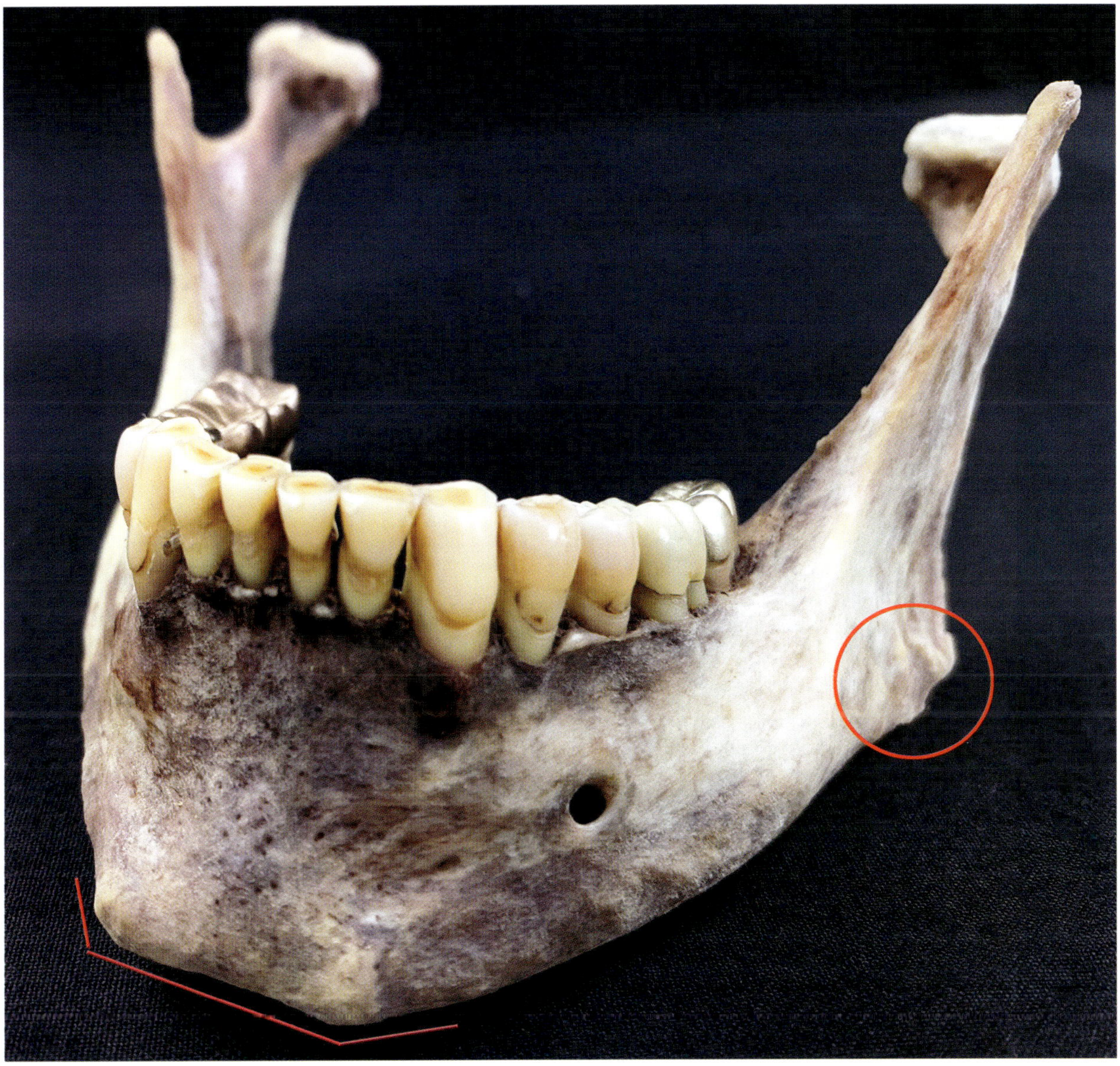

Figure 283. Left oblique view of mandible of 79-year-old European male showing a square chin (red line) and gonial eversion (circle).

TABLE 25.
Cranial Measurements in a 79-year-old European Male

Measurement	(mm)	Right	Left
Maximum Cranial Length (g-op, GOL)	181		
Maximum Cranial Breadth (eu-eu, XCB)	140		
Bizygomatic Breadth (zy-zy, ZYB)	122		
Basion-Bregma Height (ba-b, BBH)	130		
Cranial Base Length (ba-n, BNL)	96		
Basion-Prosthion (ba-pr, BPL)	93		
Maxillo-Alveolar Breadth (ecm-ecm, MAB)	–		
Maxillo-Alveolar Length (pr-alv, MAL)	54		
Biauricular Breadth (au-au, AUB)	117		
Upper Facial Height (n-pr)	72		
Minimum Frontal Breadth (ft-ft, WFB)	93		
Upper Facial Breadth (fmt-fmt)	100		
Nasal Height (n-ns, NLH)	54		
Nasal Breadth (al-al, NLB)	24		
Orbital Breadth (d-ec, OBB)	44		L
Orbital Height (OBH)	36		L
Biorbital Breadth (ec-ec, EKB)	94		
Interorbital Breadth (d-d, DKB)	19		
Frontal Chord (n-b, FRC)	109		
Parietal Chord (b-l, PAC)	111		
Occipital Chord (l-o, OCC)	98		
Foramen Magnum Length (ba-o, FOL)	38		
Foramen Magnum Breadth (FOB)	33		
Mastoid Length (MDH)			23

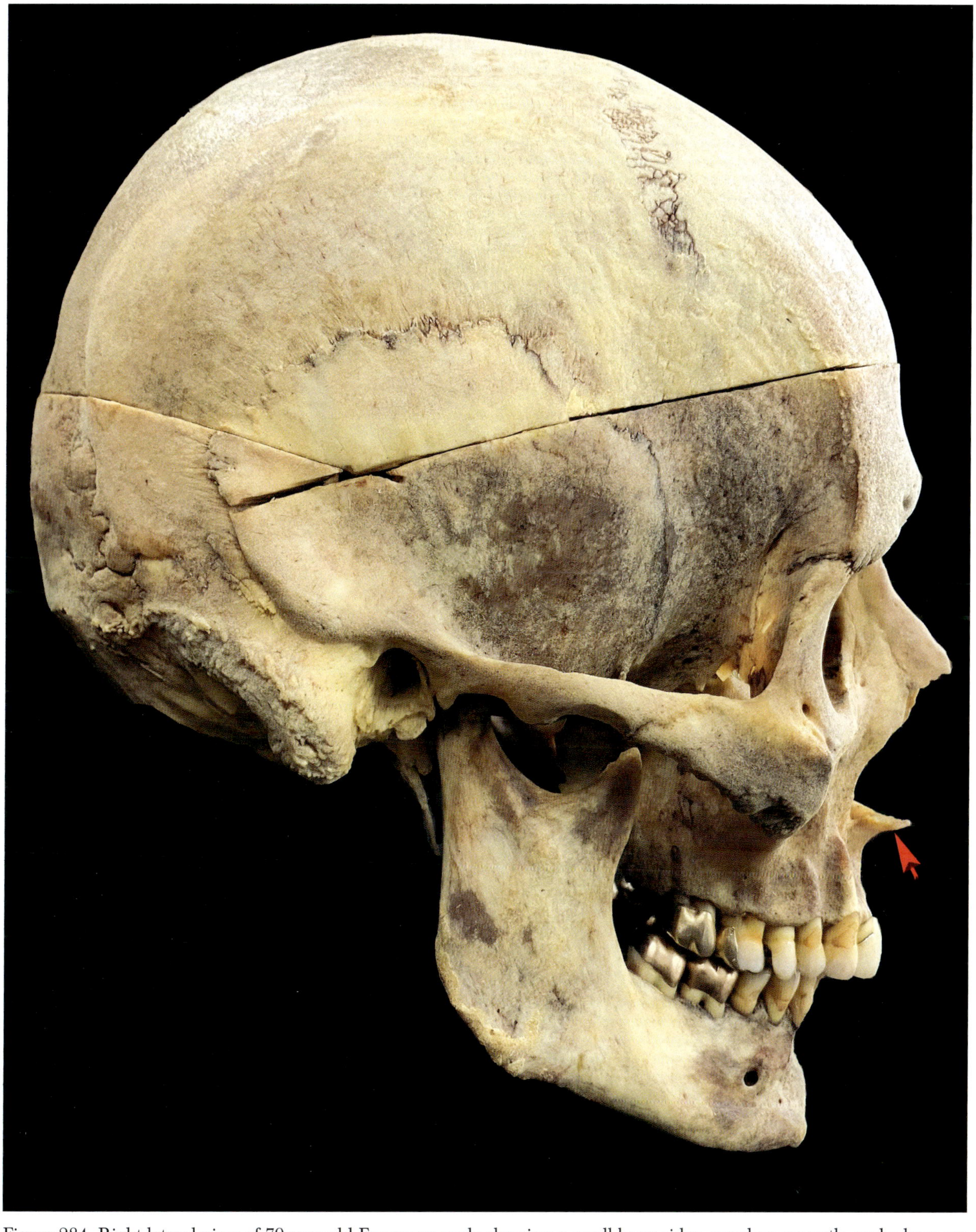

Figure 284. Right lateral view of 79-year-old European male showing a small brow ridge, nasal overgrowth, and a long anterior nasal spine (7.5 mm in length [arrow]). Note the well-developed supramastoid and nuchal crests and absence of alveolar prognathism resulting in a straight facial profile.

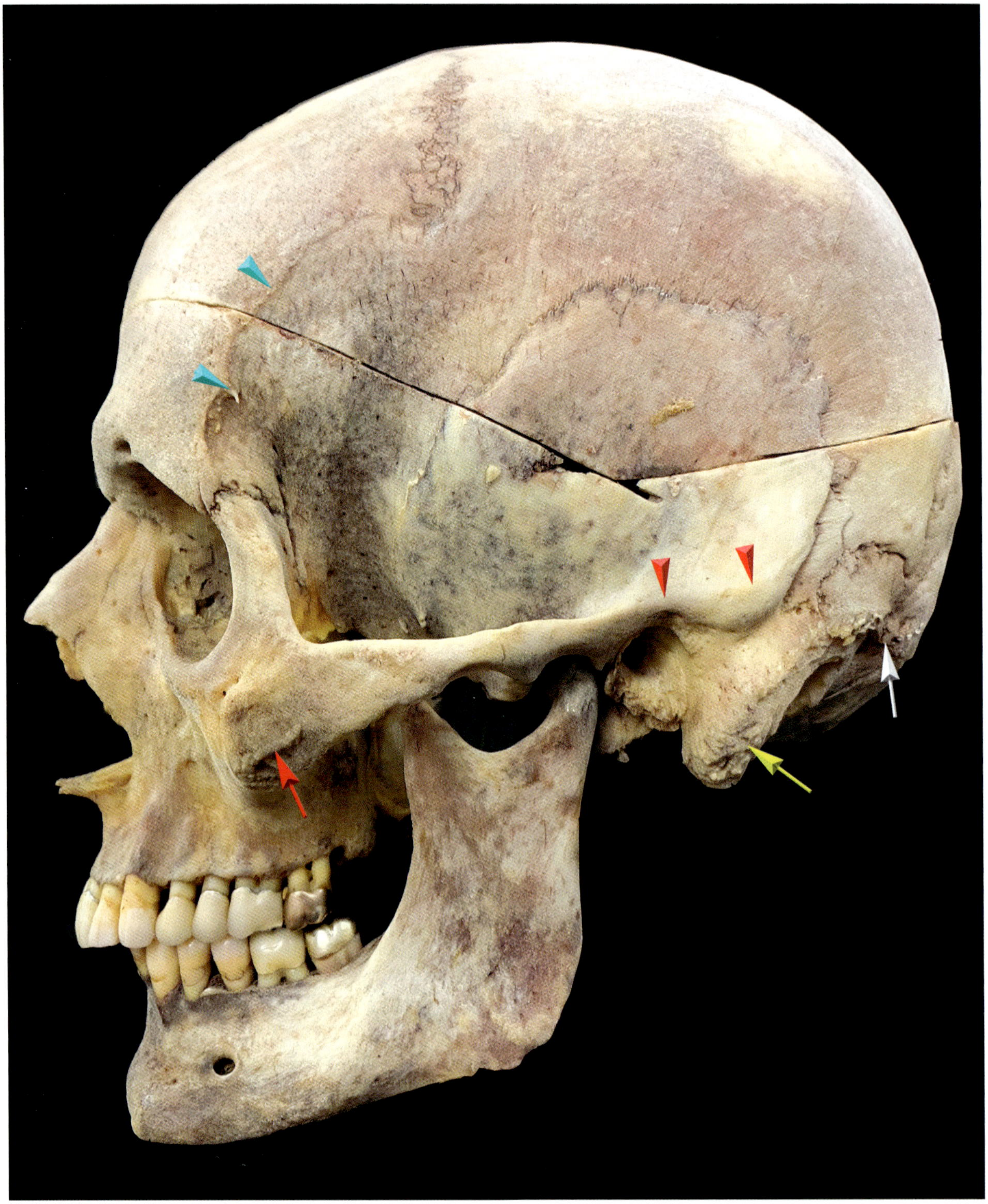

Figure 285. Left lateral view of 79-year-old European male with nasion depression, a sharp, curved projection along the frontal portion of the temporal line (blue arrowhead), and a bony ridge on the malar (red arrow) for attachment of the masseter muscle. Also note the developed supramastoid crest (red arrowhead), the mastoid process (yellow arrow) and a developed nuchal crest (white arrow).

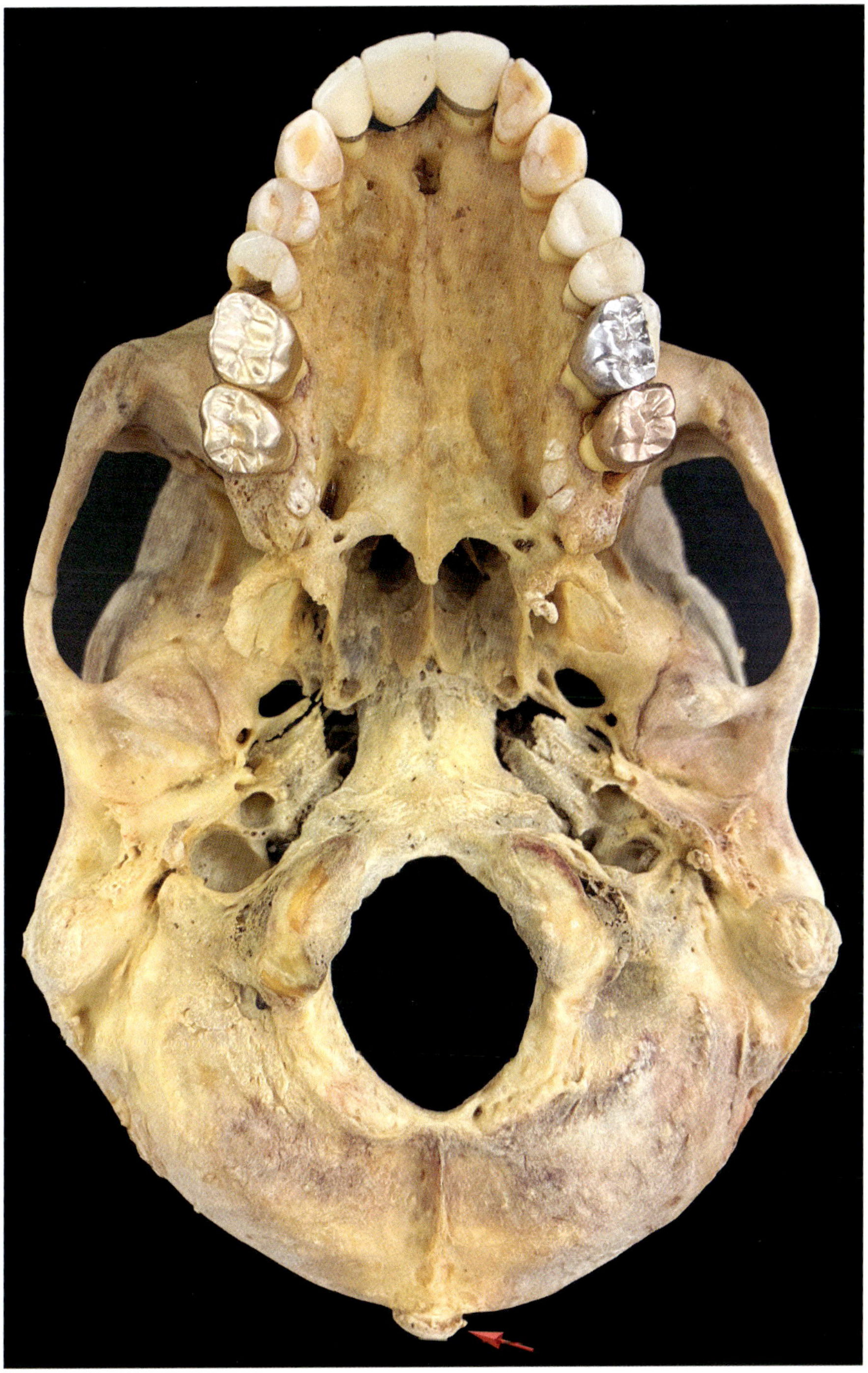

Figure 286. Inferior view of 79-year-old European male showing the size and shape of the cranium, the palate, and the presence of an inion hook (arrow).

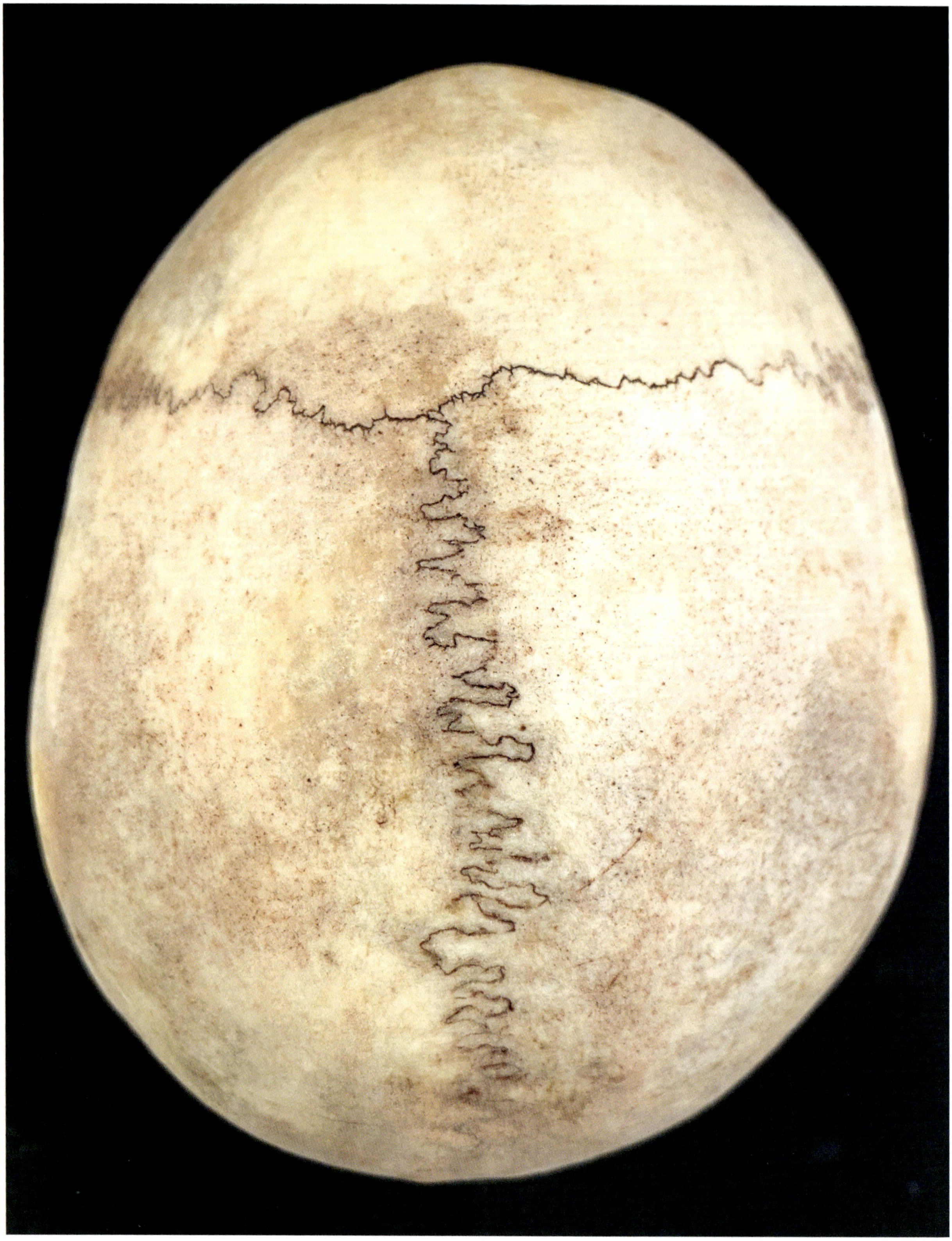

Figure 287. Superior view of 79-year-old European male with open sutures.

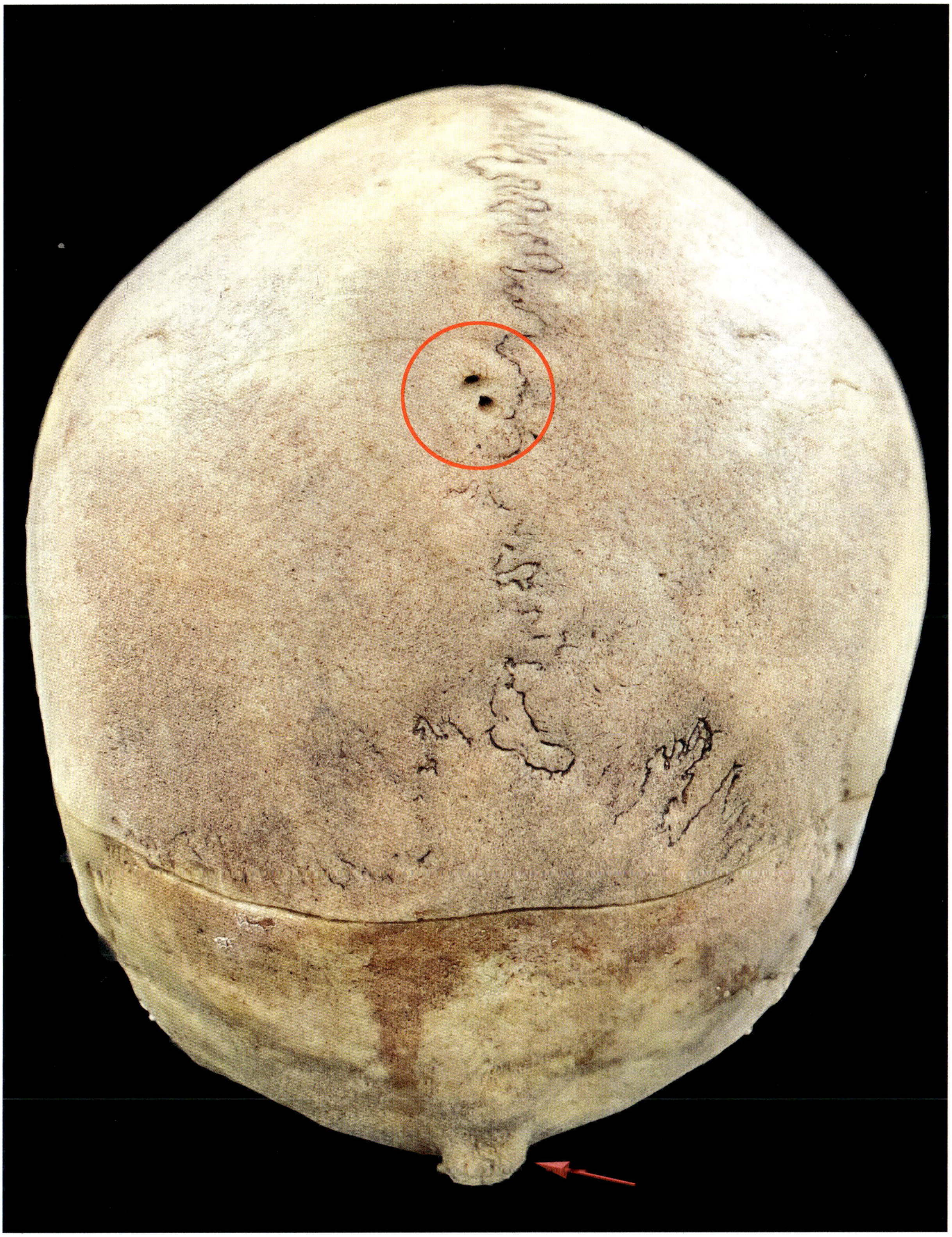

Figure 288. Posterior view of 79-year-old European male with two adjacent left parietal foramina (circle) and an inion hook (arrow).

7. EUROPEAN MALE

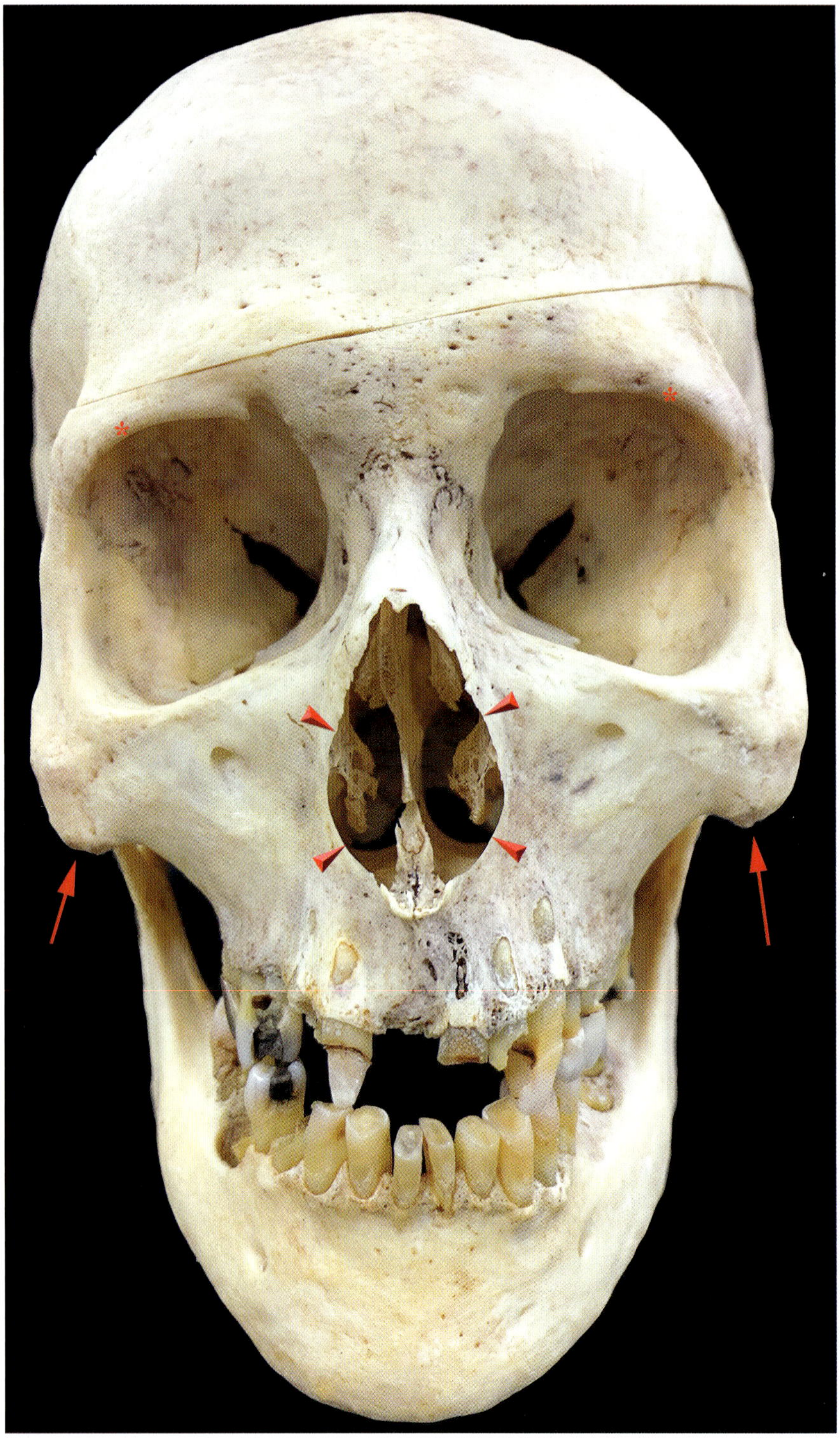

Figure 289. Anterior view 96-year-old European male with blunt superior orbital margins (*), and downward sloping (angular) inferior orbital margins. Note that the nasal aperture is narrow and tall and that its outline is sharp and continuous (arrowhead). These features are typical of European ancestry (see Figure 290). Note, also the malar tubercles (arrow), and the rounded chin.

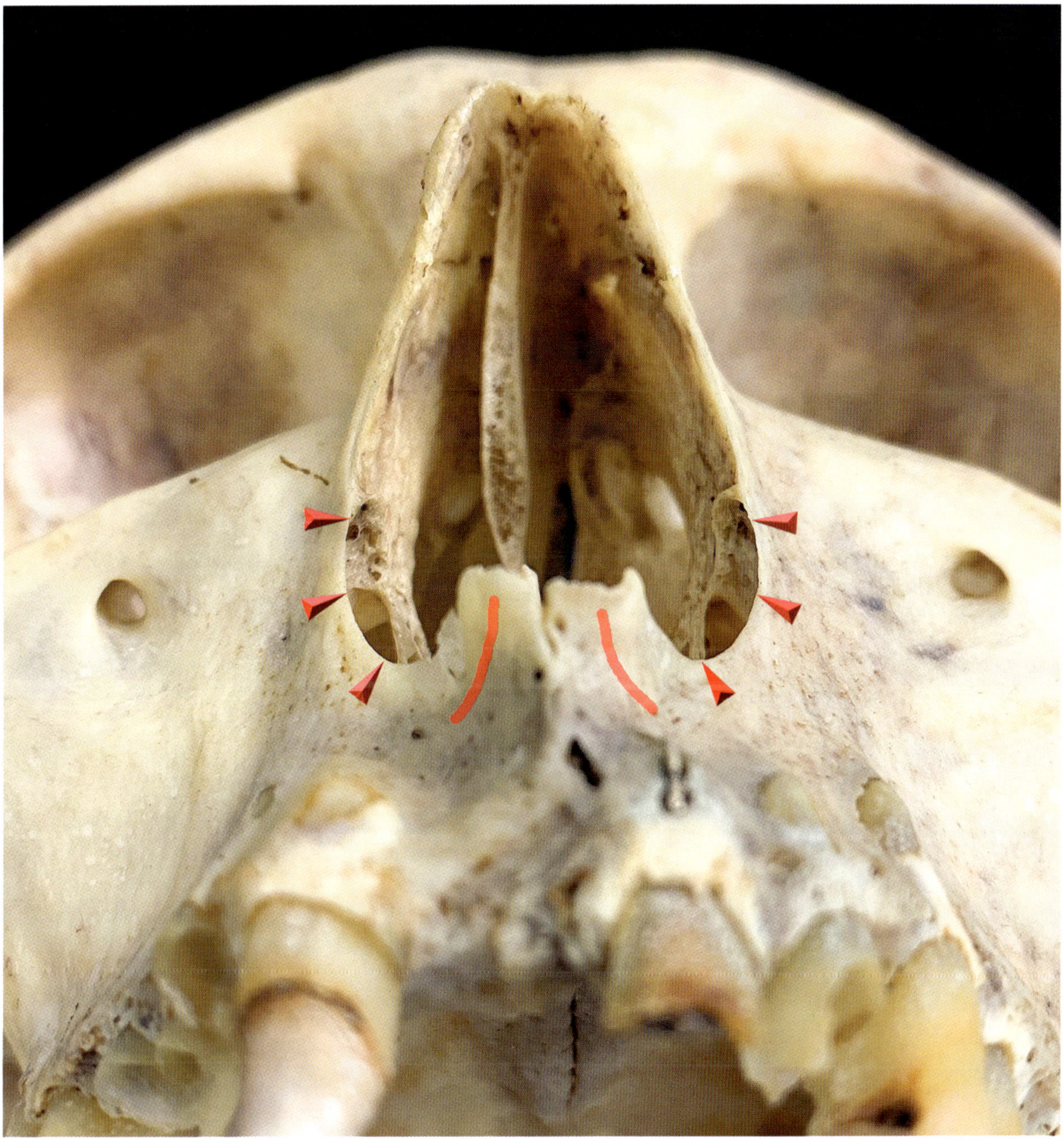

Figure 290. Inferior-superior view of 96-year-old European male showing a sharp and continuous inferior nasal rim/sill (arrowhead) and an anteriorly projecting (curved red line) anterior nasal spine.

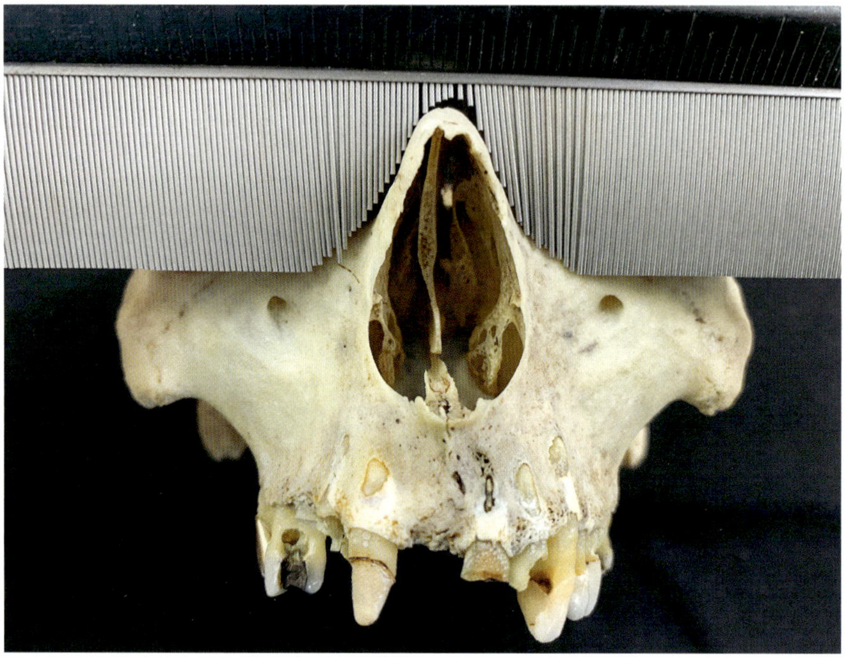

Figure 291. Nasal bone contour in a 96-year-old European male.

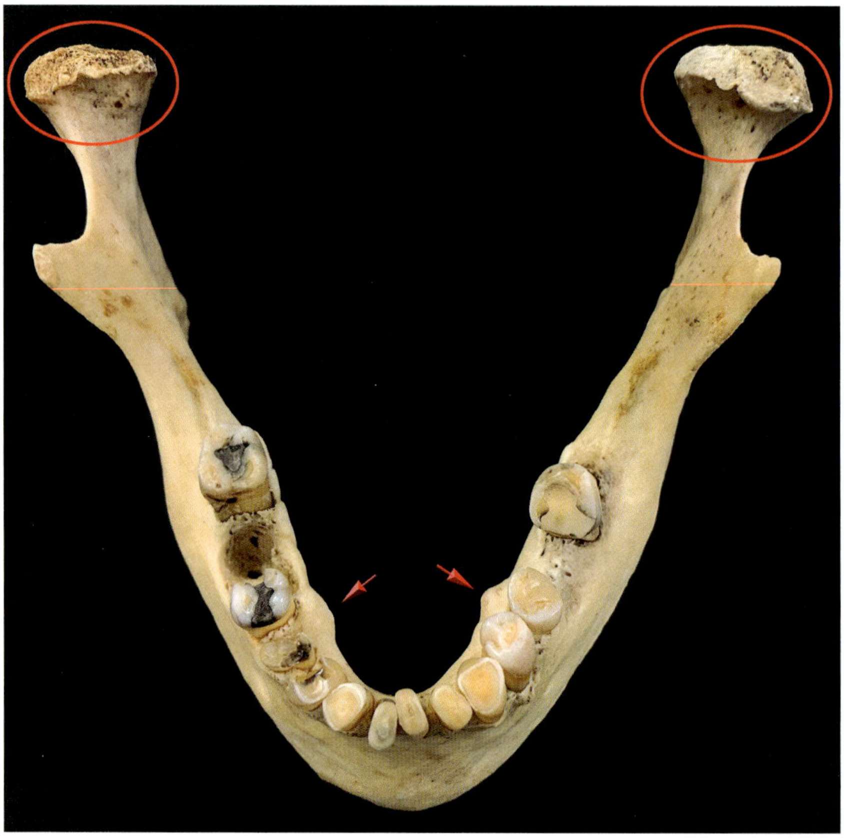

Figure 292. Mandible of 96-year-old European male with degenerative osteoarthritis of the condyles (oval), small lingual exostoses (arrow) and severe occlusal wear of the teeth.

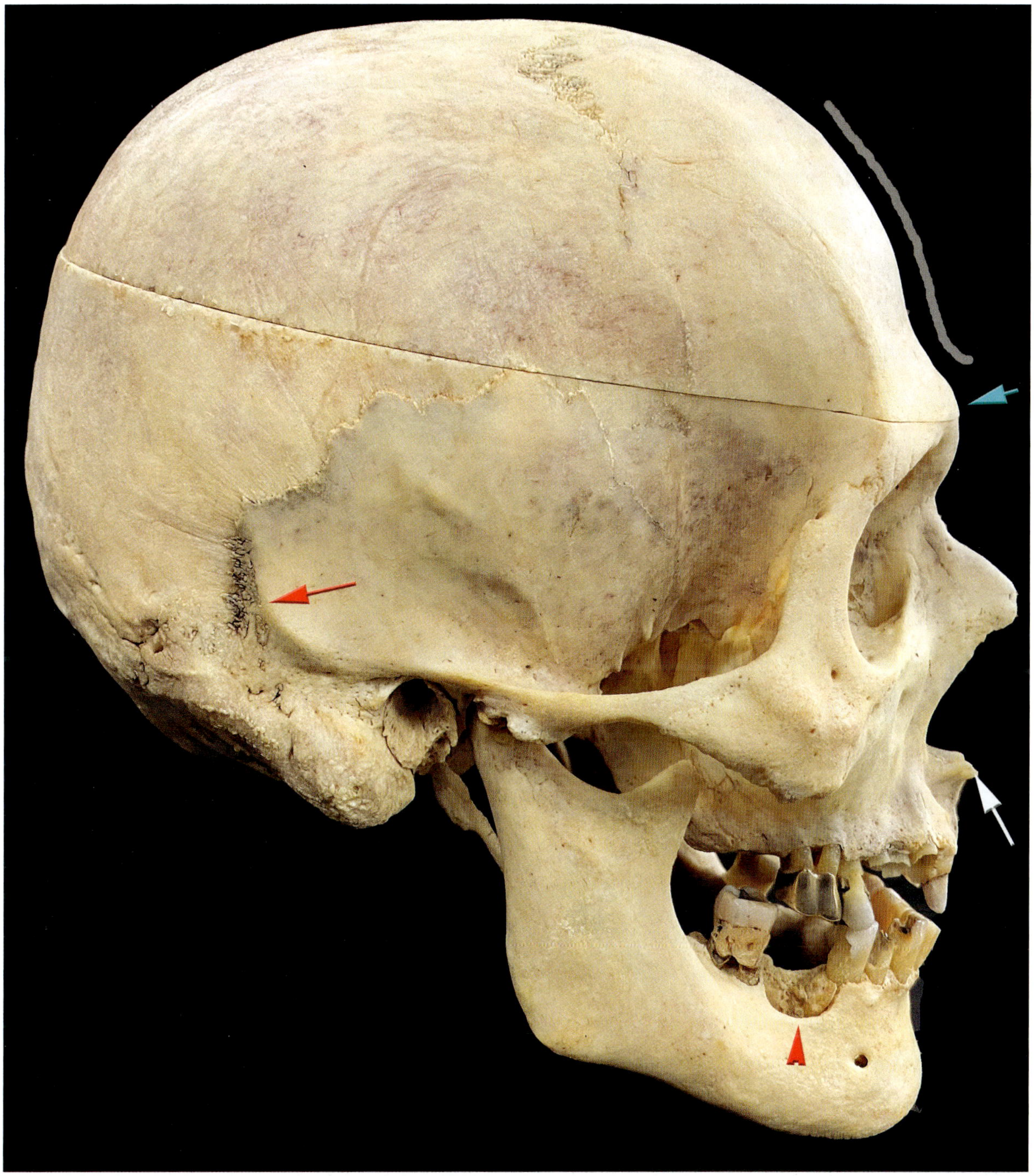

Figure 293. Right lateral view 96-year-old European male with a sloping frontal bone (silver line) and a pronounced brow ridge (blue arrow). Note the anterior nasal spine (white arrow), and a pronounced supramastoid crest (red arrow). The right hemi-mandible has a periapical lesion (arrowhead) that has eroded the mandible.

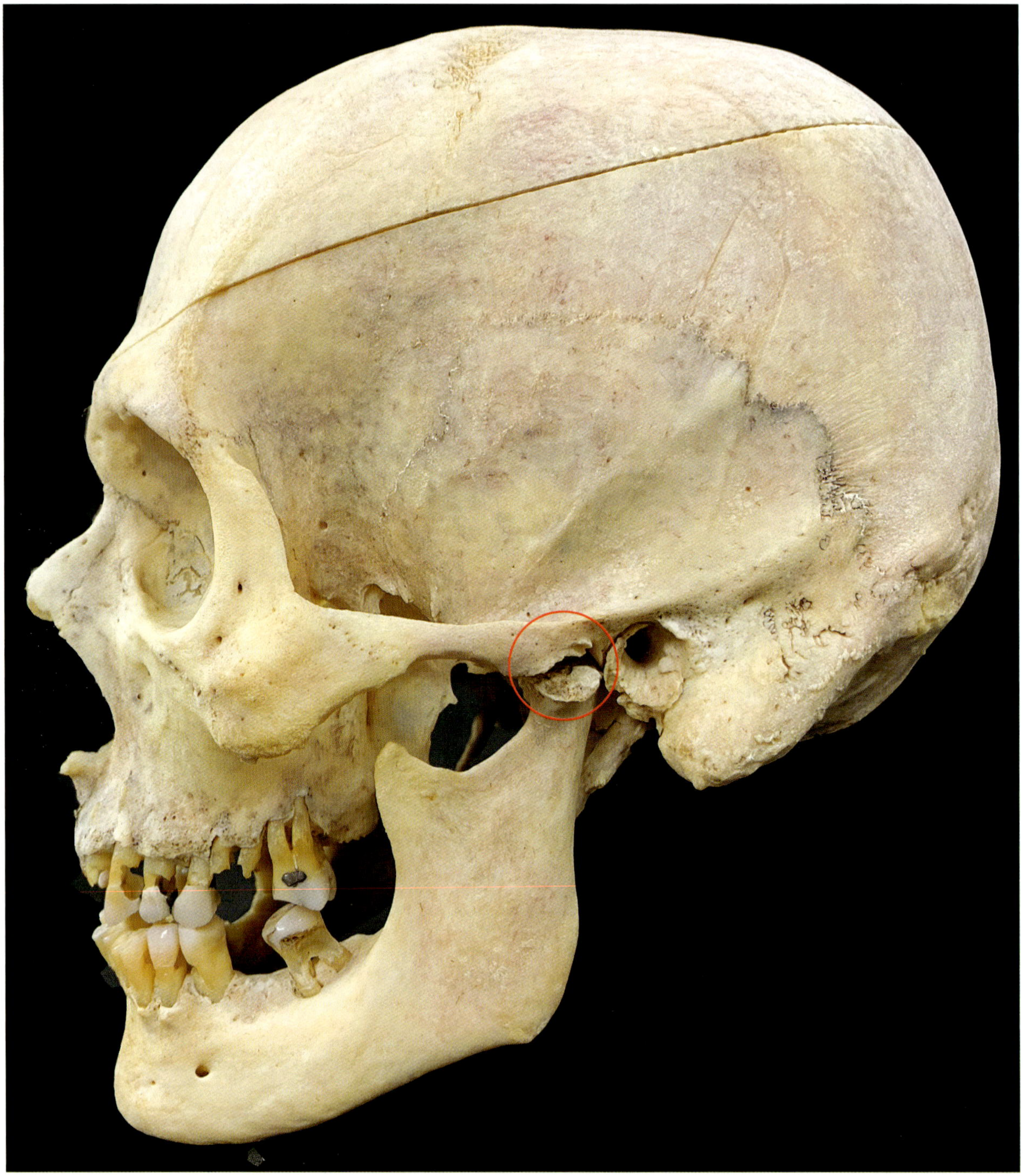

Figure 294. Left lateral view 96-year-old European male with a sloping frontal bone, degenerative joint disease of the left temporomandibular joint (circle), and several carious teeth.

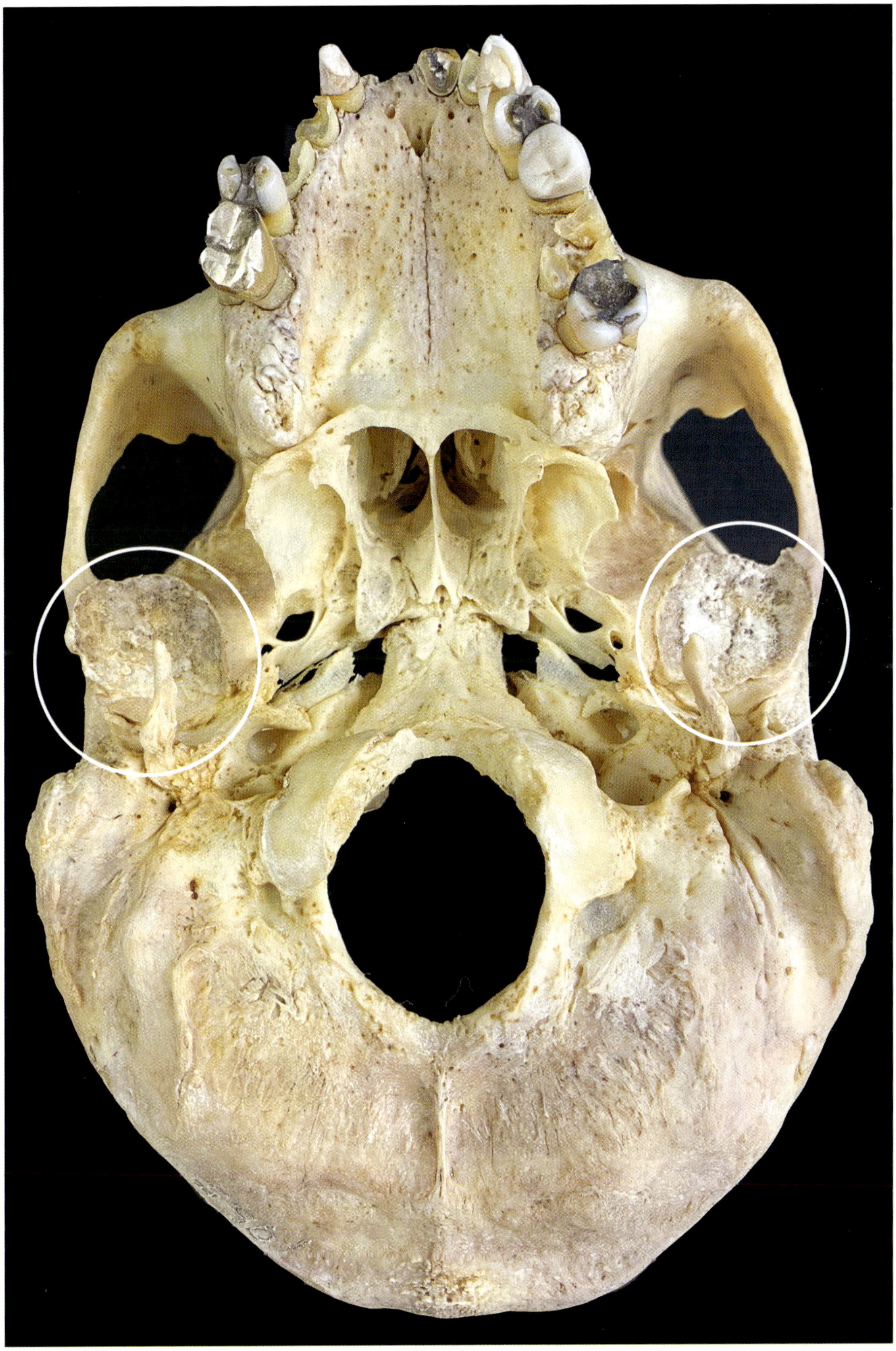

Figure 295. Inferior view of 96-year-old European male with temporomandibular joint disease (circle).

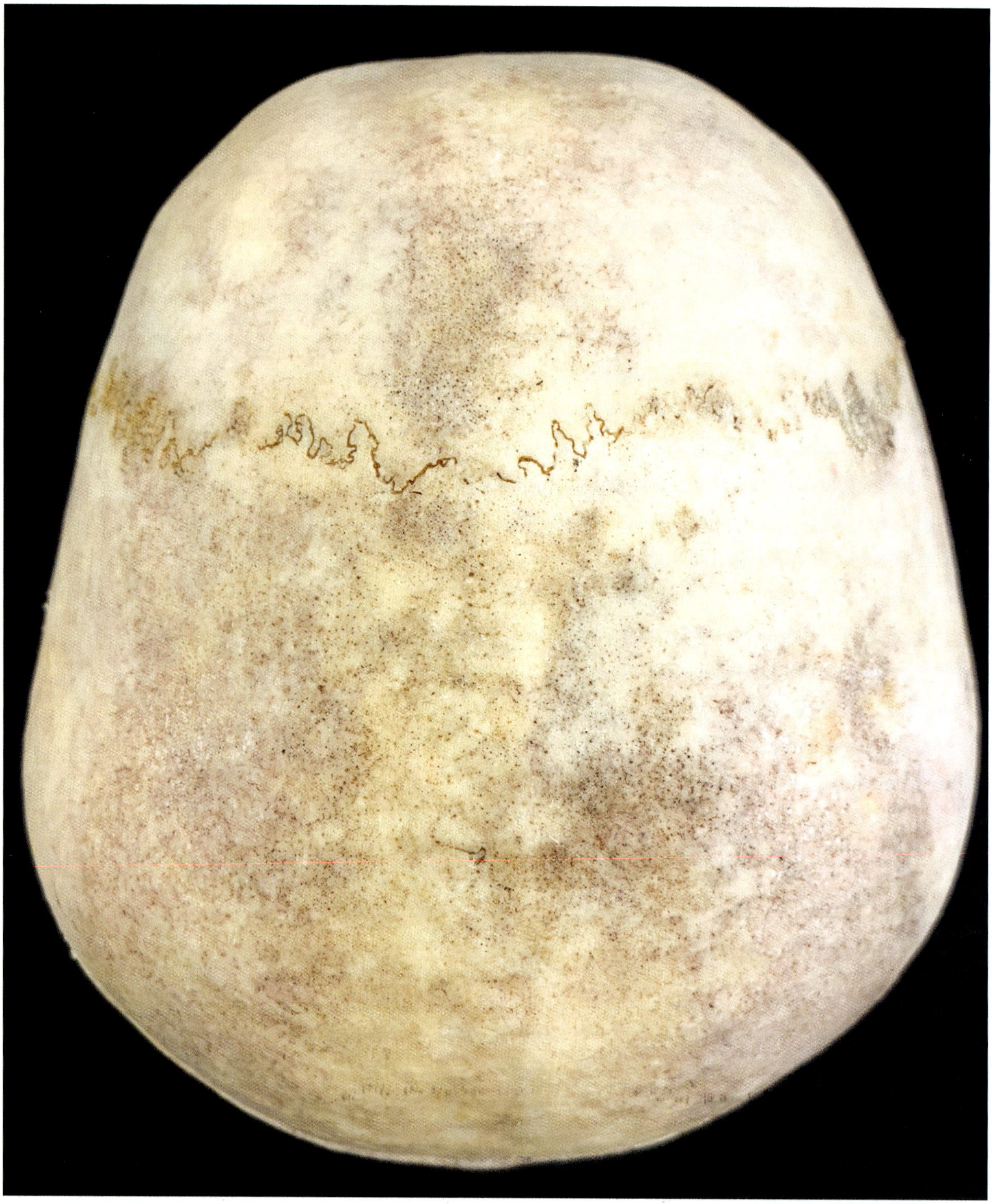

Figure 296. Superior view 96-year-old European male showing complete obliteration of the sagittal suture and partial obliteration of the coronal suture.

Figure 297. Posterior view of 96-year-old European male with complete obliteration of the sagittal and lambdoidal sutures.

8. EUROPEAN MALE

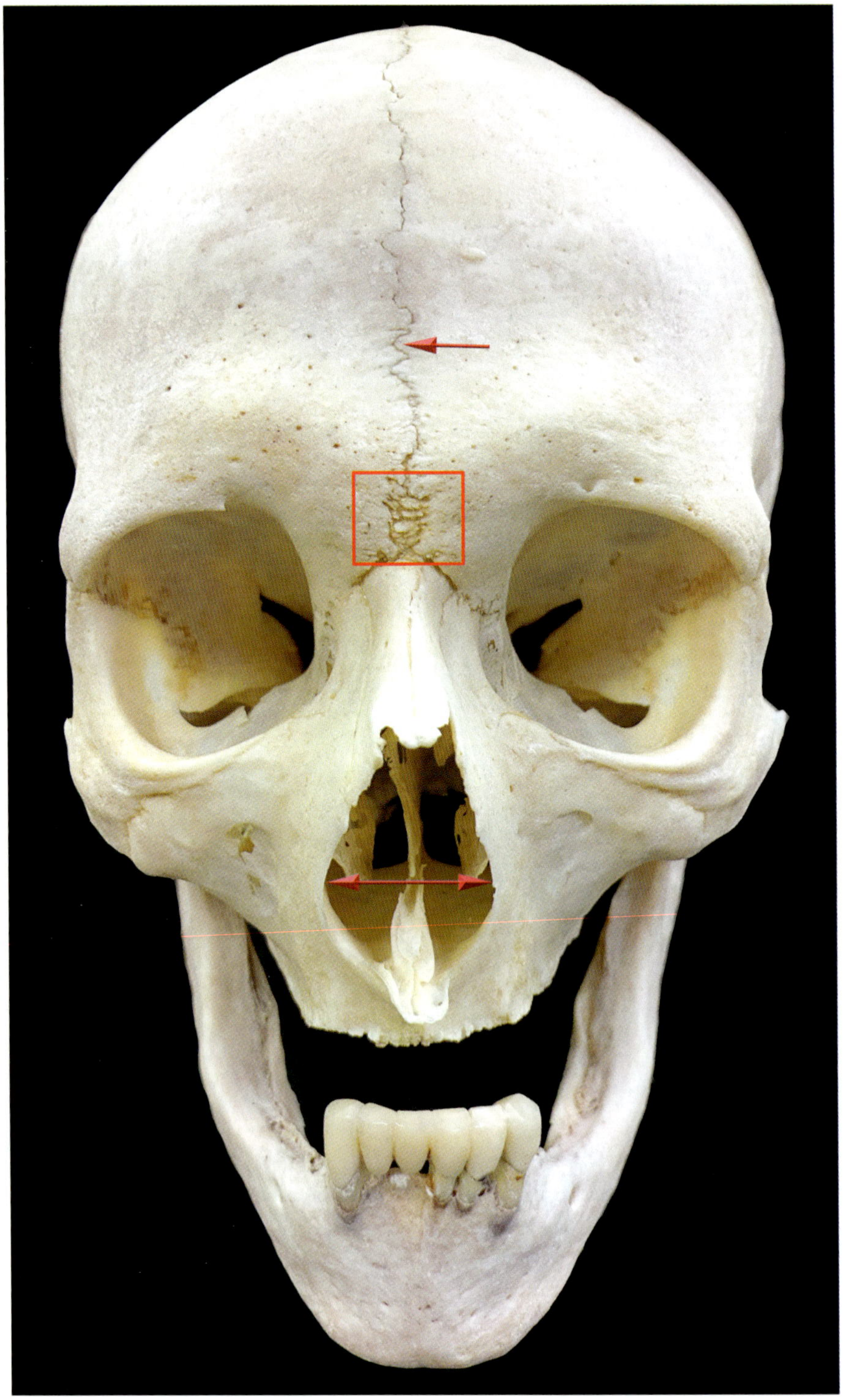

Figure 298. Anterior view of 83-year-old European male with metopism (arrow), a supranasal suture (rectangle), prominent nasal bones, rectangular orbits, a narrow nasal aperture (double arrow), and a prominent anterior nasal spine. Note the antemortem loss of all maxillary (upper) teeth and severe erosion of the maxilla.

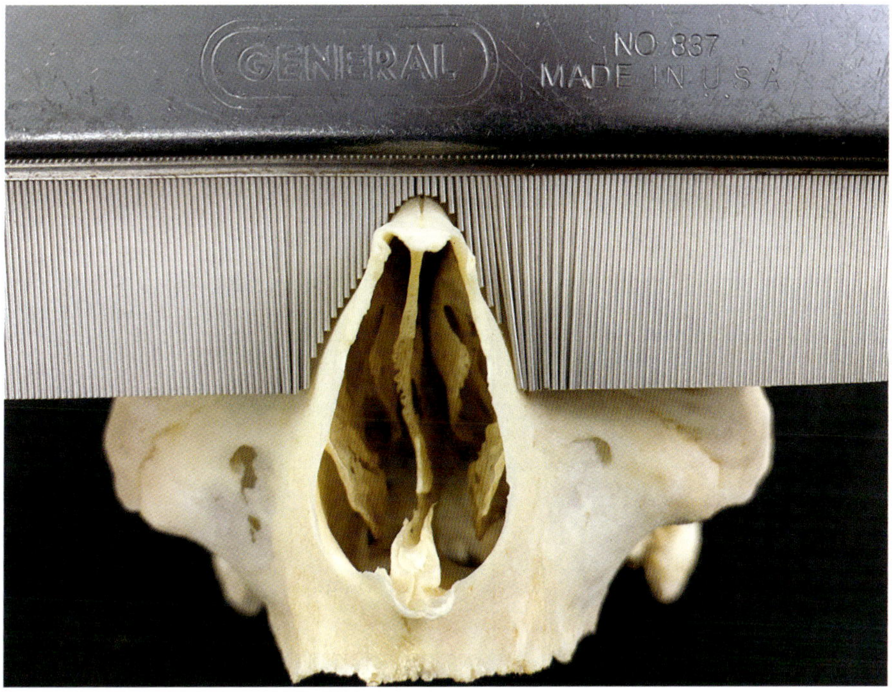

Figure 299. Nasal bone contour in an 83-year-old European male.

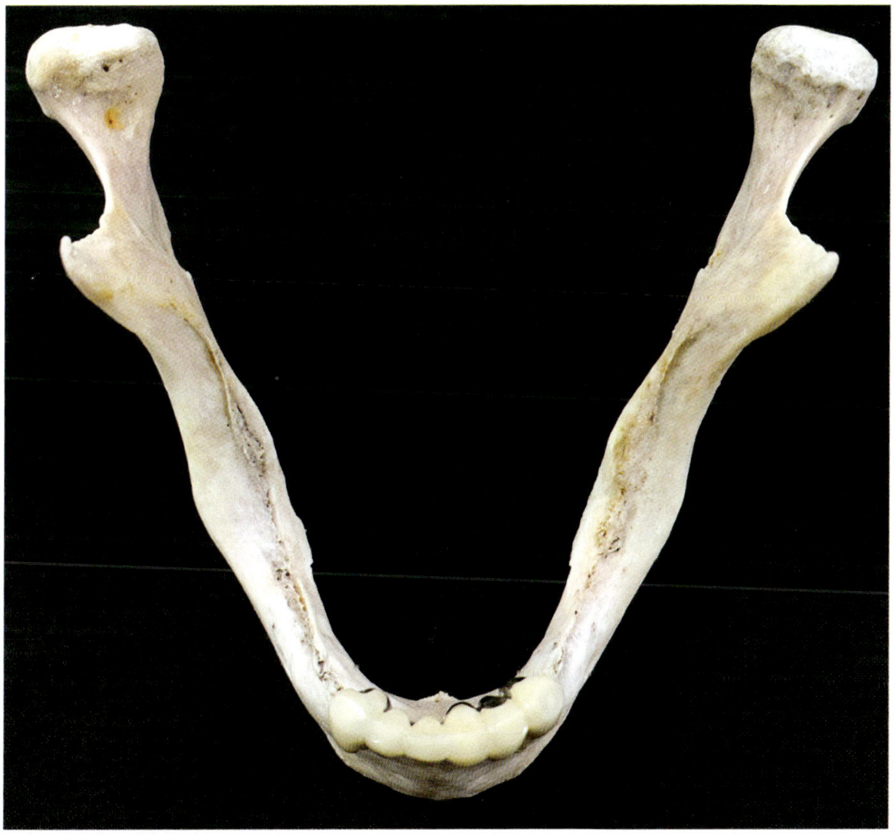

Figure 300. Mandible of 83-year-old European male showing resorption of the alveoli for the premolars and molars.

TABLE 26.
Cranial Measurements in an 83-year-old European Male

Measurement	(mm)	Right	Left
Maximum Cranial Length (g-op, GOL)	178		
Maximum Cranial Breadth (eu-eu, XCB)	137		
Bizygomatic Breadth (zy-zy, ZYB)	126		
Basion-Bregma Height (ba-b, BBH)	136		
Cranial Base Length (ba-n, BNL)	103		
Basion-Prosthion (ba-pr, BPL)	–		
Maxillo-Alveolar Breadth (ecm-ecm, MAB)	–		
Maxillo-Alveolar Length (pr-alv, MAL)	–		
Biauricular Breadth (au-au, AUB)	125		
Upper Facial Height (n-pr)	–		
Minimum Frontal Breadth (ft-ft, WFB)	96		
Upper Facial Breadth (fmt-fmt)	102		
Nasal Height (n-ns, NLH)	52		
Nasal Breadth (al-al, NLB)	23		
Orbital Breadth (d-ec, OBB)	35		L
Orbital Height (OBH)	31		L
Biorbital Breadth (ec-ec, EKB)	89		
Interorbital Breadth (d-d, DKB)	25		
Frontal Chord (n-b, FRC)	109		
Parietal Chord (b-l, PAC)	114		
Occipital Chord (l-o, OCC)	98		
Foramen Magnum Length (ba-o, FOL)	35		
Foramen Magnum Breadth (FOB)	29		
Mastoid Length (MDH)		30	28

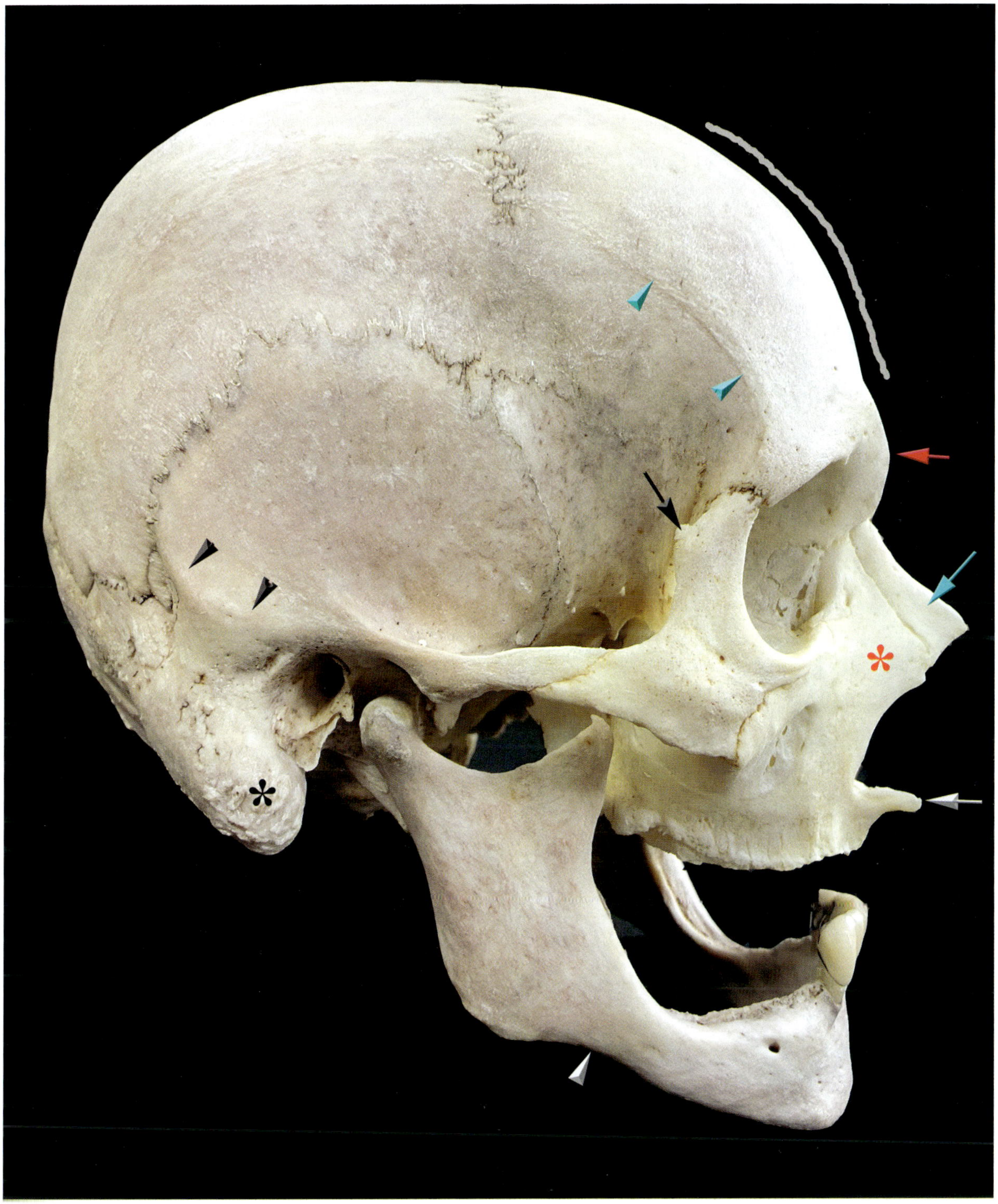

Figure 301. Right lateral view of 83-year-old European male with a sloping frontal bone (curved white line), slight temporal lines (blue arrowheads), small brow ridge (red arrow), prominent nasal bones (blue arrow), frontal process of the maxilla (red*), an anterior nasal spine (white arrow), antegonial notch (white arrowhead), moderate supramastoid crest (black arrowhead), small posterior zygomatic tubercle (black arrow), and moderate-size mastoid process (black*). The maxilla is eroded due to antemortem loss of the teeth that accentuates protrusion of the anterior nasal spine.

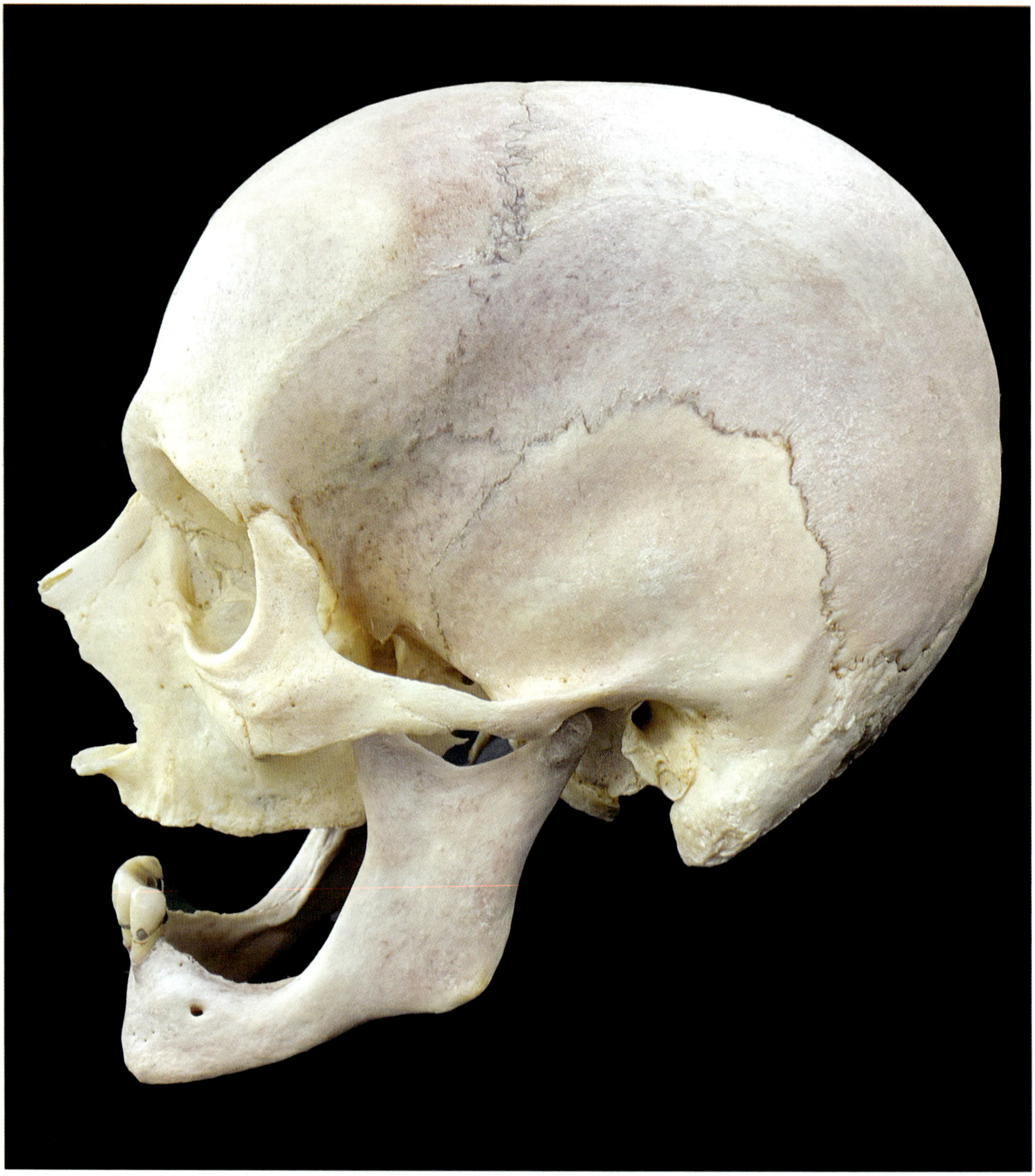

Figure 302. Left lateral view of 83-year-old European male with prominent nasal bones and frontal process of the maxilla. The maxilla/hard palate is eroded due to antemortem loss of the teeth that accentuates protrusion of the anterior nasal spine.

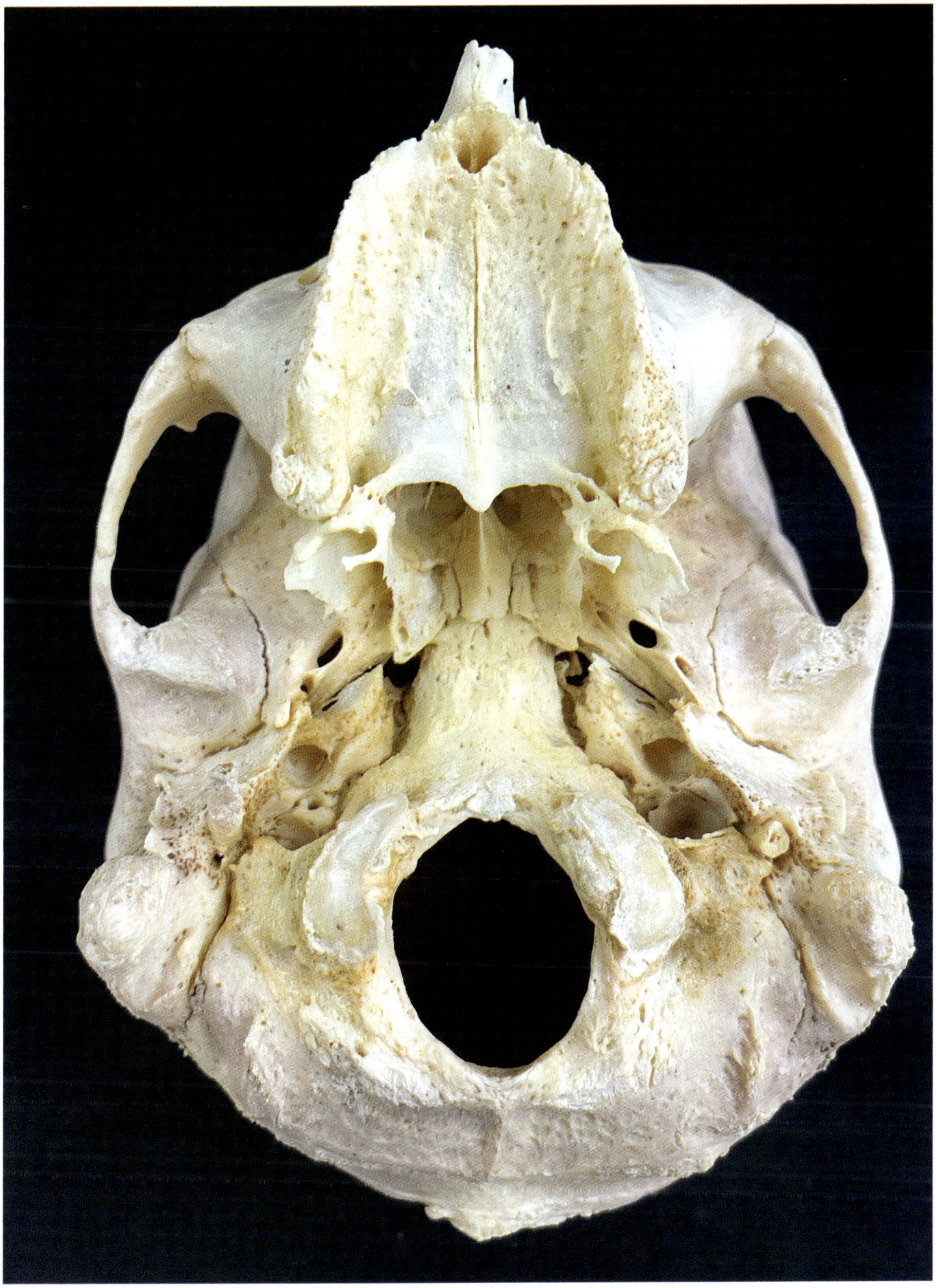

Figure 303. Inferior view of 83-year-old European male. Note the V-shaped and thin (osteoporosis) hard palate/maxilla, asymmetrical mastoid processes, protrusion of the anterior nasal spine due in part to antemortem loss of the teeth, and subsequent resorption of the maxilla.

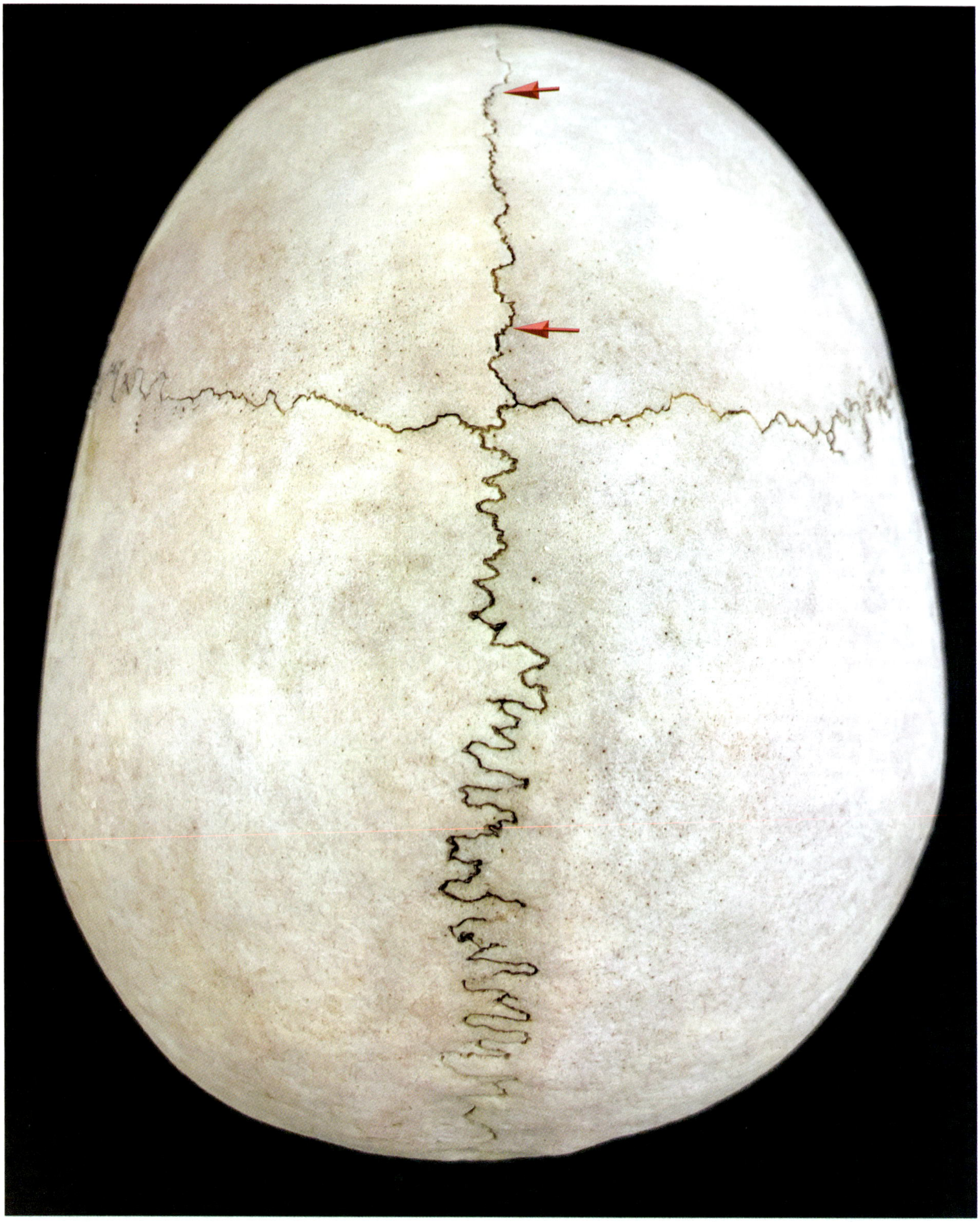

Figure 304. Superior view of 83-year-old European male with open sutures, including the metopic suture (arrow), sometimes referred to as the interfrontal suture, resulting in "metopism."

Figure 305. Posterior view of 83-year-old European male with minimal obliteration of the sutures. Note the prominent inion hook (*).

9. EUROPEAN FEMALE

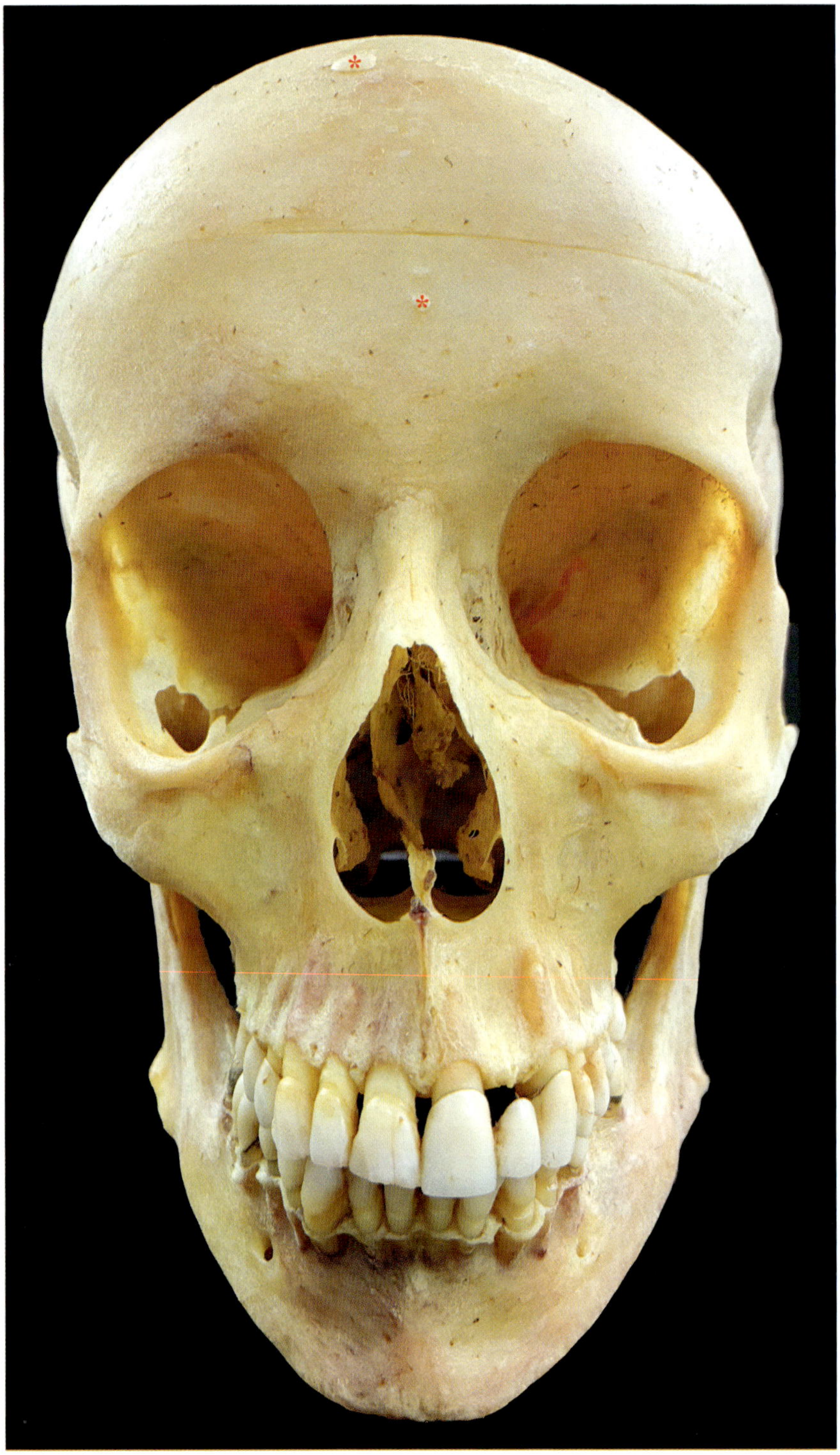

Figure 306. Anterior view of 100-year-old European female with two small button osteomas (*), benign bony tumors, in the frontal bone. Note the narrow chin, the small and posteriorly retreating/slanted zygomas, the complete nasal sill/rim, the prominent nasal bones, and mild gonial eversion. Observe also the translucent lateral orbital plates due to osteoporosis.

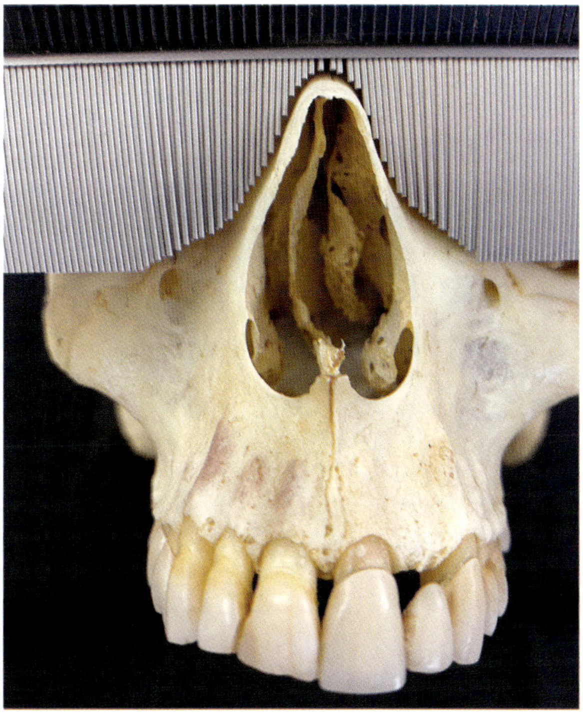

Figure 307. Nasal bone contour in a 100-year-old European female.

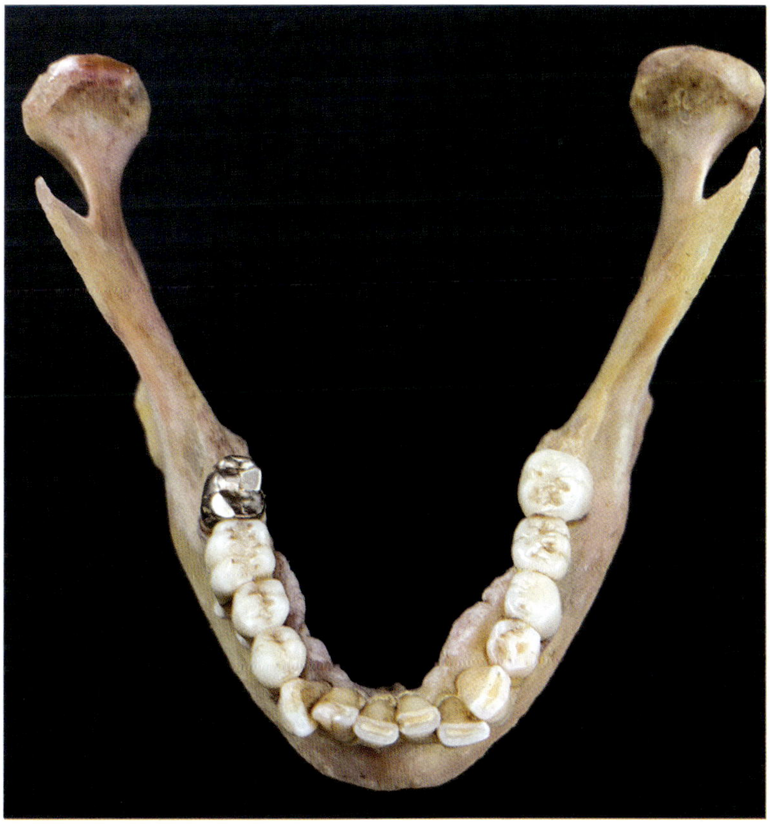

Figure 308. Mandible of 100-year-old European female.

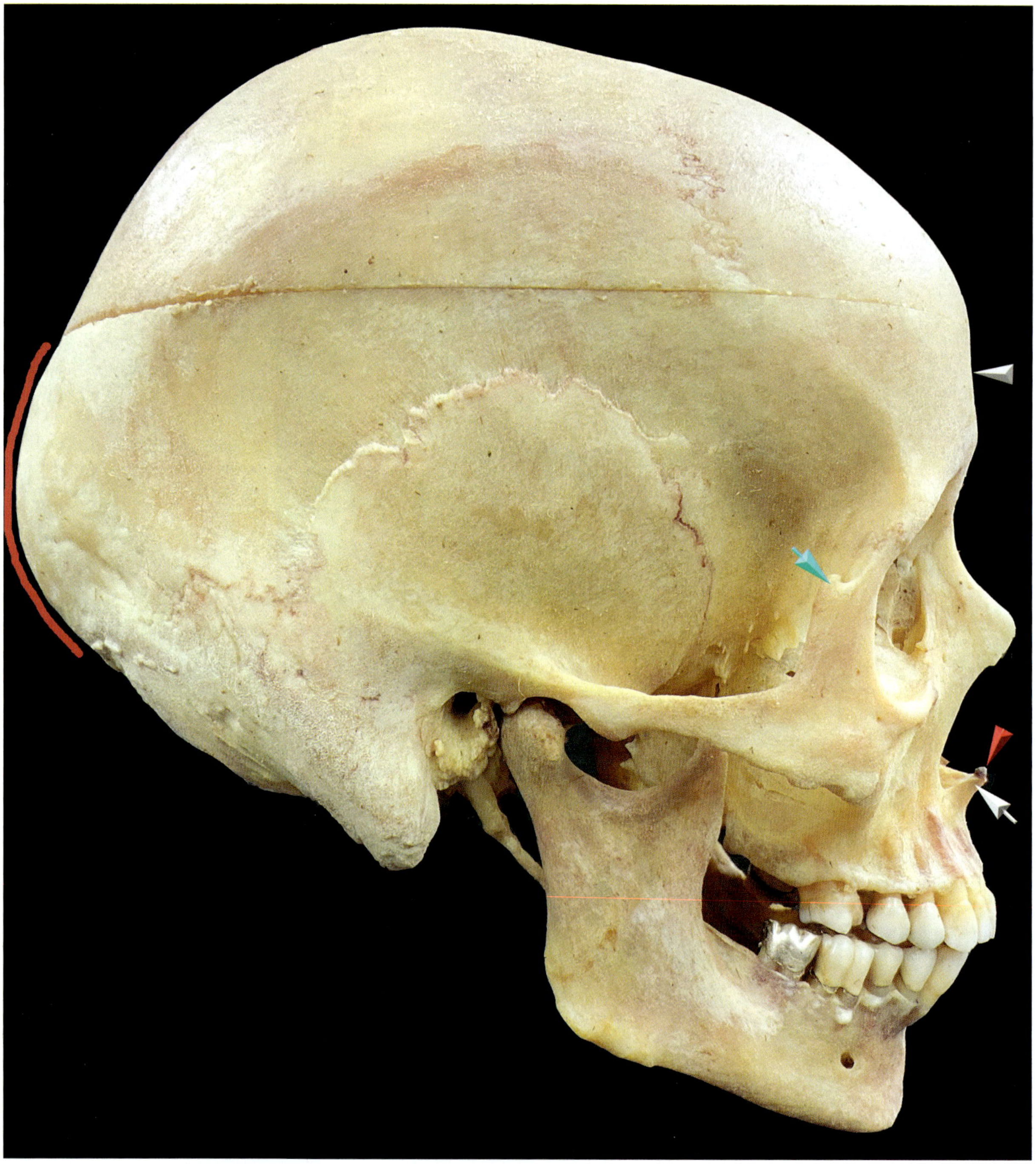

Figure 309. Right lateral view 100-year-old European female with a vertical frontal bone (white arrowhead), a developed posterior zygomatic tubercle (blue arrow), a small calcification (red arrowhead) at the tip of the anterior nasal spine (white arrow) that exhibits mild nasal overhang, a small mandible, and a protruding and rounded occipital bone (curved red line) resulting in a long (anterior-posterior) skull.

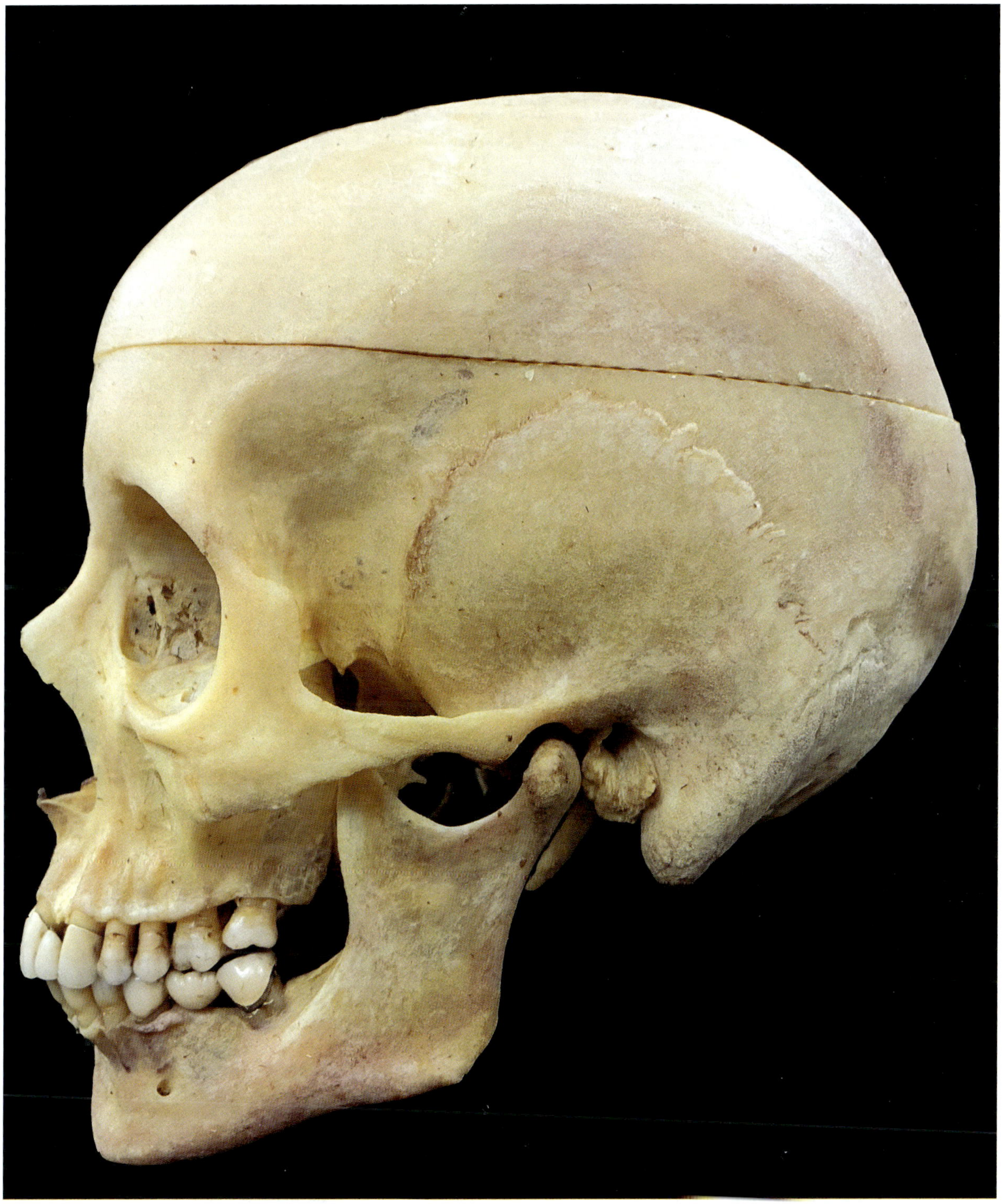

Figure 310. Left lateral view of 100-year-old European female.

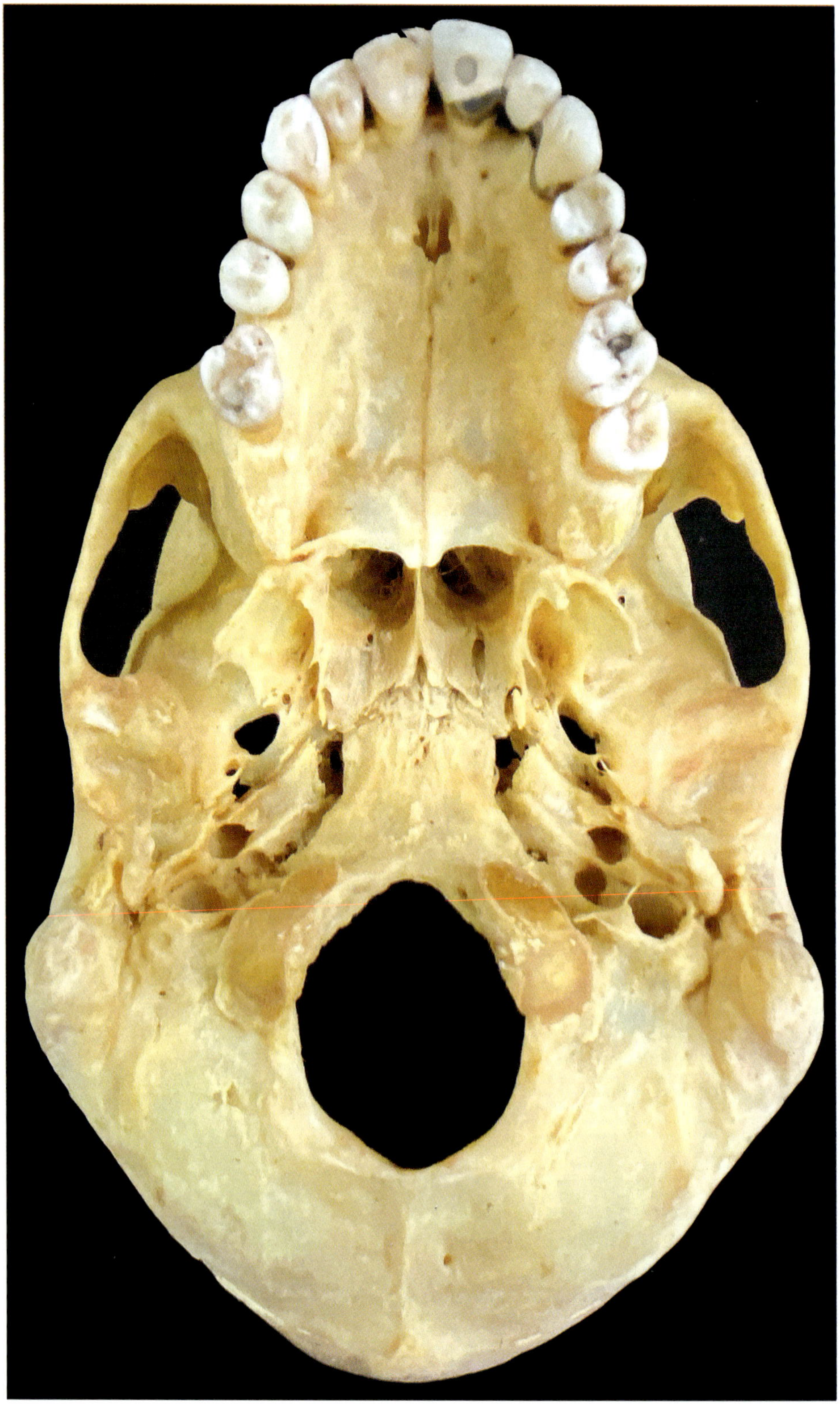

Figure 311. Inferior view of 100-year-old European female with a thin (osteoporosis) and translucent hard palate. Notable at this age is the retention of most of the teeth.

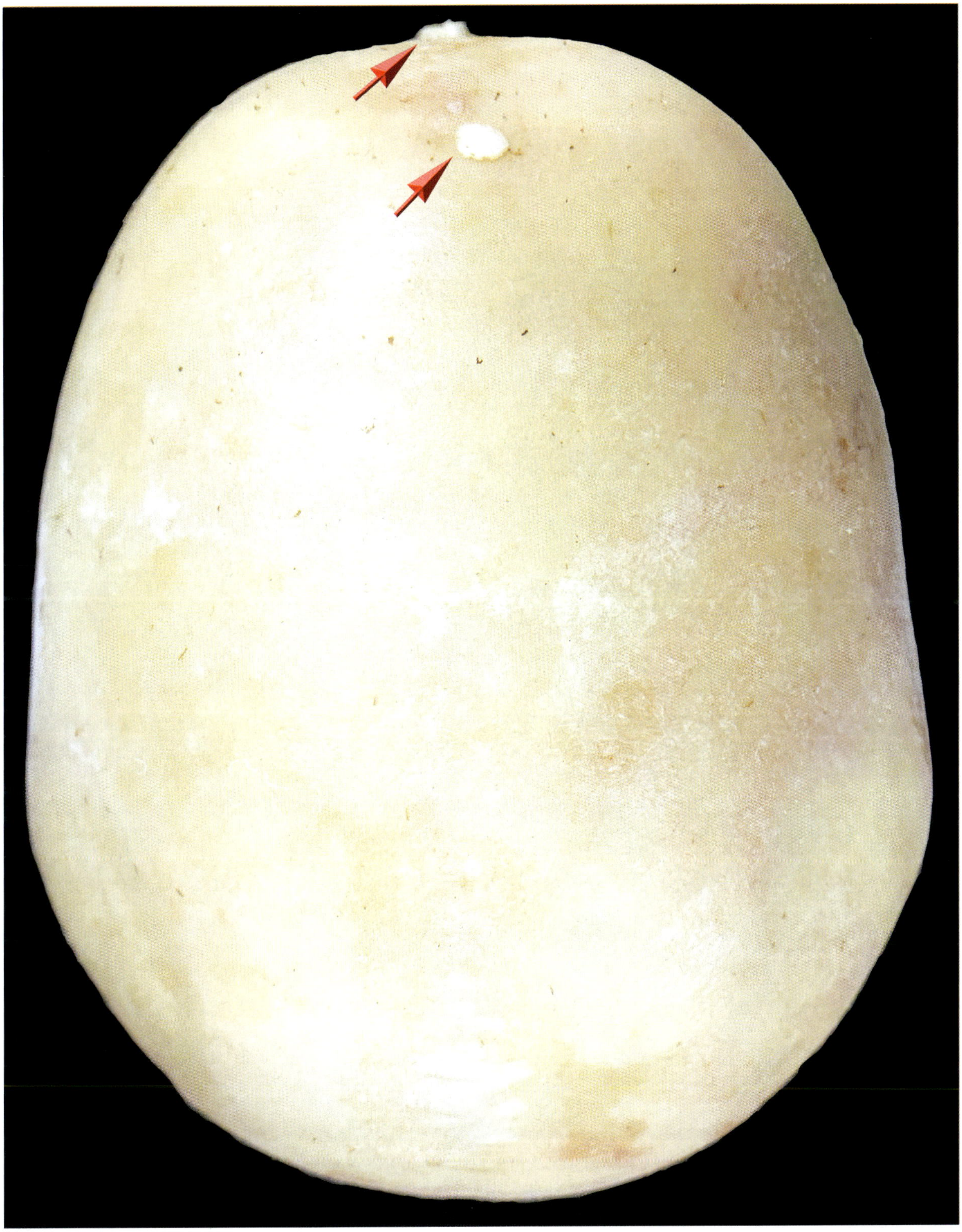

Figure 312. Superior view of 100-year-old European female with an oval-shaped cranium, complete obliteration of the sutures, and two small button osteomas (arrow) in the frontal bone.

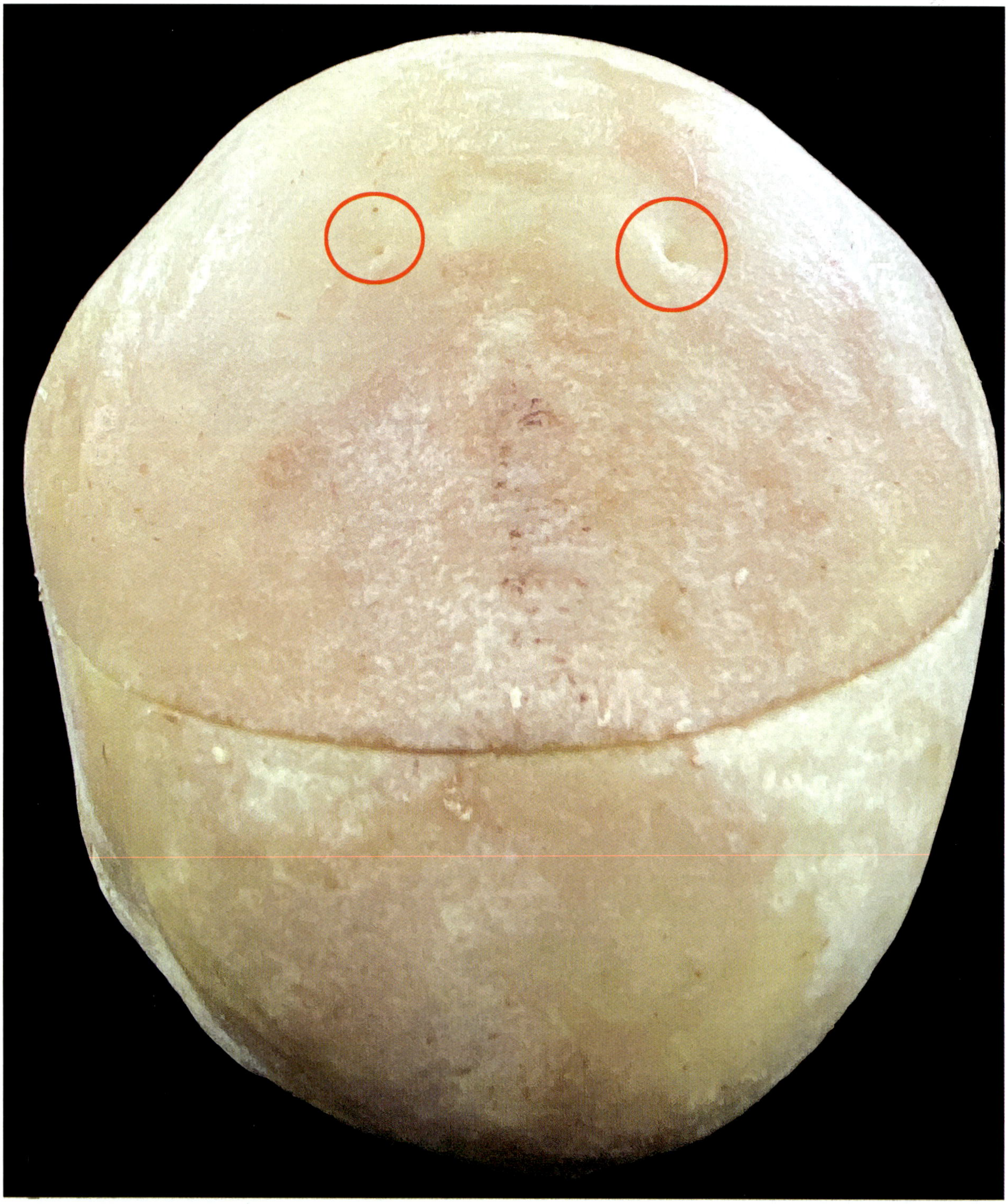

Figure 313. Posterior view of 100-year-old European female with complete obliteration of the sutures and widely spaced parietal foramina (circle) representing normal variation.

10. EUROPEAN FEMALE

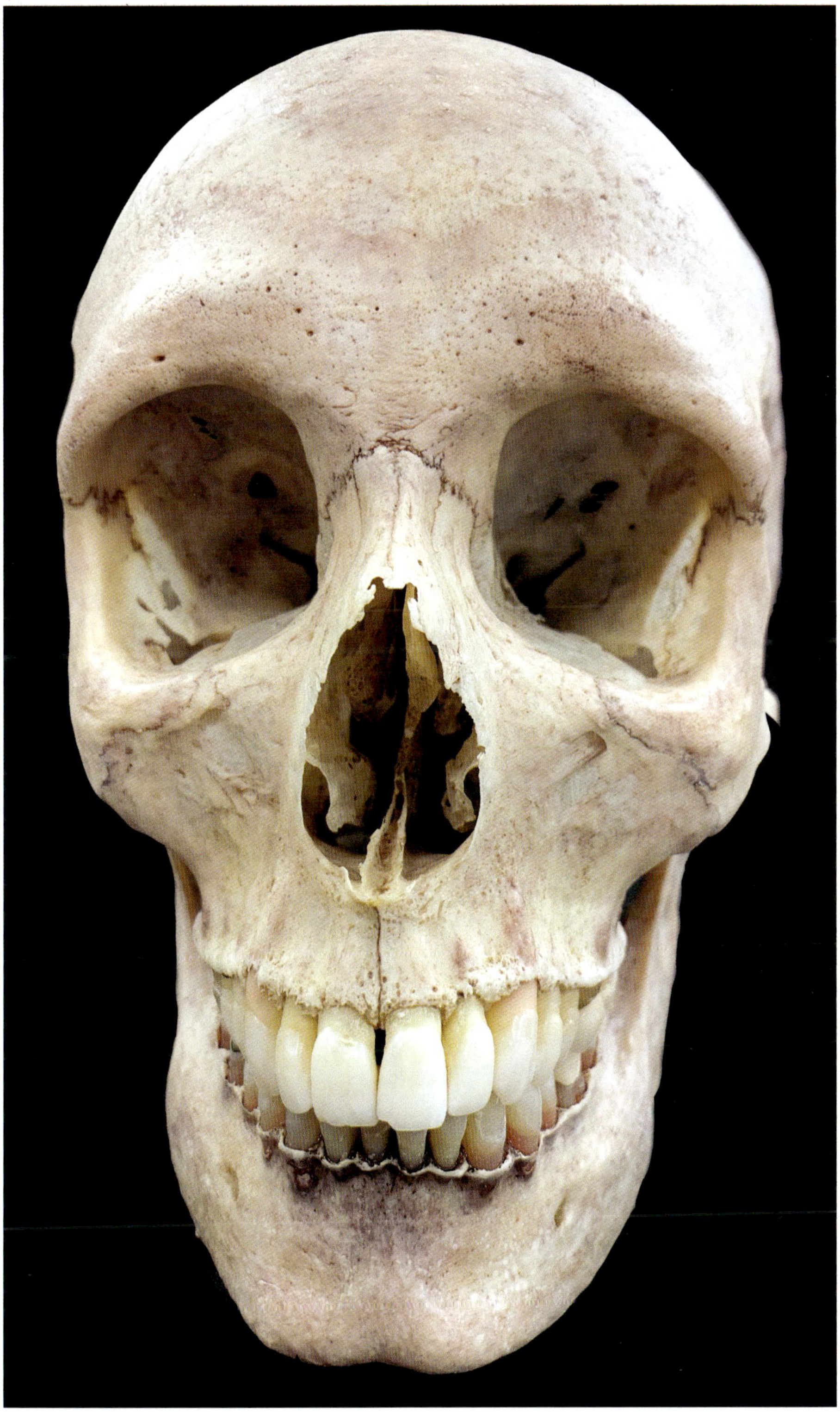

Figure 314. Anterior view 72-year-old European female with a widely separated brow ridge, pronounced nasal bones and frontal process of the maxilla, a sharply-defined nasal rim and sill, and a bilobate chin.

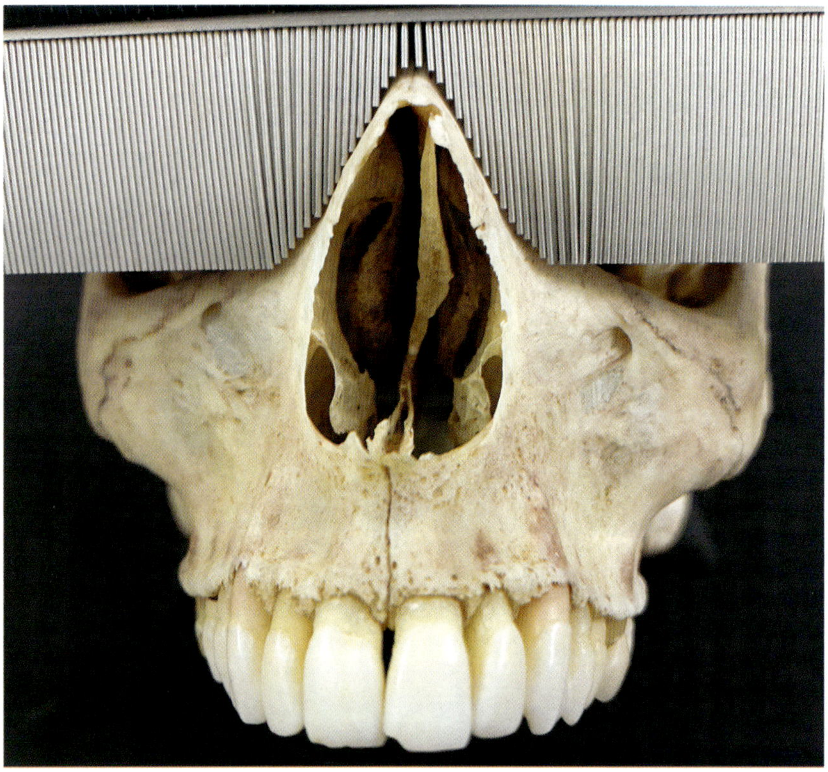

Figure 315. Nasal bone contour in a 72-year-old European female.

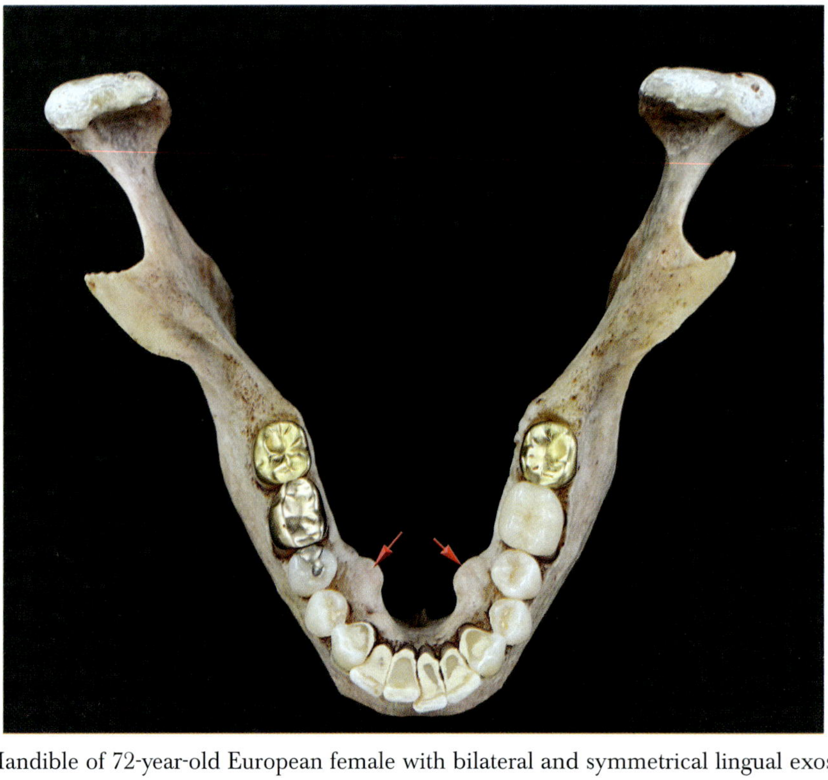

Figure 316. Mandible of 72-year-old European female with bilateral and symmetrical lingual exostoses (arrow).

TABLE 27.
Cranial Measurements in a 72-year-old European Female

Measurement	(mm)	Right	Left
Maximum Cranial Length (g-op, GOL)	168		
Maximum Cranial Breadth (eu-eu, XCB)	130		
Bizygomatic Breadth (zy-zy, ZYB)	115		
Basion-Bregma Height (ba-b, BBH)	125		
Cranial Base Length (ba-n, BNL)	94		
Basion-Prosthion (ba-pr, BPL)	83		
Maxillo-Alveolar Breadth (ecm-ecm, MAB)	60		
Maxillo-Alveolar Length (pr-alv, MAL)	52		
Biauricular Breadth (au-au, AUB)	111		
Upper Facial Height (n-pr)	63		
Minimum Frontal Breadth (ft-ft, WFB)	94		
Upper Facial Breadth (fmt-fmt)	98		
Nasal Height (n-ns, NLH)	50		
Nasal Breadth (al-al, NLB)	21		
Orbital Breadth (d-ec, OBB)	38		L
Orbital Height (OBH)	31		L
Biorbital Breadth (ec-ec, EKB)	86		
Interorbital Breadth (d-d, DKB)	21		
Frontal Chord (n-b, FRC)	109		
Parietal Chord (b-l, PAC)	104		
Occipital Chord (l-o, OCC)	88		
Foramen Magnum Length (ba-o, FOL)	38		
Foramen Magnum Breadth (FOB)	29		
Mastoid Length (MDH)		23	22

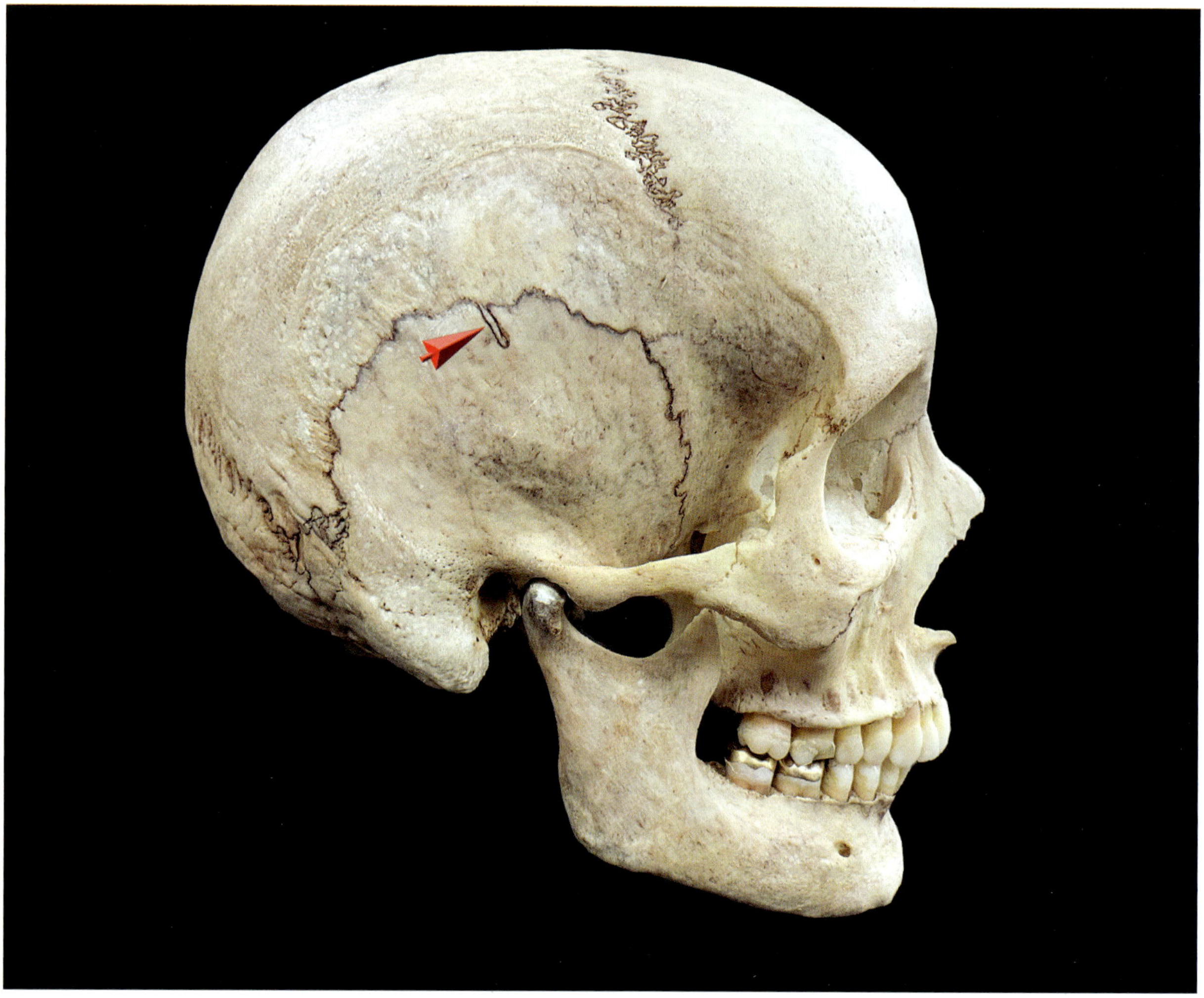

Figure 317. Right lateral view 72-year-old European female with an elongated ossicle (arrow), an uncommon normal anatomical variant, along the squamosal suture.

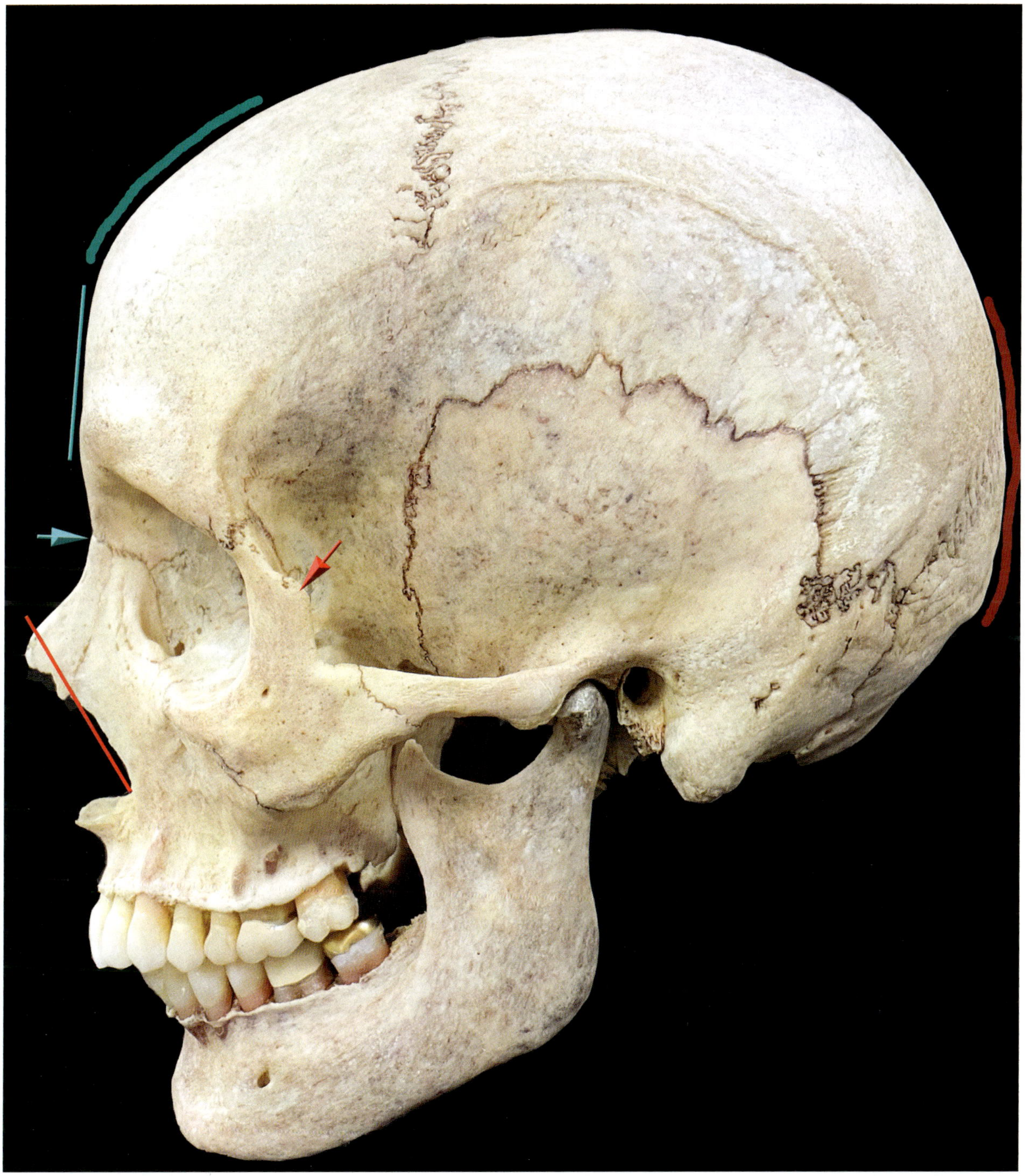

Figure 318. Left lateral view 72-year-old European female with vertical frontal bone (straight blue line) in the region of glabella leading to a sloping frontal bone (curved blue line), absence of a depression at nasion (blue arrow), and prominent, overhanging nasal bones (straight red line), and a posterior zygomatic tubercle (red arrow). Also note an intermediate-size anterior nasal spine, large zygoma, flat supramastoid crest, small mastoid process and mandible, and a rounded occipital region (curved red line).

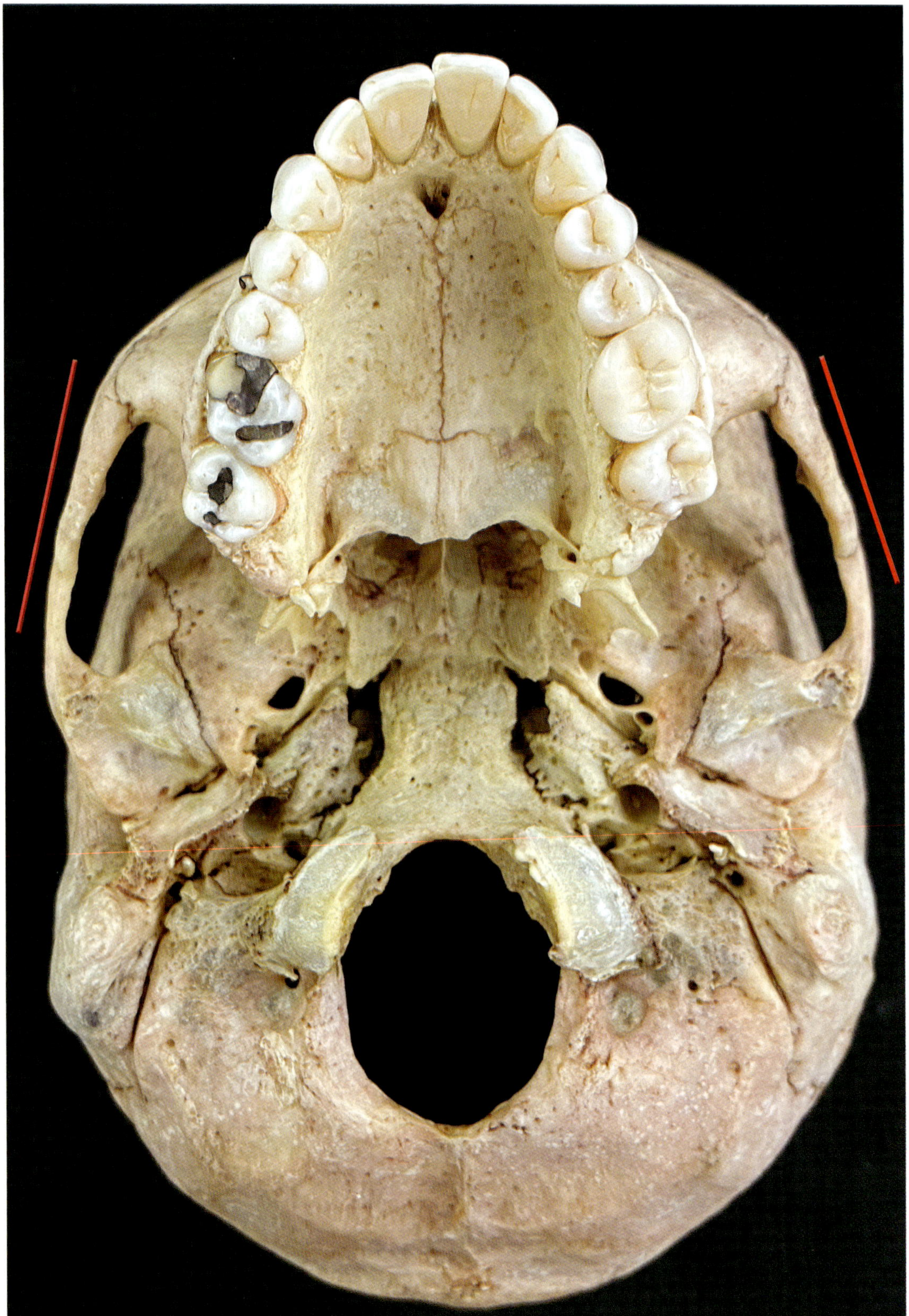

Figure 319. Inferior view 72-year-old European female with a V-shaped and deep palate and nearly flat zygomatic arches (red line).

Figure 320. Superior view 72-year-old European female with an oval-shaped cranium with minimal obliteration of the sutures.

Figure 321. Posterior view 72-year-old European female with minimal obliteration of the sutures and four parietal foramina (arrow), normal anatomical variants.

11. EUROPEAN FEMALE

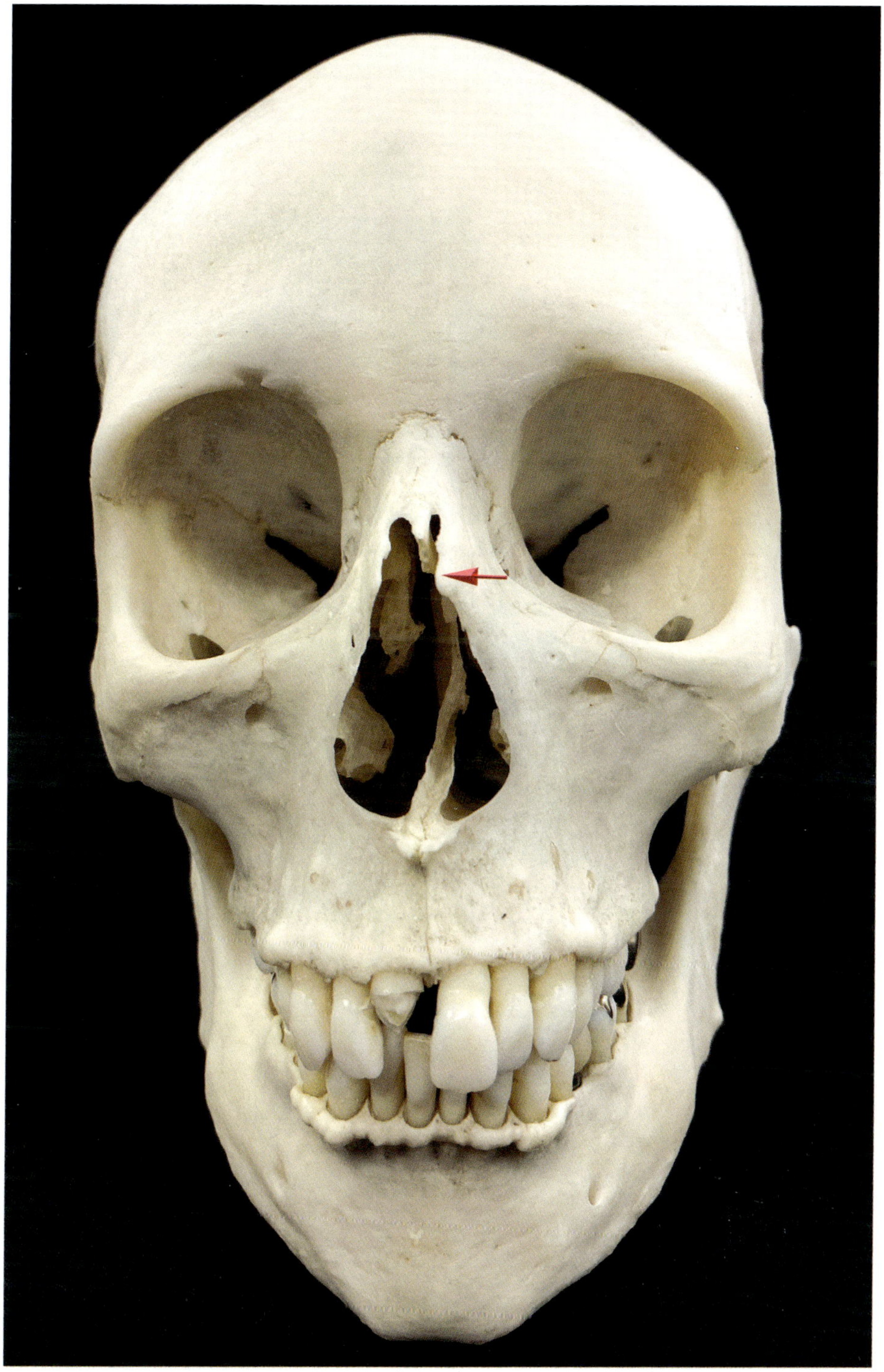

Figure 322. Anterior view of 74-year-old European female. Note the nasal septum deviated to the left and asymmetry of the left nasal aperture (arrow). This individual exhibits prominent nasal bones and maxilla, blunt supraorbital margins, small zygomas, buccal exostoses of the maxilla and mandible, and a narrow, square chin.

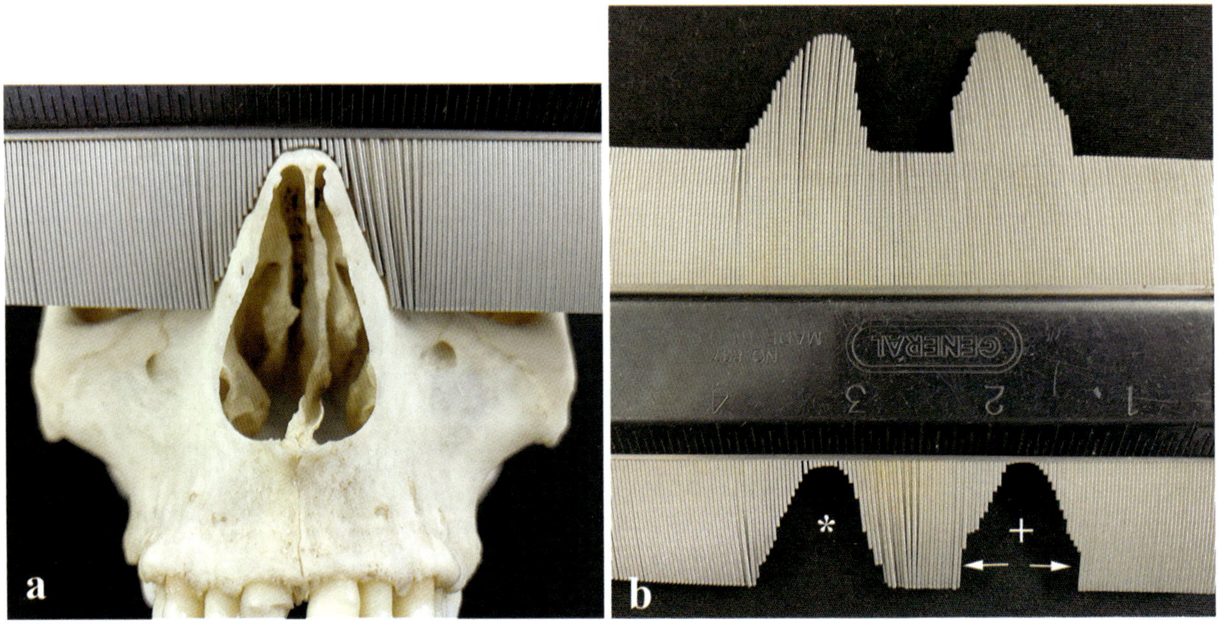

Figure 323a-b. Nasal bone contour in a 74-year-old European female with (a) the gauge held perpendicular to the hard palate and (b) a comparison of nasal bone contour with the gauge held perpendicular to the hard palate (*) and perpendicular to the nasal bones (+). (b) Note the squared sides (arrow) when including the medial orbits (+).

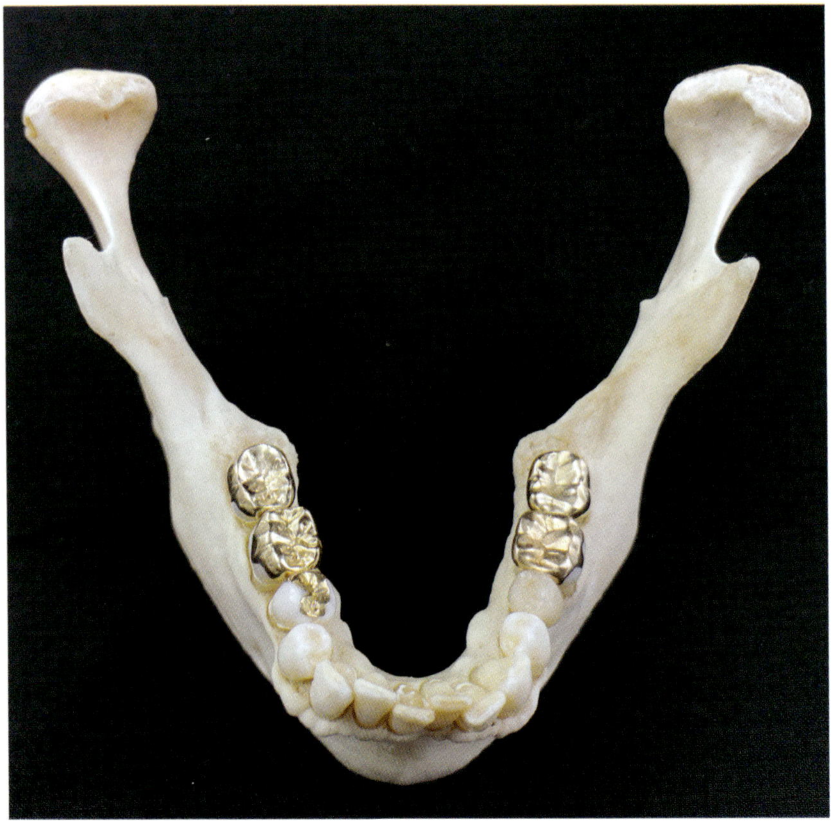

Figure 324. Mandible of 74-year-old European female with a rounded chin, and bilateral and symmetrical buccal and lingual exostoses.

TABLE 28.
Cranial Measurements in a 74-year-old European Female

Measurement	(mm)	Right	Left
Maximum Cranial Length (g-op, GOL)	182		
Maximum Cranial Breadth (eu-eu, XCB)	143		
Bizygomatic Breadth (zy-zy, ZYB)	130		
Basion-Bregma Height (ba-b, BBH)	142		
Cranial Base Length (ba-n, BNL)	107		
Basion-Prosthion (ba-pr, BPL)	100		
Maxillo-Alveolar Breadth (ecm-ecm, MAB)	68*		
Maxillo-Alveolar Length (pr-alv, MAL)	57		
Biauricular Breadth (au-au, AUB)	128		
Upper Facial Height (n-pr)	71		
Minimum Frontal Breadth (ft-ft, WFB)	102		
Upper Facial Breadth (fmt-fmt)	106		
Nasal Height (n-ns, NLH)	52		
Nasal Breadth (al-al, NLB)	24		
Orbital Breadth (d-ec, OBB)	41		L
Orbital Height (OBH)	36		L
Biorbital Breadth (ec-ec, EKB)	96		
Interorbital Breadth (d-d, DKB)	24		
Frontal Chord (n-b, FRC)	111		
Parietal Chord (b-l, PAC)	115		
Occipital Chord (l-o, OCC)	107		
Foramen Magnum Length (ba-o, FOL)	40		
Foramen Magnum Breadth (FOB)	30		
Mastoid Length (MDH)		23	25

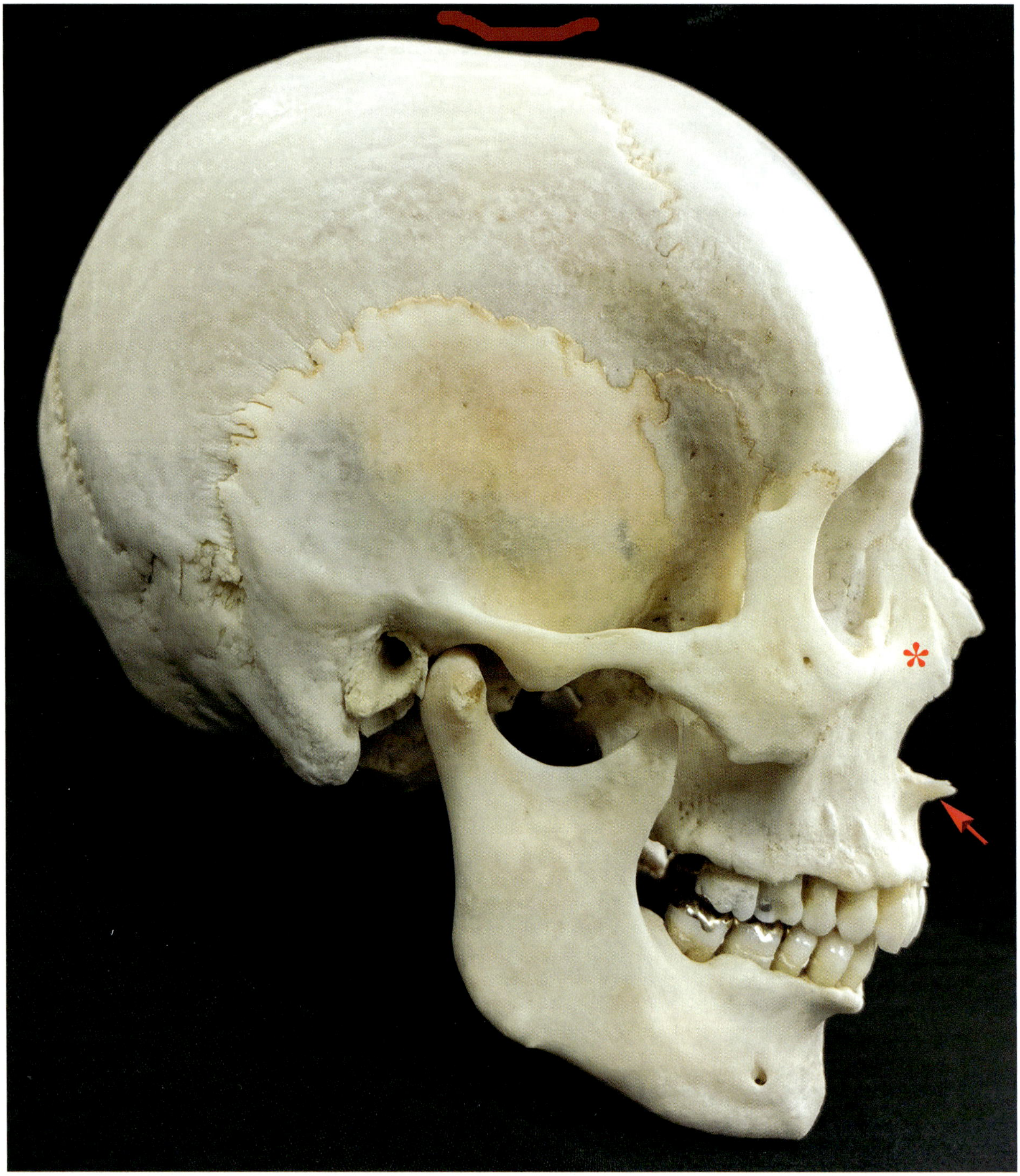

Figure 325. Right lateral view of 74-year-old European female. This individual exhibits a slight postbregmatic depression (curved red line), prominent nasal bones and frontal process of the maxilla (*), a prominent anterior nasal spine (arrow), and the absence of alveolar prognathism.

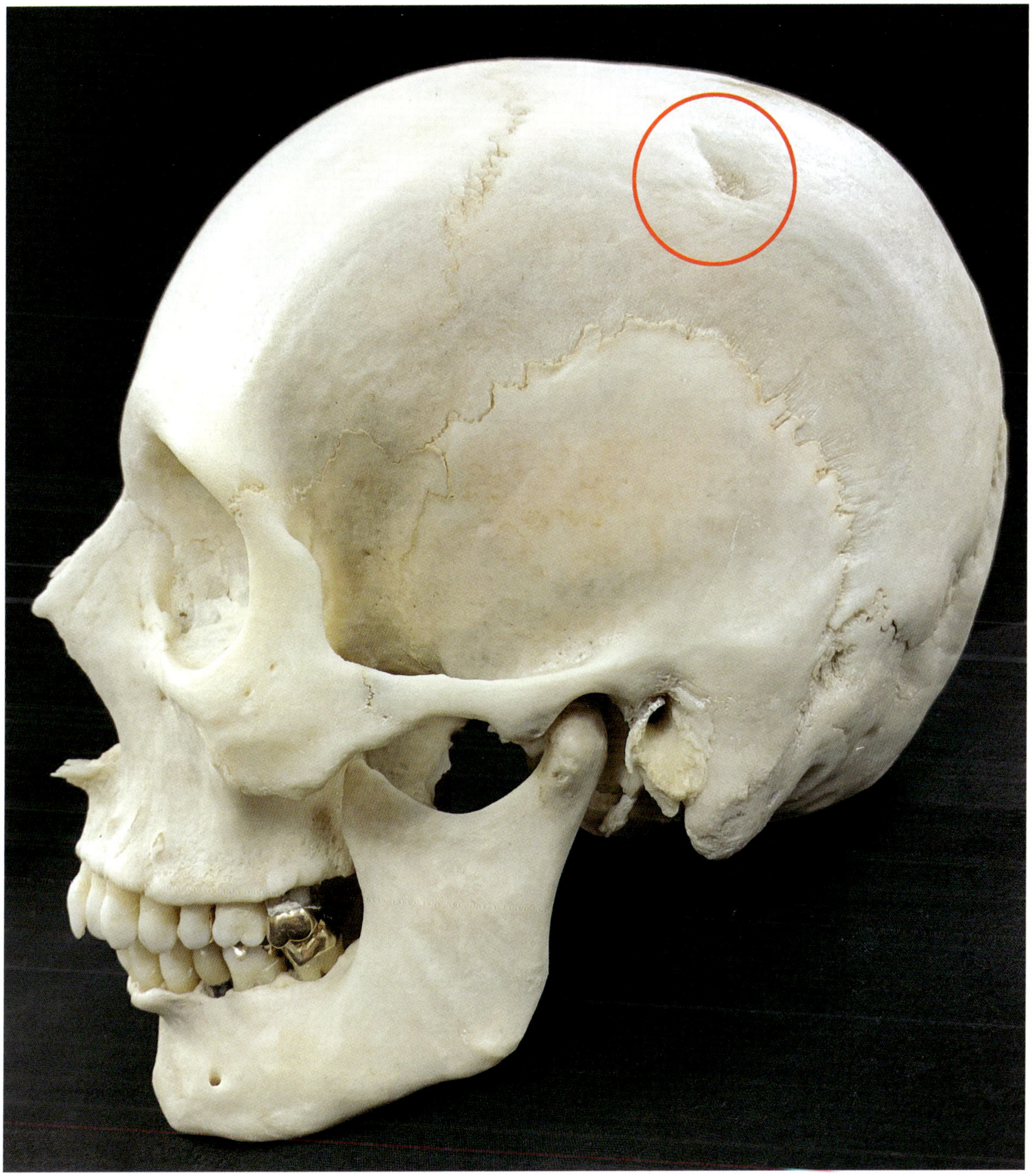

Figure 326. Left lateral view of 74-year-old European female. Note the healed blunt force trauma (circle) in the left parietal bone.

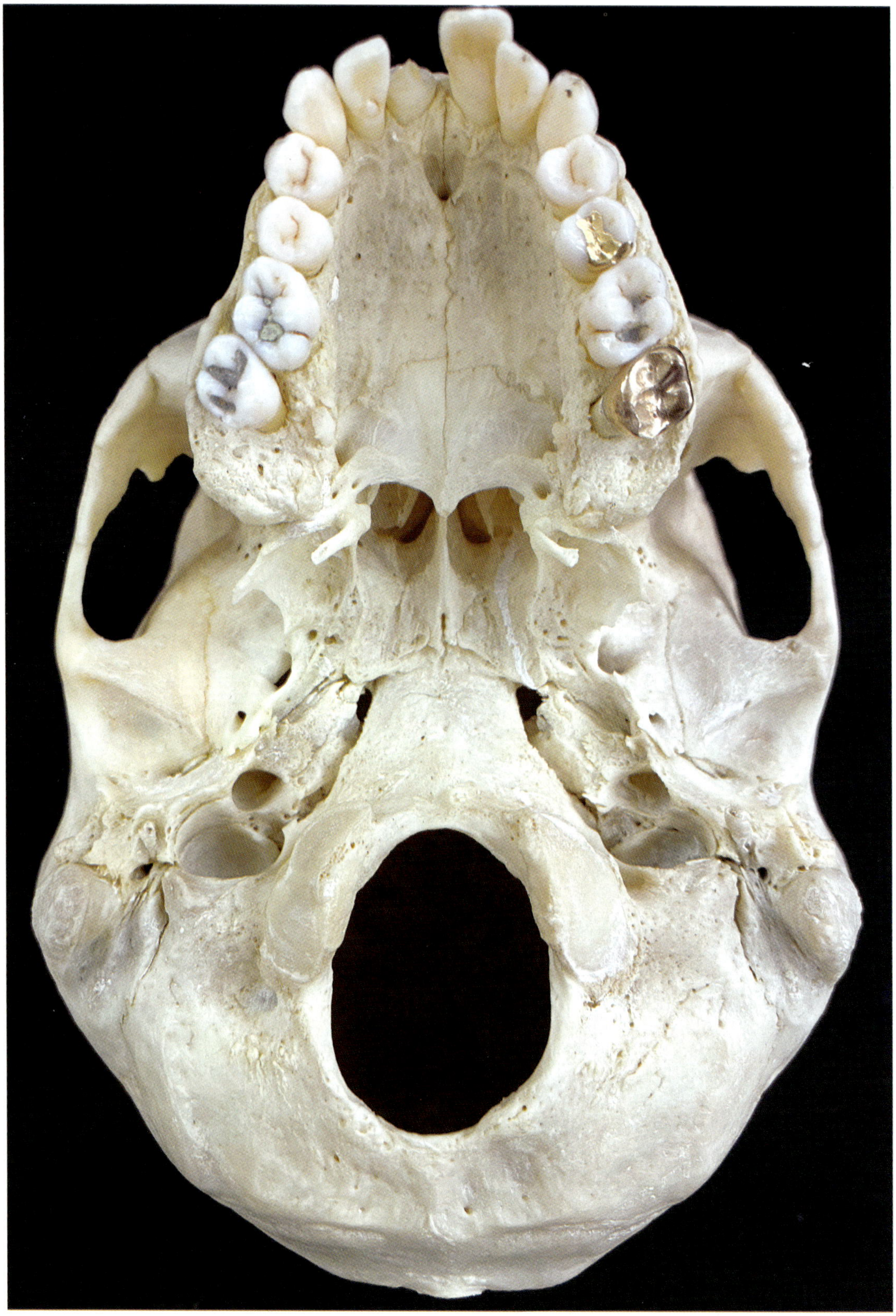

Figure 327. Inferior view of 74-year-old European female with a long and rounded occipital region.

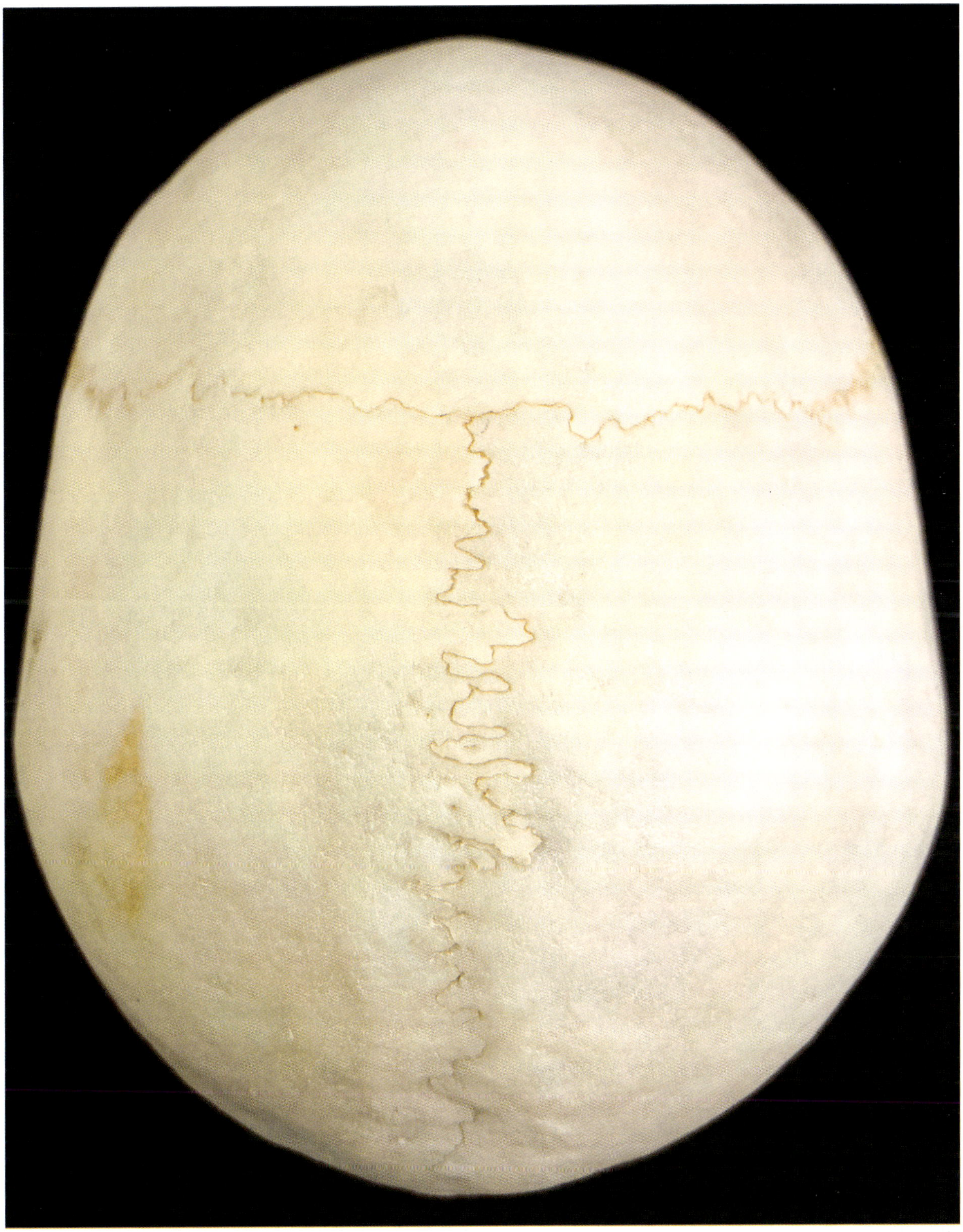

Figure 328. Superior view of 75-year-old European female. The area of blunt force trauma in the left parietal bone appears darker than the surrounding bone due to the effect of lighting and shadows when photographed.

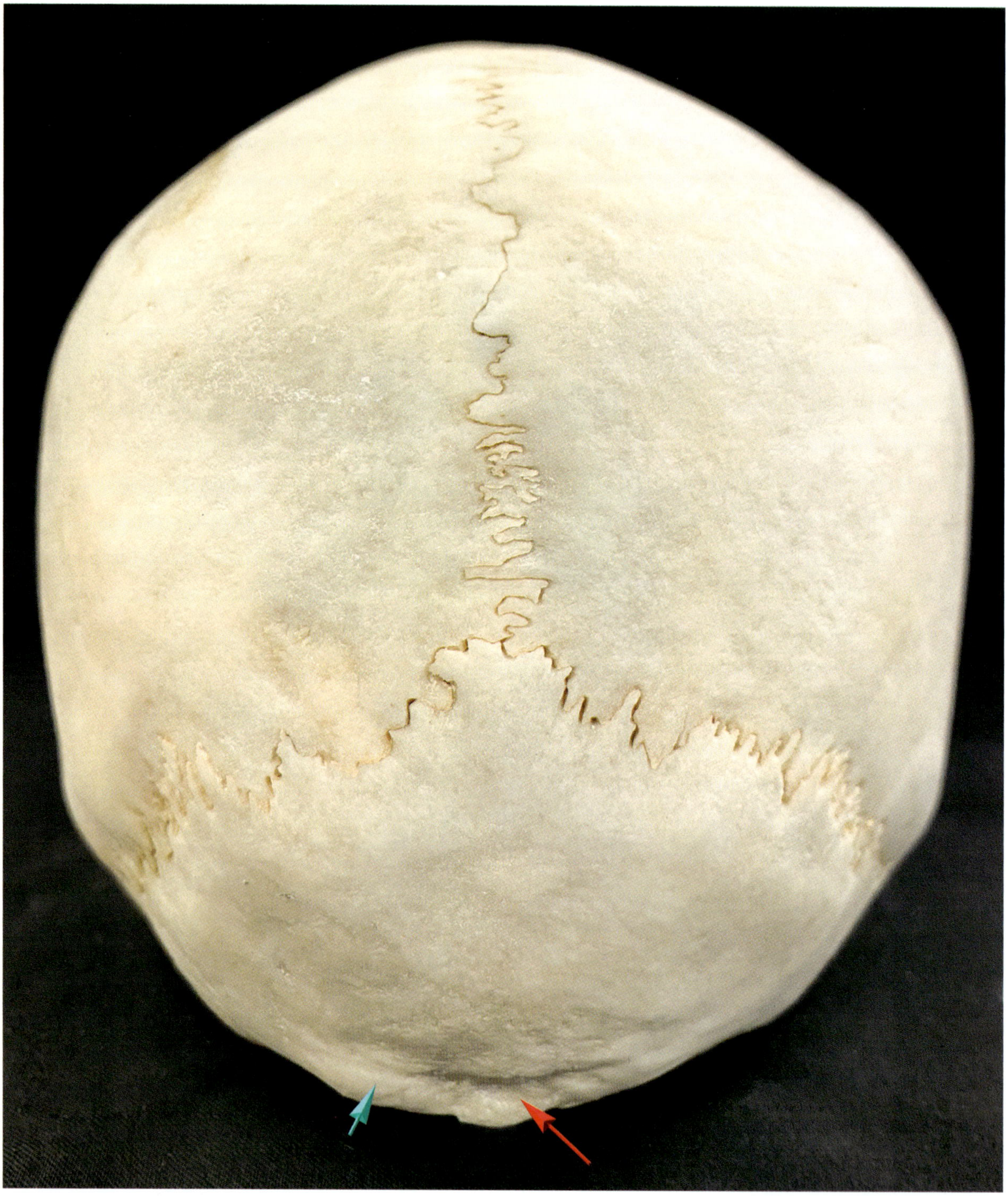

Figure 329. Posterior view of 74-year-old European female with open sutures, and moderate development of the nuchal crest (blue arrow) and inion hook (red arrow).

12. EUROPEAN FEMALE

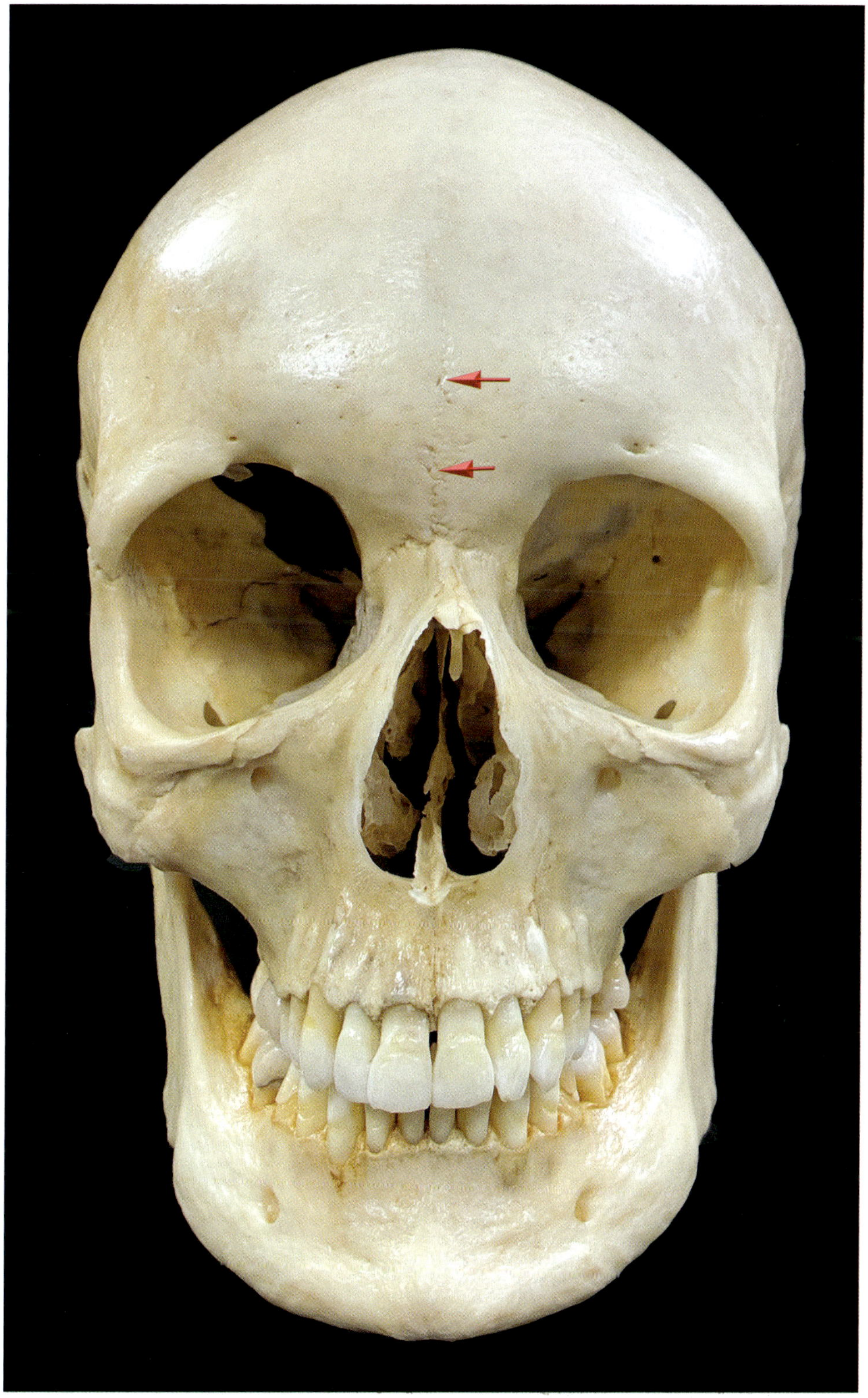

Figure 330. Anterior view 73-year-old European female with a trace of the metopic suture (arrow).

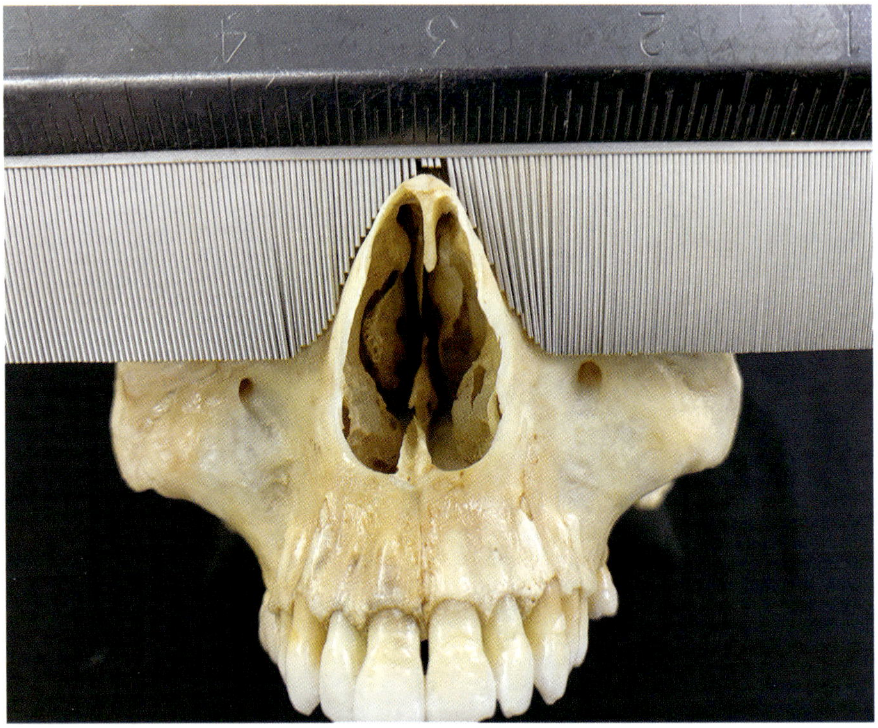

Figure 331. Nasal bone contour in a 73-year-old European female.

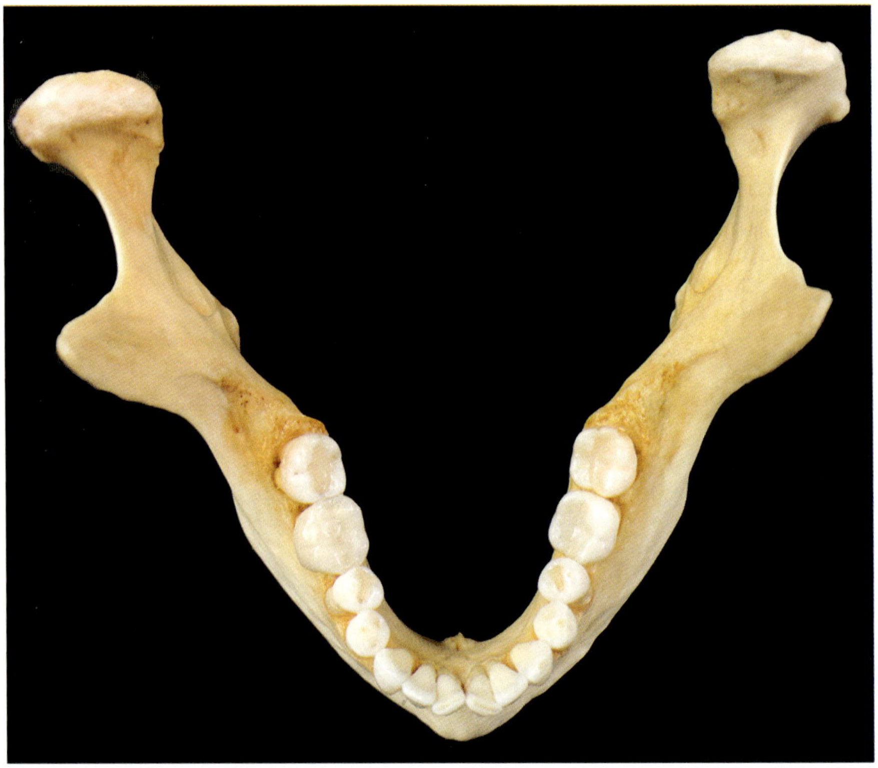

Figure 332. Mandible of 73-year-old European female. Note that the V-shaped chin is unusually pointed and the rami are widely separated.

TABLE 29.
Cranial Measurements in a 73-year-old European Female

Measurement	(mm)	Right	Left
Maximum Cranial Length (g-op, GOL)	177		
Maximum Cranial Breadth (eu-eu, XCB)	136		
Bizygomatic Breadth (zy-zy, ZYB)	119		
Basion-Bregma Height (ba-b, BBH)	136		
Cranial Base Length (ba-n, BNL)	99		
Basion-Prosthion (ba-pr, BPL)	95		
Maxillo-Alveolar Breadth (ecm-ecm, MAB)	56		
Maxillo-Alveolar Length (pr-alv, MAL)	55		
Biauricular Breadth (au-au, AUB)	117		
Upper Facial Height (n-pr)	61		
Minimum Frontal Breadth (ft-ft, WFB)	100		
Upper Facial Breadth (fmt-fmt)	103		
Nasal Height (n-ns, NLH)	43		
Nasal Breadth (al-al, NLB)	21		
Orbital Breadth (d-ec, OBB)	38		L
Orbital Height (OBH)	31		L
Biorbital Breadth (ec-ec, EKB)	91		
Interorbital Breadth (d-d, DKB)	20		
Frontal Chord (n-b, FRC)	109		
Parietal Chord (b-l, PAC)	114		
Occipital Chord (l-o, OCC)	96		
Foramen Magnum Length (ba-o, FOL)	34		
Foramen Magnum Breadth (FOB)	32		
Mastoid Length (MDH)		24	23

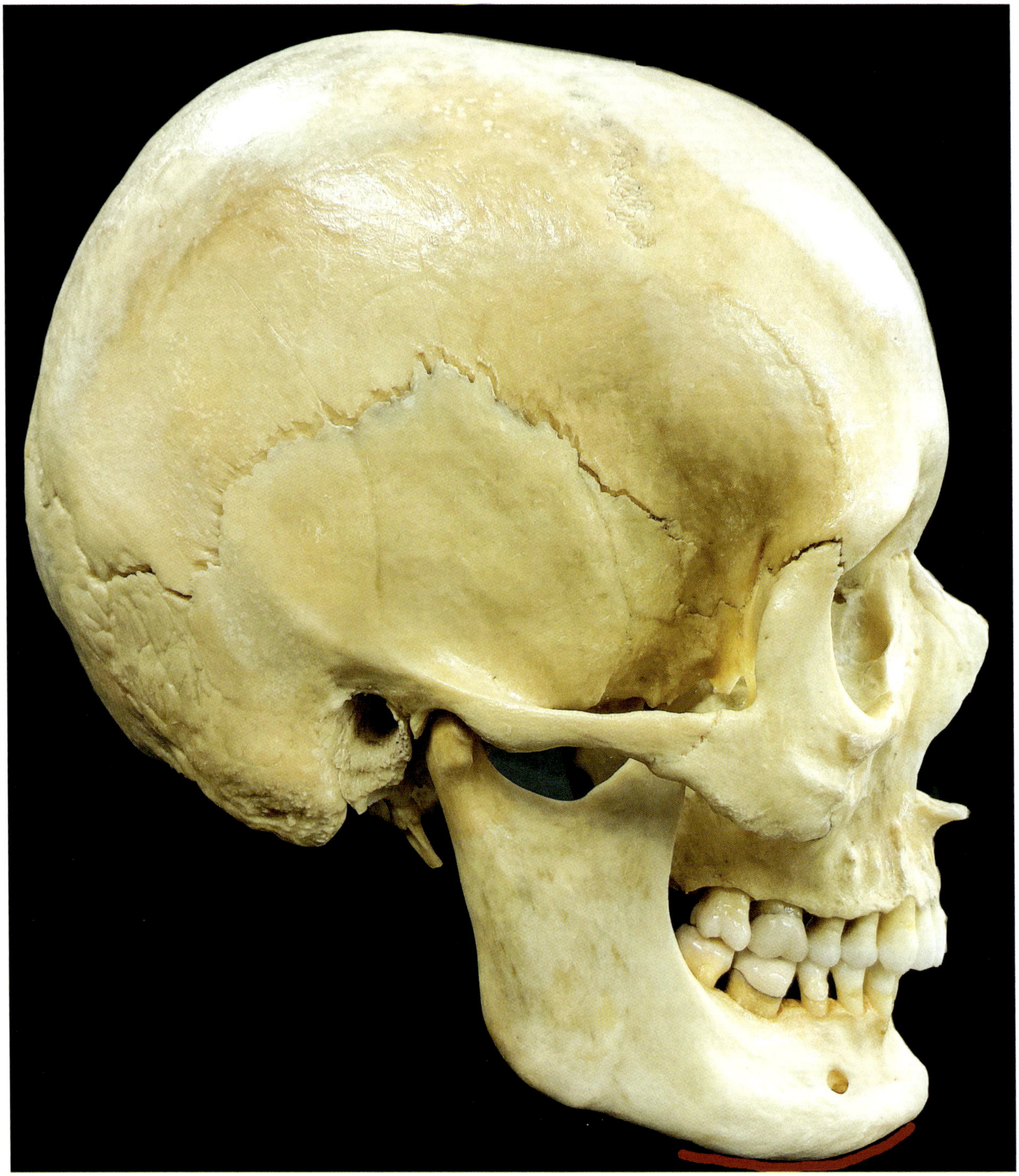

Figure 333. Right lateral view 73-year-old European female with a prominent anterior nasal spine and a convex mandibular body (curved red line).

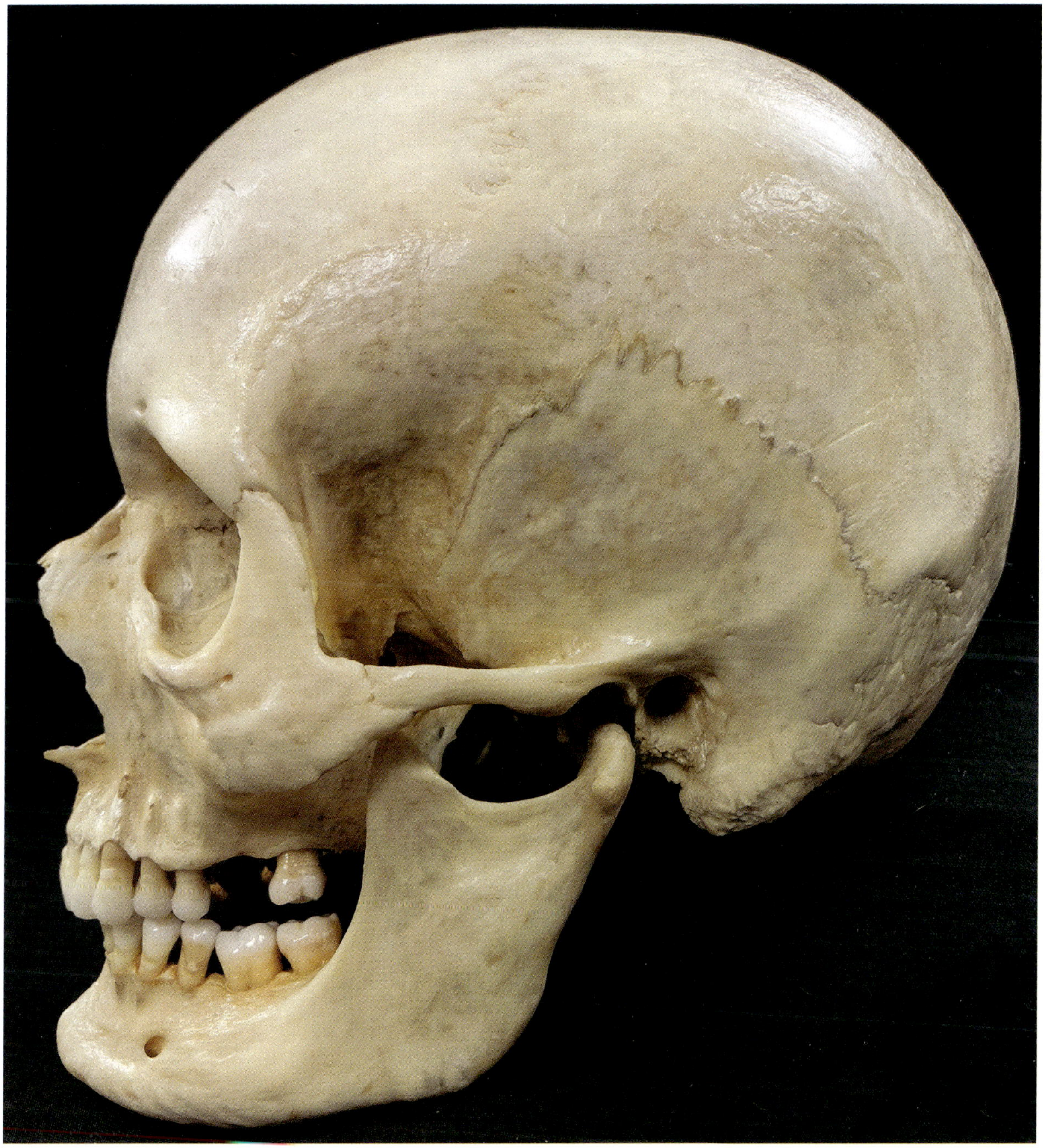

Figure 334. Left lateral view 73-year-old European female. Note the curved inferior border resulting in a "rocker jaw" typically associated with Polynesians (Houghton 1977, Snow 1974, Weisler and Swindler 2002). Rocker jaw is the result of a convex inferior border of the mandible allowing it to rock anterior-posteriorly when placed on a flat surface.

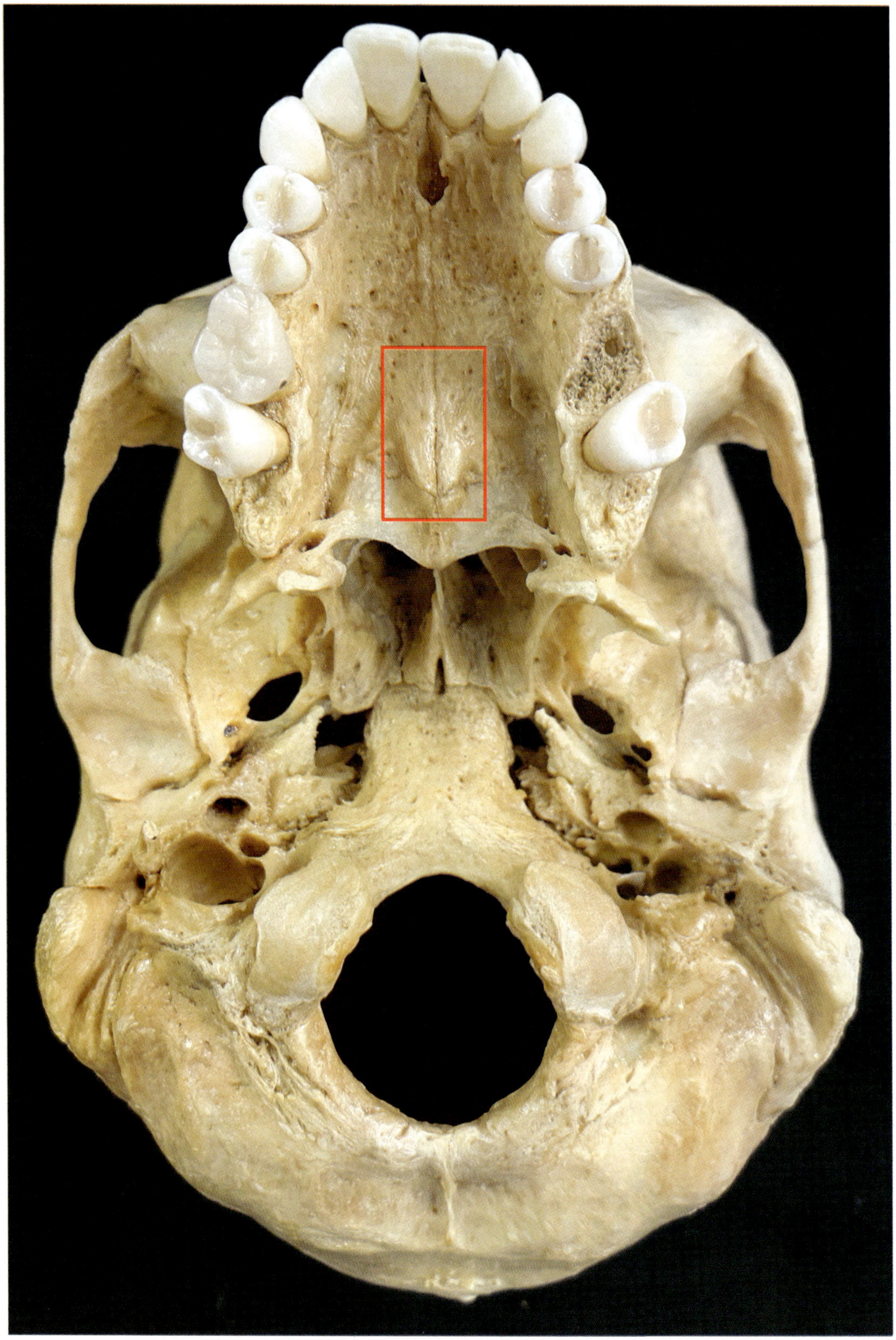

Figure 335. Inferior view 73-year-old European female with spatulate incisors, a small torus palatinus (rectangle), and a V-shaped and narrow maxillary arch.

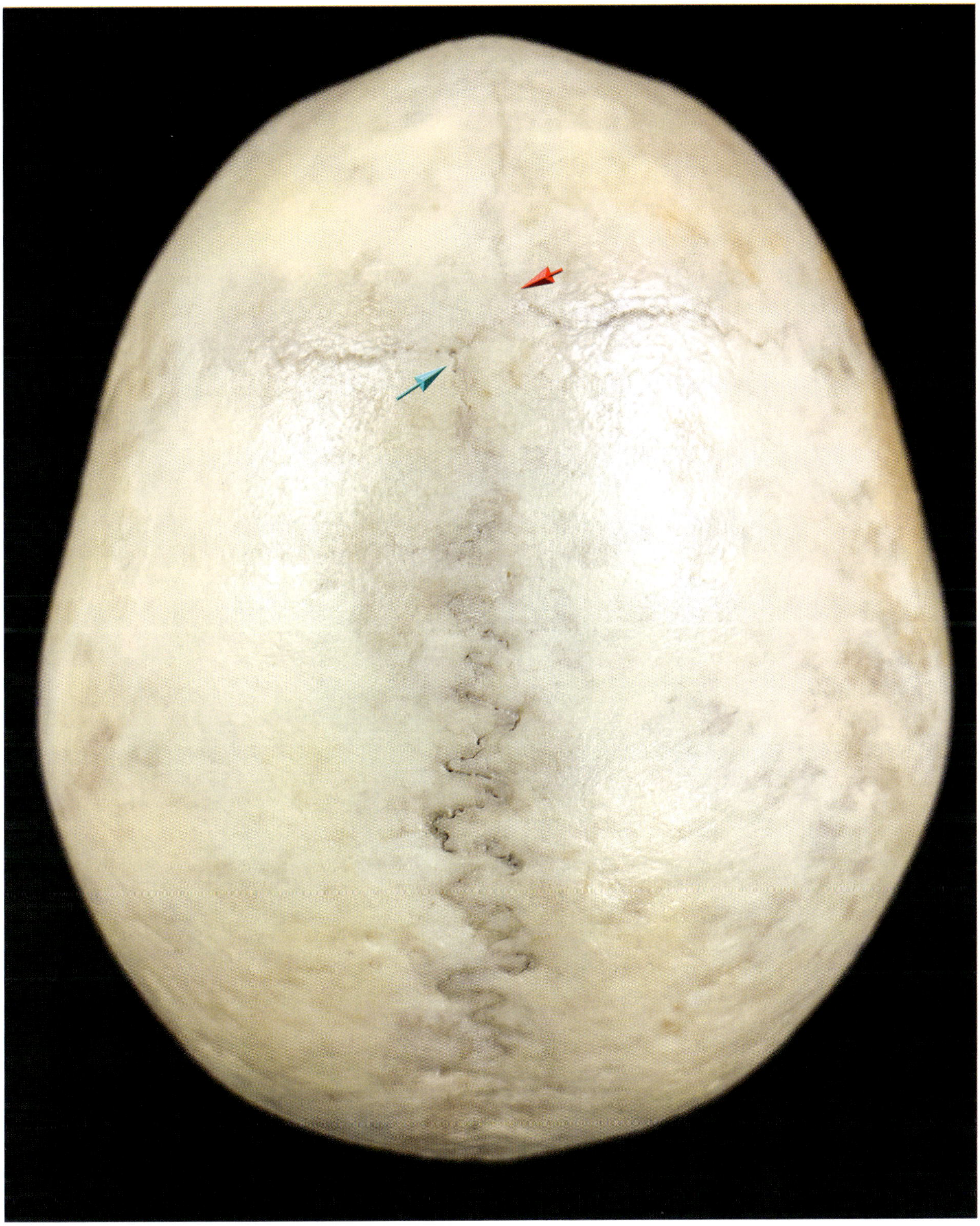

Figure 336. Superior view 73-year-old European female with asymmetry at the junction of the metopic (red arrow) and sagittal (blue arrow) sutures. Asymmetry of the sutures such as this is a common variant.

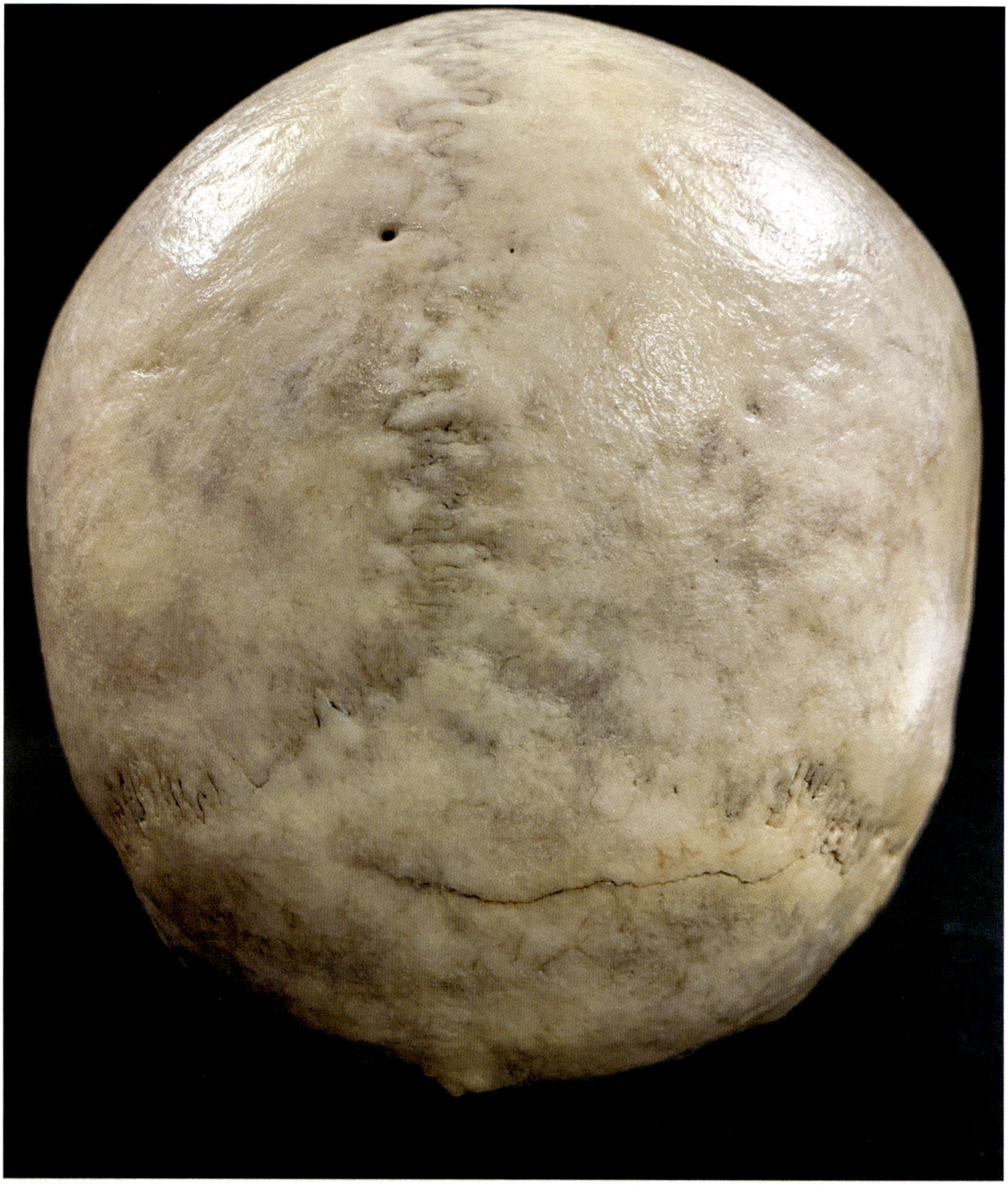

Figure 337. Posterior view 73-year-old European female. The horizontal fracture in the occipital bone is postmortem (taphonomic).

VI. HISPANIC CRANIA

(2 Female)

1. HISPANIC (MEXICAN) EUROPEAN FEMALE

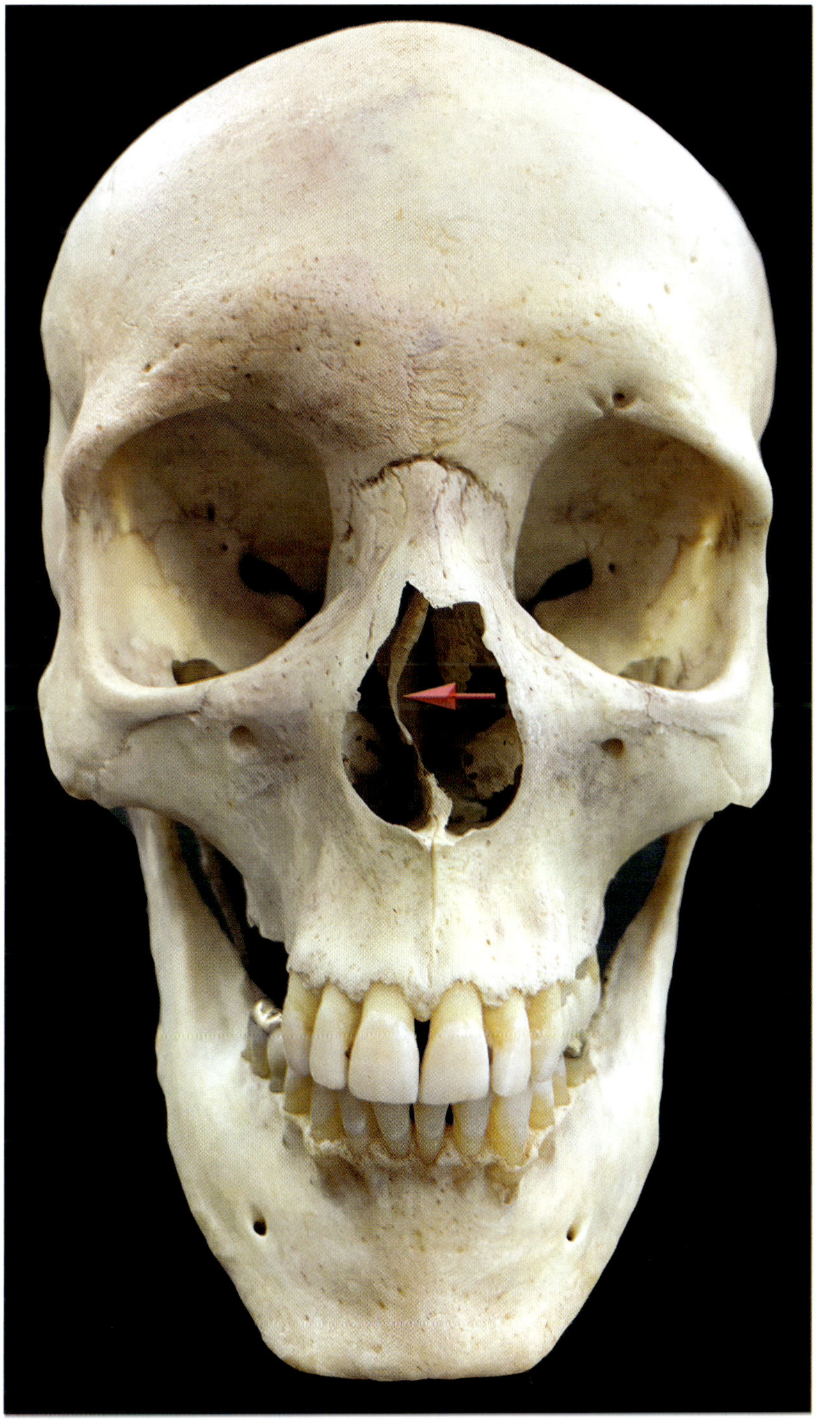

Figure 338. Anterior view 89-year-old Hispanic (Mexican) European female with a small brow ridge, trace supranasal suture, a broad interorbital and nasal aperture breadth, a deviated nasal septum (arrow), blunt supraorbital margin, and a sharp inferior nasal aperture/sill. This individual exhibits a broad and square chin. This female exhibits small zygomas, a broad and square chin, and a vermiculate pattern of pitting and short, shallow grooves along the brow ridge that is often, but not exclusively, found in males.

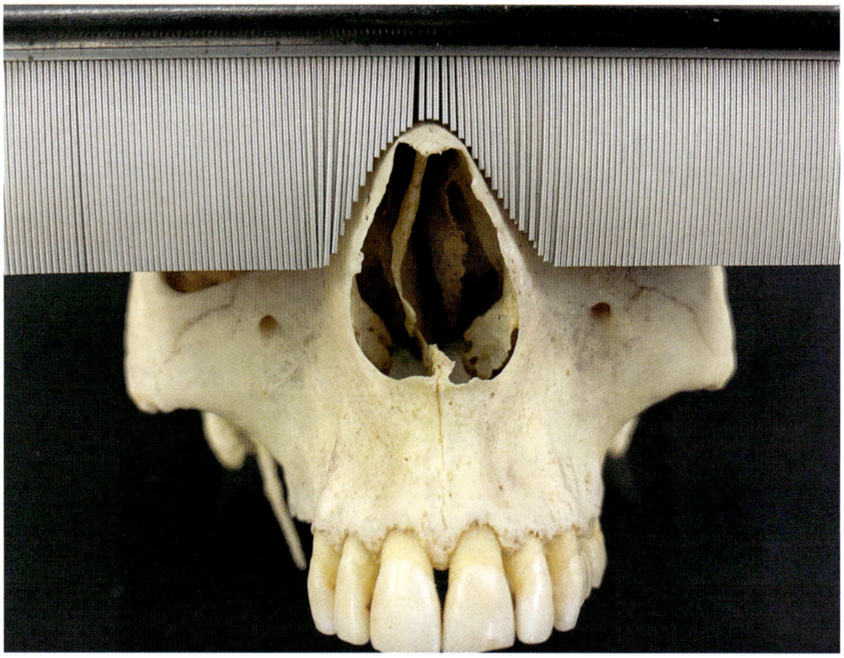

Figure 339. Nasal bone contour in an 89-year-old Hispanic (Mexican) European female.

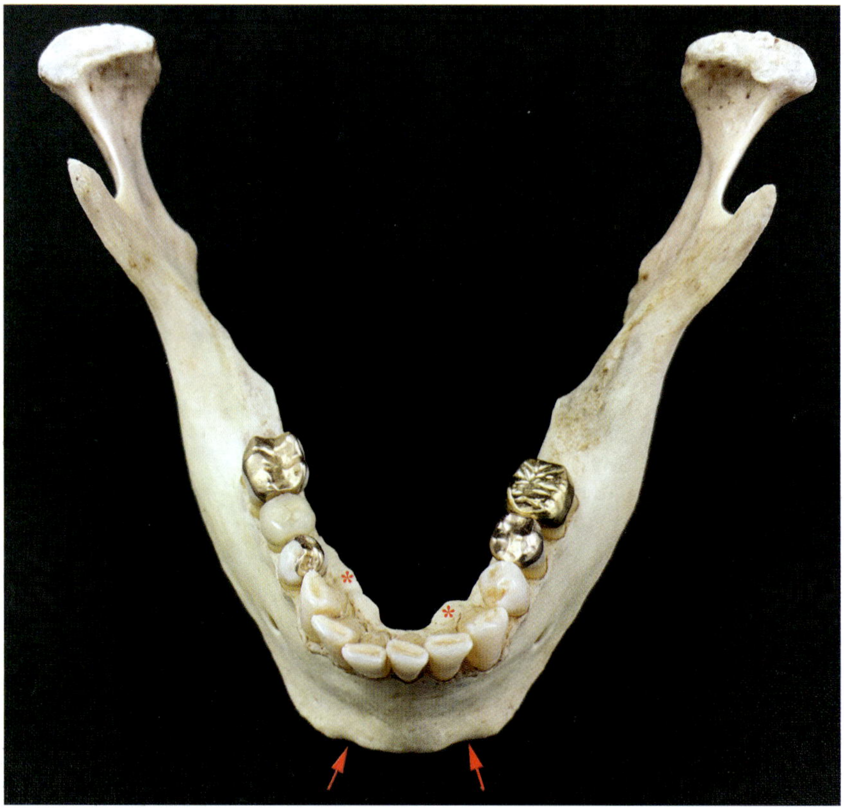

Figure 340. Mandible of 89-year-old Hispanic (Mexican) European female with lingual exostoses (*) and unusual indentations (arrow).

TABLE 30.
Cranial Measurements in an 89-year-old Hispanic (Mexican) European Female

Measurement	(mm)	Right	Left
Maximum Cranial Length (g-op, GOL)	180		
Maximum Cranial Breadth (eu-eu, XCB)	144		
Bizygomatic Breadth (zy-zy, ZYB)	125		
Basion-Bregma Height (ba-b, BBH)	137		
Cranial Base Length (ba-n, BNL)	98		
Basion-Prosthion (ba-pr, BPL)	94		
Maxillo-Alveolar Breadth (ecm-ecm, MAB)	52		
Maxillo-Alveolar Length (pr-alv, MAL)	53		
Biauricular Breadth (au-au, AUB)	121		
Upper Facial Height (n-pr)	65		
Minimum Frontal Breadth (ft-ft, WFB)	95		
Upper Facial Breadth (fmt-fmt)	101		
Nasal Height (n-ns, NLH)	45		
Nasal Breadth (al-al, NLB)	23		
Orbital Breadth (d-ec, OBB)	41		L
Orbital Height (OBH)	31		L
Biorbital Breadth (ec-ec, EKB)	92		
Interorbital Breadth (d-d, DKB)	24		
Frontal Chord (n-b, FRC)	110		
Parietal Chord (b-l, PAC)	112		
Occipital Chord (l-o, OCC)	101		
Foramen Magnum Length (ba-o, FOL)	36		
Foramen Magnum Breadth (FOB)	29		
Mastoid Length (MDH)		21	21

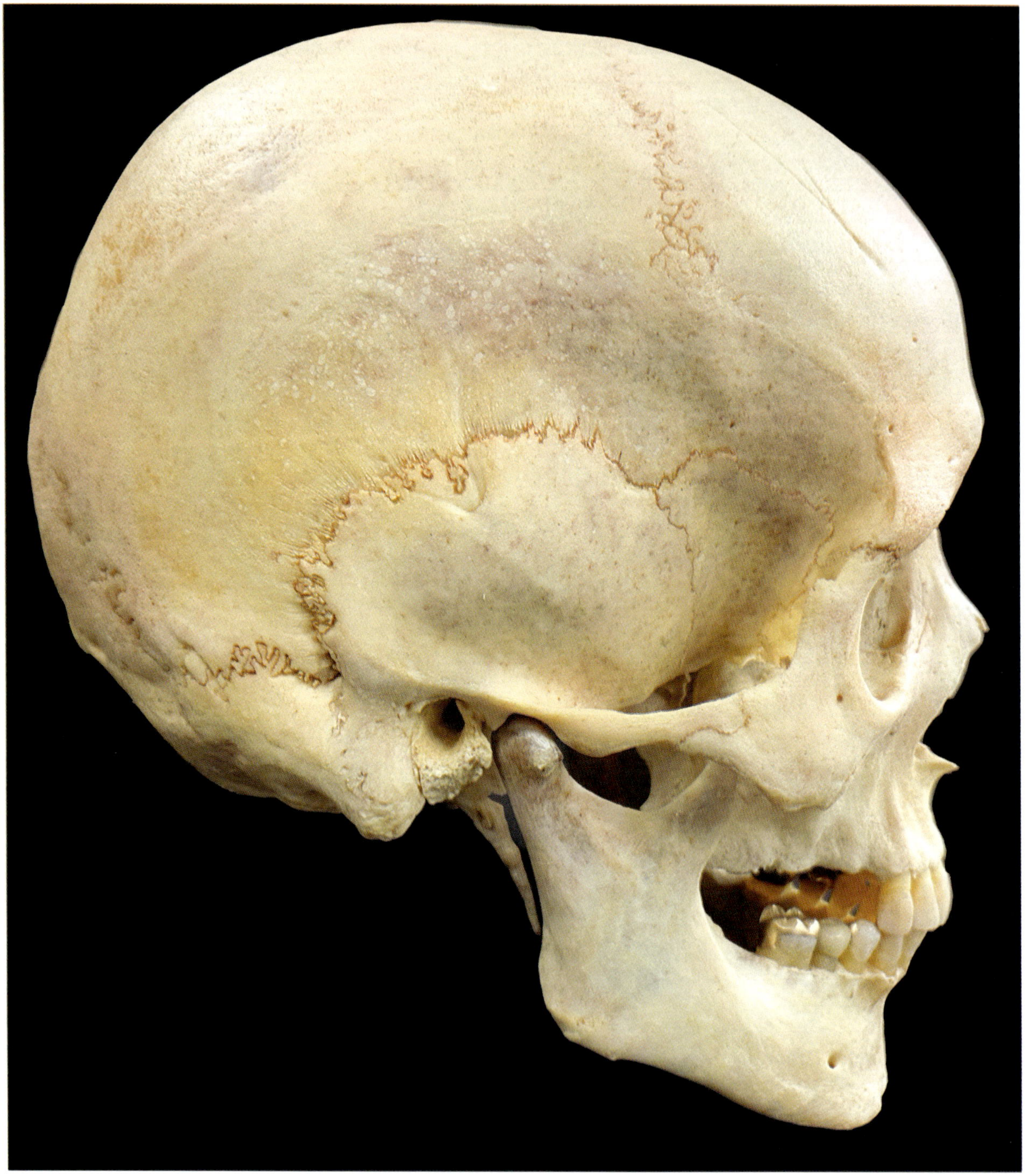

Figure 341. Right lateral view 89-year-old Hispanic (Mexican) European female with a small brow ridge with frontal grooves (arrow), a vertical frontal bone superior to glabella, small and low nasal bones, an anterior nasal spine, small supramastoid crest, and a small mastoid process. The frontal grooves (arrow) transmit vessels and could be mistaken for cuts or trauma (see Mann and Hunt 2019).

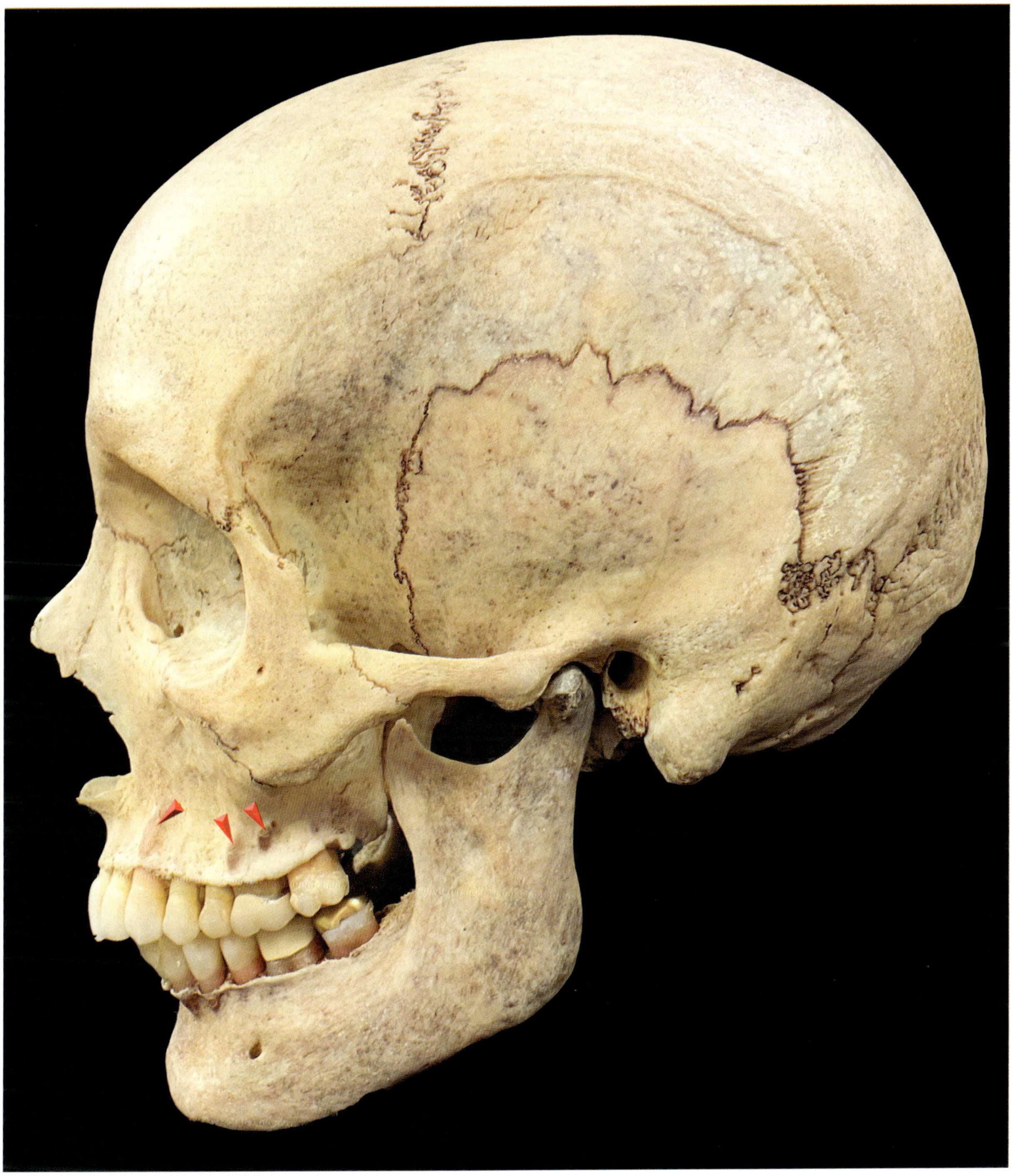

Figure 342. Left lateral view 89-year-old Hispanic (Mexican) European female. Note the areas of missing bone covering the roots of several teeth. These openings in the bone are known as fenestrations (arrowhead) and are often misinterpreted as occurring as a result of burial or postmortem handling. These fenestrations are probably the result of prominent tooth roots and thin alveolar bone (Fiorellini et al. 2006; Sun et al. 2019). The squamosal suture in the area of the parietal bone notch is unusually complex and exhibits numerous small ossicles (bone islands) representing anatomical variation.

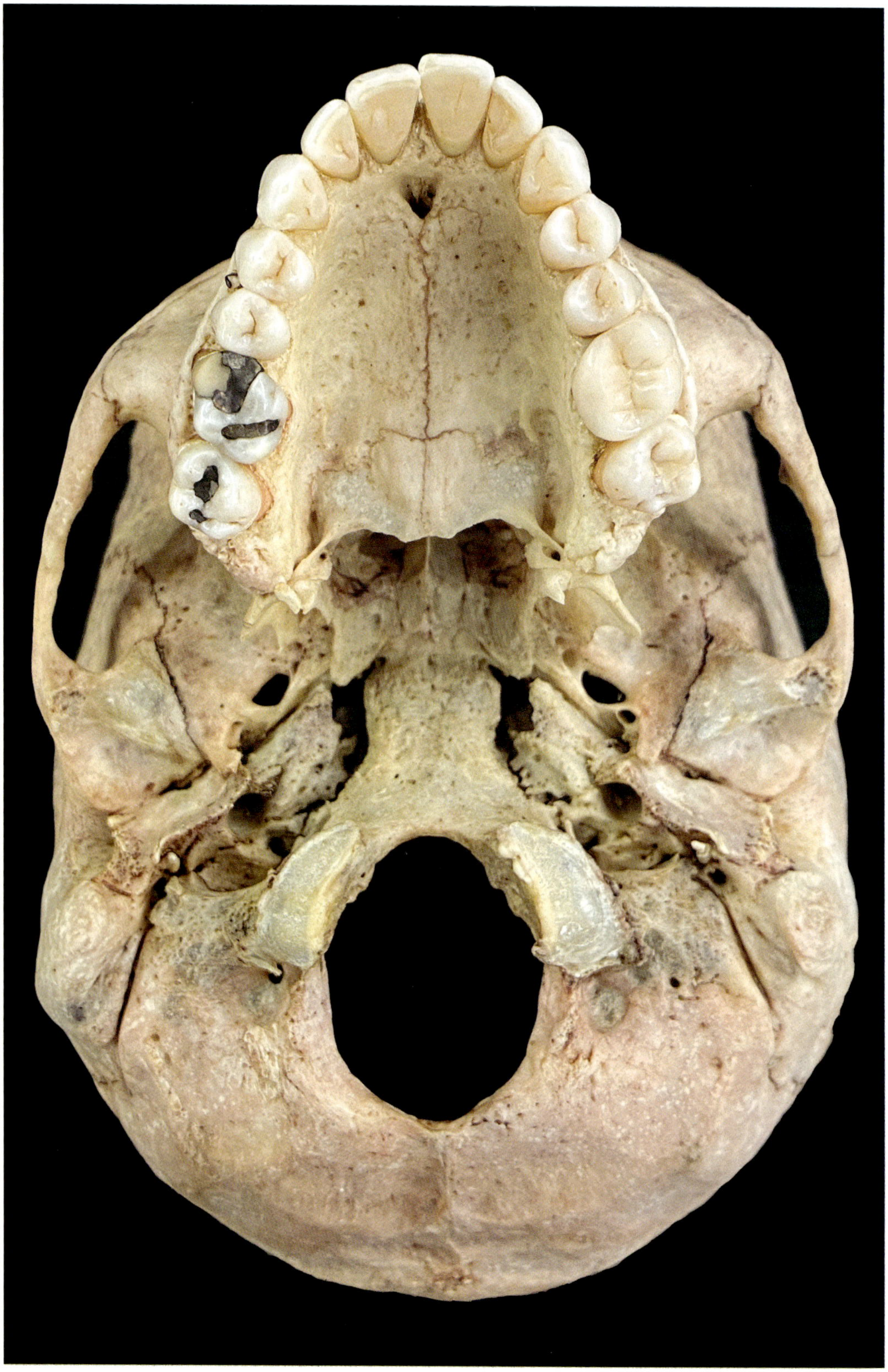

Figure 343. Inferior view 89-year-old Hispanic (Mexican) European female with a wide and deep maxilla missing only the third molars, and small mastoid processes.

Figure 344. Superior view 89-year-old Hispanic (Mexican) European female with a short and wide cranium.

Figure 345. Posterior view 89-year-old Hispanic (Mexican) European female with bilateral foramina (arrow), parietal bossing/squaring of the eminences, and nearly complete obliteration of the sutures.

2. HISPANIC (PUERTO RICAN) FEMALE

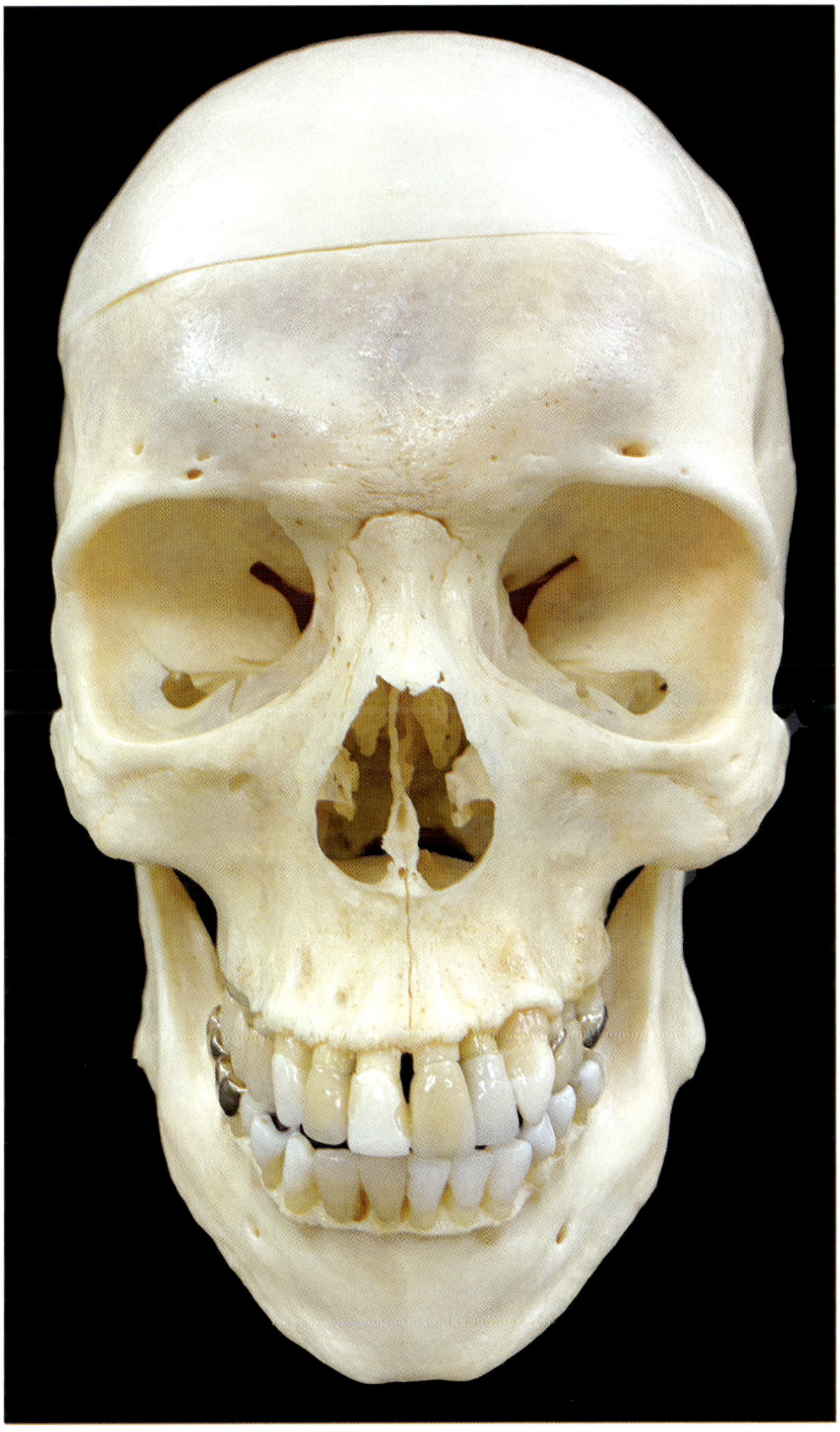

Figure 346. Anterior view 58-year-old Hispanic female (Puerto Rican) with a small brow ridge, trace supranasal suture (see Hefner 2005, 2015; Hefner et al. 2015), large nasal bones, an oval nasal aperture, well-defined nasal rim and sill, a square chin, and moderate gonial eversion.

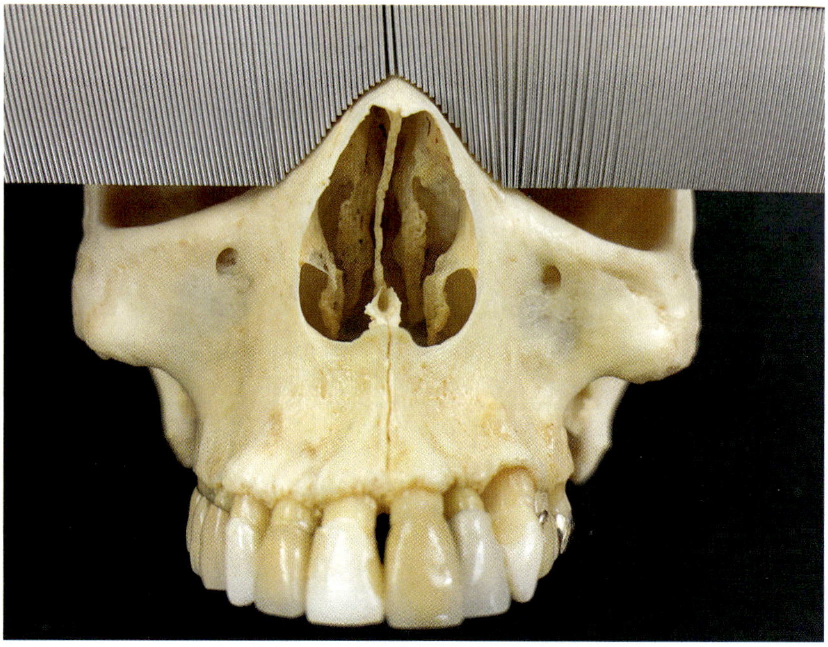

Figure 347. Nasal bone contour in a 58-year-old Hispanic female (Puerto Rican).

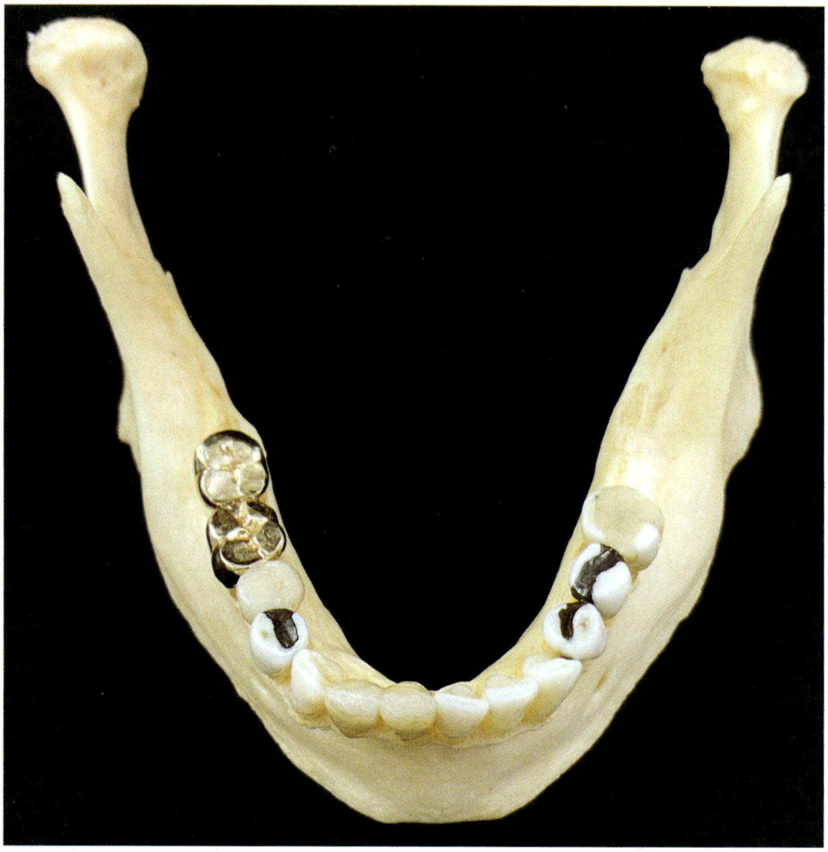

Figure 348. Mandible of 58-year-old Hispanic female (Puerto Rican) exhibiting moderate to severe arthritic destruction of the condyles.

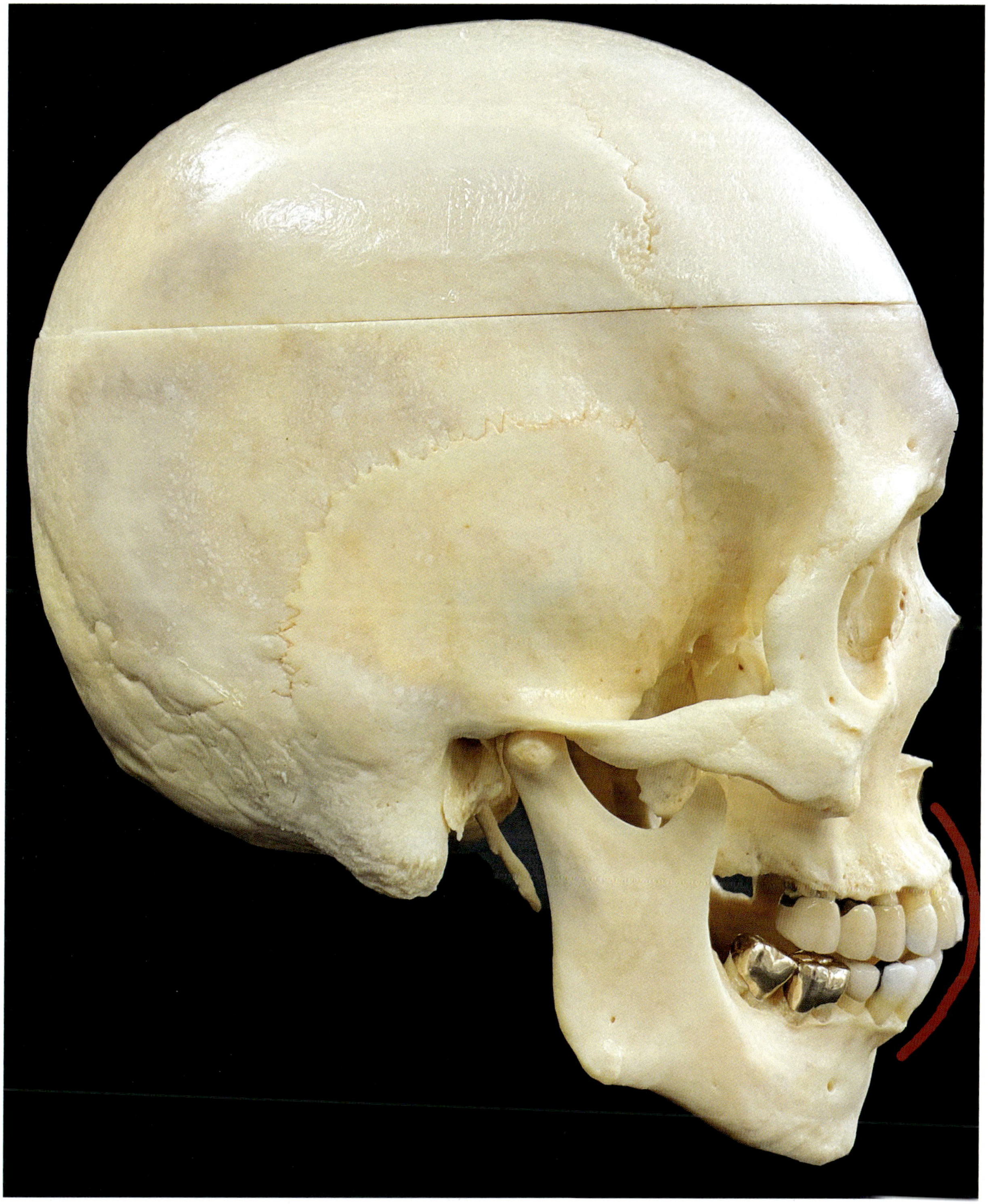

Figure 349. Right lateral view 58-year-old Hispanic female (Puerto Rican) with a rounded cranium, sloping frontal bone, a small brow ridge (glabella region), small and low nasal bones, a small anterior nasal spine, alveolar prognathism (curved red line), flat supramastoid crest, and a small mastoid process.

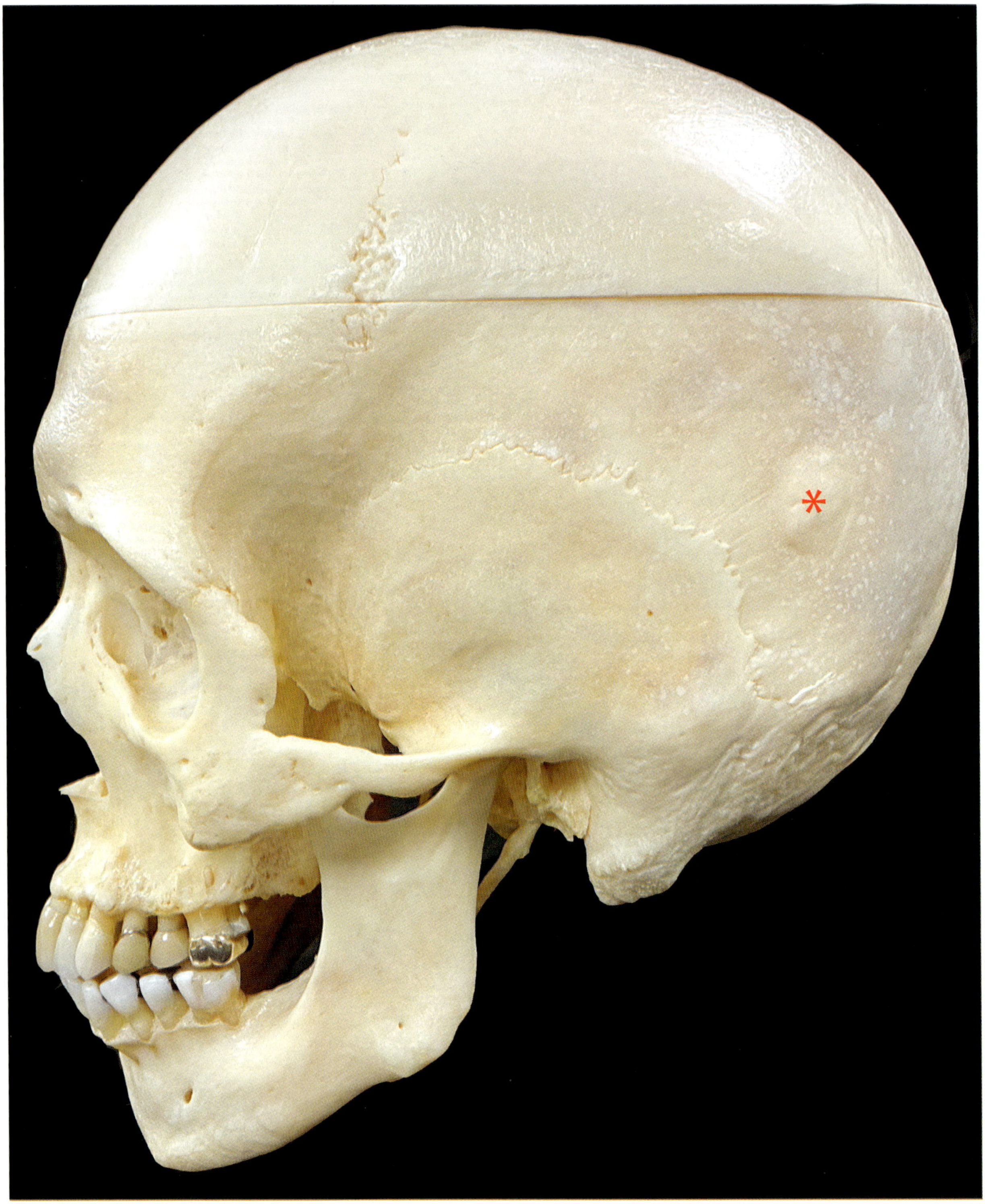

Figure 350. Left lateral view 58-year-old Hispanic female (Puerto Rican) with a small lesion (*), possibly a button osteoma, in the left parietal bone.

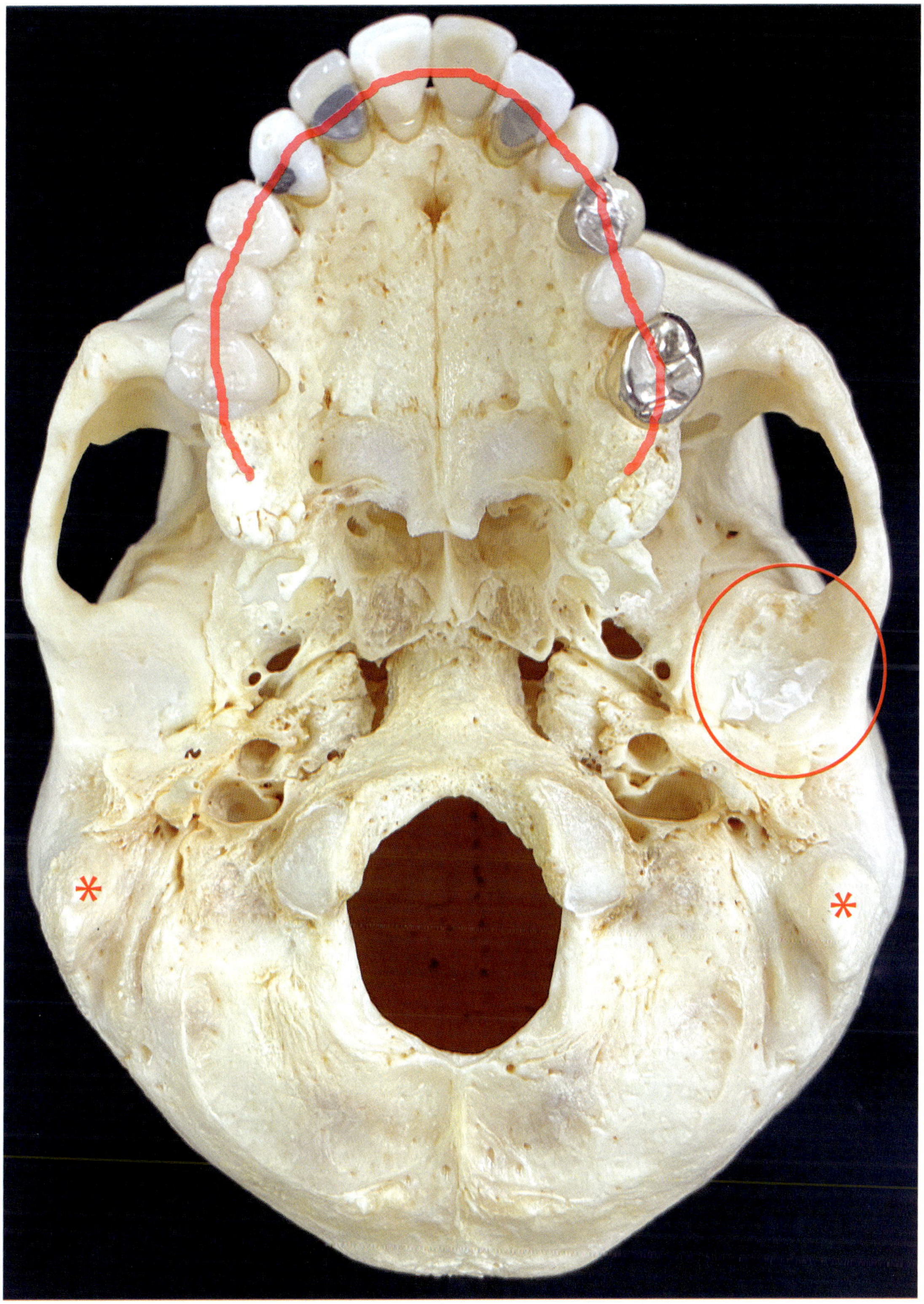

Figure 351. Inferior view 58-year-old Hispanic female (Puerto Rican) with a wide and shallow palate/maxilla that is U-shape and curves toward the midline posteriorly (curved red line), small mastoid processes (*), and degenerative joint disease of the left temporomandibular joint (circle).

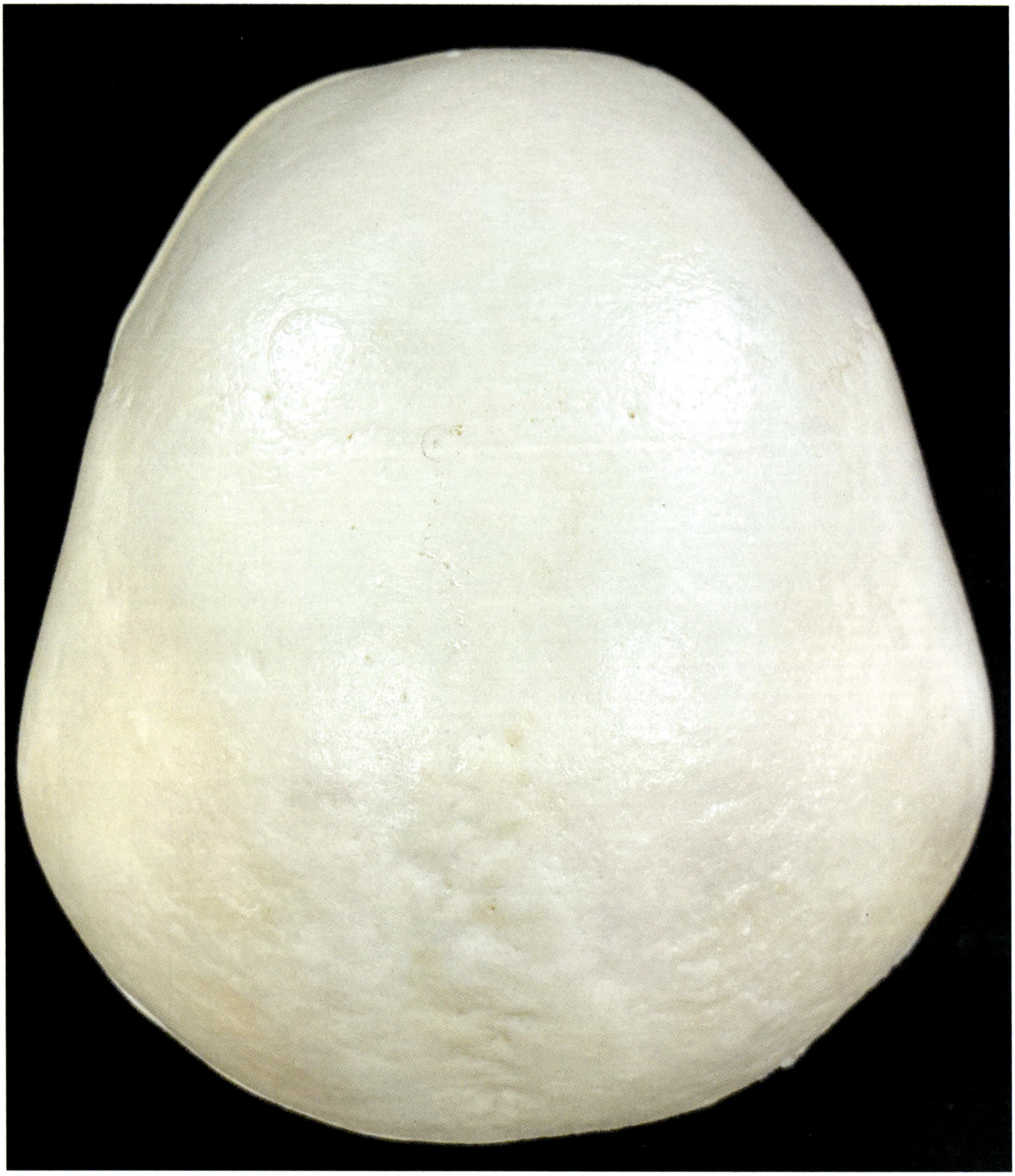

Figure 352. Superior view 58-year-old Hispanic female (Puerto Rican) with a short and wide cranium, parietal bossing, and complete obliteration of the sutures.

Figure 353. Posterior view 58-year-old Hispanic female (Puerto Rican) with prominent parietal bossing/squaring.

VII. PACIFIC ISLANDER CRANIA

(4 Male and 4 Female)

1. HAWAIIAN FEMALE

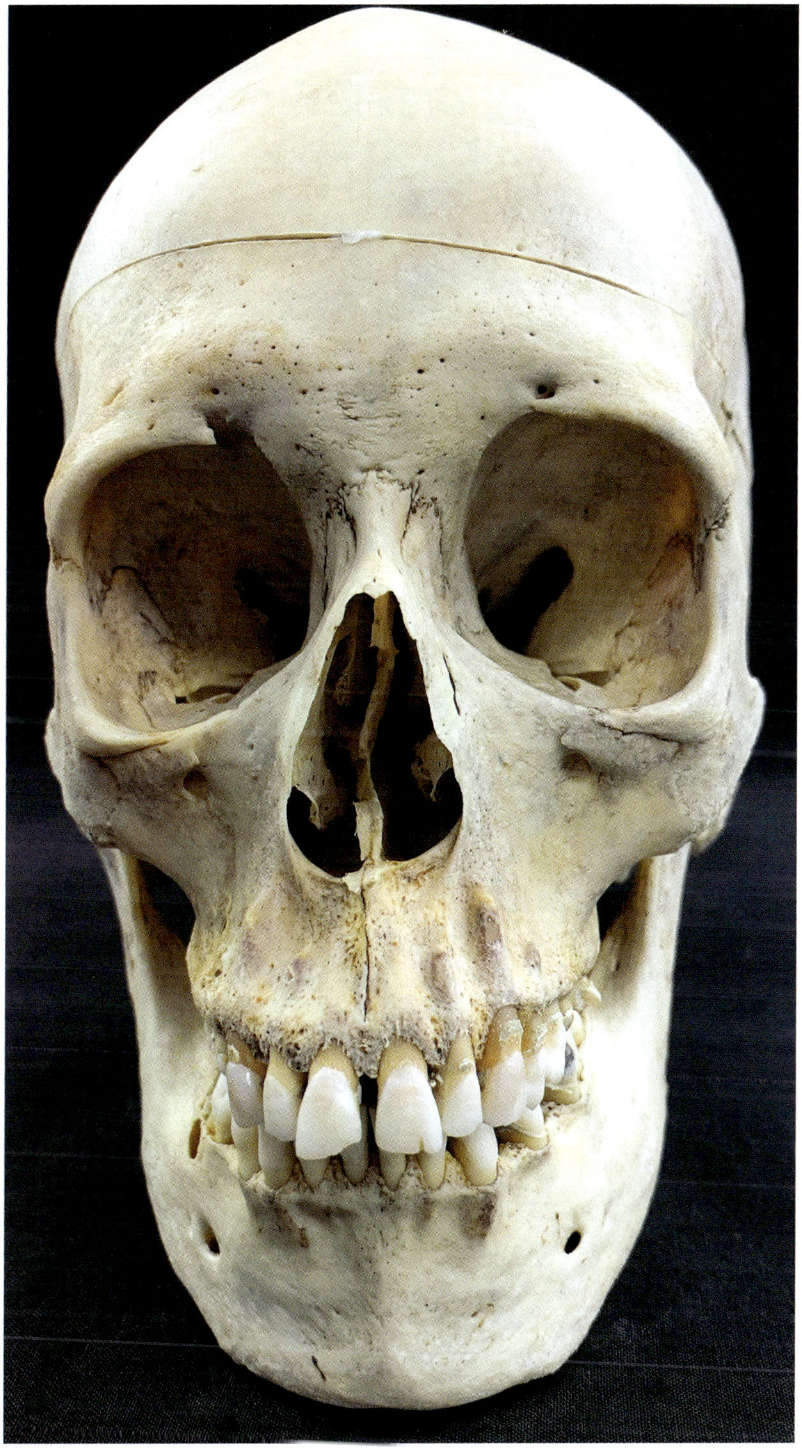

Figure 354. Anterior view of 53-year-old Hawaiian female. Note the angular orbits, the moderate interorbital breadth, the moderately wide and tall nasal aperture and the "waisted" nasal bones forming the bridge of the nose. Finally, note the nasal guttering along the inferior nasal aperture (see Figure 356), the small nasal spine, and the rounded chin.

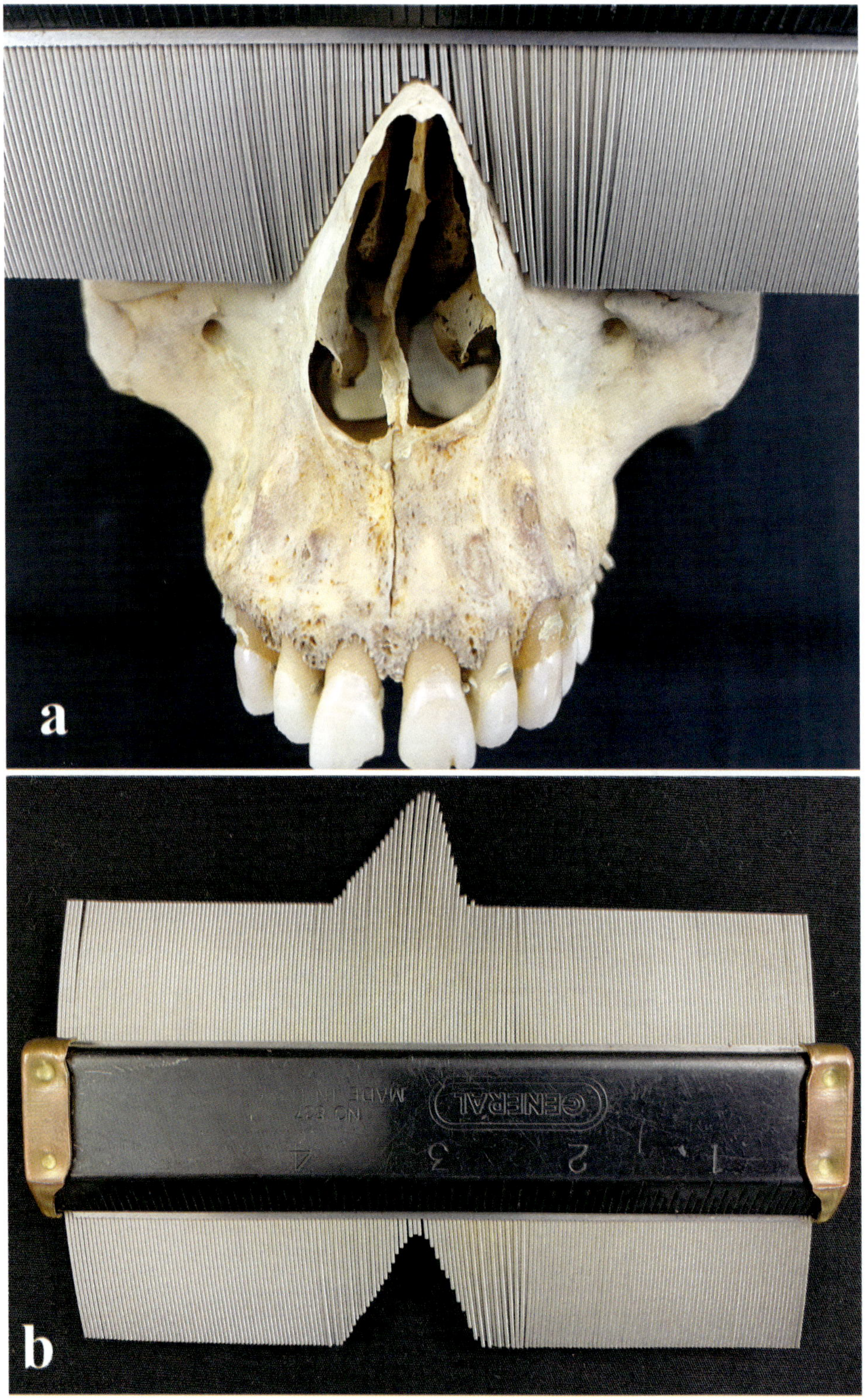

Figure 355a-b. (a) Nasal bone size (scale in inches) and contour using a contour gauge positioned perpendicular to the hard palate and approximately 1 centimeter from nasion. (b) The size and contour of the nasal bones and frontal process of the maxilla as revealed with the contour gauge. Care must be exercised when applying a contour gauge as the nasal bones and maxilla can be easily broken.

TABLE 31.
Cranial Measurements in a 53-year-old Hawaiian Female

Measurement	(mm)	Right	Left
Maximum Cranial Length (g-op, GOL)	180		
Maximum Cranial Breadth (eu-eu, XCB)	131		
Bizygomatic Breadth (zy-zy, ZYB)	120		
Basion-Bregma Height (ba-b, BBH)	144		
Cranial Base Length (ba-n, BNL)	102		
Basion-Prosthion (ba-pr, BPL)	100		
Maxillo-Alveolar Breadth (ecm-ecm, MAB)	61		
Maxillo-Alveolar Length (pr-alv, MAL)	58		
Biauricular Breadth (au-au, AUB)	119		
Upper Facial Height (n-pr)	71		
Minimum Frontal Breadth (ft-ft, WFB)	85		
Upper Facial Breadth (fmt-fmt)	96		
Nasal Height (n-ns, NLH)	51		
Nasal Breadth (al-al, NLB)	22		
Orbital Breadth (d-ec, OBB)	38		L
Orbital Height (OBH)	34		L
Biorbital Breadth (ec-ec, EKB)	90		
Interorbital Breadth (d-d, DKB)	16		
Frontal Chord (n-b, FRC)	111		
Parietal Chord (b-l, PAC)	121		
Occipital Chord (l-o, OCC)	102		
Foramen Magnum Length (ba-o, FOL)	37		
Foramen Magnum Breadth (FOB)	30		
Mastoid Length (MDH)		23	21

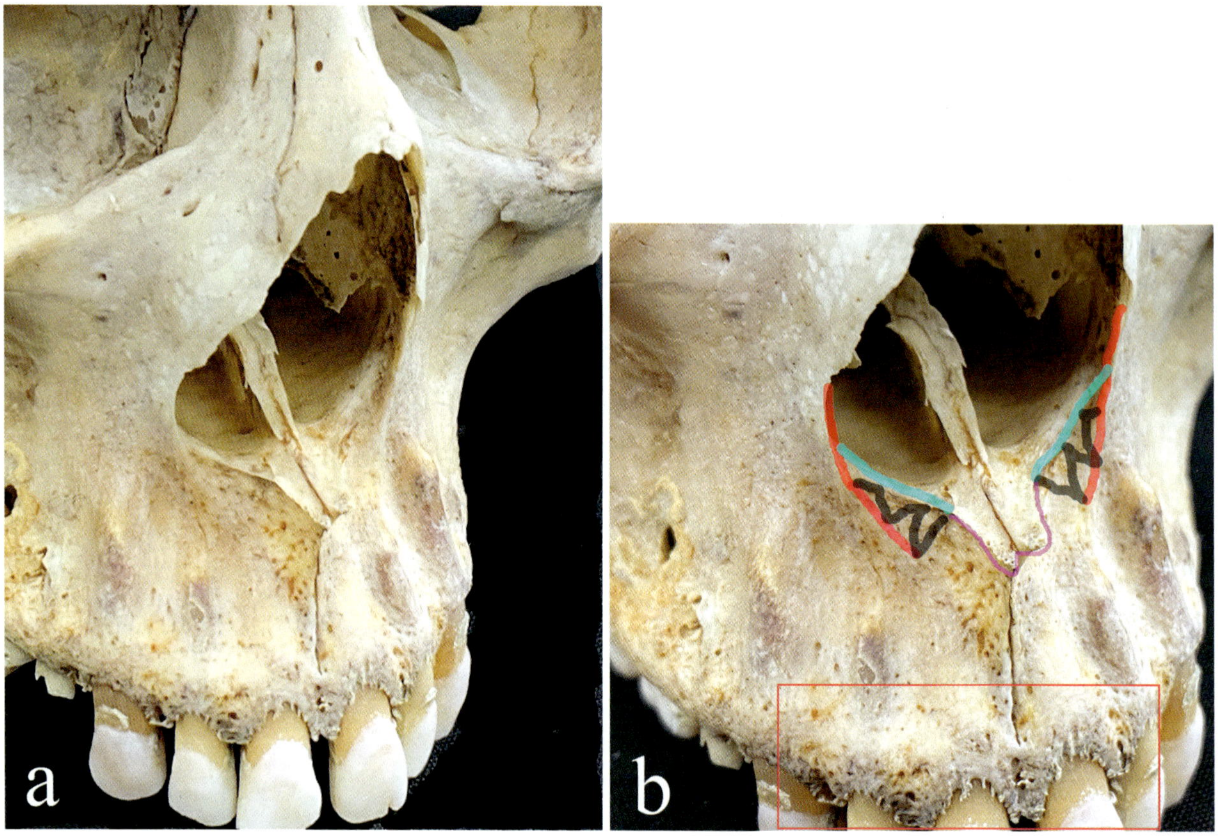

Figure 356a-b. (a) Right superior-oblique view of 53-year-old Hawaiian female showing a developed anterior nasal spine and nasal guttering. (b) Note that nasal guttering consists of two components/features. The first component is the "gutter" (black line): a downward sloping area of bone that the author refers to as "internal and external" rims (green=internal and red=external). Nasal guttering, in varying expressions, is typically present in African, African American, and Asian groups, but may be present in any ancestral group and at any age. The second component is the anterior nasal spine (outlined in purple). In this individual the anterior nasal spine is an extension of the internal rim. Note that periodontal disease (red rectangle) has resulted in erosion/resorption of the alveolar bone.

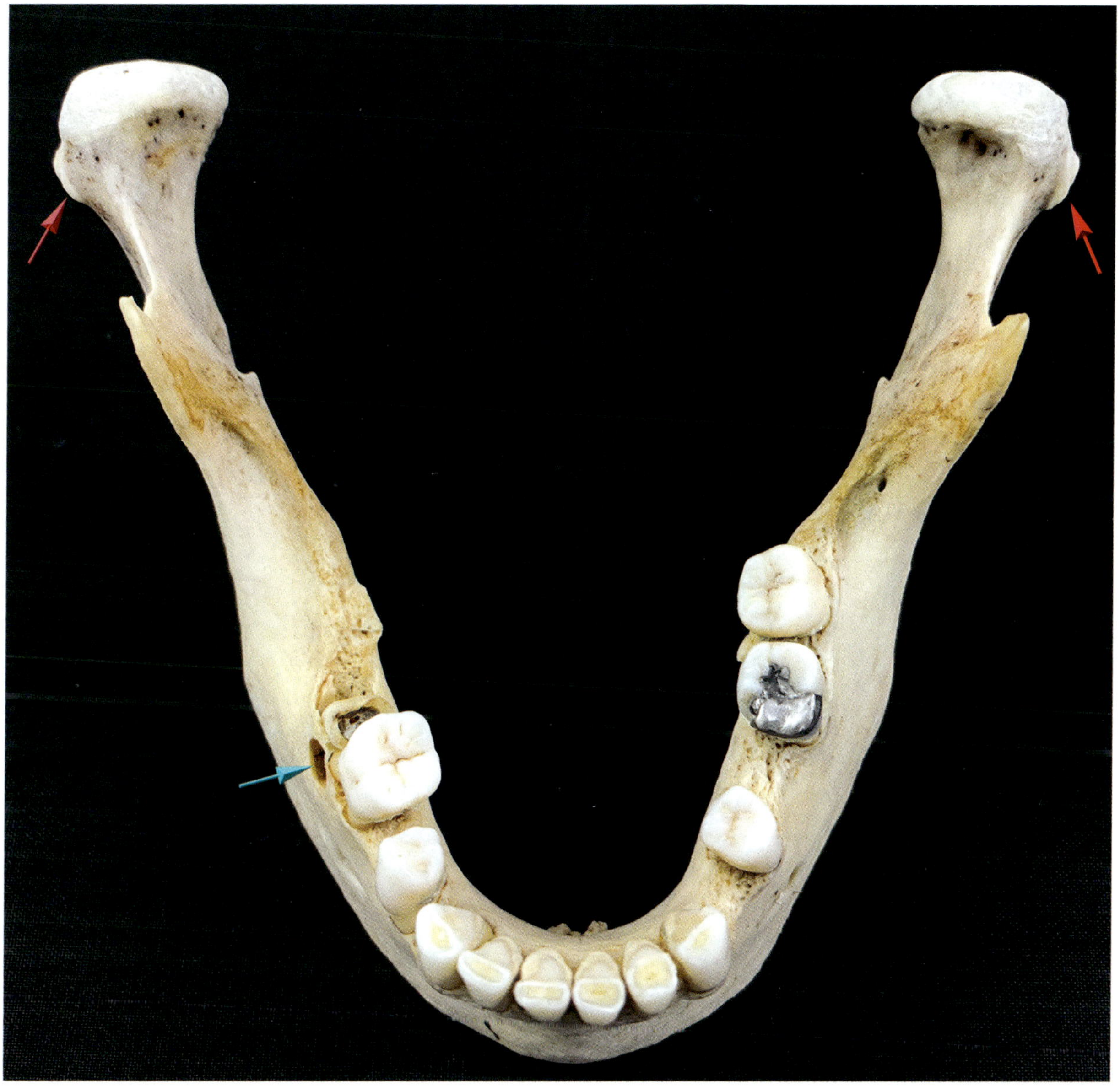

Figure 357. Mandible of 53-year-old Hawaiian female. The lateral condylar processes exhibit development at the site of the articular capsule (red arrow) and the right molar tooth exhibits a periapical lesion/abscess (blue arrow).

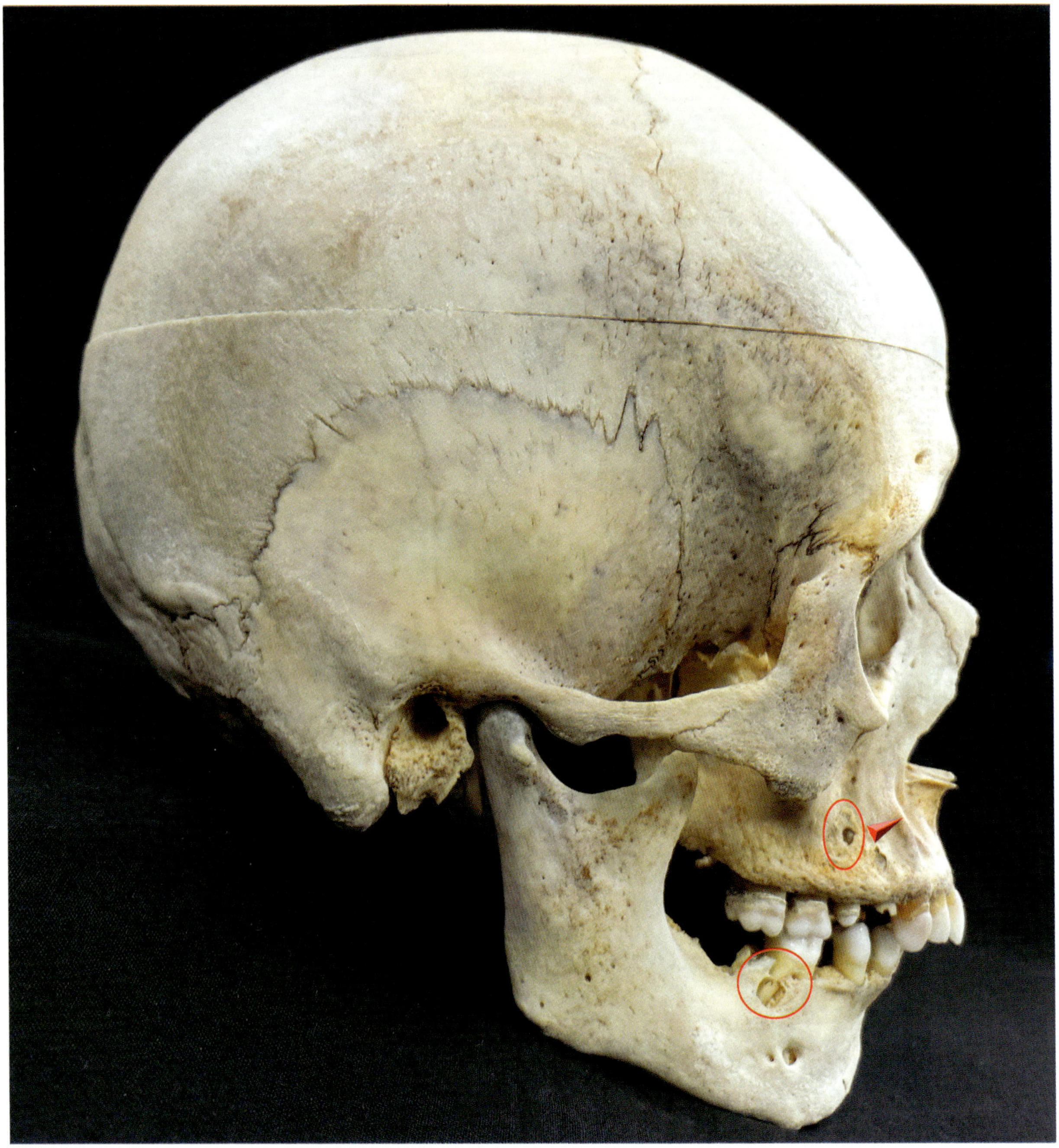

Figure 358. Right lateral view of 53-year-old Hawaiian female. Note the small brow ridge and nasal bones, the vertical frontal bone, the small anterior nasal spine, the external rim (arrowhead) creating part of the nasal guttering, absence of nasal overgrowth; presence of alveolar prognathism, a small malar tubercle and mastoid process, and a developed supramastoid crest. A lower right molar and upper right premolar exhibit periapical lesions/abscesses (circle).

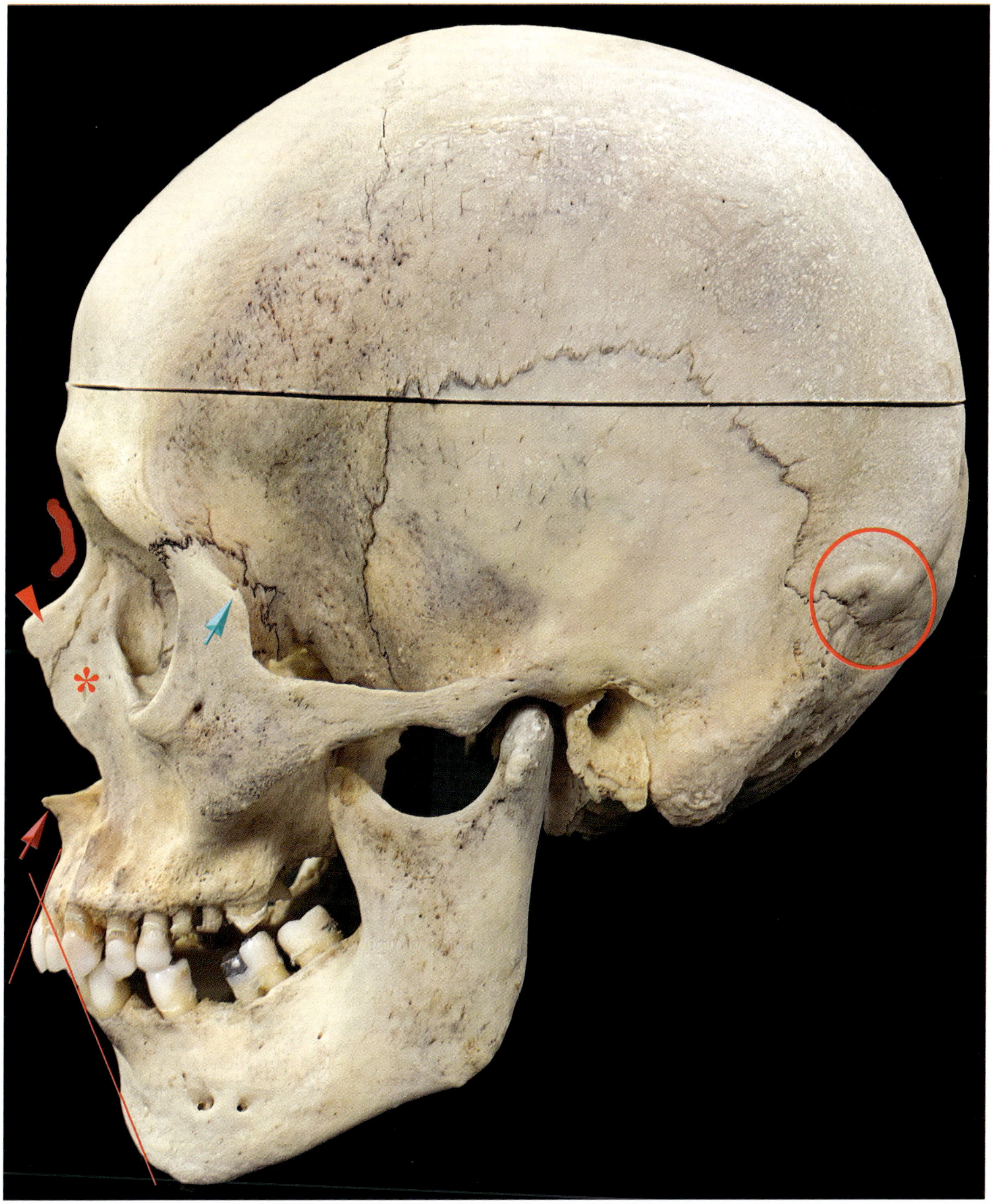

Figure 359. Left lateral view of 53-year-old Hawaiian female showing the nasion depression (curved red line), the projecting nasal bones (arrowhead) and frontal process (*) of the maxilla, the long, sharp anterior nasal spine (red arrow), alveolar prognathism (intersecting red lines), and the small posterior zygomatic tubercle (blue arrow). Note the occipital torus tubercle (circle) for muscle attachments on the parietal bone. Occipital torus consists of solid/dense bone without diploe (see Figure 362). This symmetrical and bilateral feature is seen almost exclusively in Oceania/Polynesia (Heathcote et al. 2014). The mandible exhibits double mental foramina, a normal anatomical variant, in both the right and left hemi-mandibles.

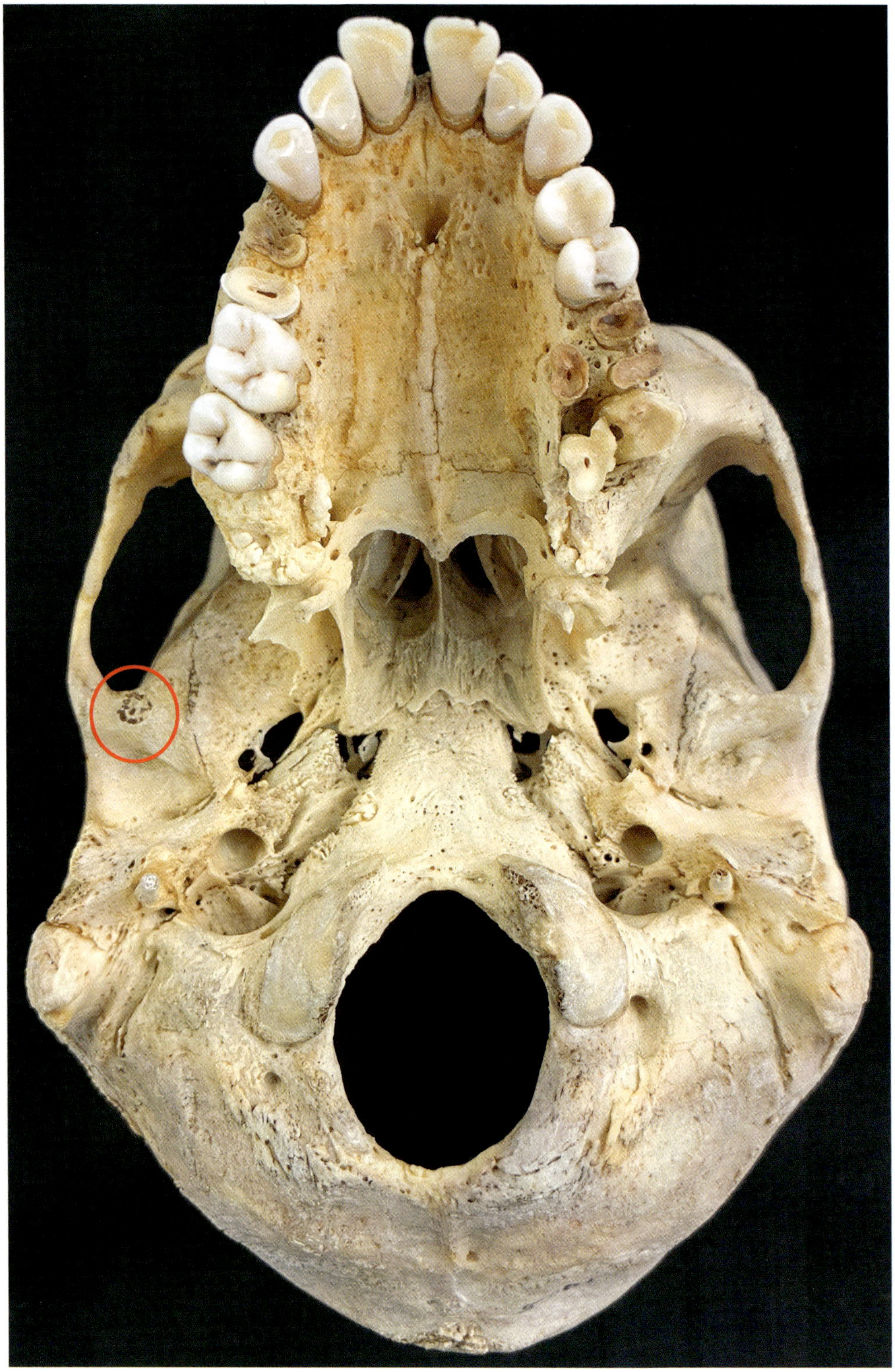

Figure 360. Inferior view of 53-year-old Hawaiian female. The lingual surfaces of the incisors are spatulate shaped; the hard palate is narrow and long with a small torus palatinus along the midline; the inion hook is mildly developed, and the mastoid processes are small. The articular eminence of the right temporomandibular joint exhibits arthritic pitting (circle) and several teeth are carious/decayed leaving only the roots.

Figure 361. Superior view of 53-year-old Hawaiian female. The sutures are simple and exhibit minimal obliteration. See Mann et al. (2009) for a study on suture complexity and ancestry, which found that the most complex sutures (sagittal) were in Caucasians and not Asians, as generally believed.

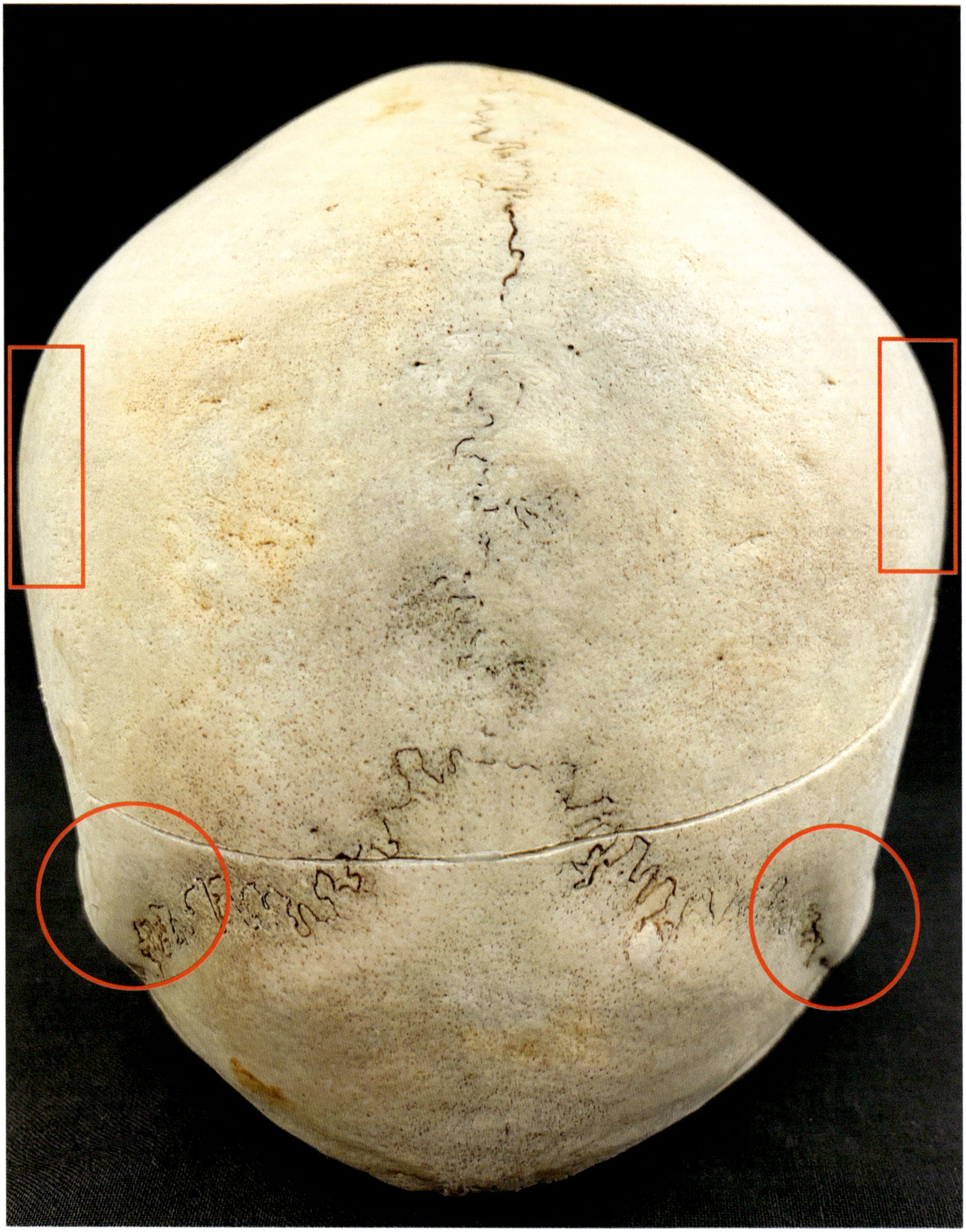

Figure 362. Posterior view of 53-year-old Hawaiian female. Note the symmetrical occipital tori tubercles (circle) in the parietal bones immediately superior to the lambdoidal suture, and the parietal bossing/squaring of the eminences (rectangle). Occipital tori such as these have been reported in Chamarro individuals from Guam (see Heathcote et al. 2014 and 2019).

2. HAWAIIAN CHINESE SPANISH MALE

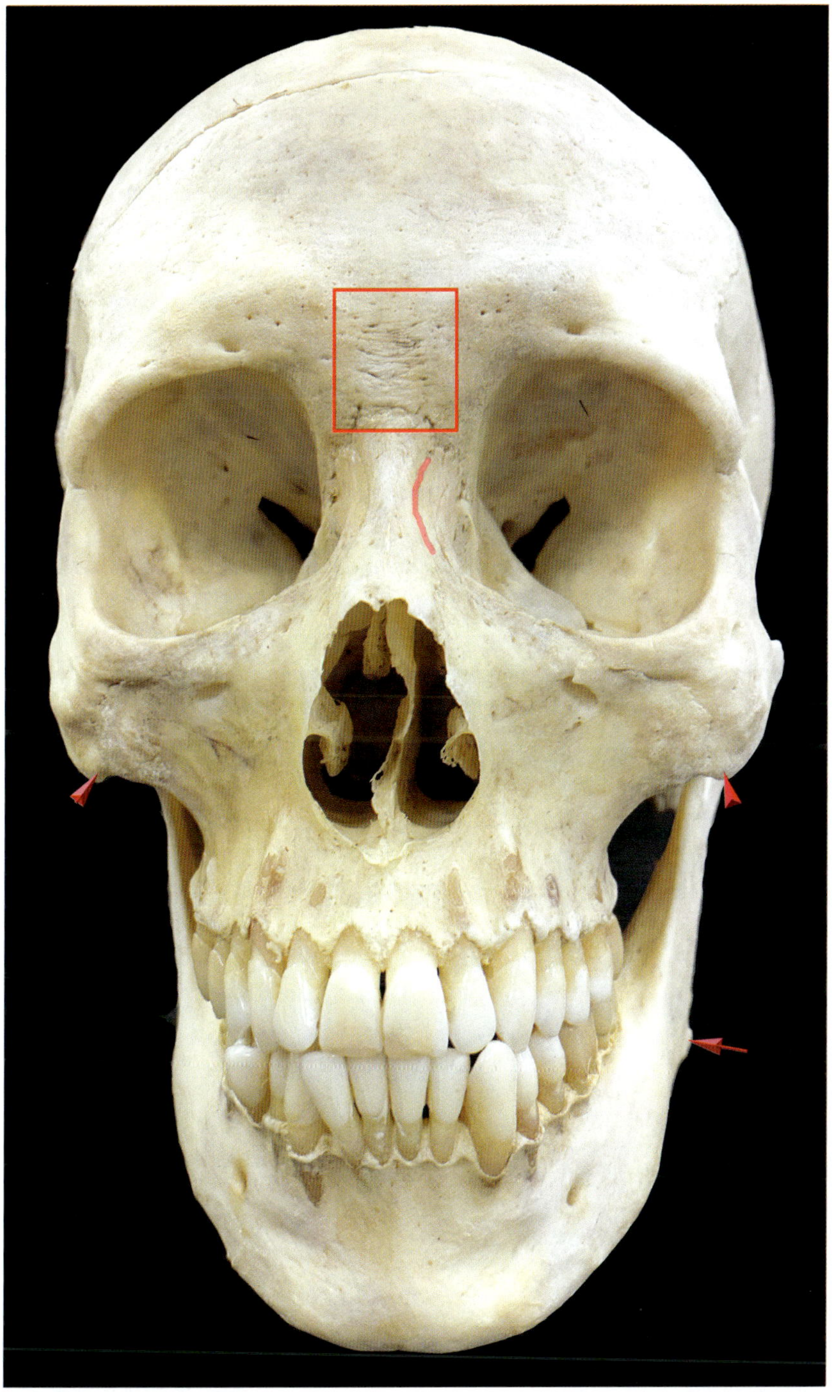

Figure 363. Anterior view 80-year-old Hawaiian Chinese Spanish male. This cranium exhibits a rounded vault, remnant of the supranasal suture (square; Mann et al. 2016); a moderate brow ridge; malar tubercles (arrowhead); a nearly complete nasal sill; a developed and projecting anterior nasal spine; narrow and waisted nasal bones (curved red line); a medium width nasal aperture; moderate interorbital breadth; a blunt superior margin of the orbits; angular orbits; a wide and rounded mandible, and gonial eversion (red arrow).

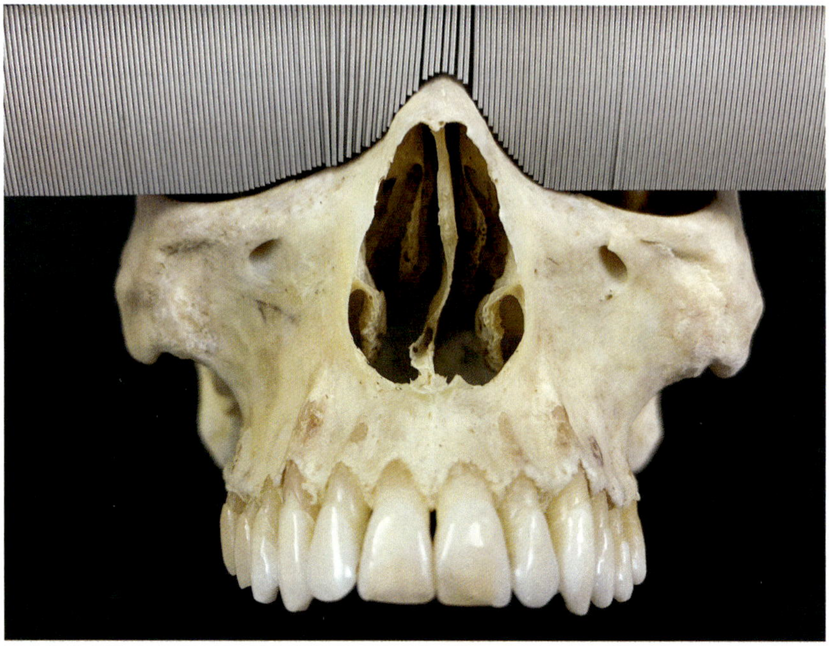

Figure 364. Nasal bone contour in an 80-year-old Hawaiian Chinese Spanish male.

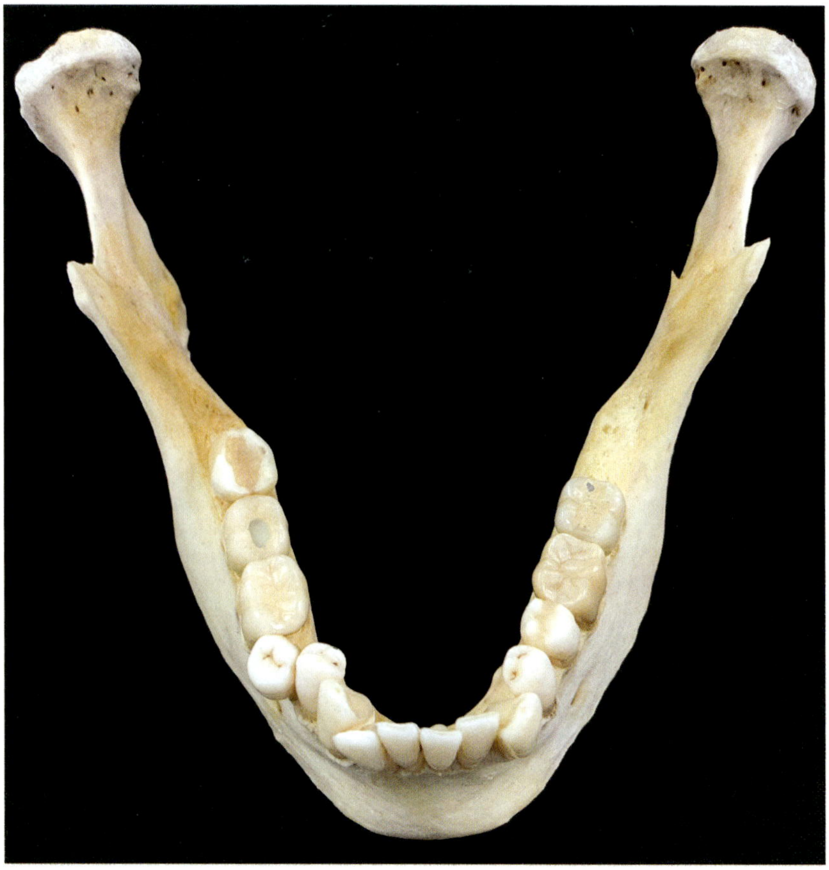

Figure 365. Mandible of 80-year-old Hawaiian Chinese Spanish male showing a rounded chin and a crowded and rotated right second premolar (bicuspid) tooth.

TABLE 32.
Cranial Measurements in an 80-year-old Hawaiian Chinese Spanish Male

Measurement	(mm)	Right	Left
Maximum Cranial Length (g-op, GOL)	174		
Maximum Cranial Breadth (eu-eu, XCB)	146		
Bizygomatic Breadth (zy-zy, ZYB)	133		
Basion-Bregma Height (ba-b, BBH)	131		
Cranial Base Length (ba-n, BNL)	98		
Basion-Prosthion (ba-pr, BPL)	96		
Maxillo-Alveolar Breadth (ecm-ecm, MAB)	68		
Maxillo-Alveolar Length (pr-alv, MAL)	54		
Biauricular Breadth (au-au, AUB)	132		
Upper Facial Height (n-pr)	70		
Minimum Frontal Breadth (ft-ft, WFB)	95		
Upper Facial Breadth (fmt-fmt)	101		
Nasal Height (n-ns, NLH)	59		
Nasal Breadth (al-al, NLB)	25		
Orbital Breadth (d-ec, OBB)	38		L
Orbital Height (OBH)	34		L
Biorbital Breadth (ec-ec, EKB)	94		
Interorbital Breadth (d-d, DKB)	21		
Frontal Chord (n-b, FRC)	112		
Parietal Chord (b-l, PAC)	111		
Occipital Chord (l-o, OCC)	99		
Foramen Magnum Length (ba-o, FOL)	36		
Foramen Magnum Breadth (FOB)	31		
Mastoid Length (MDH)		30	30

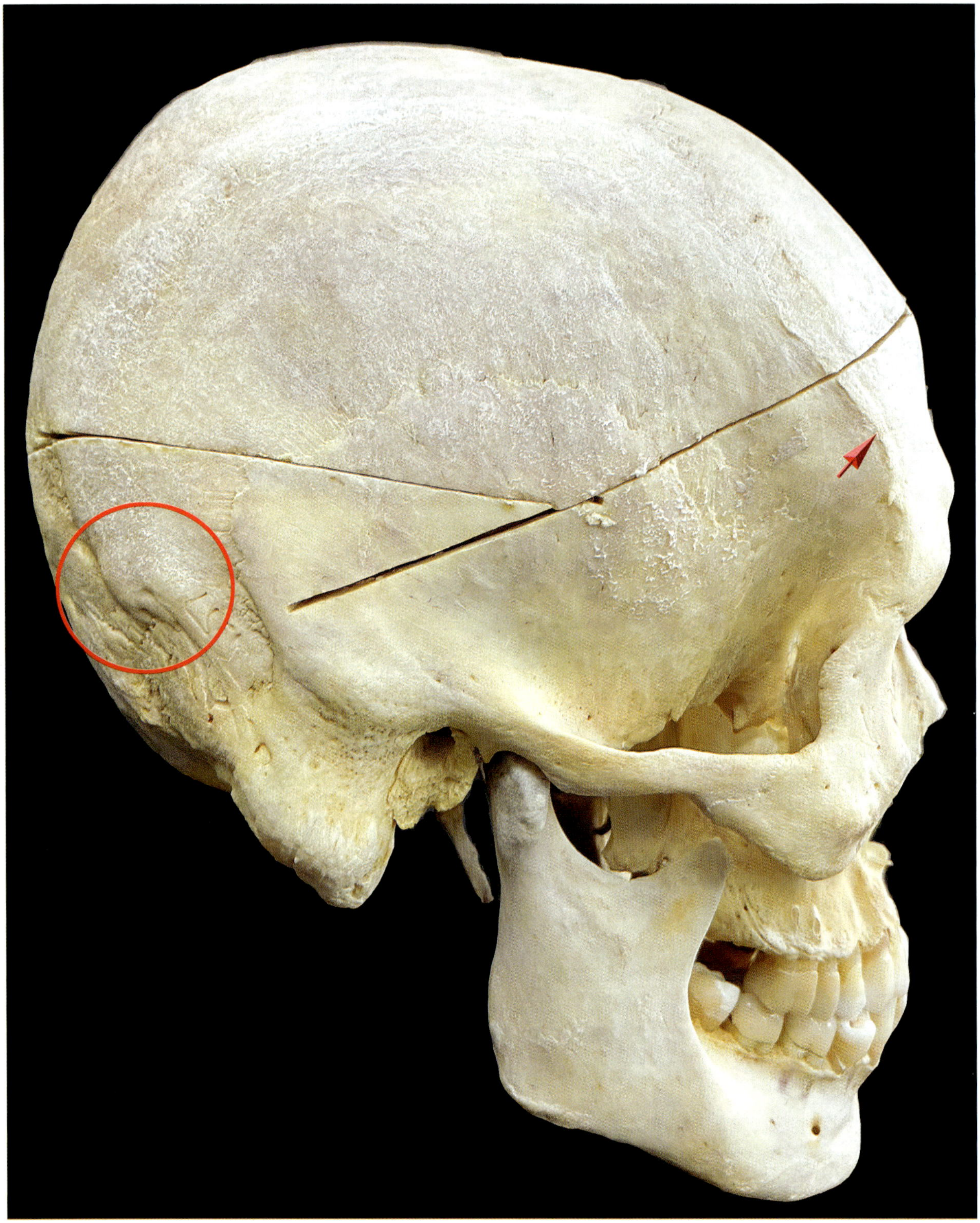

Figure 366. Right lateral view 80-year-old Hawaiian Chinese Spanish male. Note a developed brow ridge, sloping frontal bone, small nasal bones with minimal overgrowth; large mastoid process and supramastoid crest; medium posterior zygomatic tubercle; a robust mandible, developed superior temporal line (arrow); and a small occipital torus tubercle (circle) (Heathcote et al. 2014 and 2019).

Figure 367. Left lateral view 80-year-old Hawaiian Chinese Spanish male with minimal alveolar prognathism, small and short nasal bones, a robust zygomatic arch (*), and a short, rounded vault and occipital region.

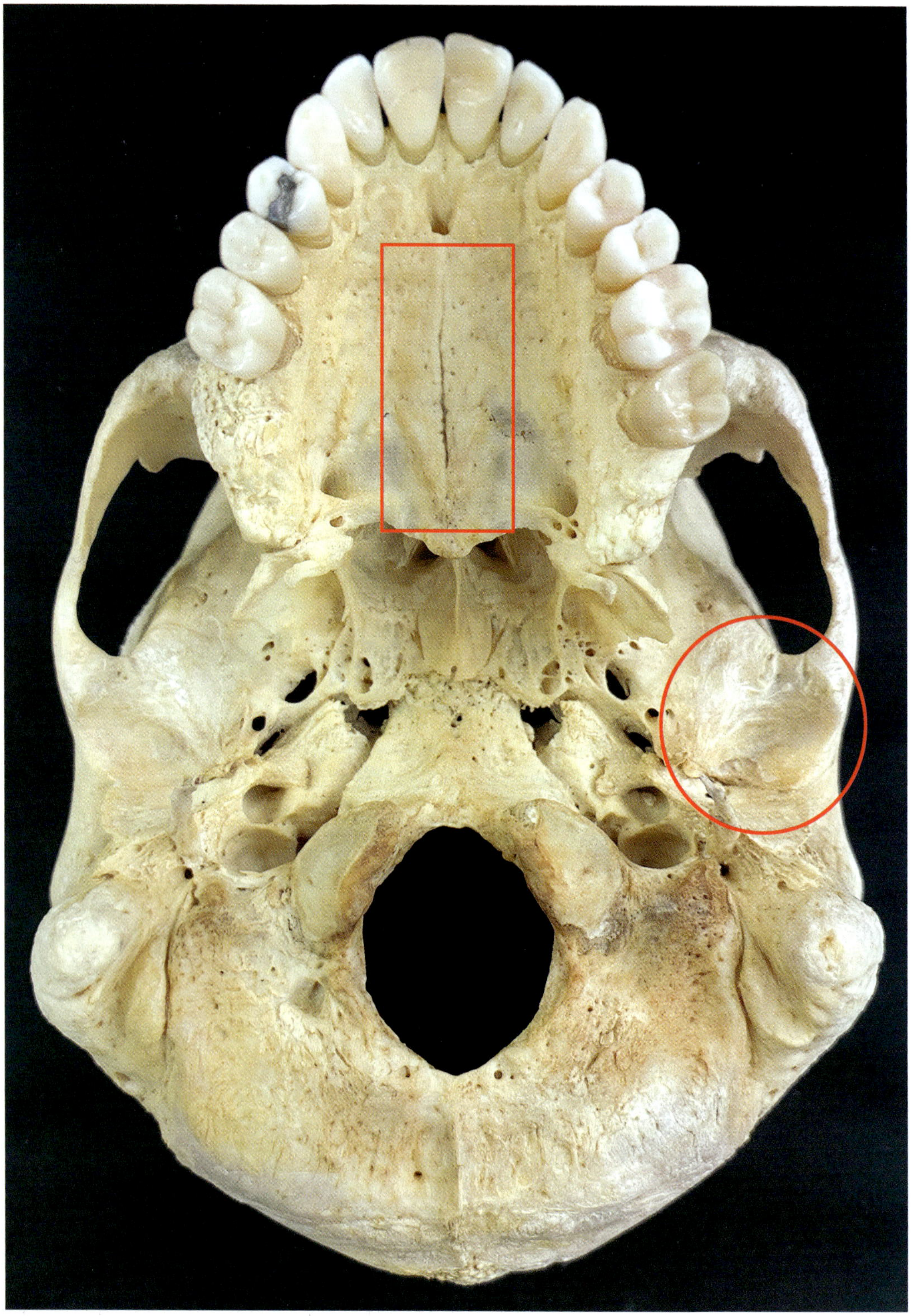

Figure 368. Inferior view 80-year-old Hawaiian Chinese Spanish male with a large, wide, and shallow hard palate with spatulate incisor teeth, a small torus palatinus (rectangle), and intermediate-size mastoid processes. The enlarged and flattened left temporomandibular joint (circle) indicates osteoarthritis, but did not affect the left mandibular condyle (see Figure 365). The posterior portion of the hard palate is thin and translucent due to osteoporosis.

Figure 369. Superior view 80-year-old Hawaiian Chinese Spanish male with incomplete obliteration of the sutures. Note the pitting in the outer vault, possibly indicative of iron-deficiency anemia.

Figure 370. Posterior view 80-year-old Hawaiian Chinese Spanish male. The developed nuchal crest area (where the neck muscles attach to the occipital) has an irregular and pitted surface, consistent with porotic hyperostosis possibly due to iron-deficiency anemia.

3. HAWAIIAN EUROPEAN MALE

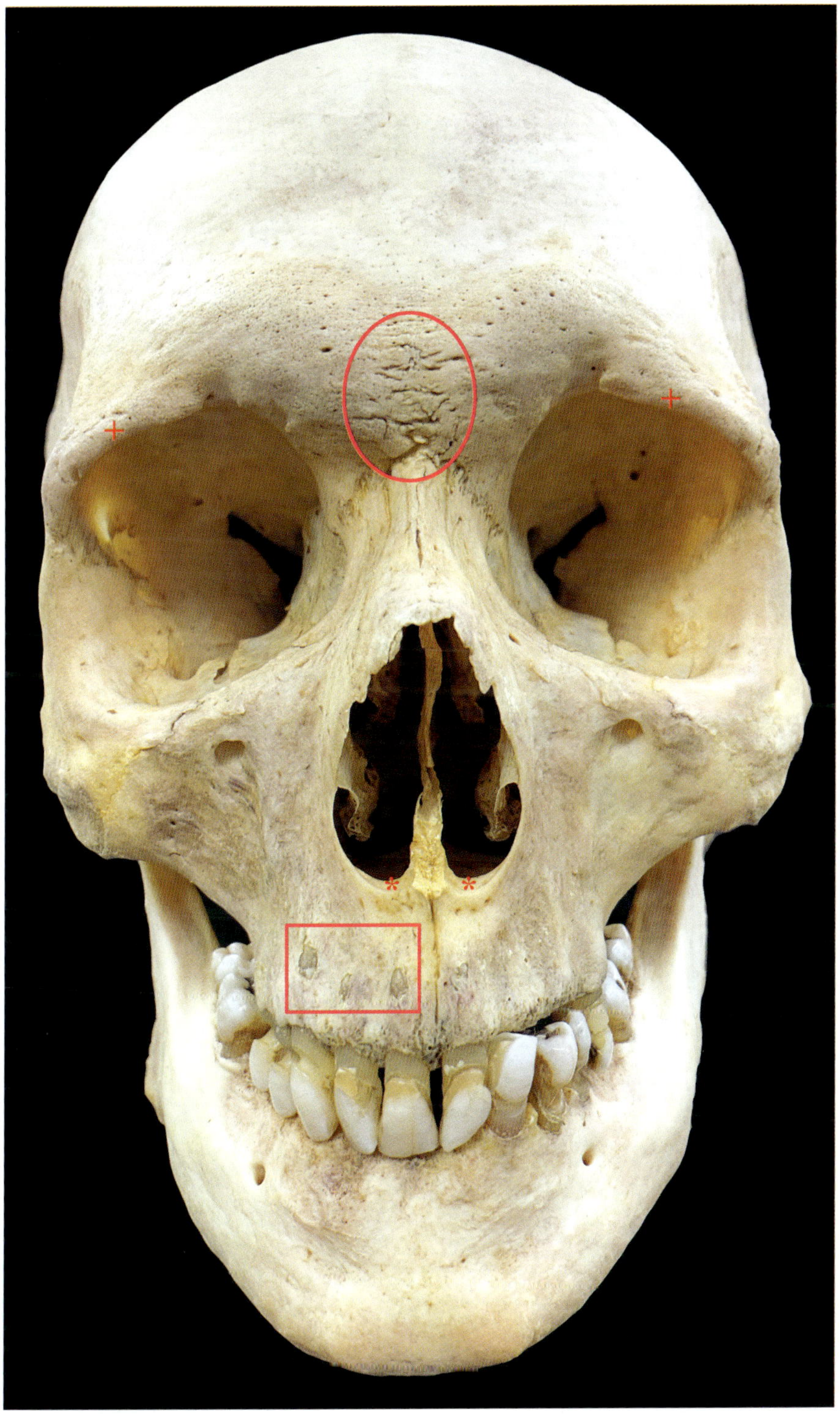

Figure 371. Anterior view 83-year-old Hawaiian European male. Note trace of the supranasal suture (oval), the blunt orbital rims (+), mild nasal guttering (*), and fenestrations (rectangle) indicating thin bone covering the tooth roots.

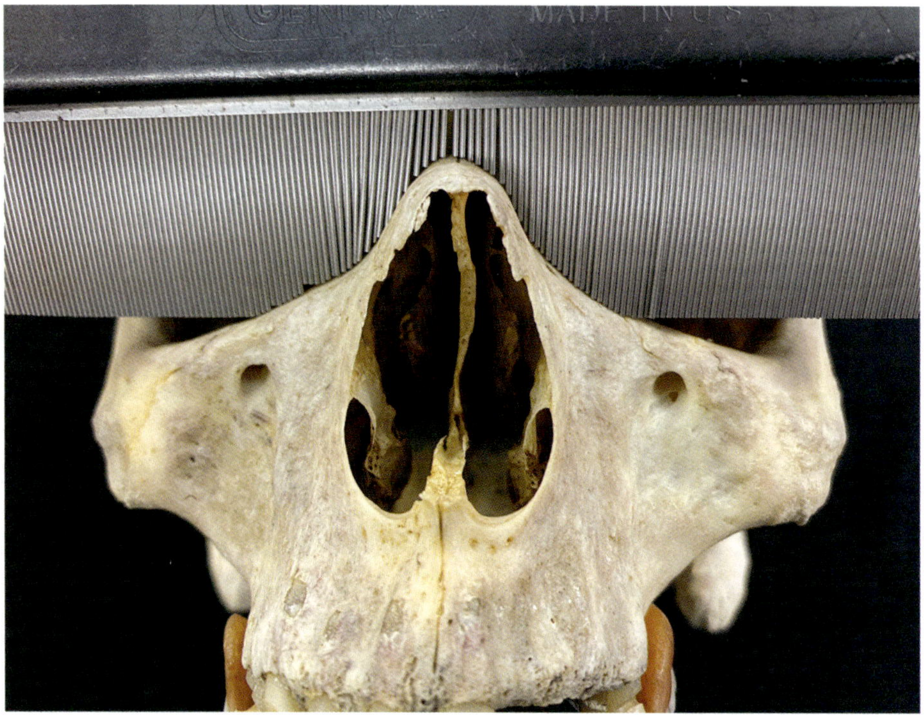

Figure 372. Nasal bone contour in an 83-year-old Hawaiian European male.

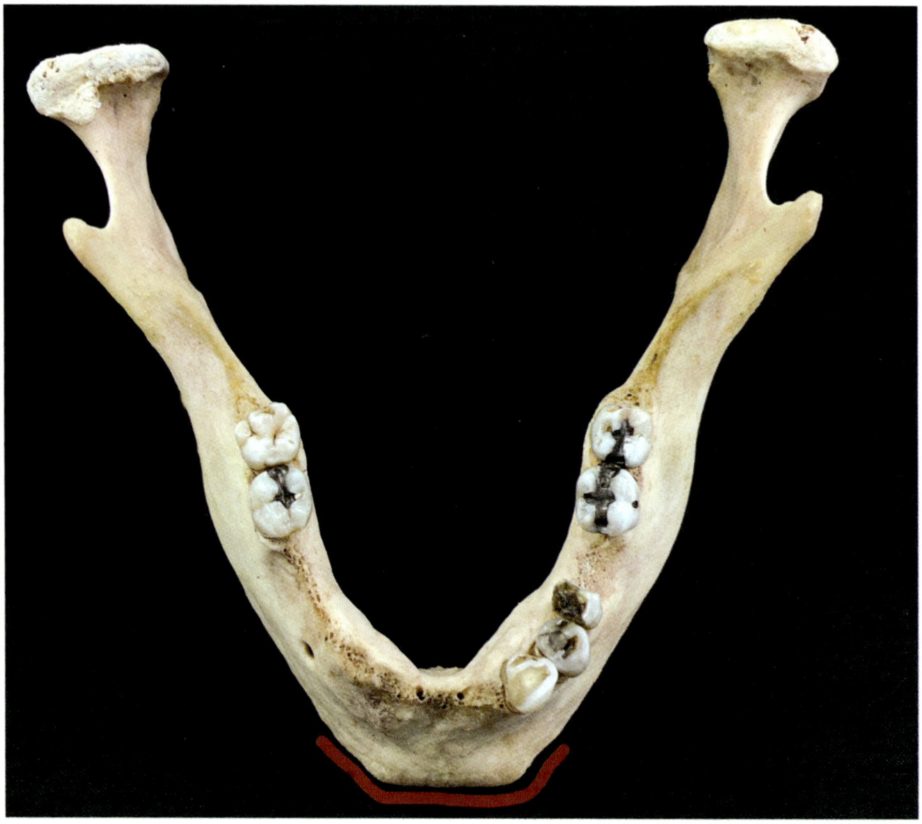

Figure 373. Mandible of 83-year-old Hawaiian European male with a square chin.

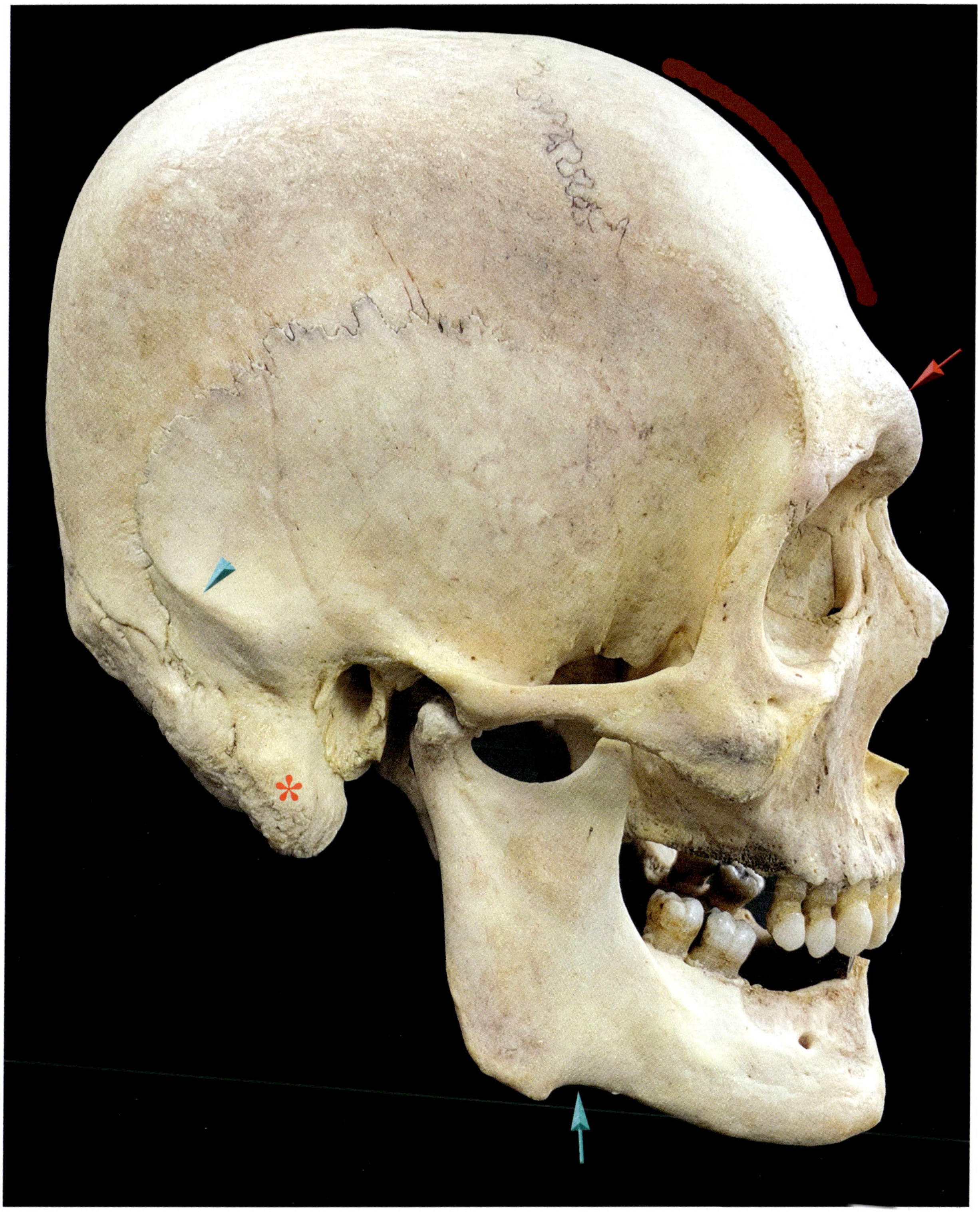

Figure 374. Right lateral view 83-year-old Hawaiian European male. This skull exhibits a sloping frontal bone (curved red line), prominent brow ridge (red arrow), a prominent supramastoid crest (arrowhead), large mastoid process (*), a small postbregmatic depression, and a small anterior nasal spine, and antegonial notch (blue arrow).

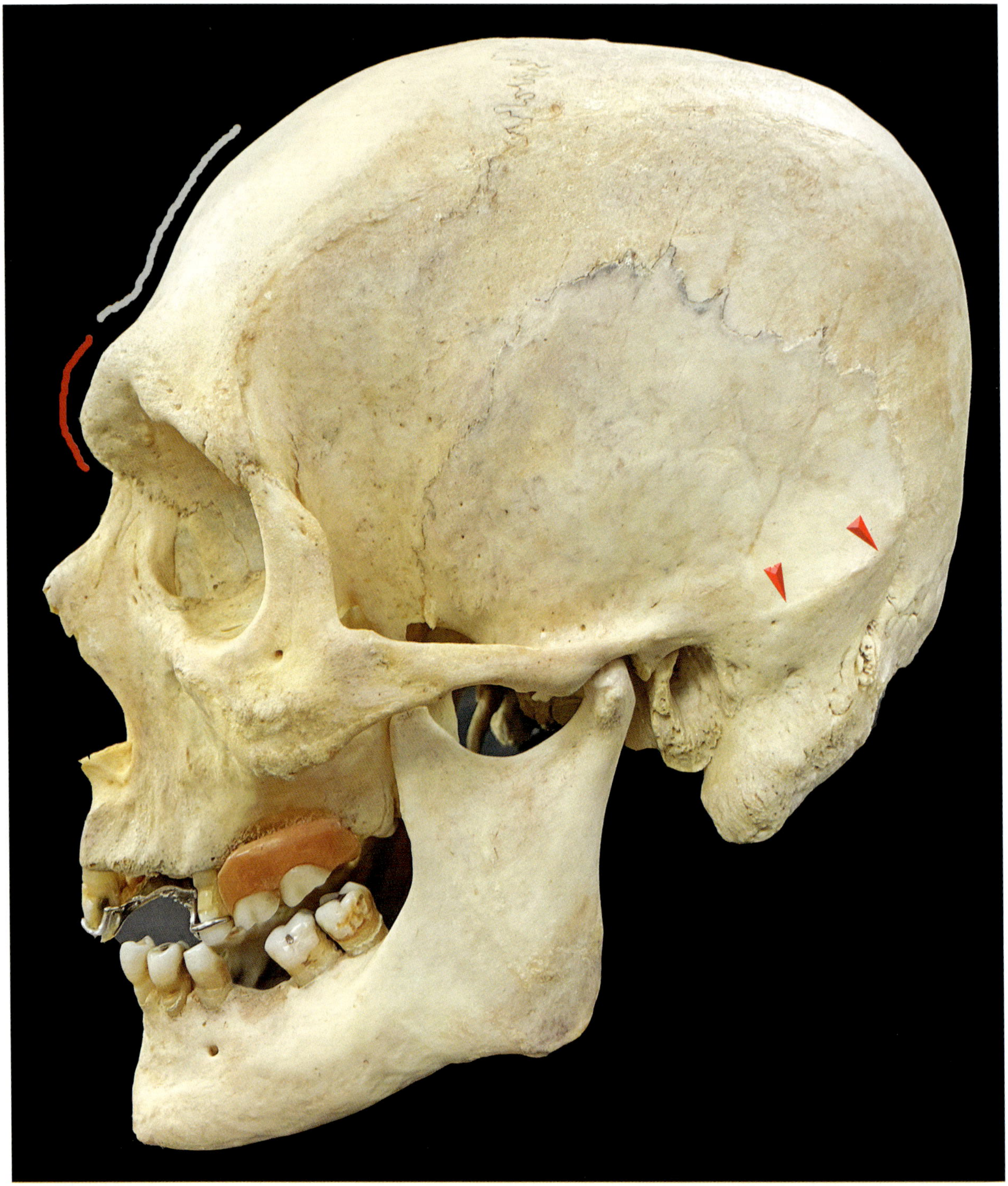

Figure 375. Left lateral view of 83-year-old Hawaiian European male with a large brow ridge (curved red line), sloping frontal bone (curved white line), concave/convex nasal bones, a small anterior nasal spine, and a well-developed supramastoid crest (arrowhead) positioned low on the cranium. The supramastoid crest is usually positioned a little higher in the temporal bone. In this image the cranium was photographed with its upper denture/plate.

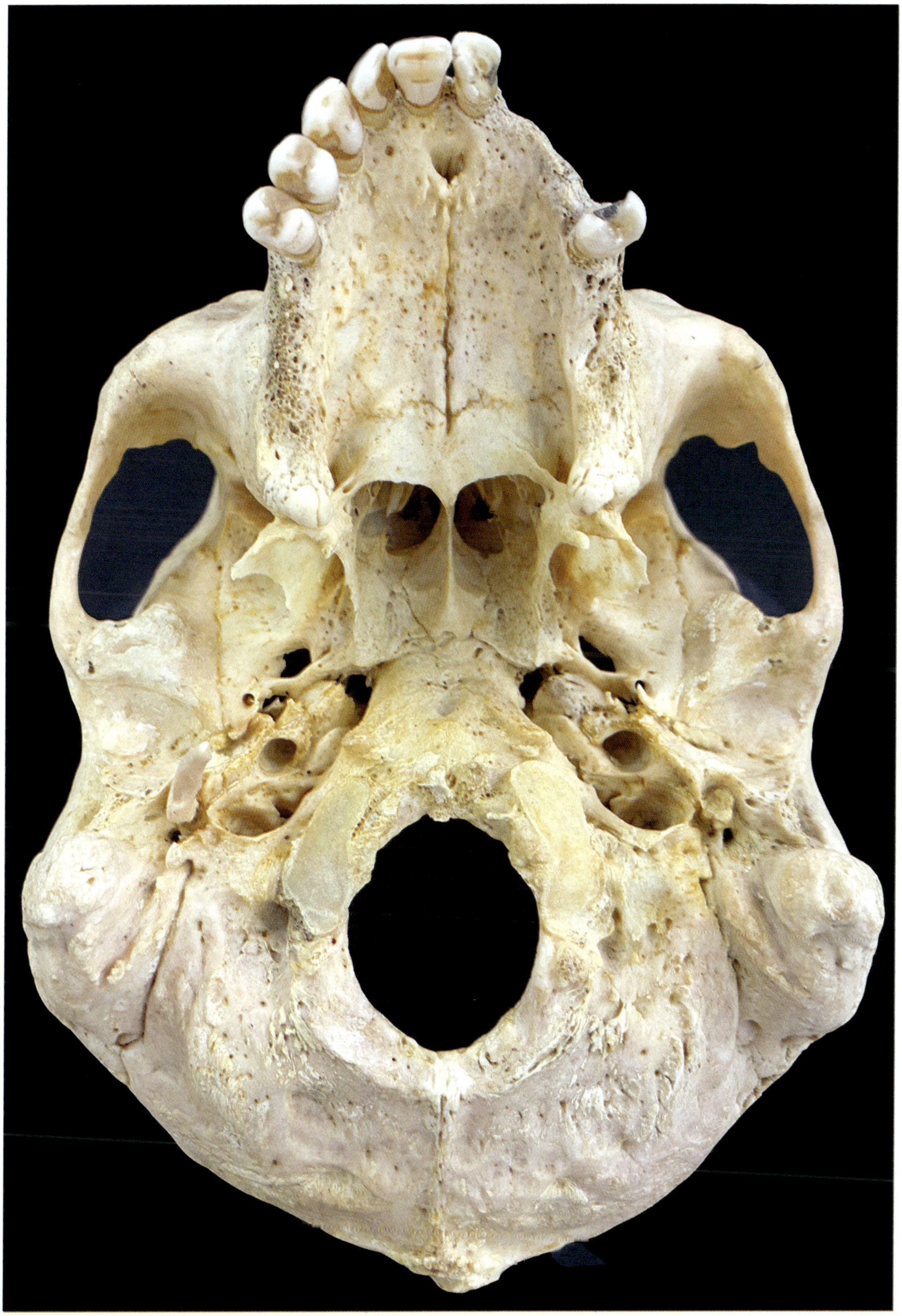

Figure 376. Inferior view of 83-year-old Hawaiian European male showing the shape of the palate, wide maximum cranial breadth (at the mastoid processes), and large mastoid processes.

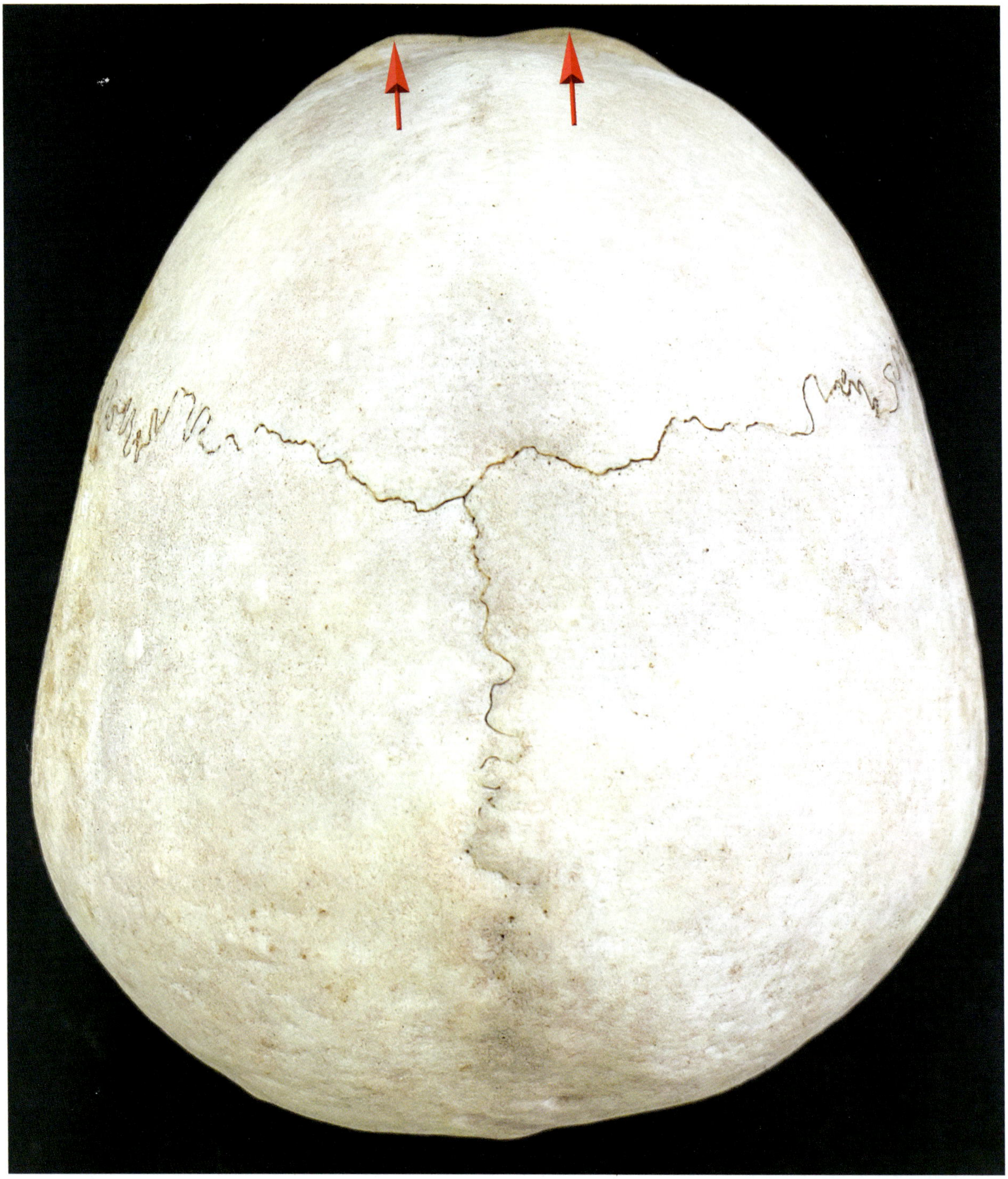

Figure 377. Superior view of 83-year-old Hawaiian European showing a short and wide vault, a prominent brow ridge (arrow), and partial obliteration of the coronal and sagittal sutures.

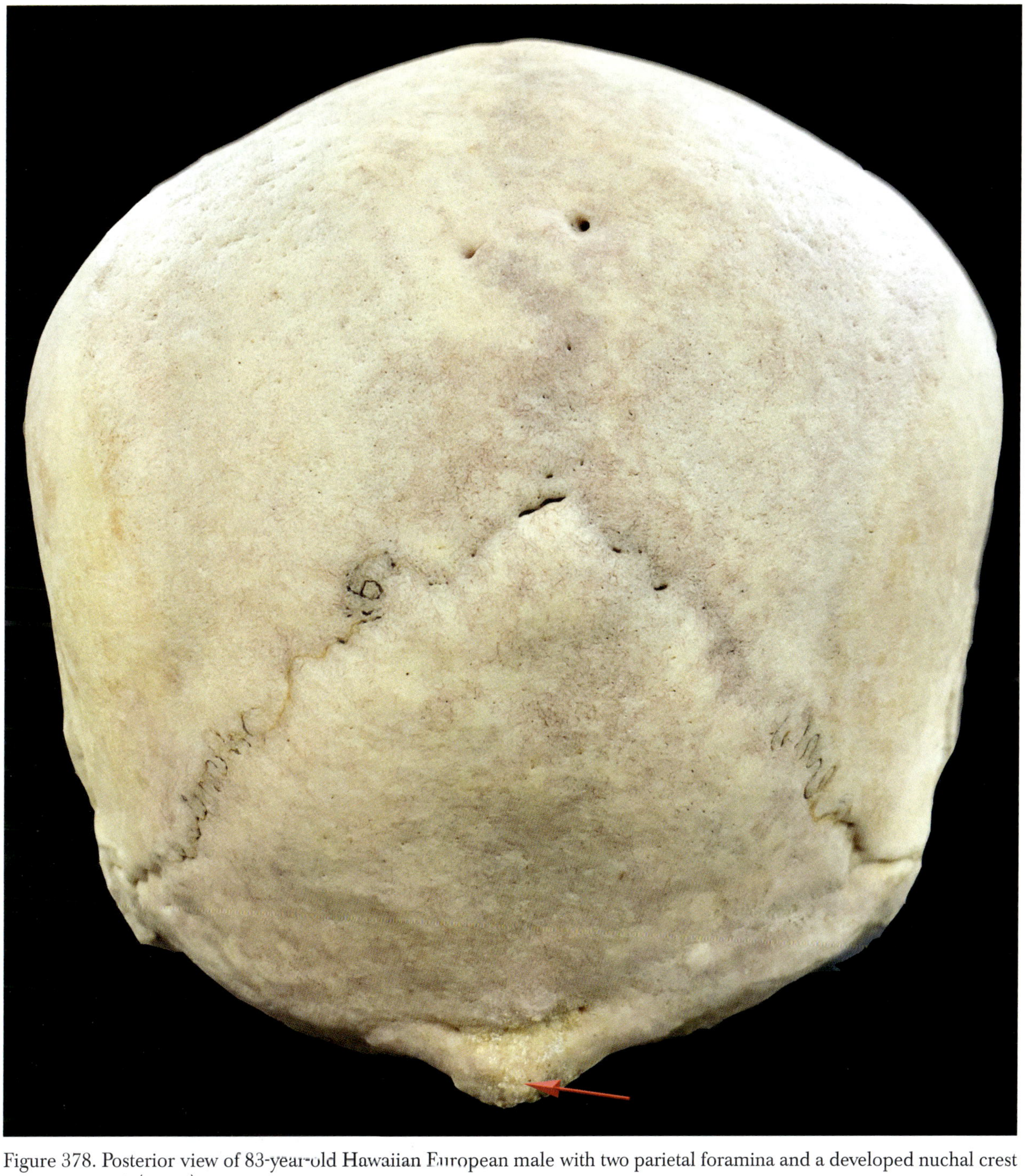

Figure 378. Posterior view of 83-year-old Hawaiian European male with two parietal foramina and a developed nuchal crest and inion hook (arrow).

4. HAWAIIAN EUROPEAN MALE

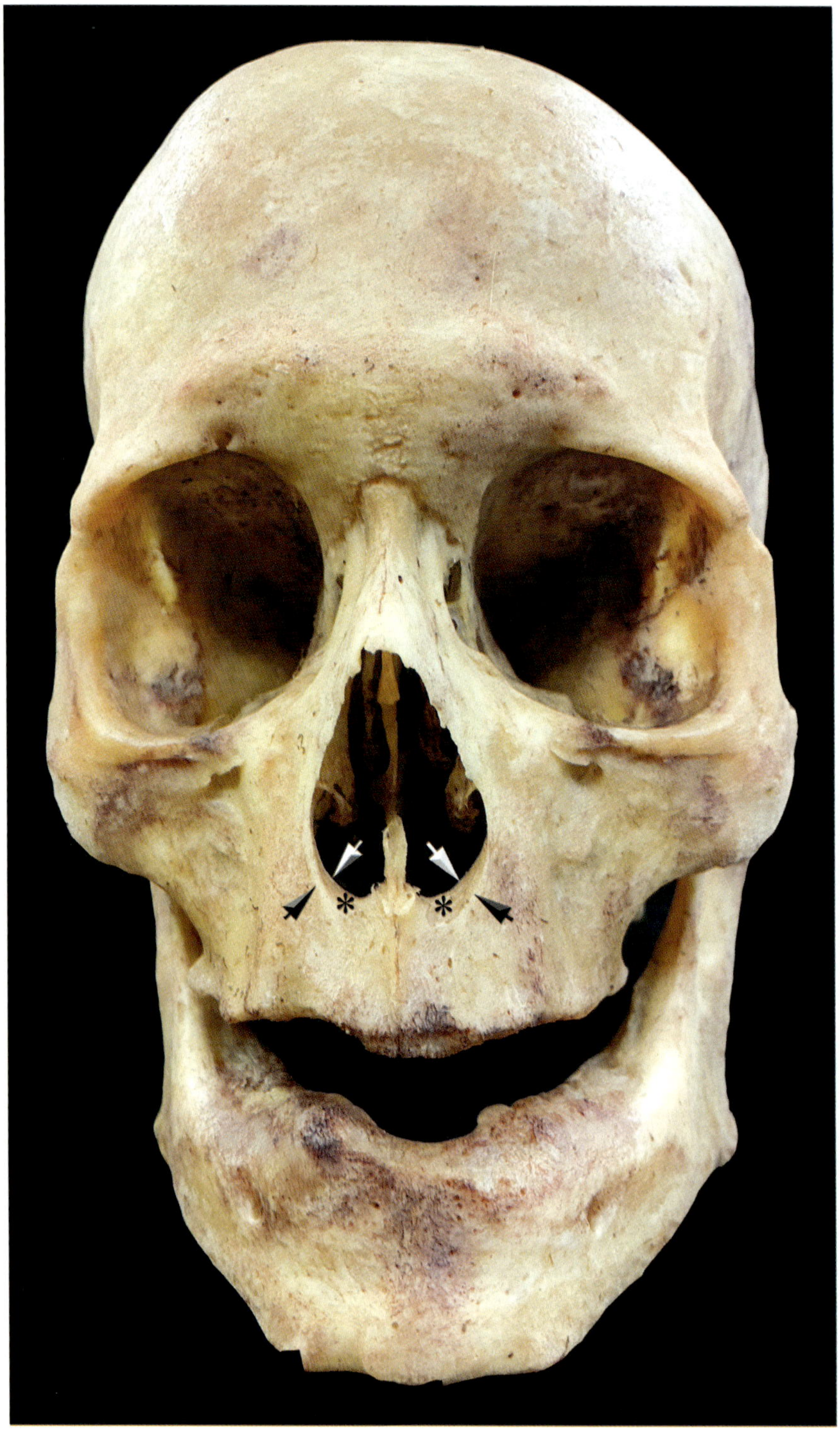

Figure 379. Anterior view of an edentulous 66-year-old Hawaiian European male with a medium-size brow-ridge that is best visualized from a lateral view (Figures 382-383). Note the internal (white arrow) and external (black arrow) ridges and anteriorly sloping nasal aperture (*) that forms nasal guttering.

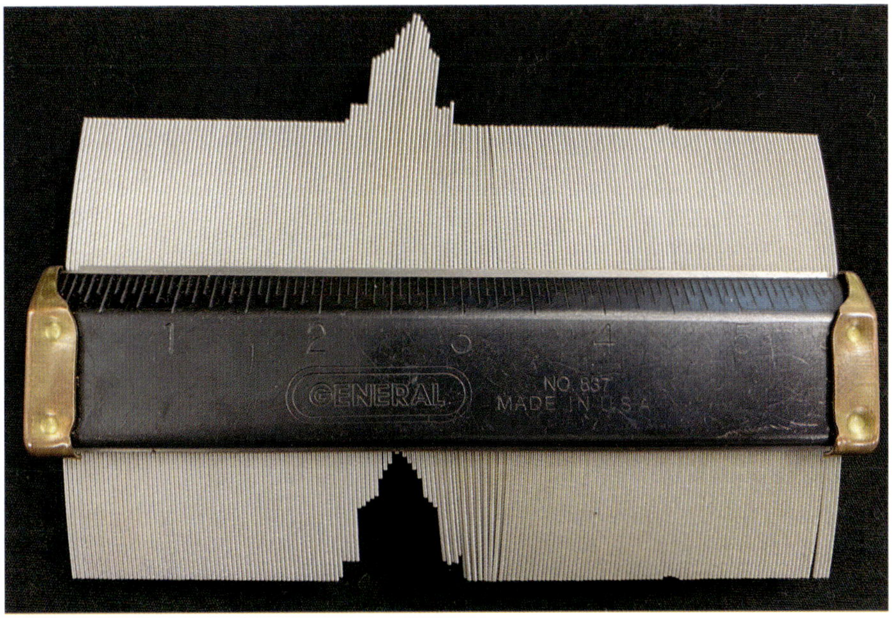

Figure 380. Nasal bone contour in a 66-year-old Hawaiian European male.

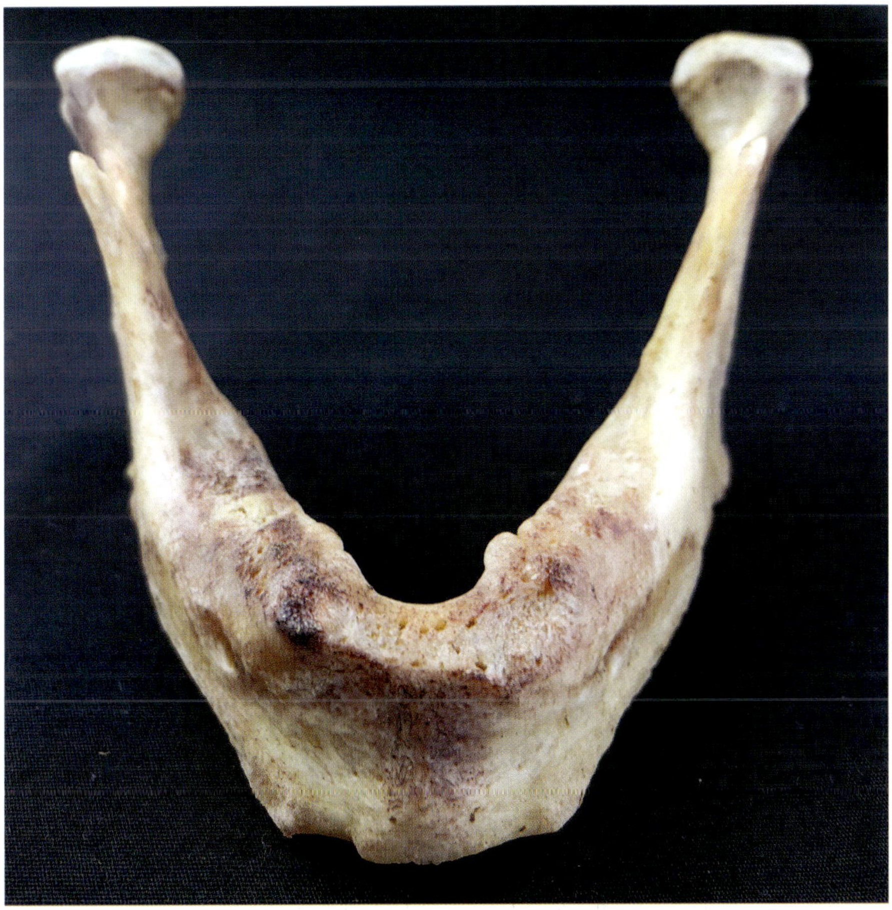

Figure 381. Anterior view of 66-year-old edentulous Hawaiian European male with a wide and irregularly shaped chin.

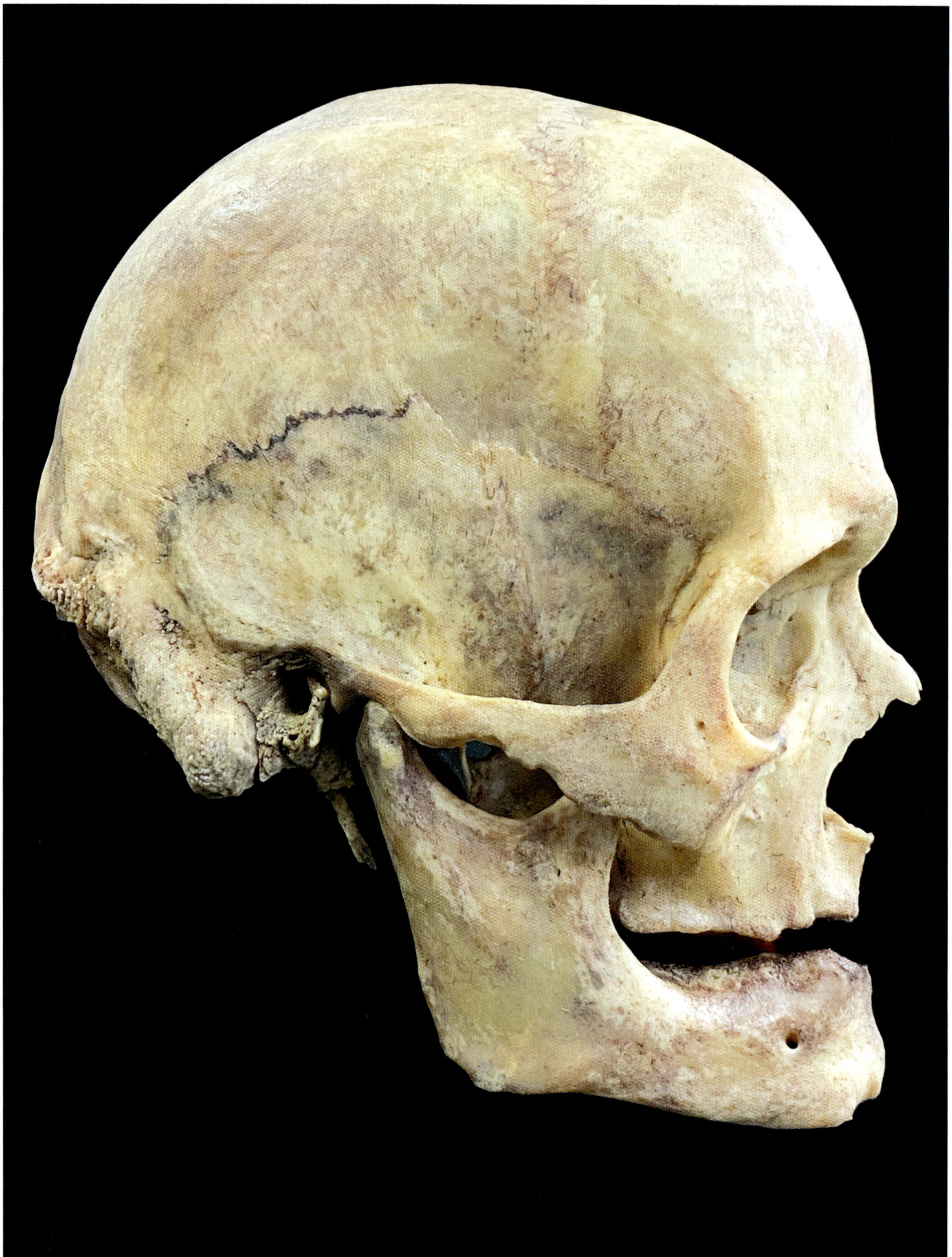

Figure 382. Right lateral view of 66-year-old edentulous Hawaiian European male with a tall and rounded vault, medium-size brow ridge, developed nuchal and supramastoid crests, large and overhanging nasal bones, and absent anterior nasal spine.

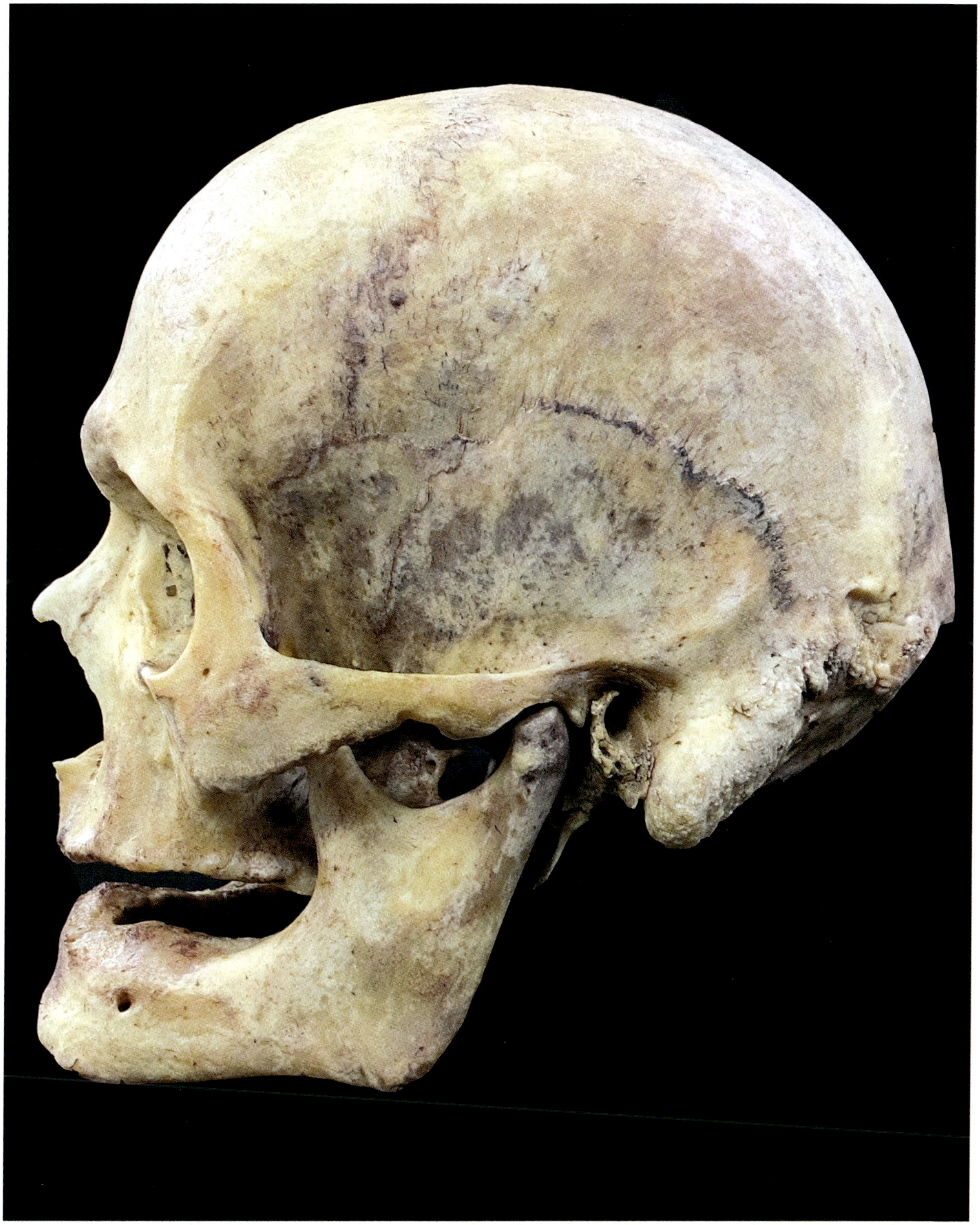

Figure 383. Left lateral view of an edentulous 66-year-old Hawaiian European male.

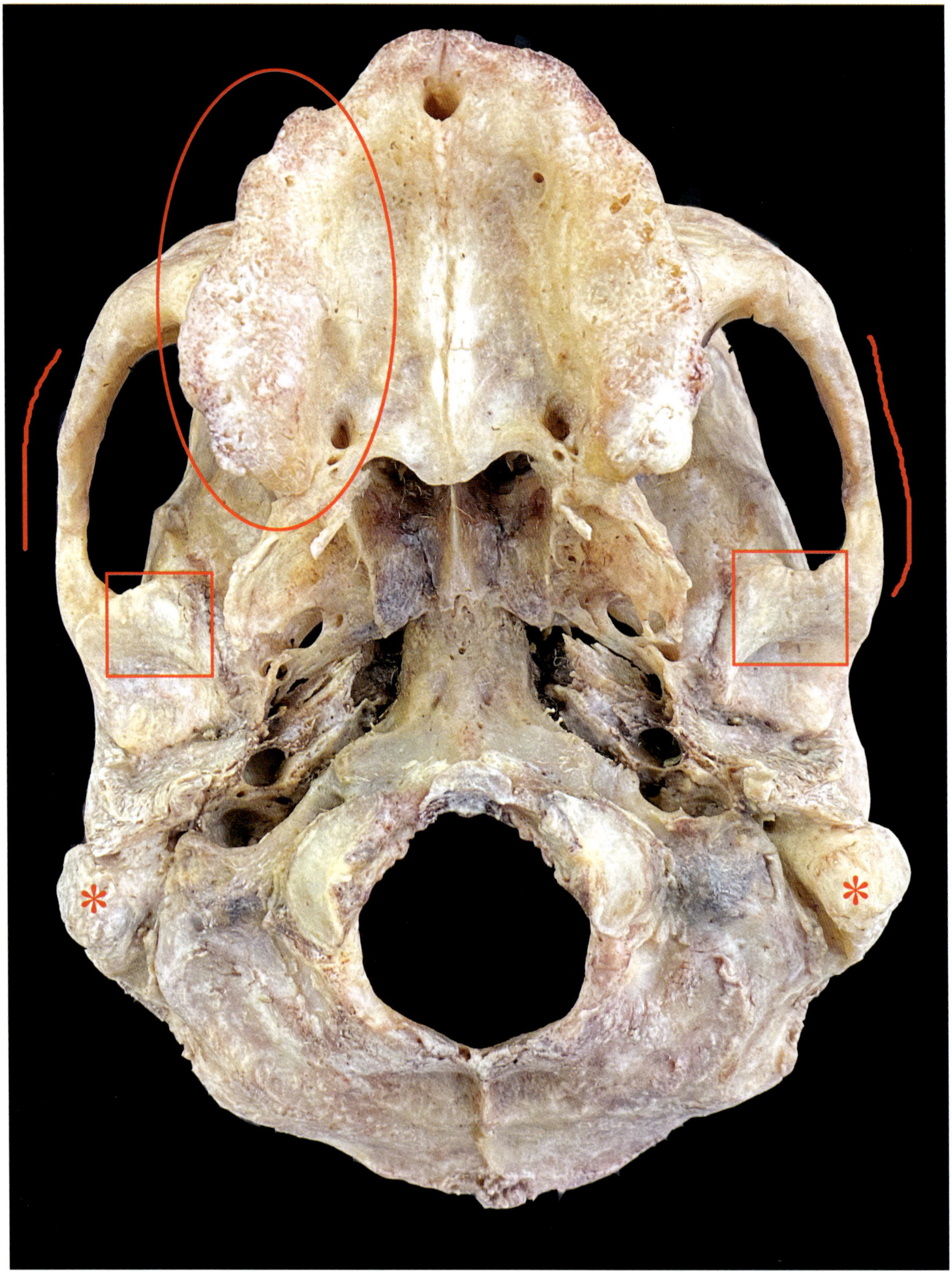

Figure 384. Inferior view of 66-year-old Hawaiian European male with a wide and shallow hard palate that has been altered/remodeled (oval) due to loss of the teeth long before death, convex zygomatic arches (curved red line), osteoarthritis of the temporomandibular joints (square) and asymmetrical mastoid processes (*).

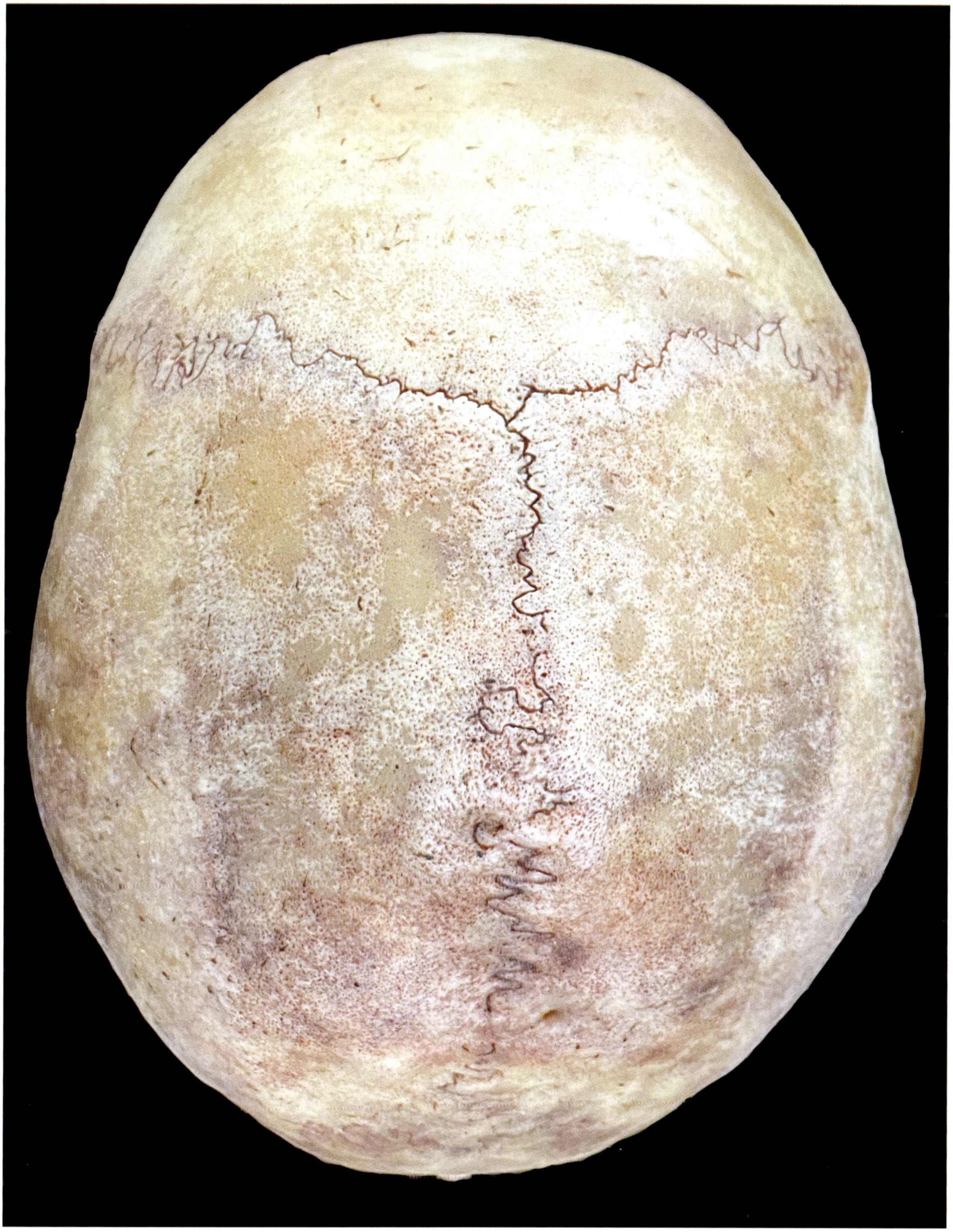

Figure 385. Superior view of 66-year-old Hawaiian European male with minimal obliteration of the sutures.

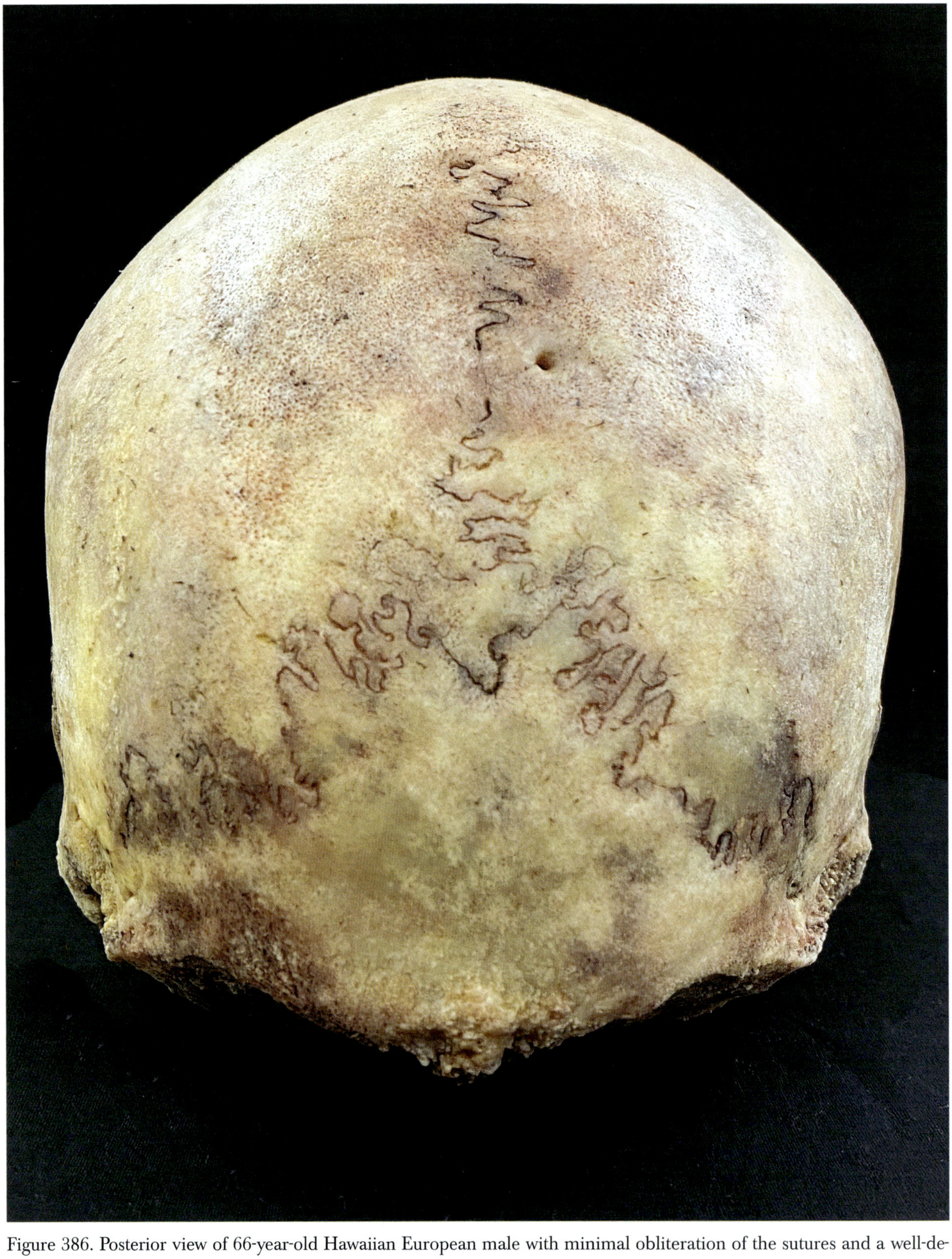

Figure 386. Posterior view of 66-year-old Hawaiian European male with minimal obliteration of the sutures and a well-developed nuchal crest.

5. HAWAIIAN EUROPEAN FEMALE

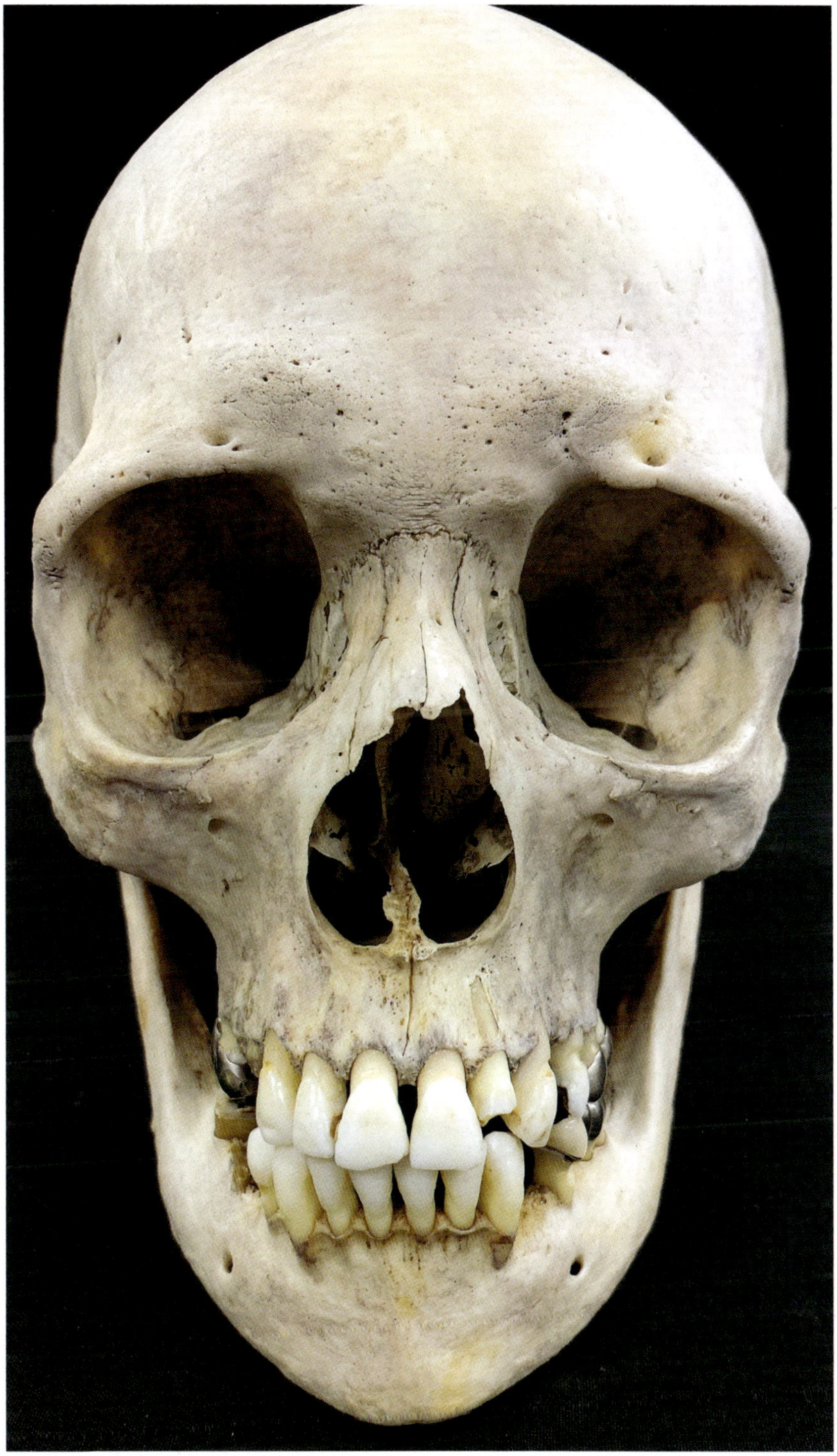

Figure 387. Anterior view of 57-year-old Hawaiian European female. Note the small brow ridge, the wide interorbital breadth, round orbits, and the blunt supraorbital margin. Notice the wide and circular nasal aperture, trace/minimal nasal guttering (represented by internal and external rims), and the "waisted" nasal bones. The malar bones are small, and the chin is V-shaped and rounded.

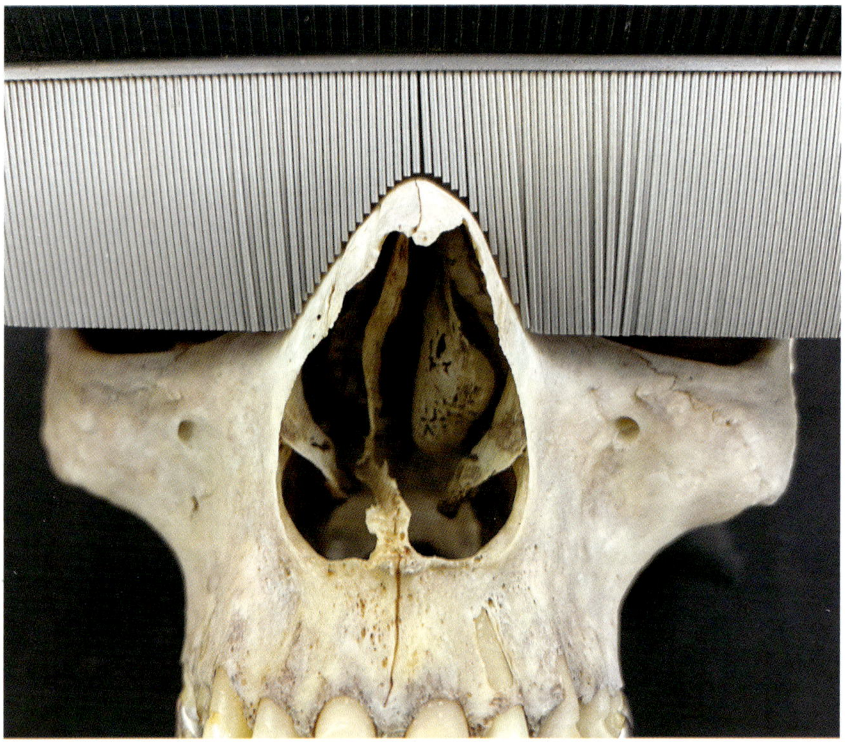

Figure 388. Nasal bone contour in a 57-year-old Hawaiian European female.

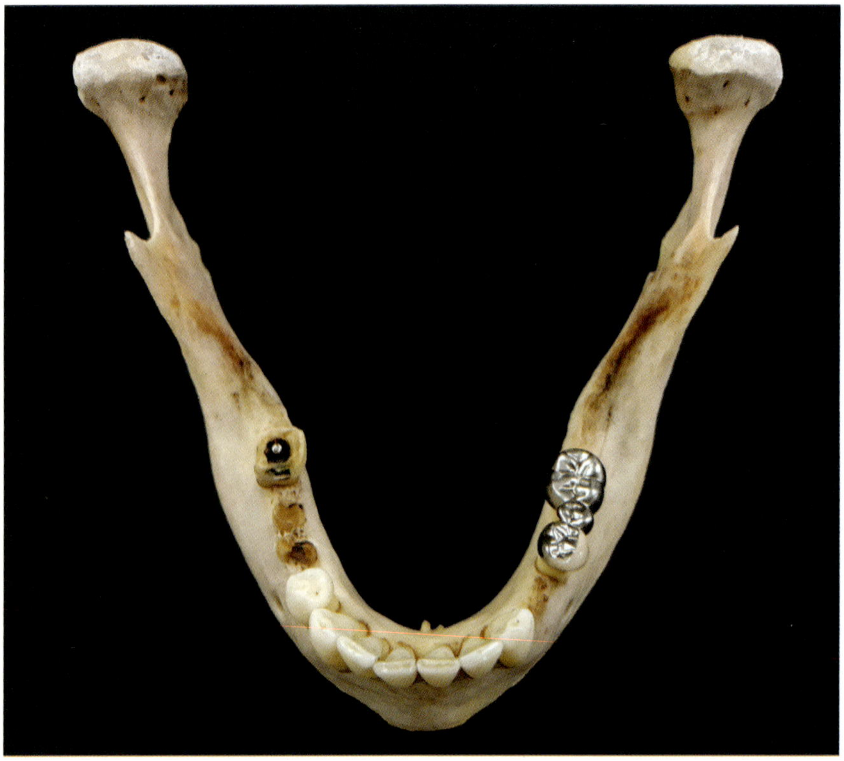

Figure 389. Mandible of 57-year-old Hawaiian and European female. The chin is narrow and V-shaped. Several of the teeth exhibit carious lesions that have destroyed the crowns.

TABLE 33.
Cranial Measurements in a 57-year-old Hawaiian European Female

Measurement	(mm)	Right	Left
Maximum Cranial Length (g-op, GOL)	173		
Maximum Cranial Breadth (eu-eu, XCB)	136		
Bizygomatic Breadth (zy-zy, ZYB)	125		
Basion-Bregma Height (ba-b, BBH)	143		
Cranial Base Length (ba-n, BNL)	103		
Basion-Prosthion (ba-pr, BPL)	96		
Maxillo-Alveolar Breadth (ecm-ecm, MAB)	56		
Maxillo-Alveolar Length (pr-alv, MAL)	45		
Biauricular Breadth (au-au, AUB)	120		
Upper Facial Height (n-pr)	63		
Minimum Frontal Breadth (ft-ft, WFB)	91		
Upper Facial Breadth (fmt-fmt)	105		
Nasal Height (n-ns, NLH)	50		
Nasal Breadth (al-al, NLB)	26		
Orbital Breadth (d-ec, OBB)	40		L
Orbital Height (OBH)	33		L
Biorbital Breadth (ec-ec, EKB)	97		
Interorbital Breadth (d-d, DKB)	23		
Frontal Chord (n-b, FRC)	112		
Parietal Chord (b-l, PAC)	109		
Occipital Chord (l-o, OCC)	96		
Foramen Magnum Length (ba-o, FOL)	36		
Foramen Magnum Breadth (FOB)	29		
Mastoid Length (MDH)		21	26

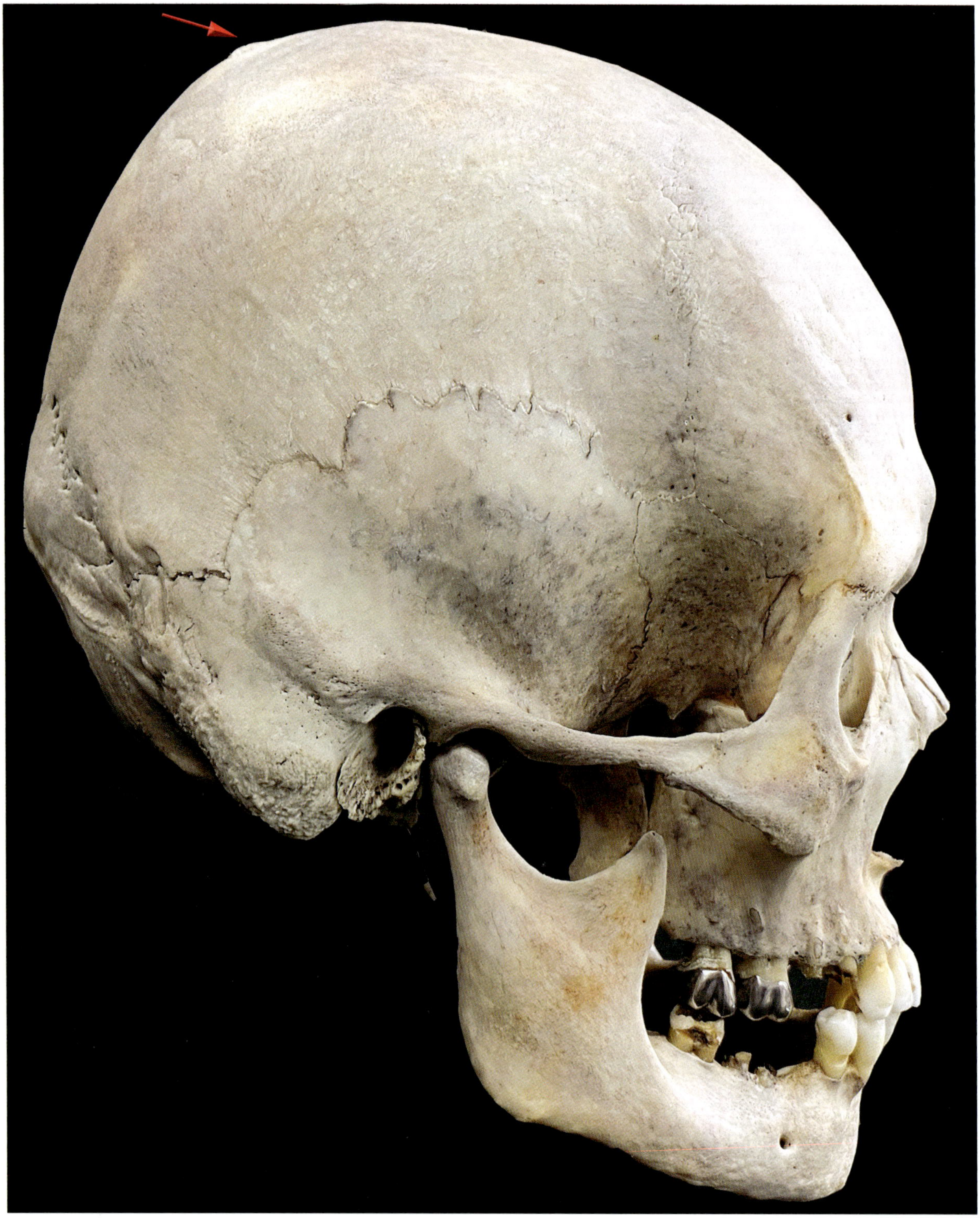

Figure 390. Right lateral view of 57-year-old Hawaiian European female. Note the small brow ridge, the small nasal bones with mild nasal overgrowth, minimal alveolar prognathism, depression at nasion, the absent posterior zygomatic tubercle, the gracile (small and slender) mandibular body, the moderate anterior nasal spine, developed supramastoid crest, and absence of malar tubercle. Also note the small mastoid process, a rounded vault, and a small, benign button osteoma (arrow) in the right parietal bone.

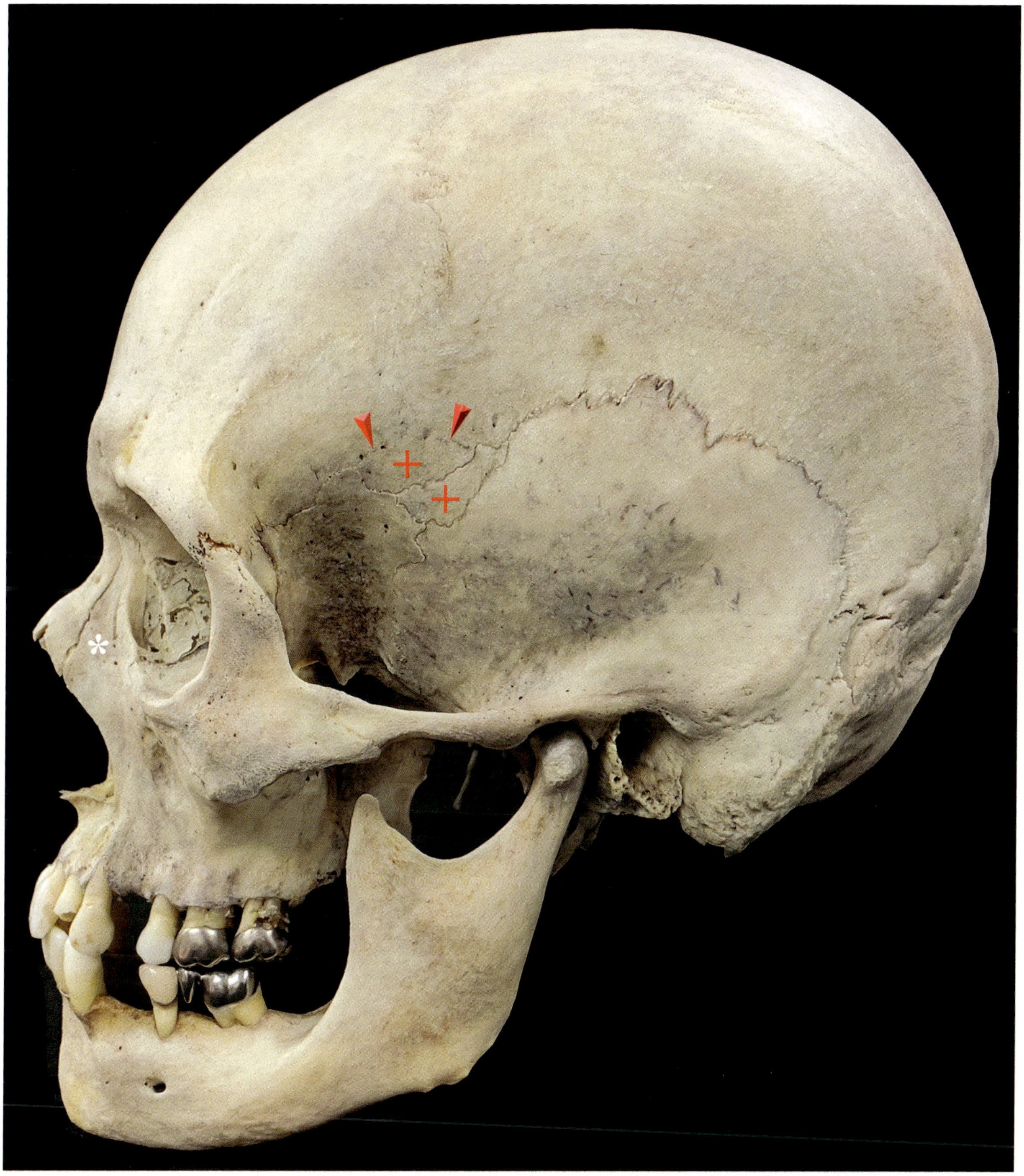

Figure 391. Left lateral view of 57-year old Hawaiian European female. Note the forward projection of the nasal bones (no nasal overgrowth) and the frontal process of the maxilla (*), the presence of an anterior nasal spine, and a divided/bipartite epipteric or squamosal ossicle (+) (arrowhead indicates mostly obliterated superior margin of the ossicle).

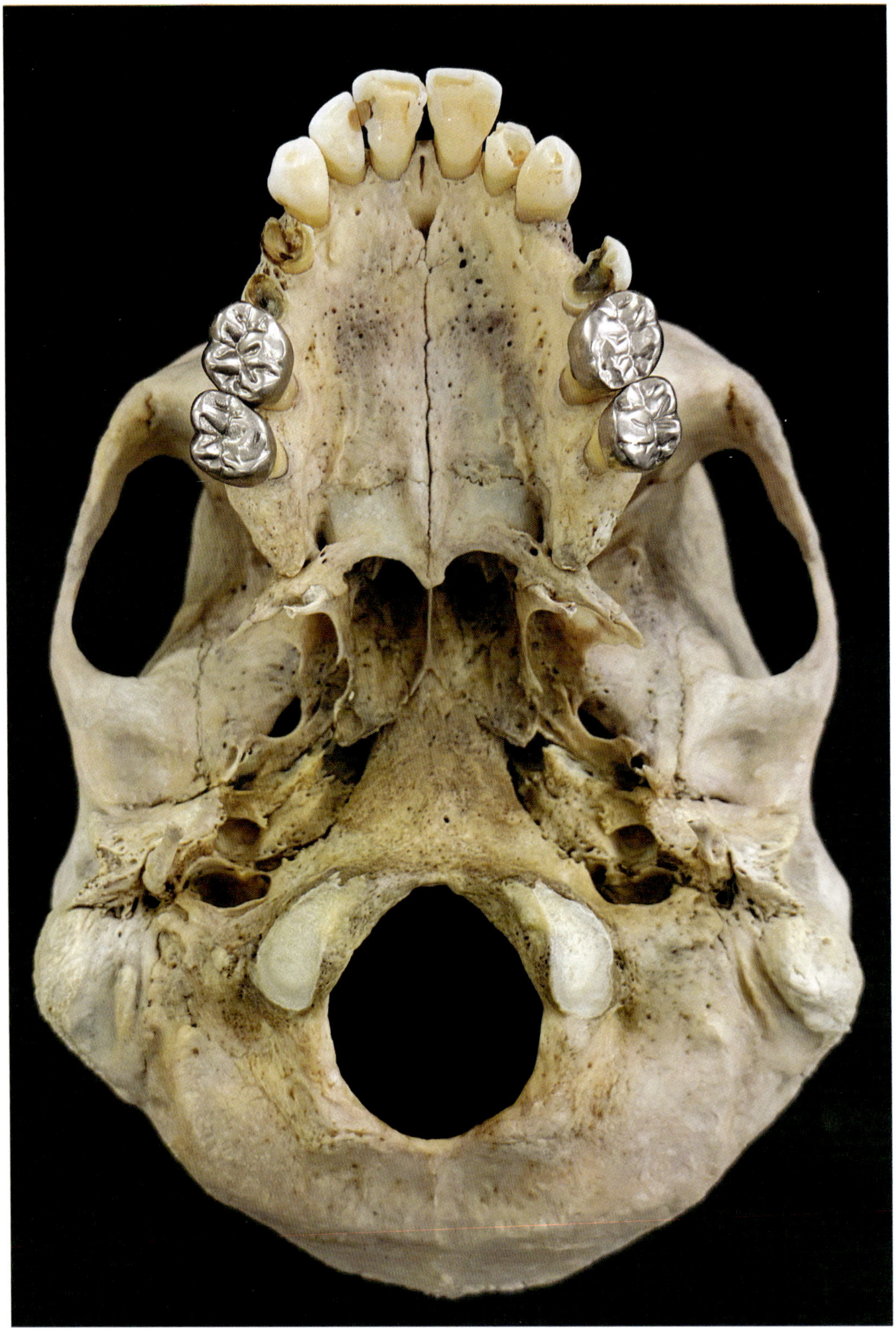

Figure 392. Inferior view of 57-year-old Hawaiian European female. This individual has a V-shaped maxilla/dental arcade, shovel-shaped incisors, small and asymmetrical mastoid processes, a small torus palatinus along the midline, and carious lesions in several teeth.

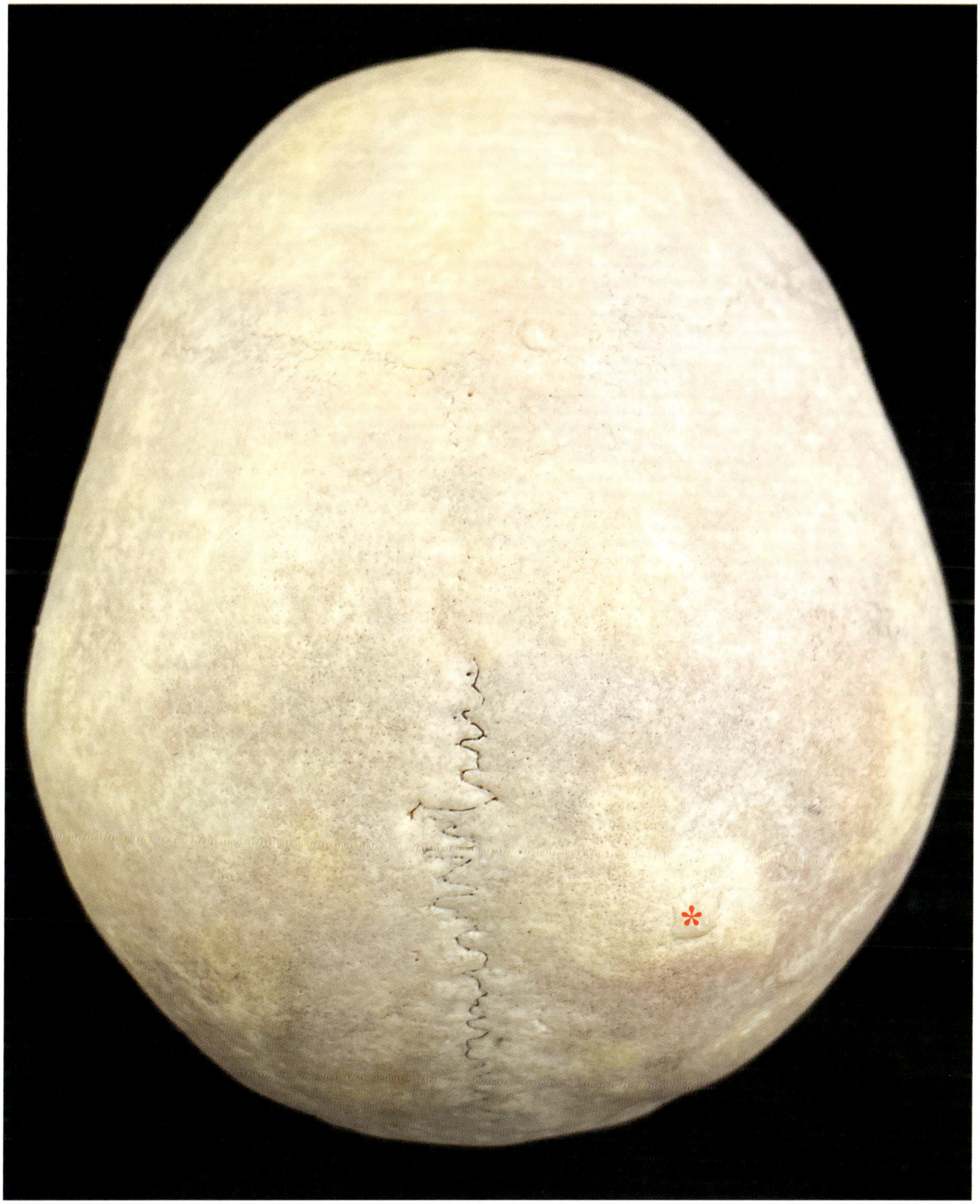

Figure 393. Superior view of 57-year-old Hawaiian European female. Note the parietal bossing/squaring, the obliteration of most of the coronal suture and less than half of the sagittal suture, and a small (benign) button osteoma (*).

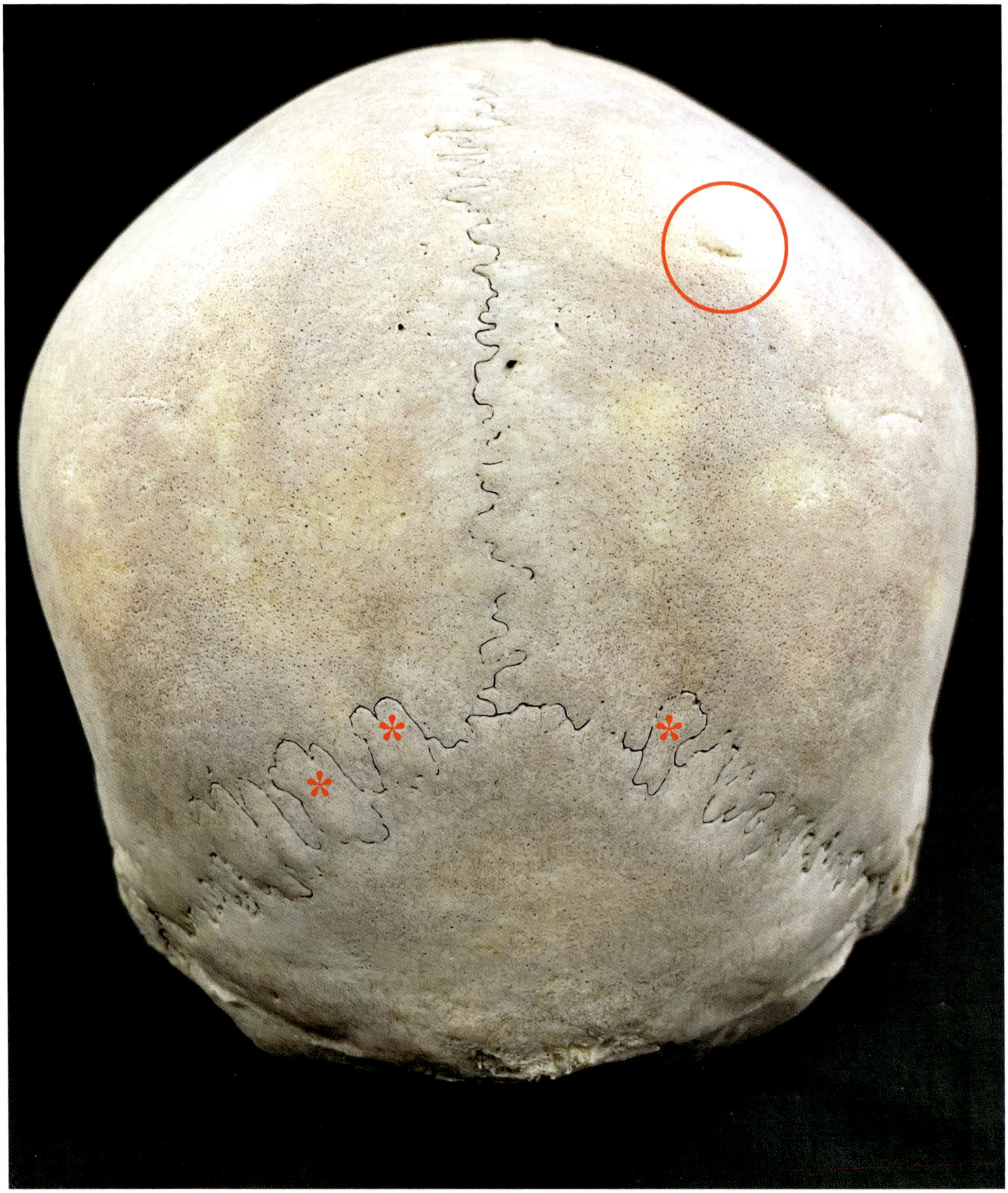

Figure 394. Posterior view of 57-year-old Hawaiian European female. This cranium exhibits parietal bossing/squaring, mild development of the nuchal crest, numerous lambdoidal ossicles (*), a benign button osteoma (circle), and minimal closure/obliteration of the sutures. Button osteomas are usually round or slightly oval mounds of bone usually present in the parietal and frontal bones.

6. HAWAIIAN FILIPINO CHINESE FEMALE

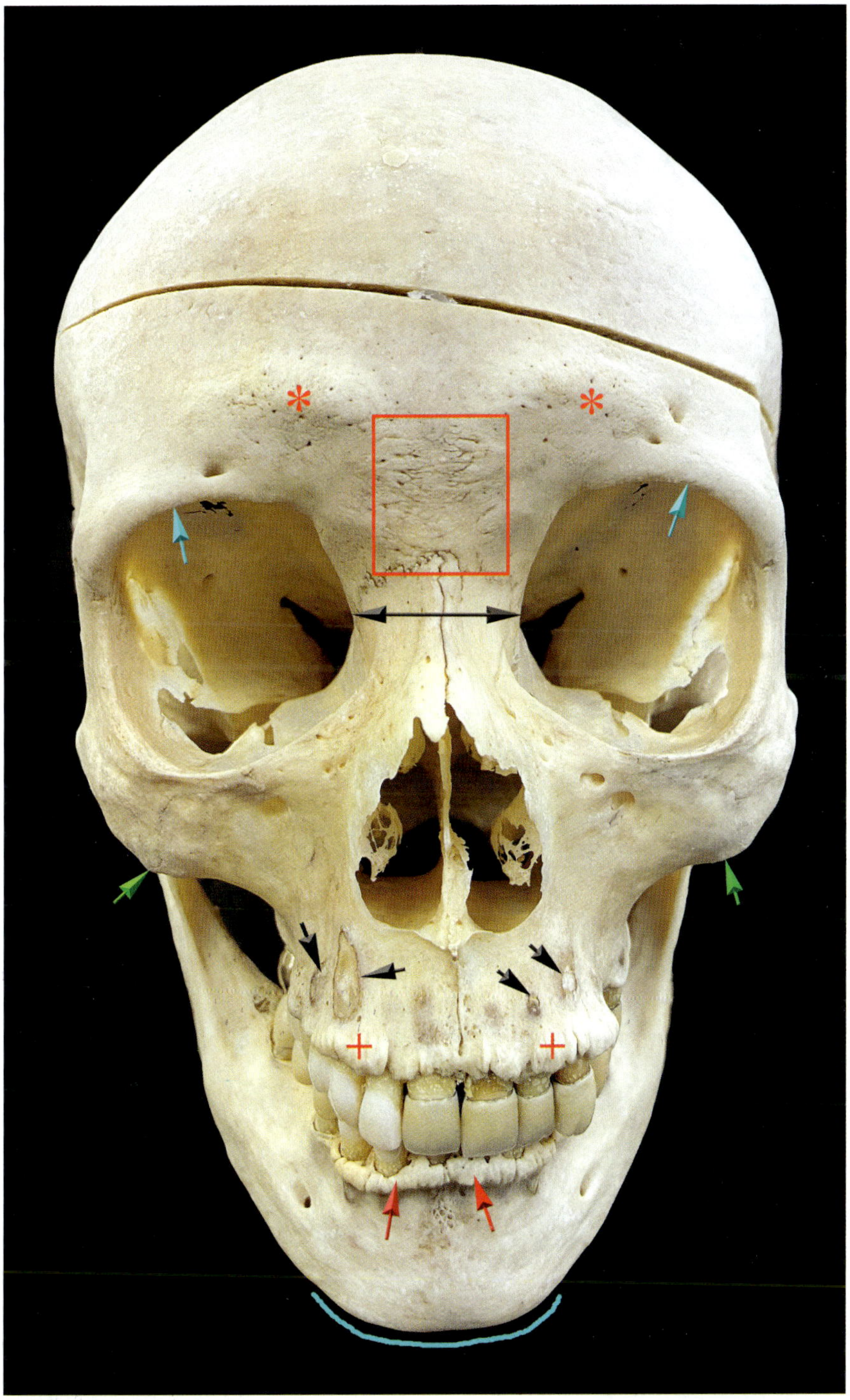

Figure 395. Anterior view of 90-year-old Hawaiian Filipino Chinese female with a small and separate brow ridge (*), trace supranasal suture (rectangle), blunt superior orbital margin (blue arrow), a wide interorbital breadth (double black arrow), absent inferior malar tubercle (green arrow), maxillary fenestrations (black arrow), maxillary buccal exostoses (+), mandibular buccal exostoses (red arrow), and a wide and rounded chin (curved blue line).

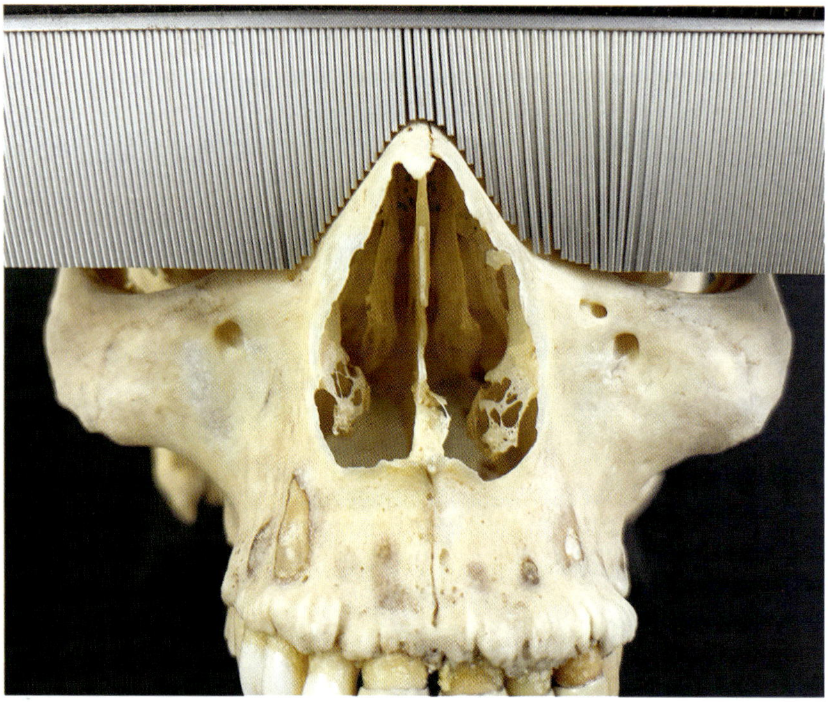

Figure 396. Nasal bone contour in a 90-year-old Hawaiian Filipino Chinese female.

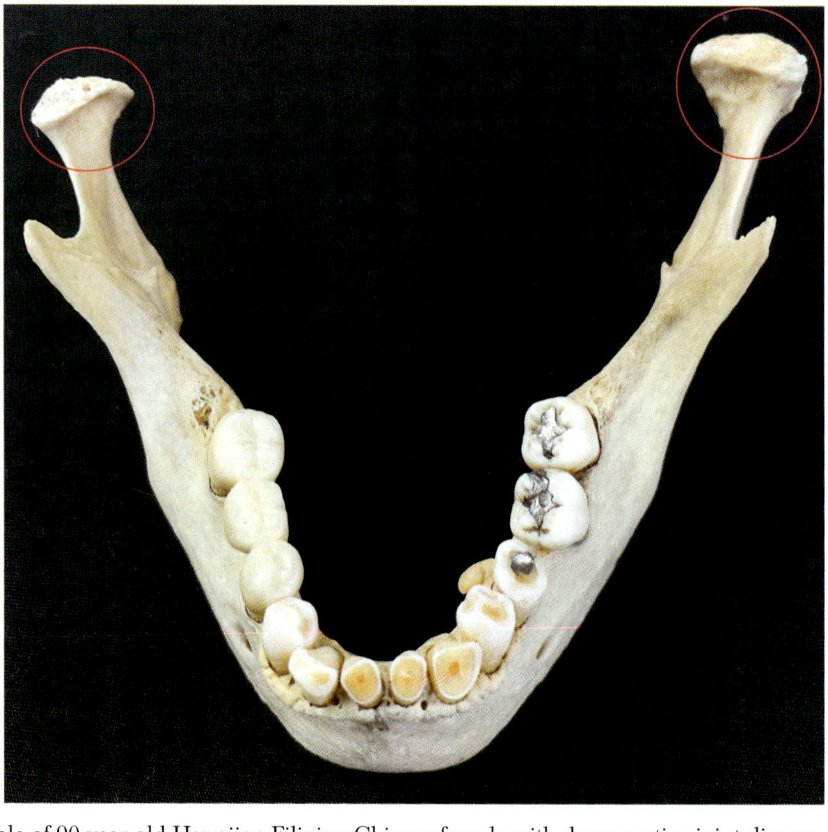

Figure 397. Mandible of 90-year-old Hawaiian Filipino Chinese female with degenerative joint disease of the condyles (circle).

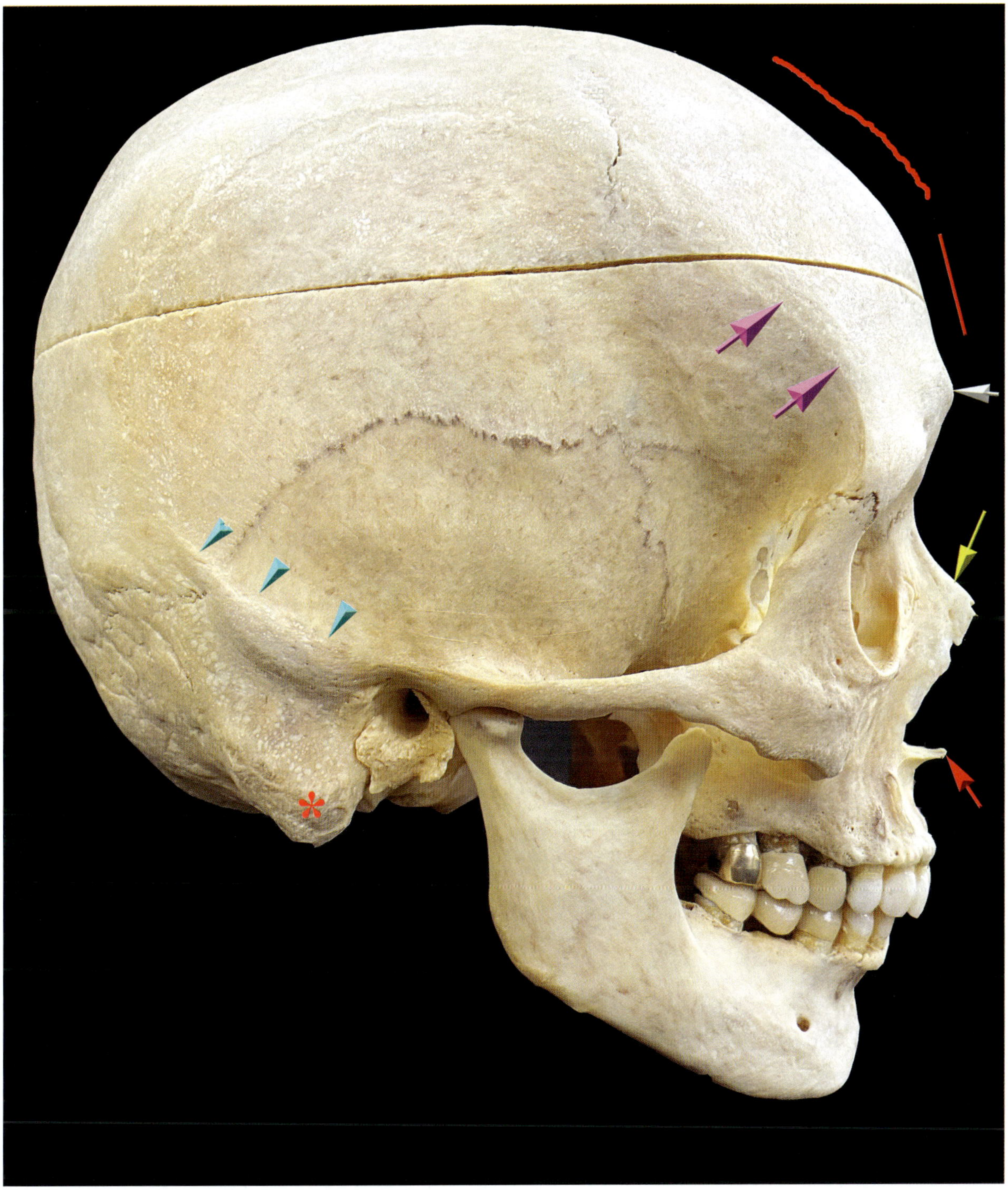

Figure 398. Right lateral view of 90-year-old Hawaiian Filipino Chinese female with mild posterior sloping of the frontal bone (curved line), nearly vertical frontal bone (straight line) superior to a projecting glabella (straight line), small nasal bones (yellow arrow), prominent anterior nasal spine (red arrow), minimal development of the temporal line (purple arrow), a developed supramastoid crest (blue arrowhead) that extends onto the parietal bone, and a small mastoid process (*).

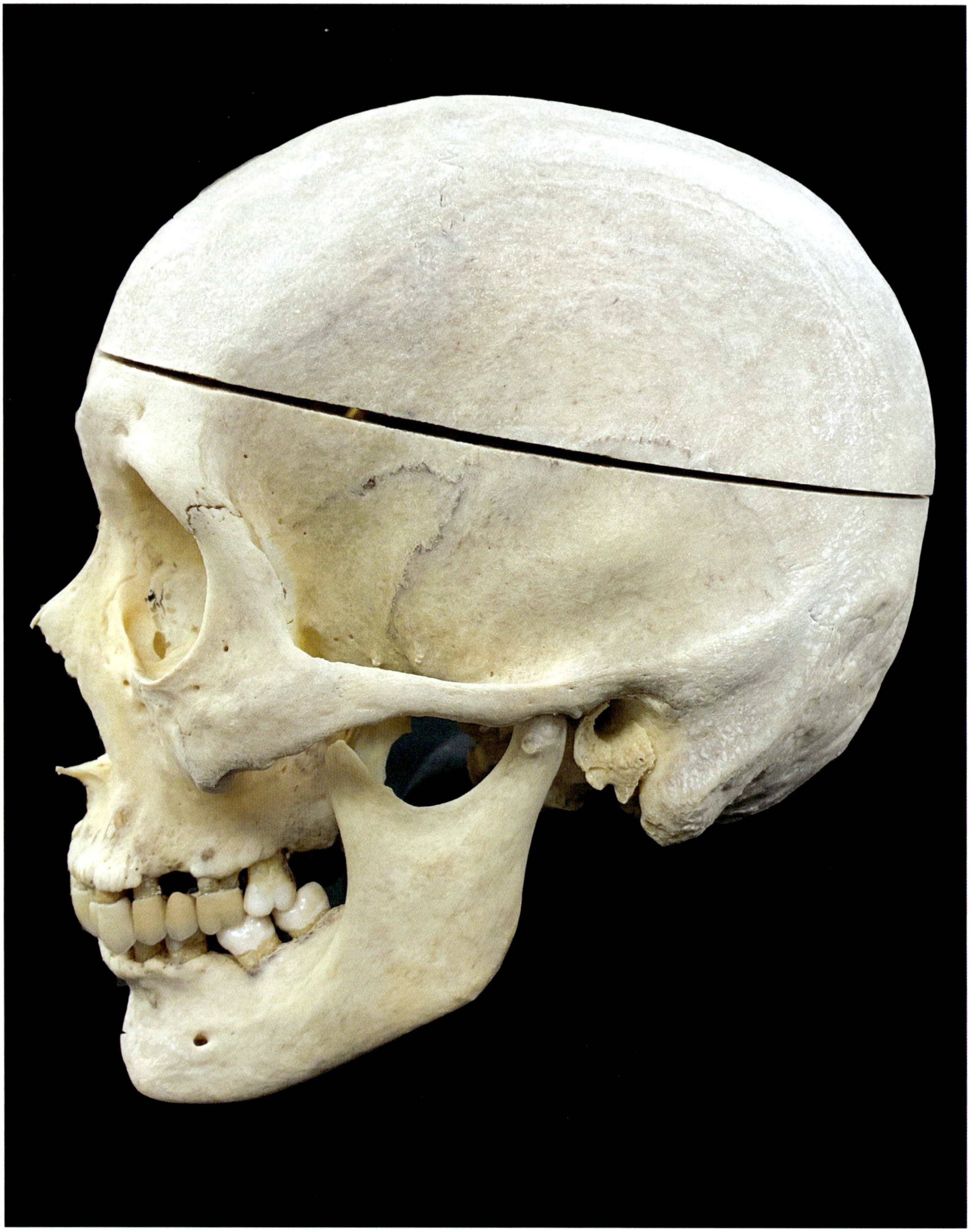

Figure 399. Left lateral view of 90-year-old Hawaiian Filipino Chinese female.

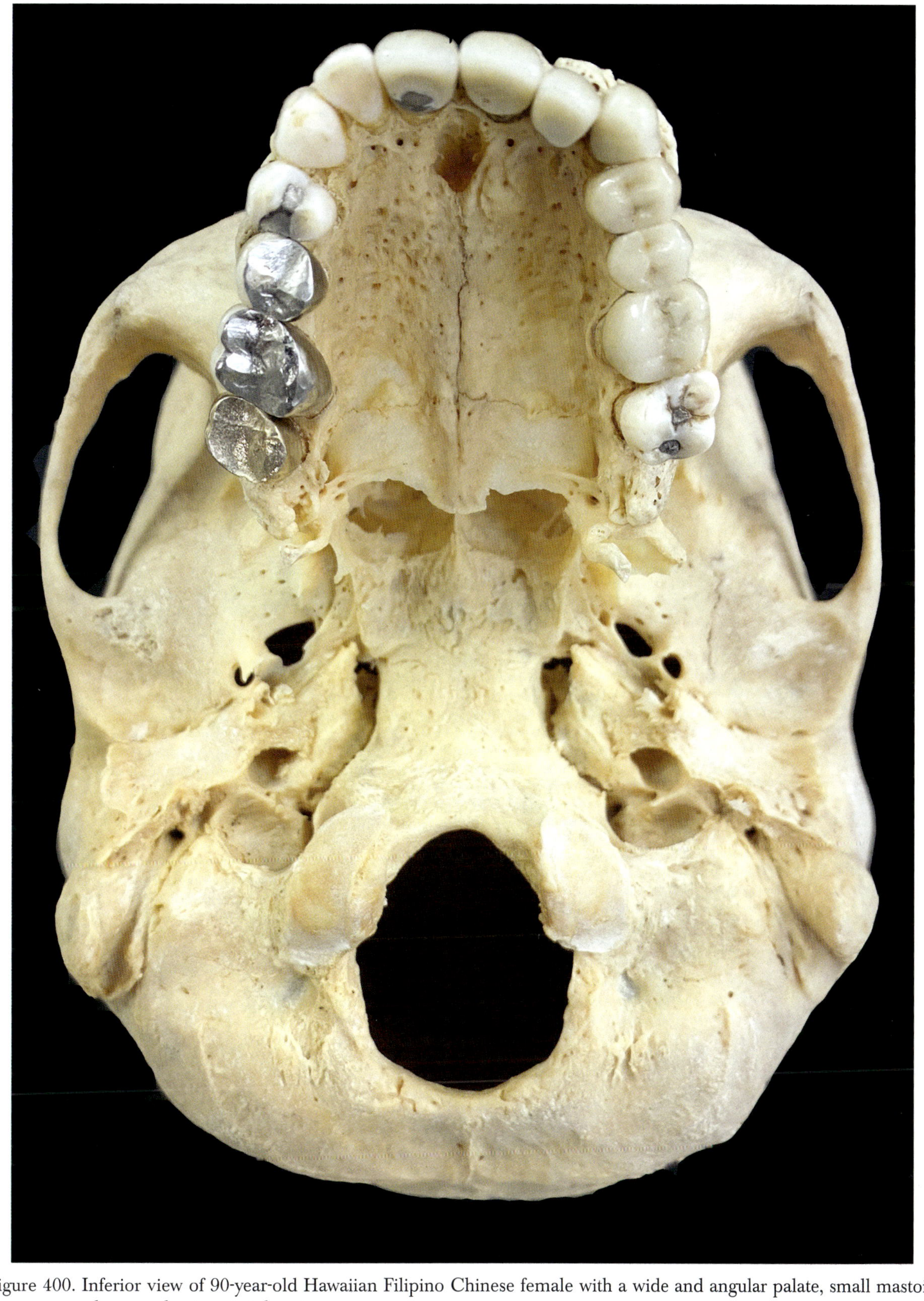

Figure 400. Inferior view of 90-year-old Hawaiian Filipino Chinese female with a wide and angular palate, small mastoid processes, and a very short occipital region.

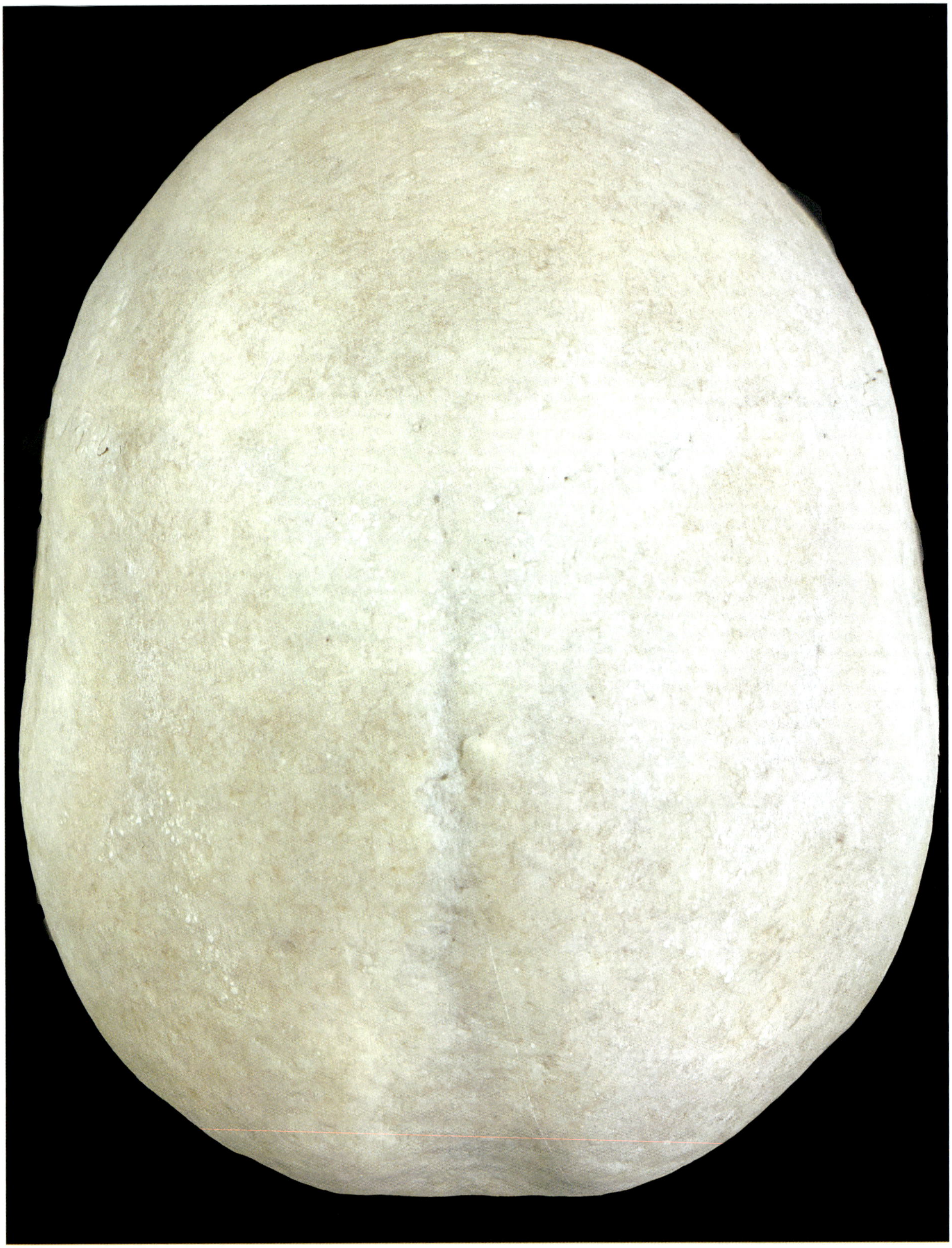

Figure 401. Superior view of 90-year-old Hawaiian Filipino Chinese female with complete obliteration of all sutures.

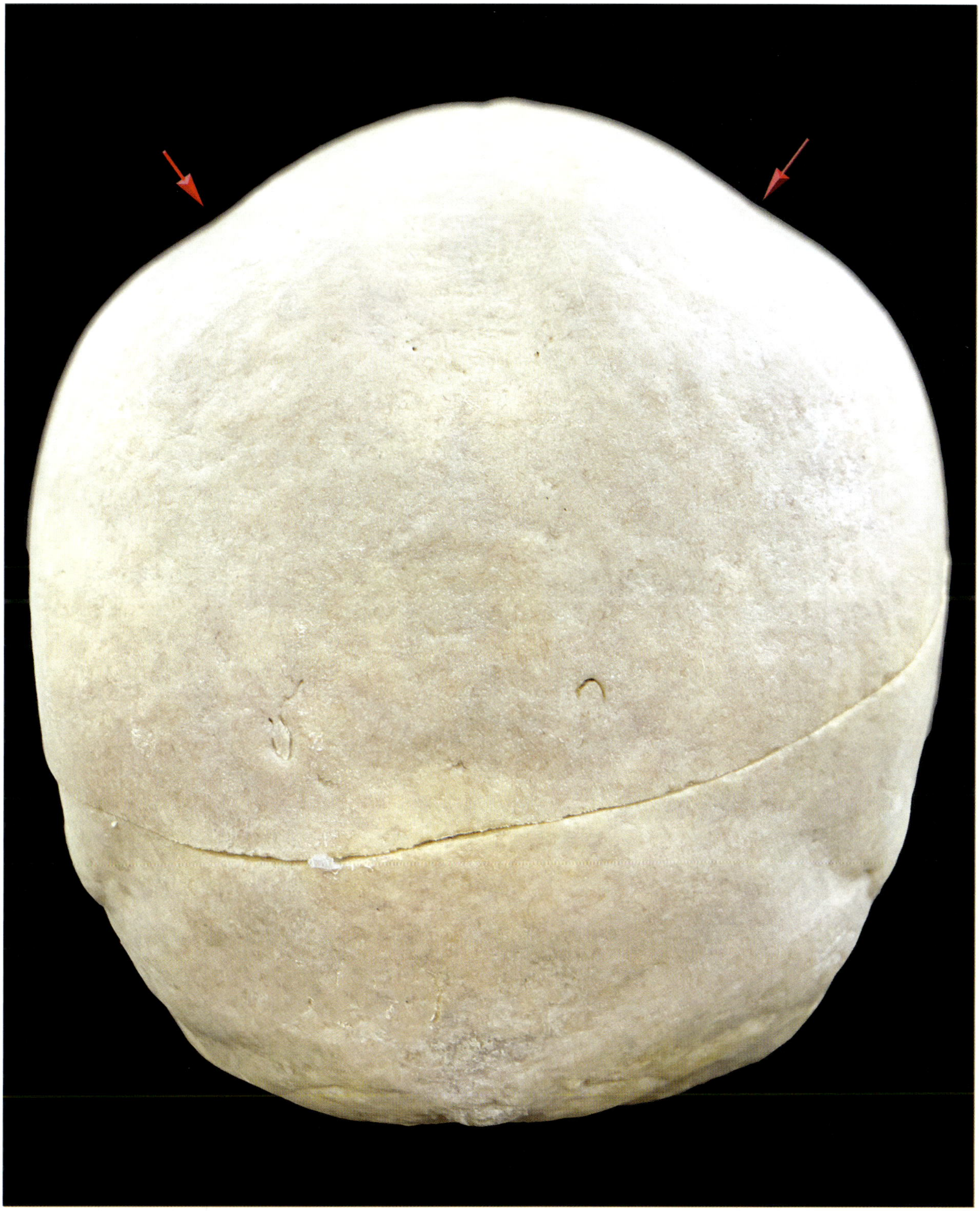

Figure 402. Posterior view of 90-year-old Hawaiian Filipino Chinese female with bilateral depressions (arrow; but no biparietal thinning) and complete obliteration of the sutures.

7. MICRONESIAN FEMALE

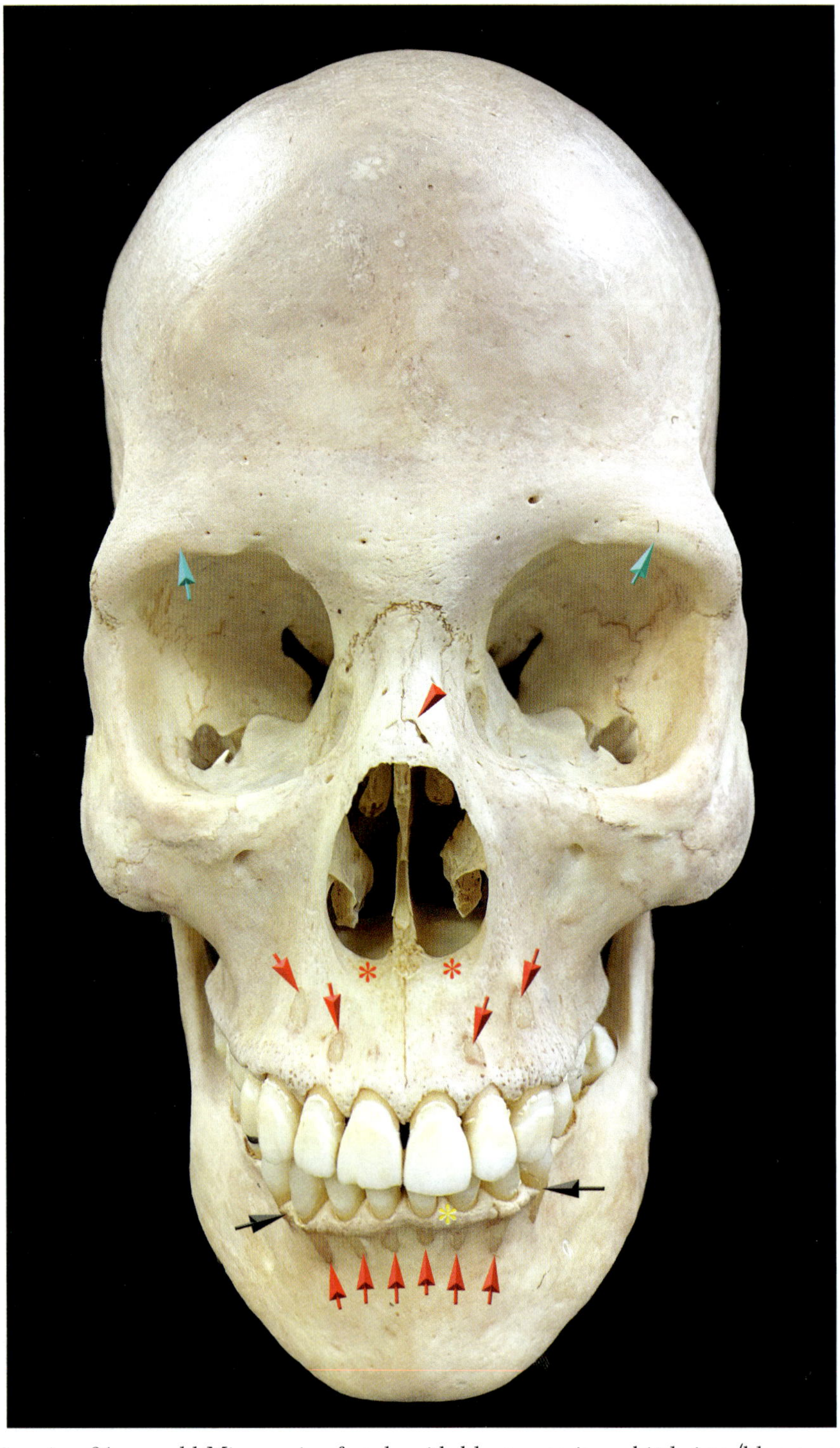

Figure 403. Anterior view 31-year-old Micronesian female with blunt superior orbital rims (blue arrow), healed fracture (red arrowhead) of the nasal bones, fenestrations (red arrow) of the roots of the maxillary and mandibular teeth, dehiscences (black arrow) that expose the roots of the mandibular first premolars, buccal exostoses (yellow*) of the mandible, and nasal guttering (red*).

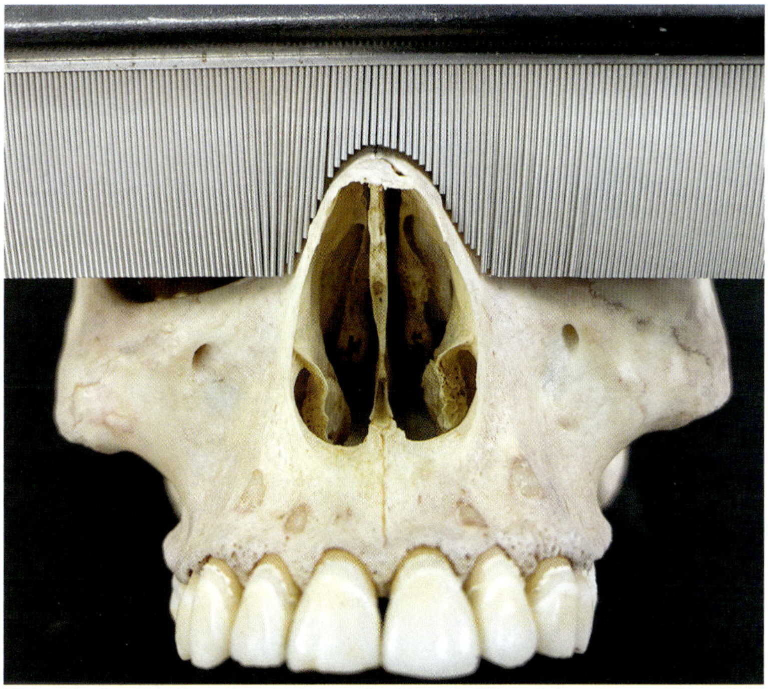

Figure 404. Nasal bone shape in a 31-year-old Micronesian female.

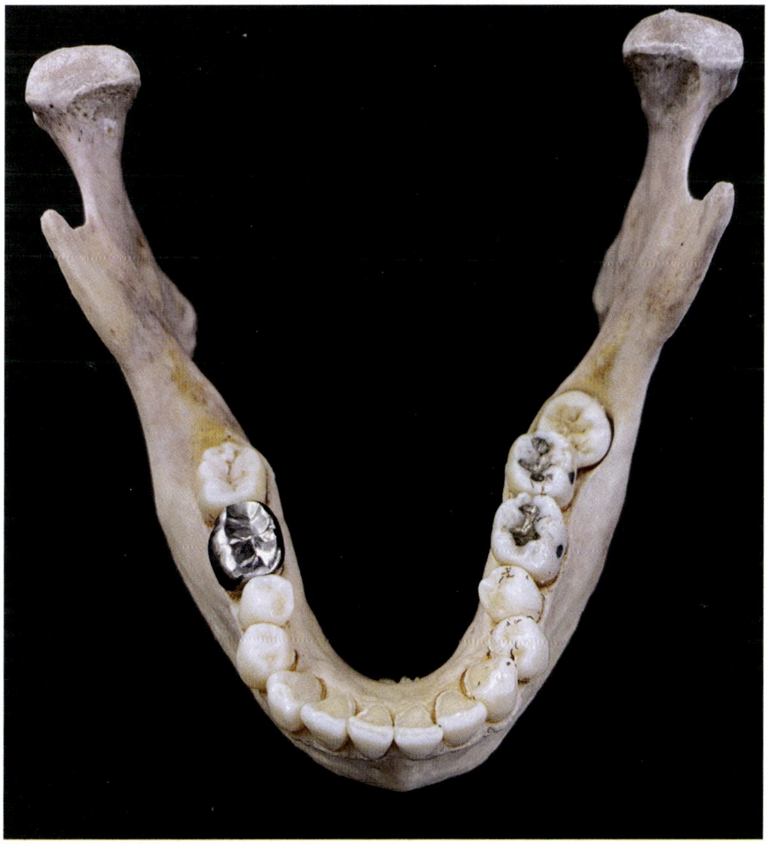

Figure 405. Mandible of 31-year-old Micronesian female with a small and rounded chin.

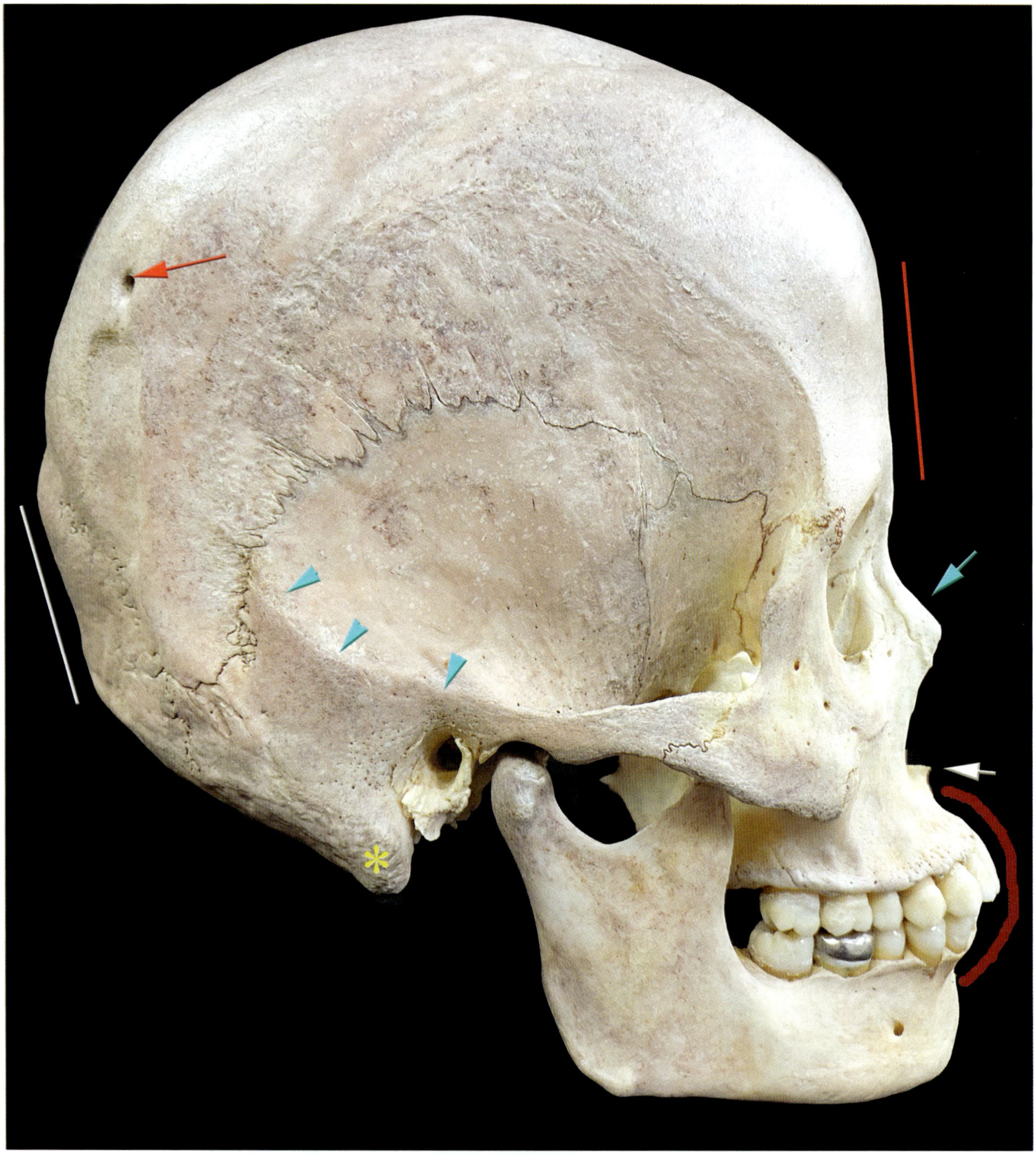

Figure 406. Right lateral view of 31-year-old Micronesian female with a vertical frontal bone (red line), small nasal bones (blue arrow), a small anterior nasal spine (white arrow), a small mastoid process (*), flat occipital region (white line), and a shunt hole (red arrow) in the right parietal bone to drain cerebral spinal fluid, and a moderate supramastoid crest (blue arrowhead). Also note the moderate alveolar prognathism (curved red line) resulting in a forward protrusion of the mouth.

Figure 407. Left lateral view of 31-year-old Micronesian female showing features of the skull including a relatively short (anterior posterior) vault.

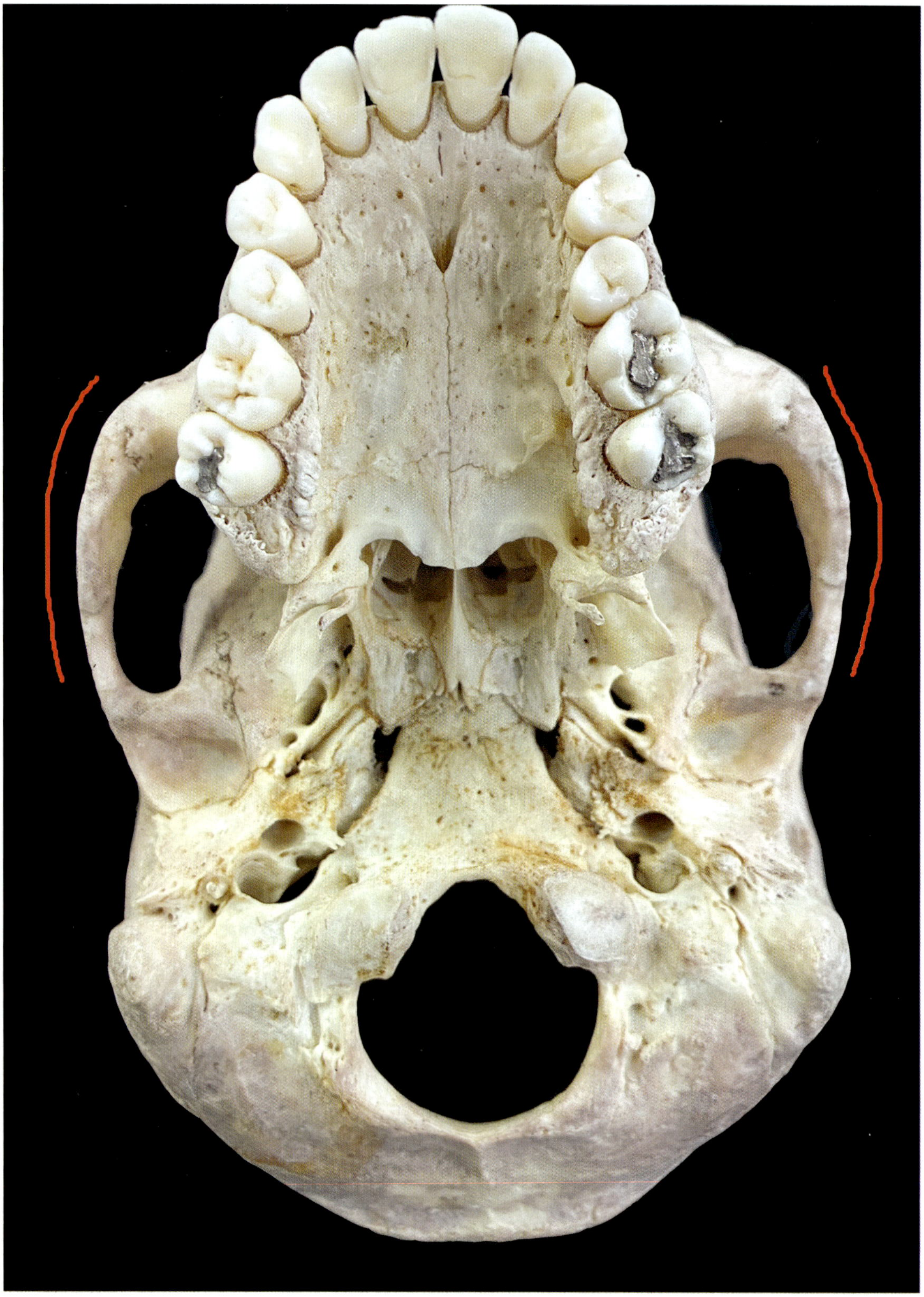

Figure 408. Inferior view of 31-year-old Micronesian female with a large, wide, and shallow hard palate. Note the convex zygomatic arches (curved red line), a feature not usually included in the assessment of ancestry, but one that with future research may hold potential as an ancestry or sex trait.

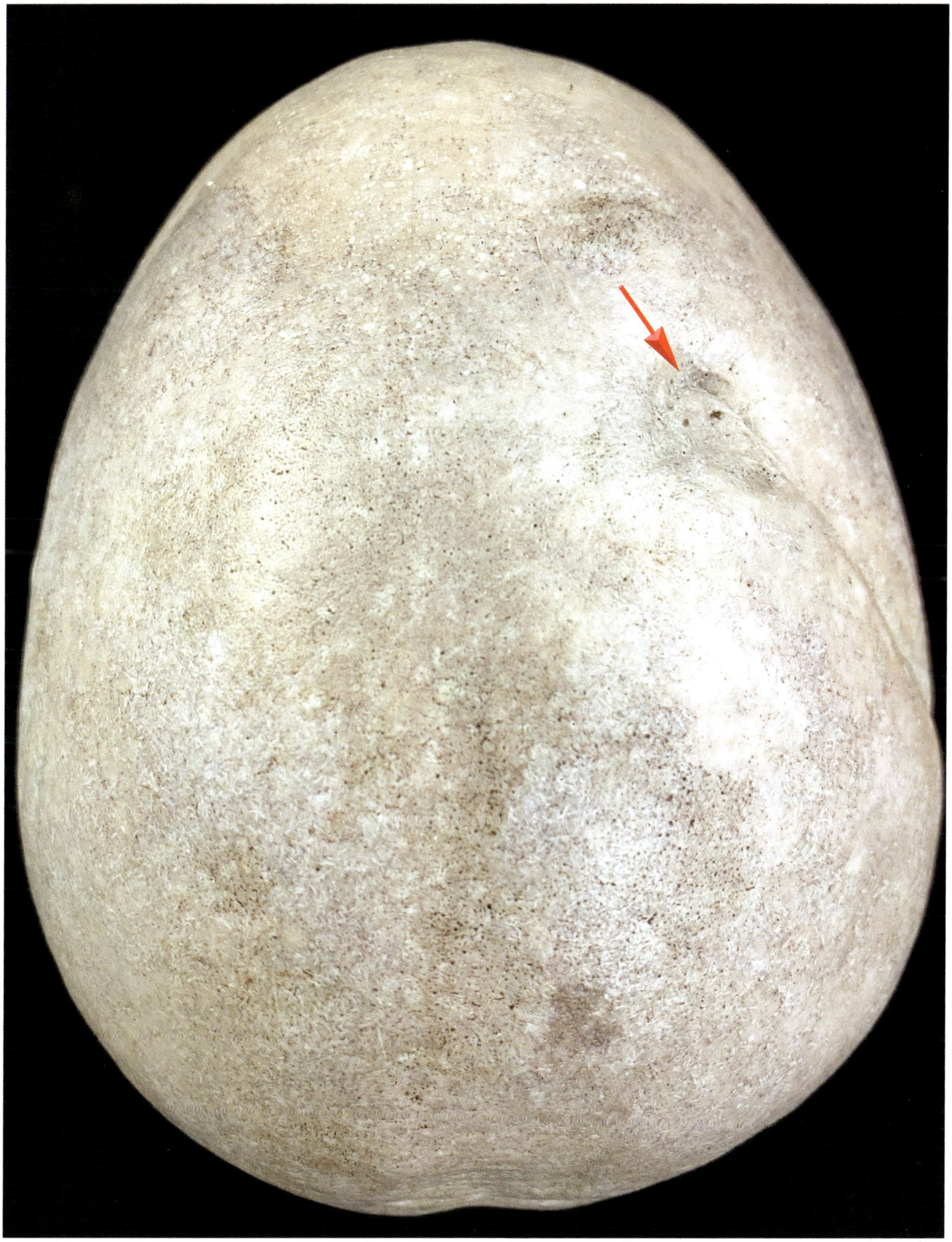

Figure 409. Superior view of 31-year-old Micronesian female with an oval-shaped skull and a healed, surgical shunt hole (arrow) used to drain cerebral spinal fluid associated with disease or trauma.

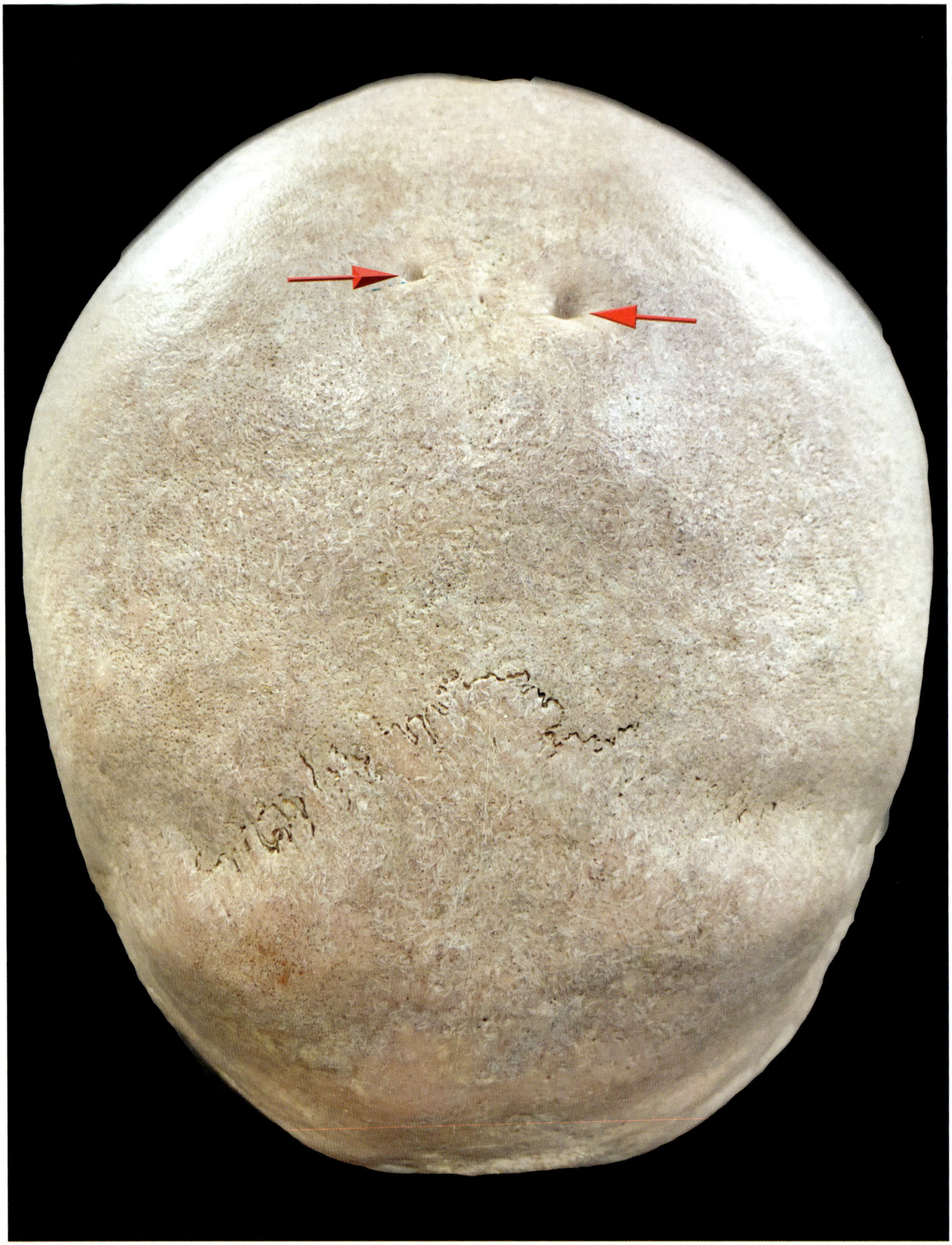

Figure 410. Posterior view of 31-year-old Micronesian female with different sized, but normal, parietal foramina (arrow) that transmit emissary veins. Individuals may have no or multiple parietal foramina representing normal variation.

8. SAMOAN MALE

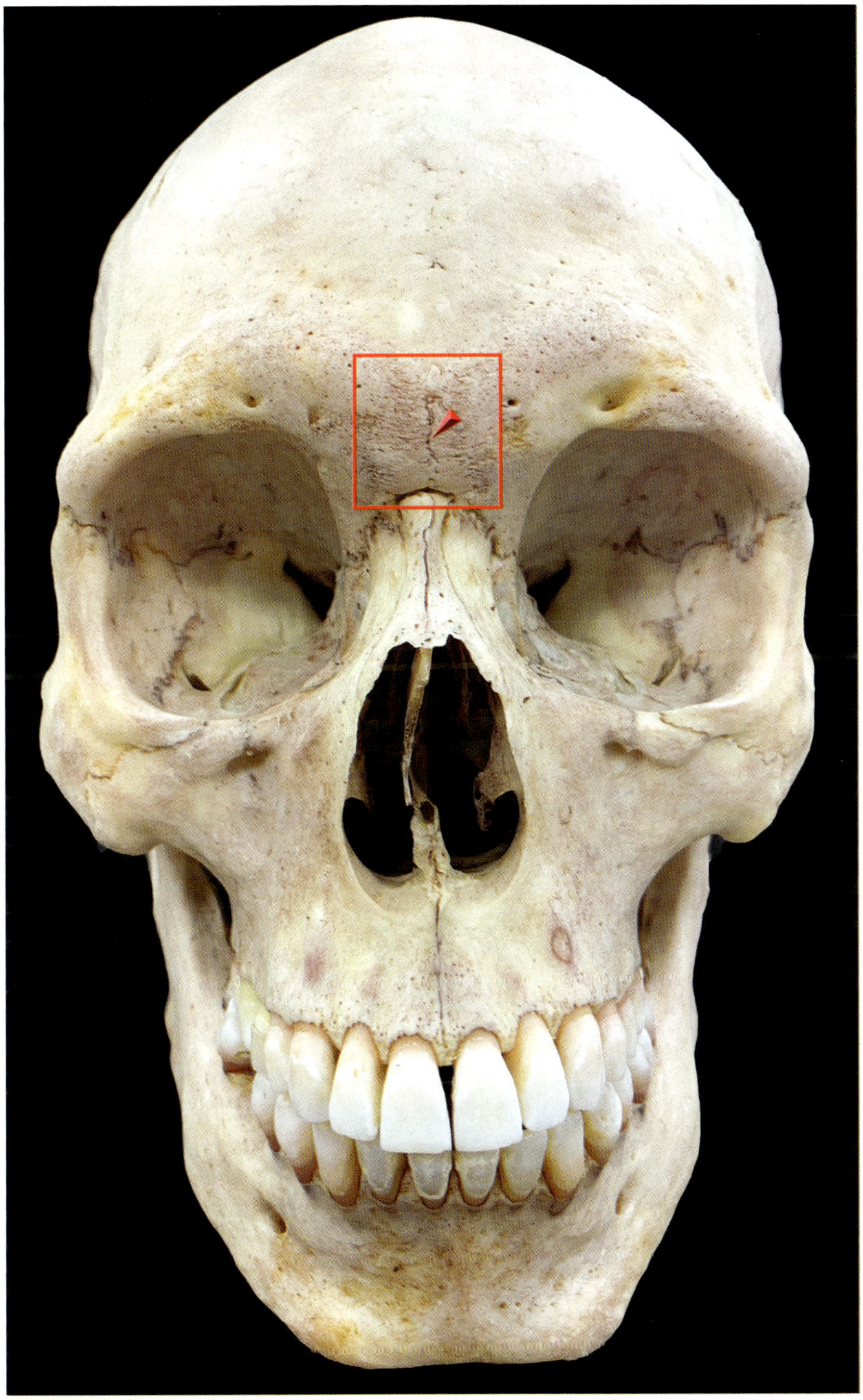

Figure 411. Anterior view 32-year-old Samoan male exhibits nasal guttering, a broad and square chin, blunt superior orbital margins, small malar tubercles, and trace metopic (arrow) and supranasal (rectangle) sutures. A metopic suture, as in this individual, is vertically oriented in comparison to a supranasal suture which is oriented in a horizontal zig-zag pattern.

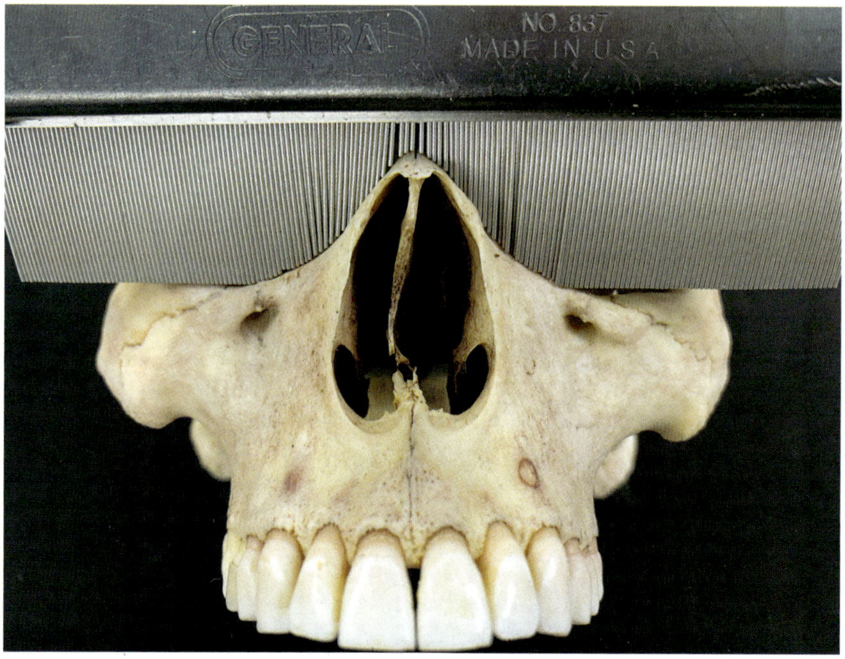

Figure 412. Nasal bone contour in a 32-year-old Samoan male.

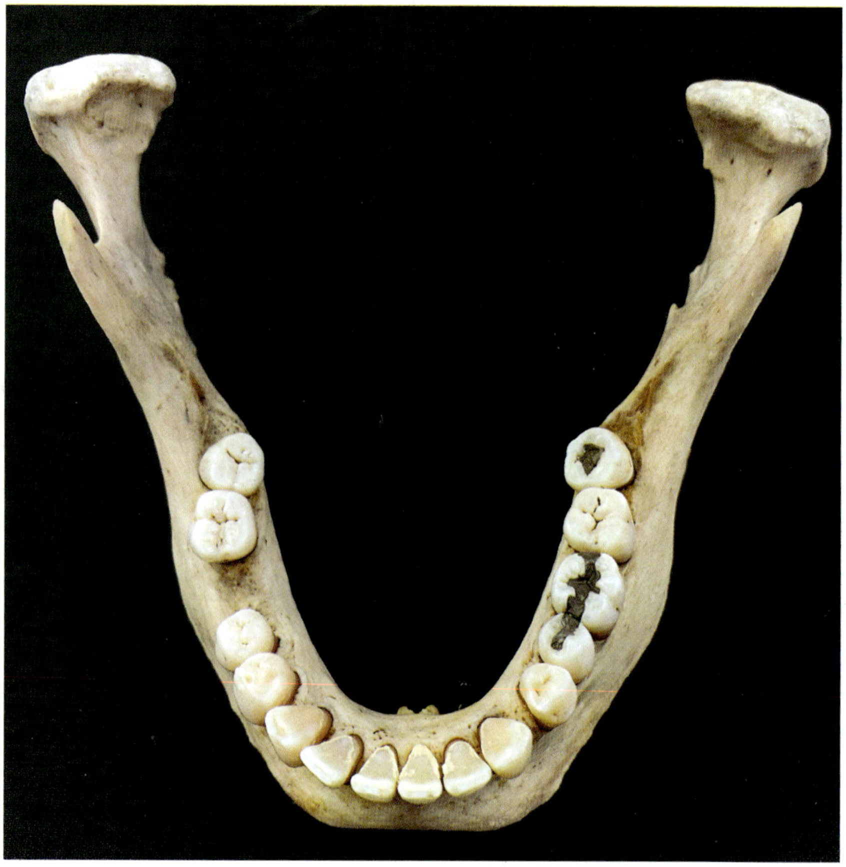

Figure 413. Mandible of 32-year-old Samoan male with an unusually wide chin (rectangle).

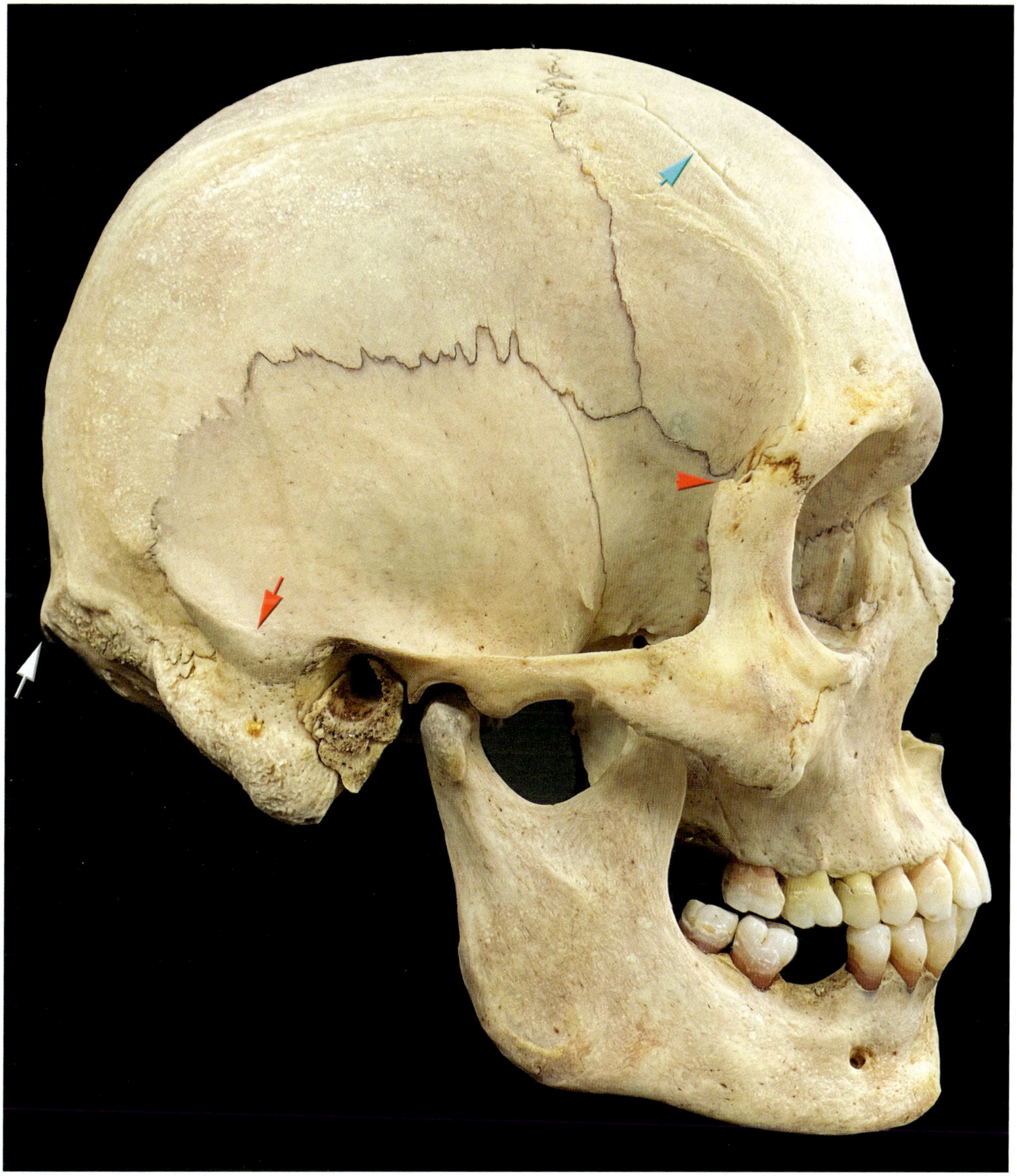

Figure 414. Right lateral view 32-year-old Samoan male with accessory frontal grooves (blue arrow), small and low nasal bones without overhang, a large nuchal crest (white arrow), prominent supramastoid crest (red arrow), alveolar prognathism, and a posterior zygomatic tubercle (arrowhead). This individual exhibits a rounded inferior border of the mandible, but not a "rocker jaw," a feature often (50-90% [Weisler and Swindler 2002]) seen in individuals of Pacific Islander ancestry (Houghton 1977, Snow 1974). Rocker jaw is due to a convex inferior border of the mandible that allows it to rock anterior-posteriorly when placed on a flat surface.

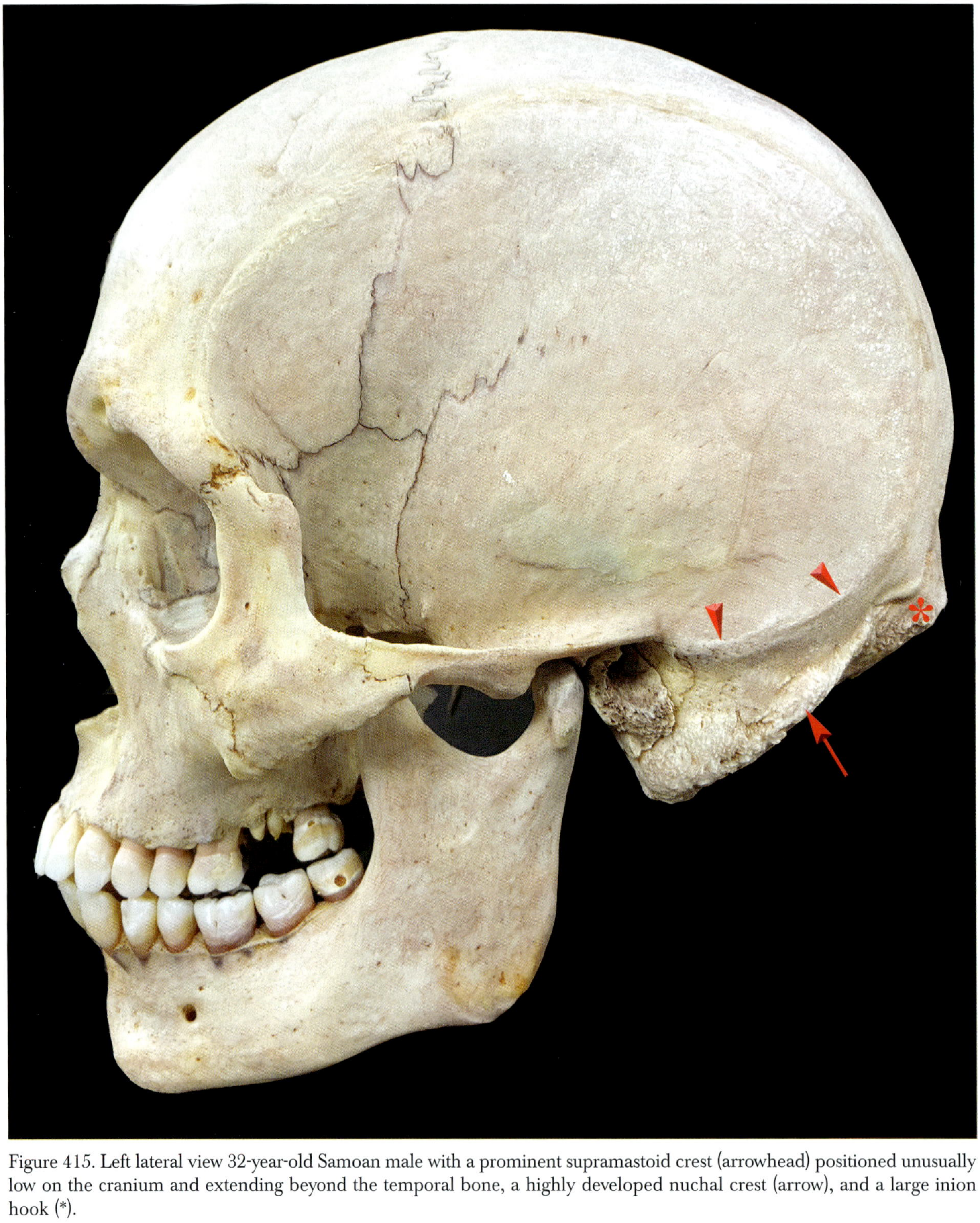

Figure 415. Left lateral view 32-year-old Samoan male with a prominent supramastoid crest (arrowhead) positioned unusually low on the cranium and extending beyond the temporal bone, a highly developed nuchal crest (arrow), and a large inion hook (*).

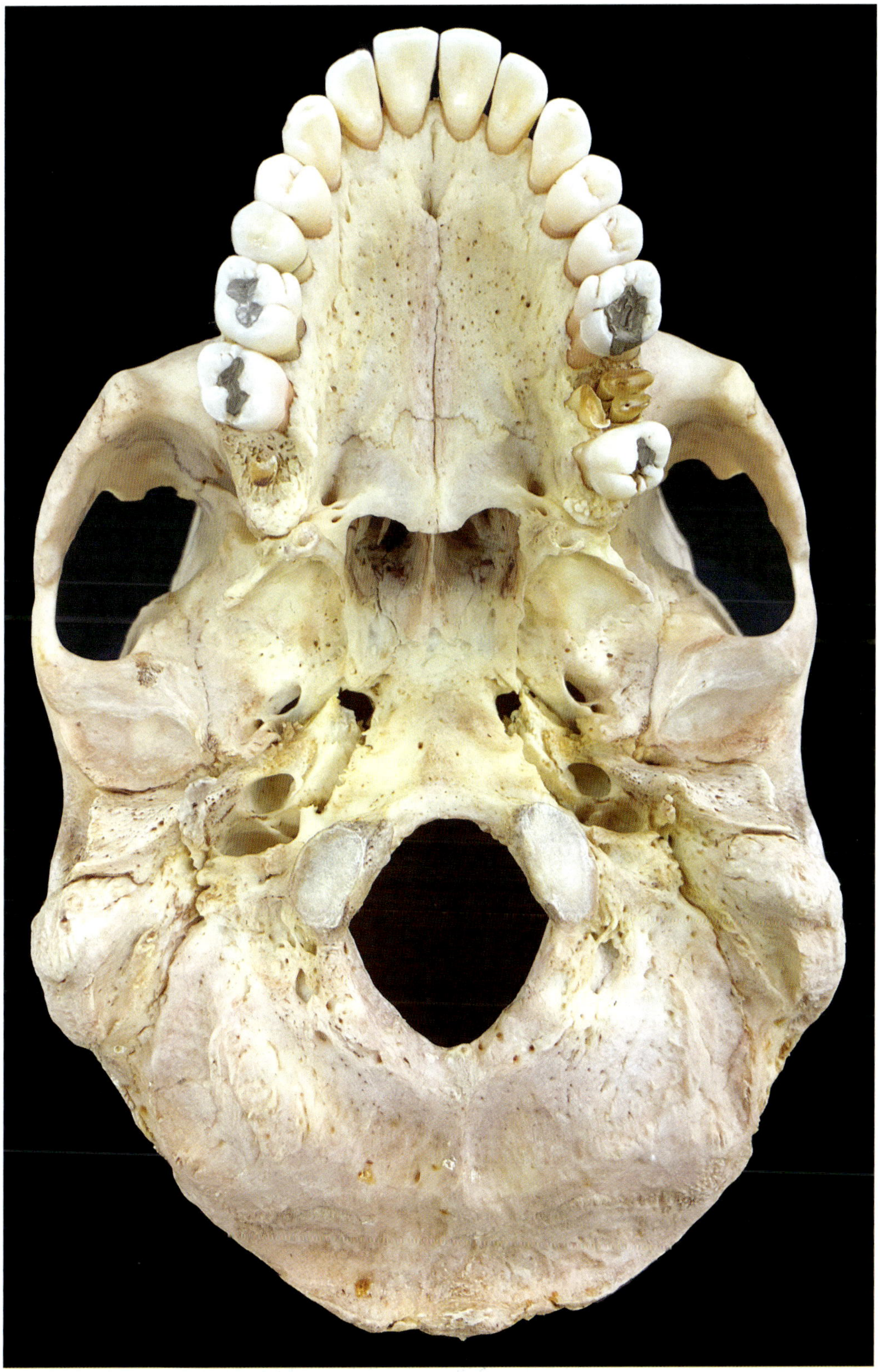

Figure 416. Inferior view 32-year-old Samoan male with a wide, shallow palate and a long occipital region.

Figure 417. Superior view of 32-year-old Samoan male with an oval-shaped cranium and partial obliteration of the sagittal and coronal sutures.

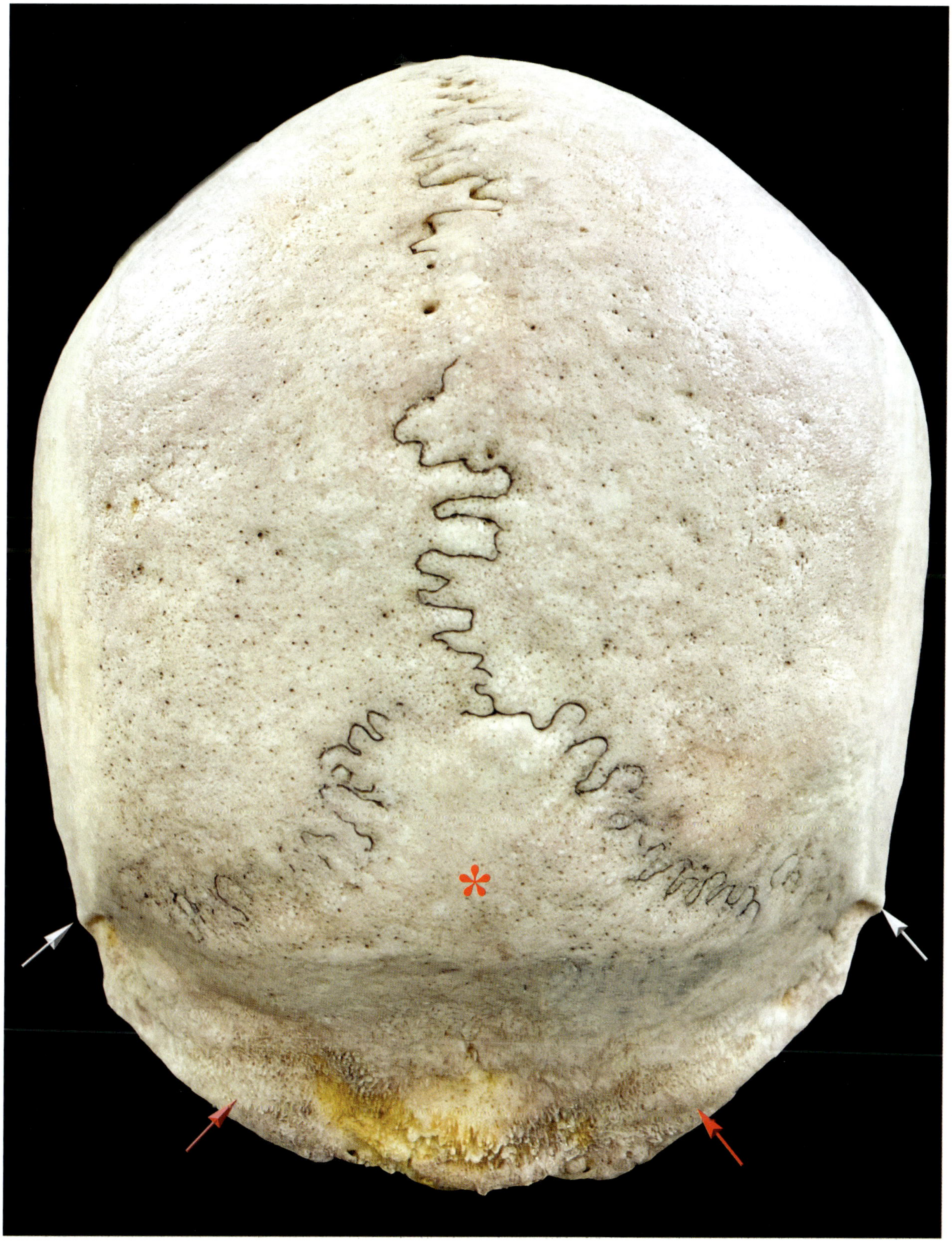

Figure 418. Posterior view 32-year-old Samoan male with a tall cranium, protrusion of the superior portion of the occipital bone (*), a very developed nuchal crest (red arrow), and a highly-developed supramastoid crest (white arrow) that is positioned low on the cranium.

APPENDICES

I. CRANIOMETRIC POINTS AND LANDMARKS USED IN THIS ATLAS

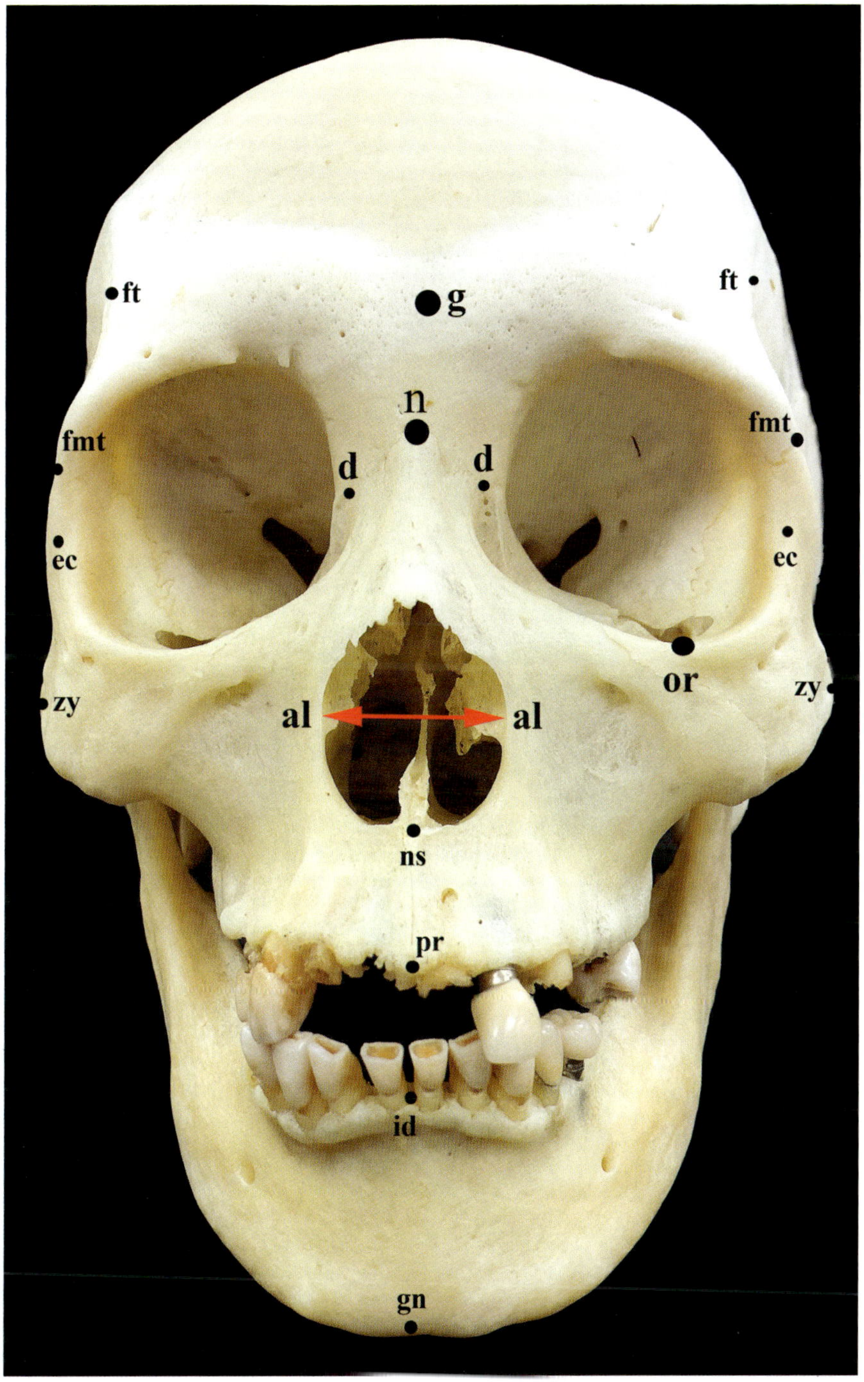

Figure 419. Anterior view 95-year-old Chinese female showing selected osteometric points and landmarks (see Bass 2005; Langley et al. 2016) which can vary slightly depending on factors such as: old age, antemortem tooth loss and alveolar resorption, disease, trauma, cultural practices, developmental conditions, and taphonomy. Ft=frontotemporale, g=glabella, fmt=frontomalare temporale, d=dacryon, n=nasion, ec=ectochoncion, zy=zygion, al=alare, or=orbitale, ns=nasospinale, pr=prosthion, id=infradentale, gn=gnathion. The red double arrow depicts the nasal breadth measured from the most lateral points on the nasal aperture, alare (al).

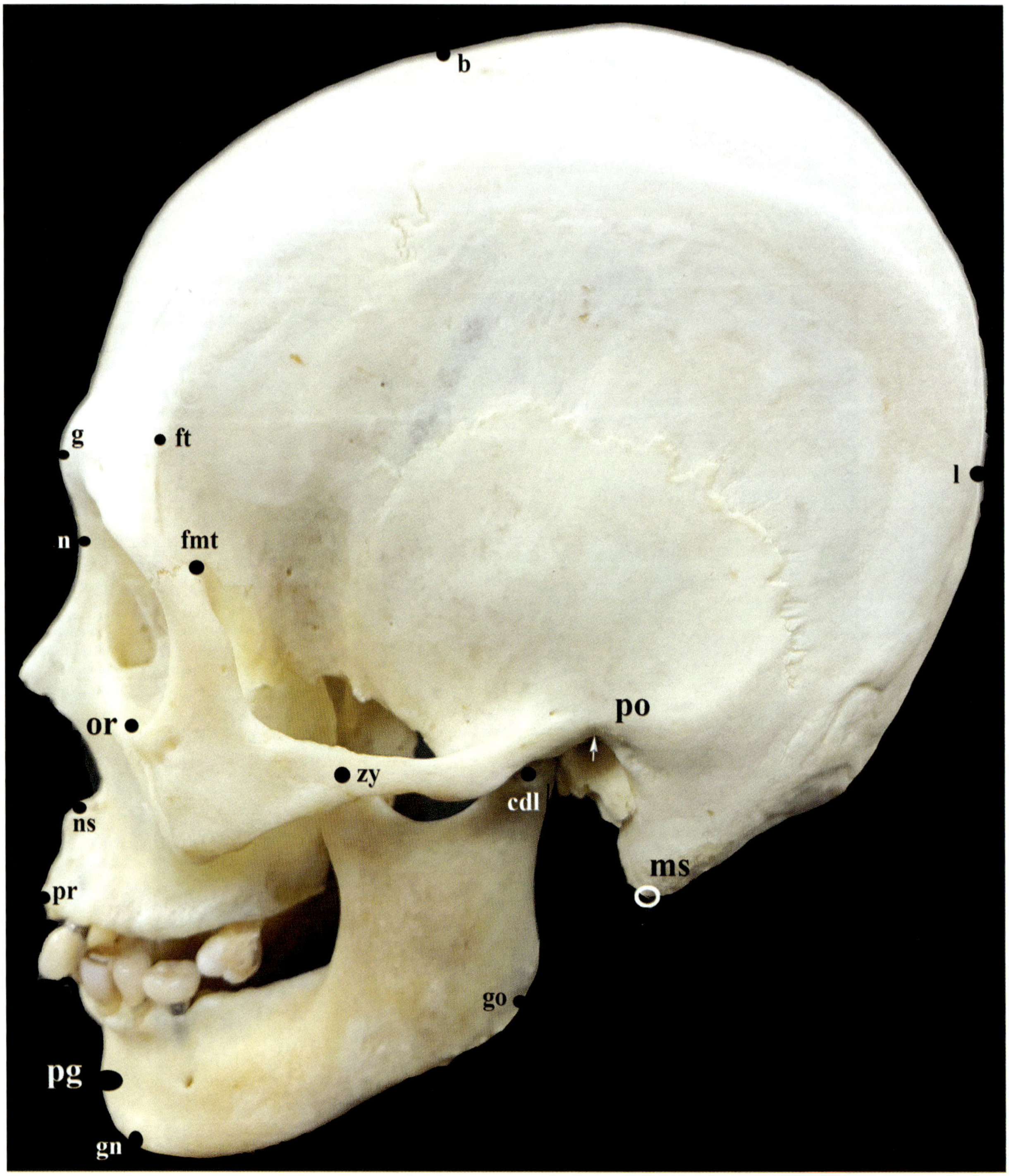

Figure 420. Left lateral view 95-year-old Chinese female showing selected osteometric points and landmarks which, depending on the morphology of skulls, can vary slightly in their location. Landmarks are: b=bregma, cdl-condylion, fmt=frontomalare temporale, ft=frontotemporale, g=glabella, gn=gnathion, go=gonion, l=lambda, ms=mastoidale, n=nasion, ns=nasospinale, or=orbitale, pg=pogonion, po=porion, pr=prosthion, zy=zygion (see Bass 2005; Langley et al. 2016).

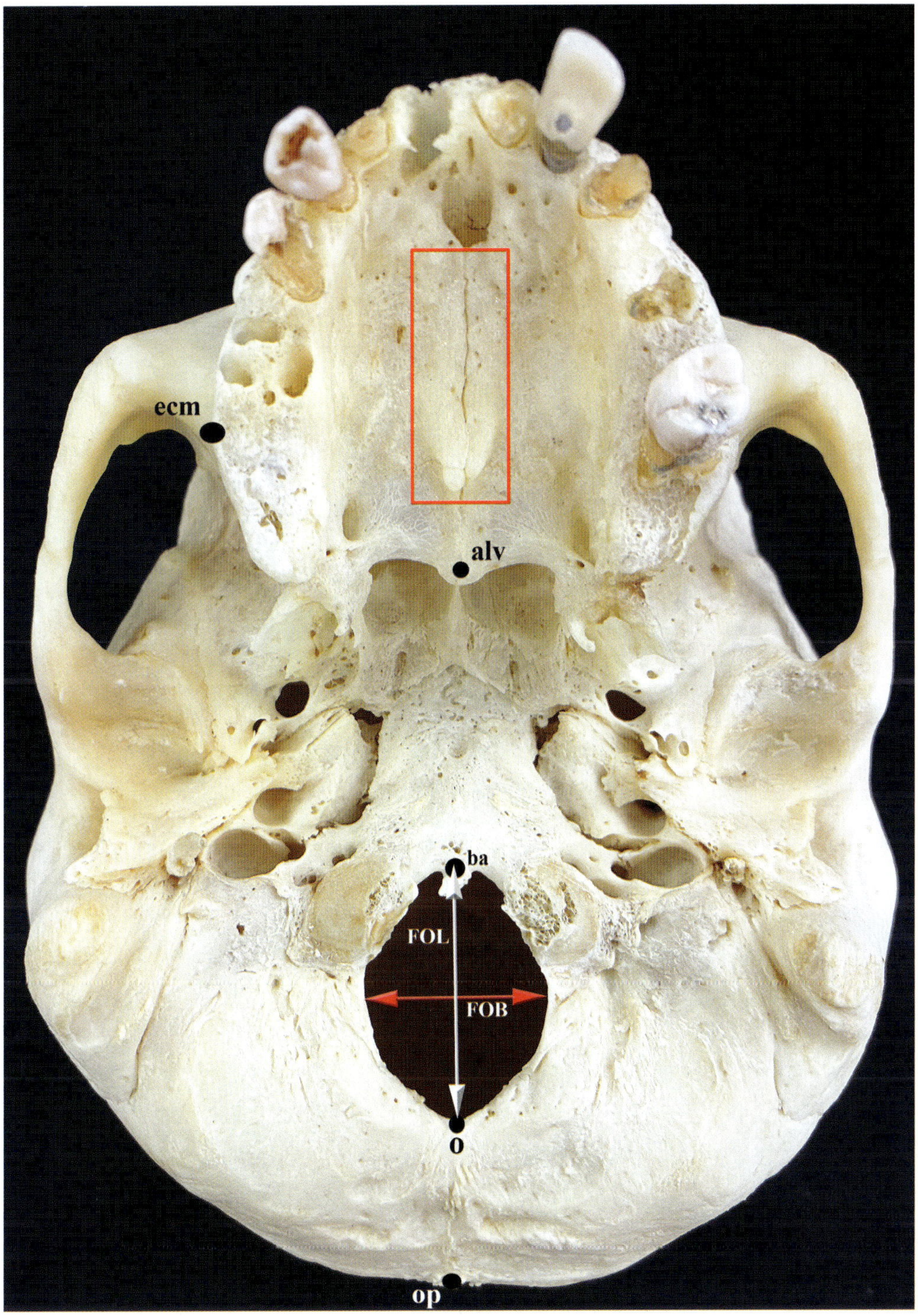

Figure 421. Inferior view 95-year-old Chinese female showing selected osteometric points and landmarks. Maximum breadth of the foramen magnum (FOB) is indicated by the red double arrow and maximum length (FOL) by the white double arrow. Basion (ba) in this individual was estimated because of the bony projection obscuring its precise location. Also estimated was ectomolare (ecm) due to loss of the maxillary second molars and the concomitant alveolar exostoses. Alveolon (alv) is the point on the palate where a line drawn from the termini of the alveolar ridges crosses the midline of the hard palate (see Bass 2005). Opistocranion (op) is the most posterior point on the cranium in the midline. Note the torus palatinus (rectangle). This cranium was chosen despite these "imprecisions" because they are frequently seen in elderly individuals.

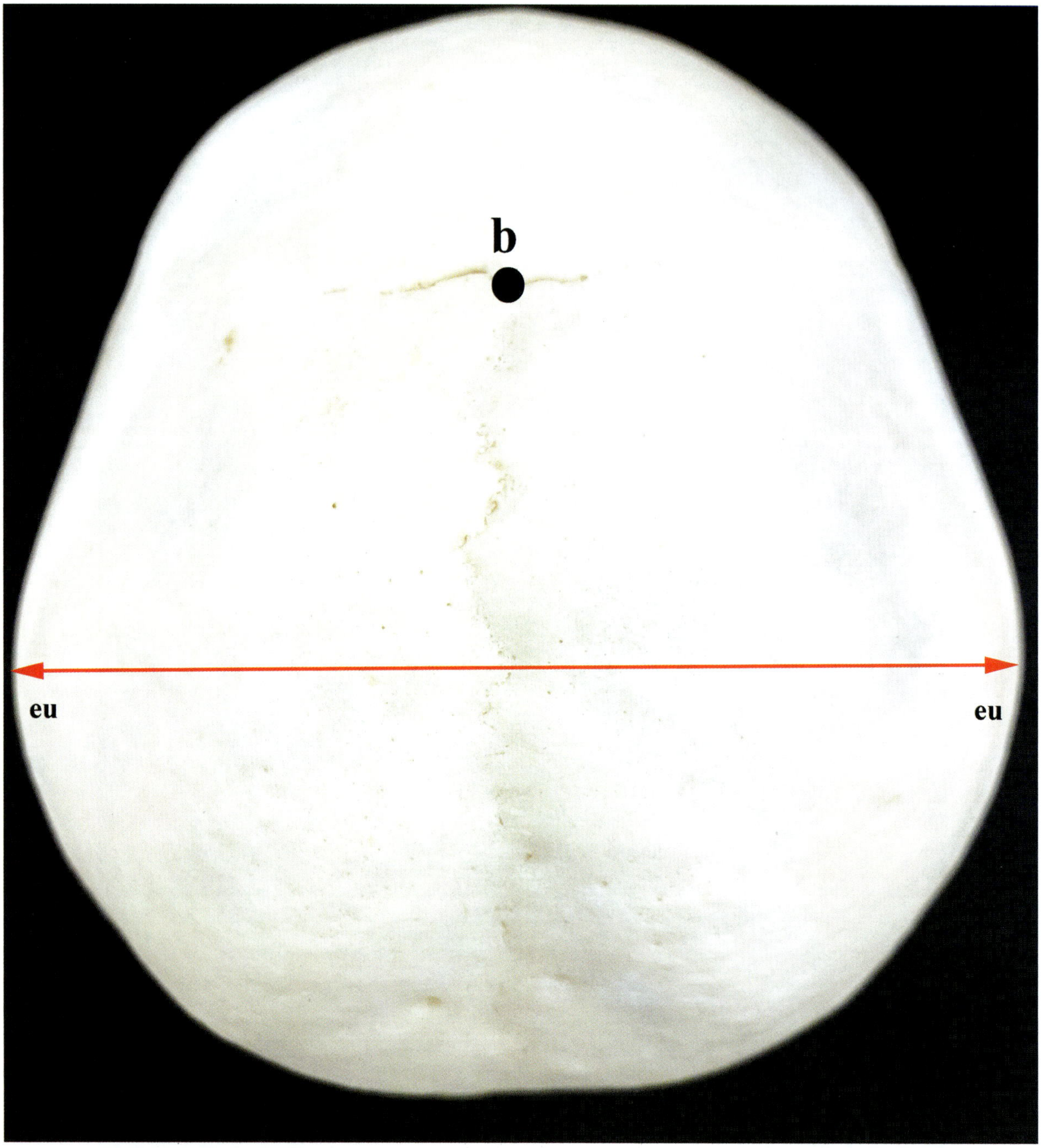

Figure 422. Superior view 95-year-old Chinese female showing bregma (b) at the intersection of the coronal and sagittal sutures, and the maximum cranial breadth (euryon to euryon/eu-eu, indicated by the red double arrow) at the parietal eminences. Note the resulting bossing/squaring of the skull (see Bass 2005).

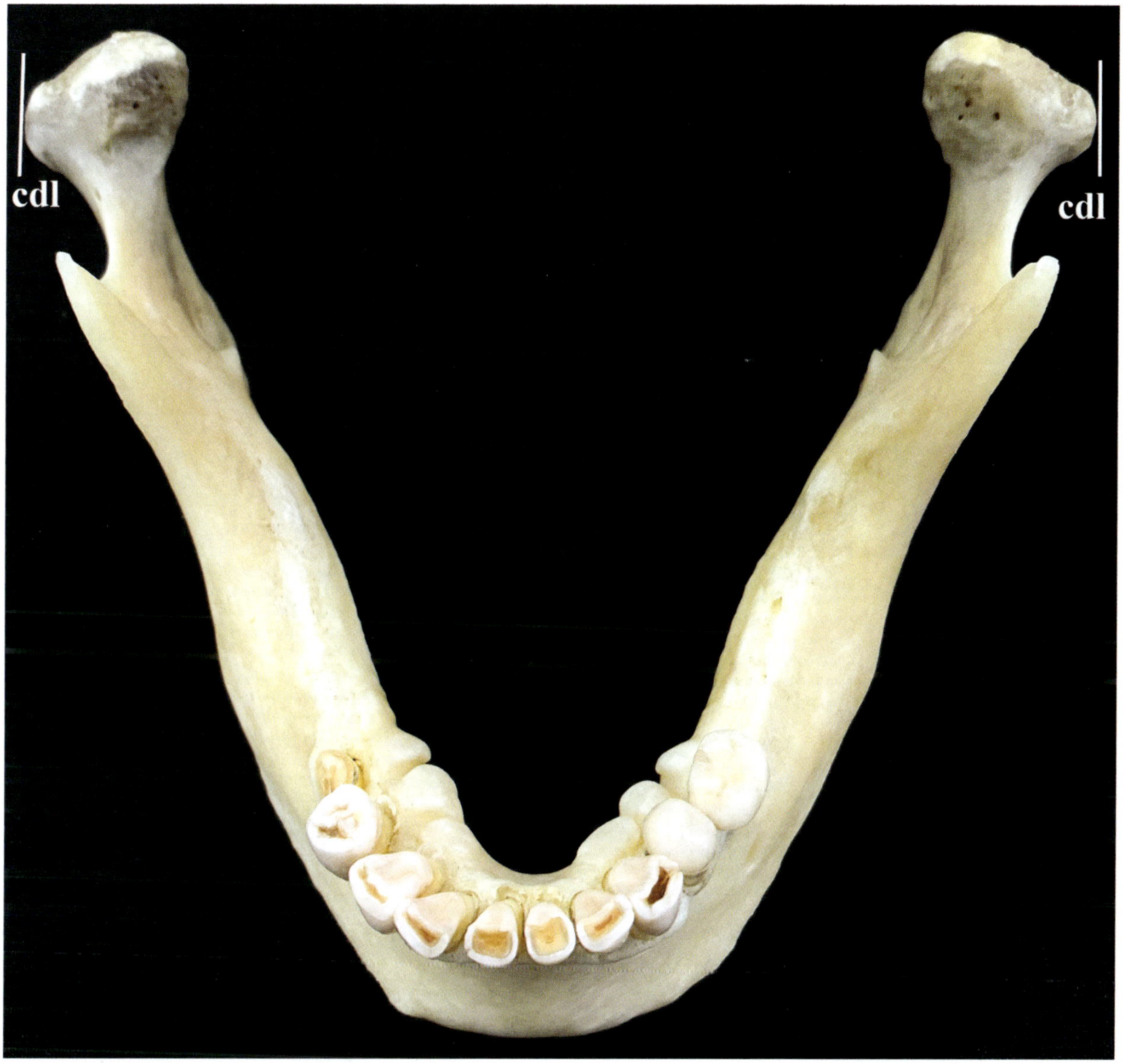

Figure 423. Maximum breadth of the mandibular condyles (condylion-condylion) (see Bass 2005).

II. SELECTED MORPHOLOGICAL FEATURES USED TO ASSESS ANCESTRY, SEX, AND VARIANTS IN THIS ATLAS

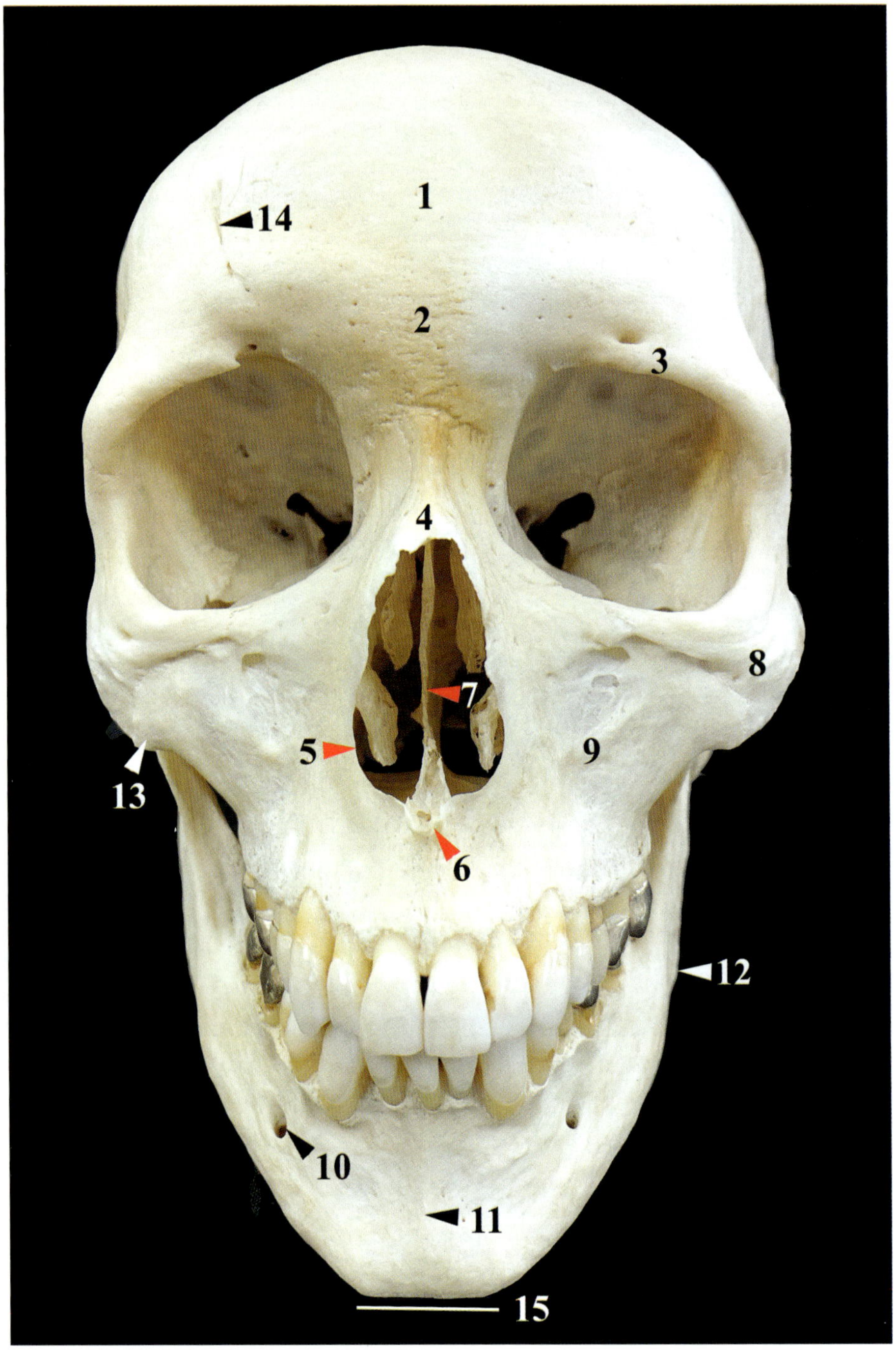

Figure 424. Anterior view of selected features used to estimate ancestry, sex, and variation. 1=frontal bone, 2=glabella, 3=superior orbital rim, 4=nasal bones, 5=nasal sill/aperture, 6=anterior nasal spine, 7=nasal septum, 8=zygomatic/malar bone, 9=maxilla and frontal process of the maxilla, 10=mental foramen, 11=mental eminence, 12=gonial eversion, 13=malar tubercle, 14=frontal groove, and 15=chin shape (See Bass 2005; Langley et al 2016).

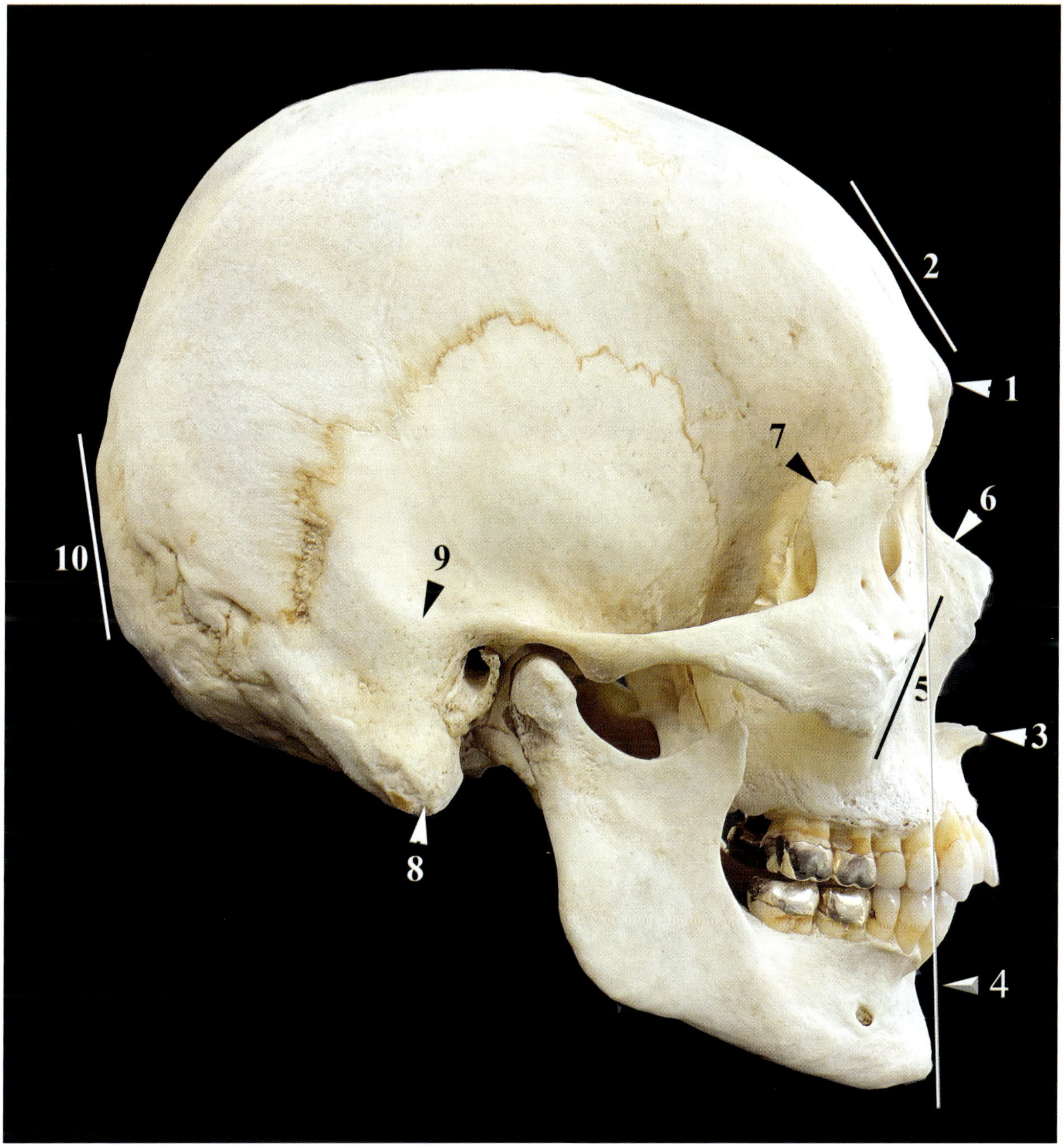

Figure 425. Right lateral view of selected features used to estimate ancestry and sex. Care must be exercised when estimating ancestry based on features such as alveolar prognathism (protrusion of the facial region) that include angles formed by a combination of the teeth and surrounding bony structures. Alveolar prognathism may also vary based on which method is used to calculate prognathism: for example, based on a straight line from nasion to pogonion, the most anterior point of the chin among the midline, as in this illustration. 1=glabella/brow ridge, 2=frontal bone shape, 3=anterior nasal spine, 4=alveolar prognathism, 5=superior-inferior angle of the zygoma, 6=nasal bone, 7=posterior zygomatic tubercle, 8=mastoid process, 9=supramastoid crest, and 10=occipital bone shape (See Bass 2005; Langley et al. 2016).

III. COMPARISON OF ALVEOLAR PROGNATHISM IN FIVE INDIVIDUALS

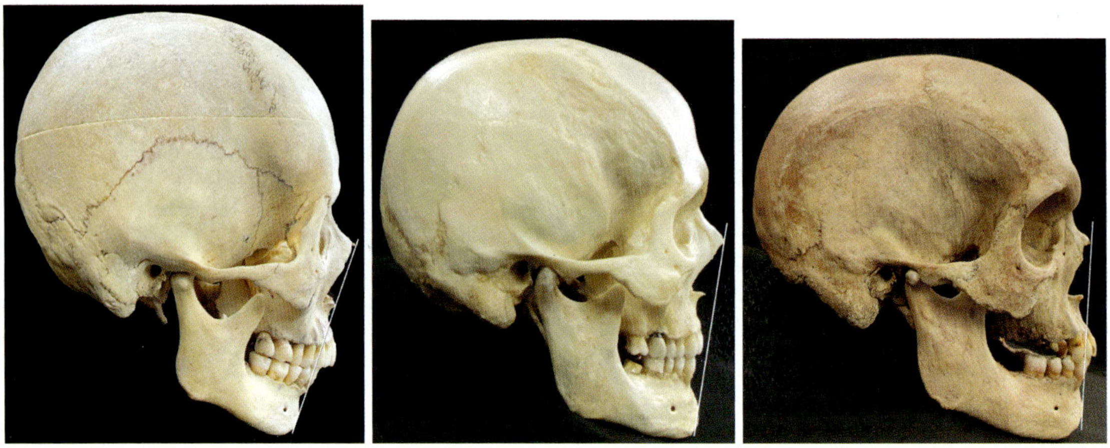

Figure 426. Alveolar prognathism in three randomly selected individuals consisting of an Asian female (left), European male (middle) and African male (right). The relative amount of prognathism or anterior projection of the midfacial region varies greatly within and between ancestral groups. One method to visualize the degree/amount of prognathism is to draw a straight line from the tip of the chin to the tip of the nasal bones. Note the relative distance of the protrusion of the maxilla/teeth and anterior nasal spine to the white line. Individuals of Asian and African ancestry typically have greater protrusion of the maxilla.

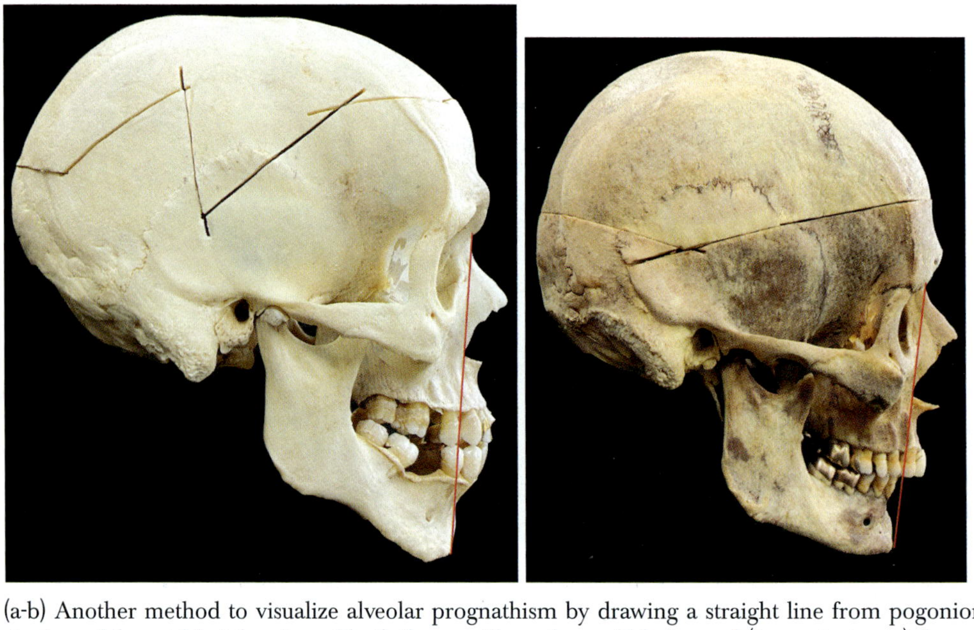

Figure 427. (a-b) Another method to visualize alveolar prognathism by drawing a straight line from pogonion to nasion. Note the projection of the mid-facial region of both the maxilla and mandible beyond (i.e., anterior to) the red line in (a) an African American male and (b) a European male.

IV. COMPARISON OF FOUR CRANIA (INFERIOR VIEW)

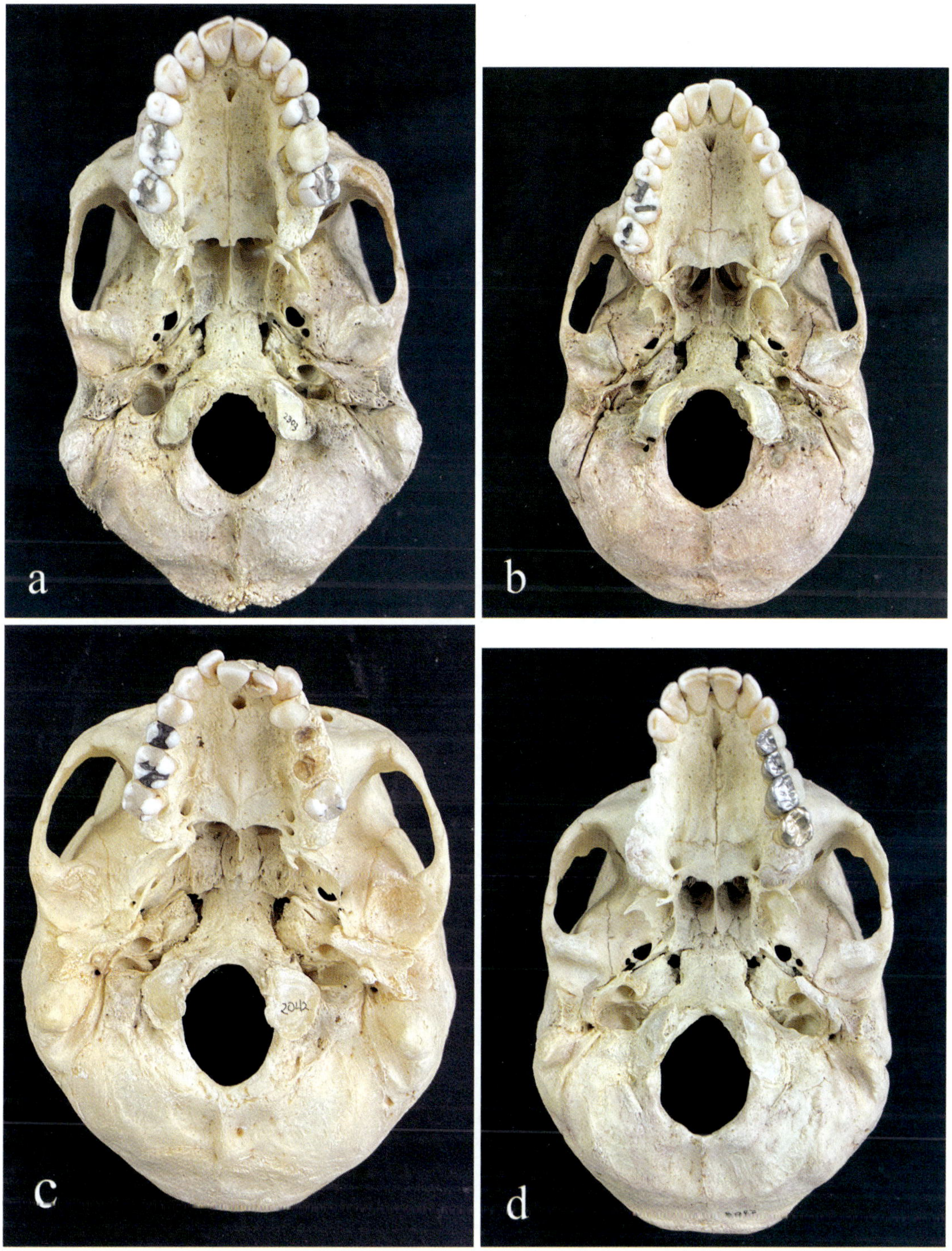

Figure 428a-d. Inferior view of four crania showing differences in size and shape. (a) European male (b) European female (c) Asian (Japanese) female, and (d) Hispanic (Mexican) European female. Note, for example, that the European male (a) and female (b) are long anterior-posterior and have flat zygomatic arches in comparison to the Asian female (c) and Hispanic female (d) with more rounded crania and convex zygomatic arches. (The four crania were photographed from the same camera focal length.)

V. SIZE AND SHAPE DIFFERENCES IN THREE EUROPEAN AND ONE ASIAN FEMALE MANDIBLES

Figure 429. A comparison of four randomly selected elderly female mandibles: three Europeans and one Asian. The Asian mandible is third from the top. Note the lingual exostoses (arrows) on two of the mandibles. This photograph illustrates size and shape differences that can make it difficult to estimate ancestry solely from the mandible.

VI. SIZE AND SHAPE DIFFERENCES IN FOUR ADULT MALE MANDIBLES

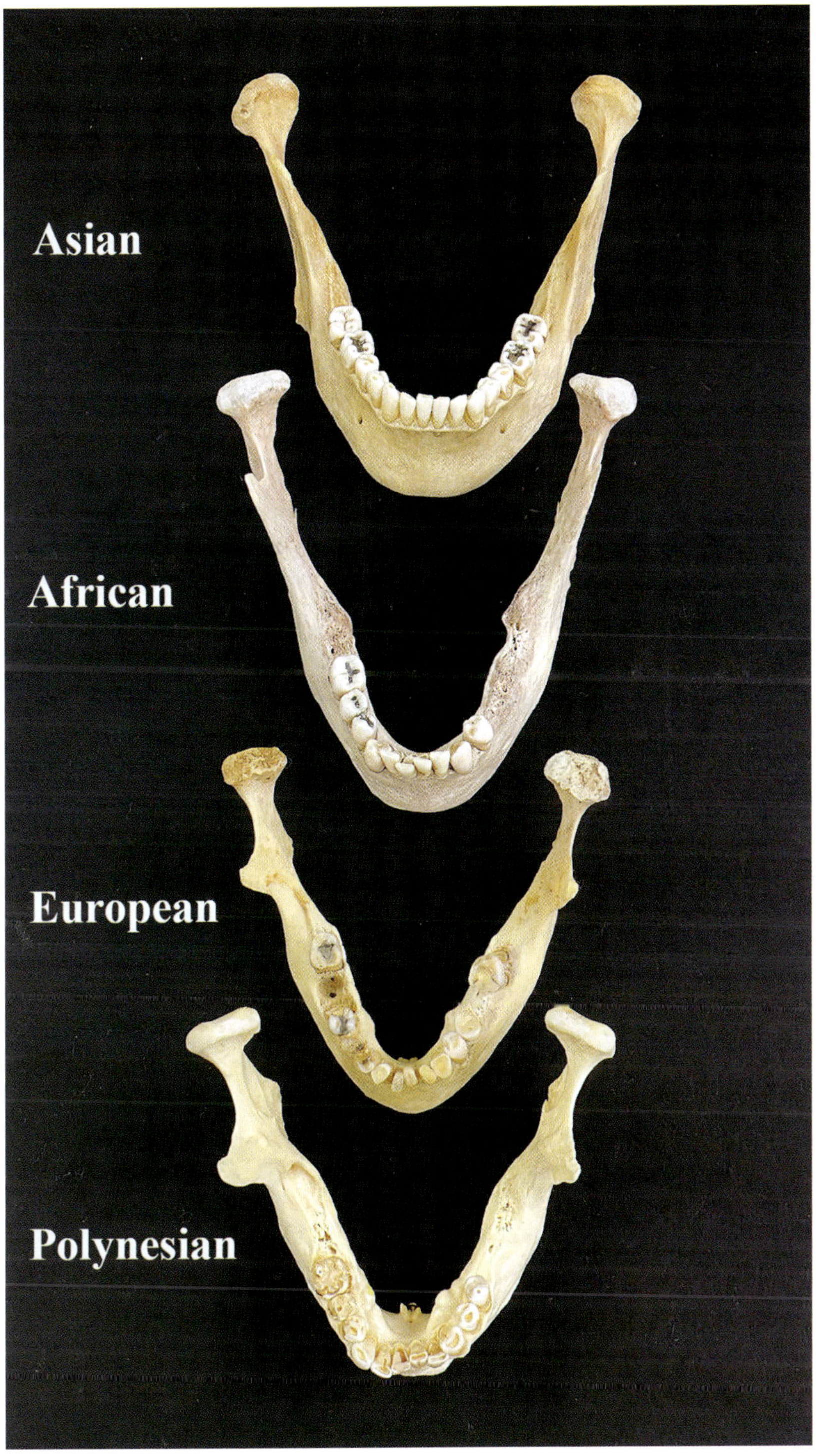

Figure 430. A comparison of four randomly selected elderly male mandibles: one Asian, one African American, one European, and one Polynesian (Pacific Islander). This photograph illustrates size and shape differences that can make it difficult to estimate ancestry solely from the mandible.

VII. ILLUSTRATION OF MOST TRAITS USED IN THIS ATLAS

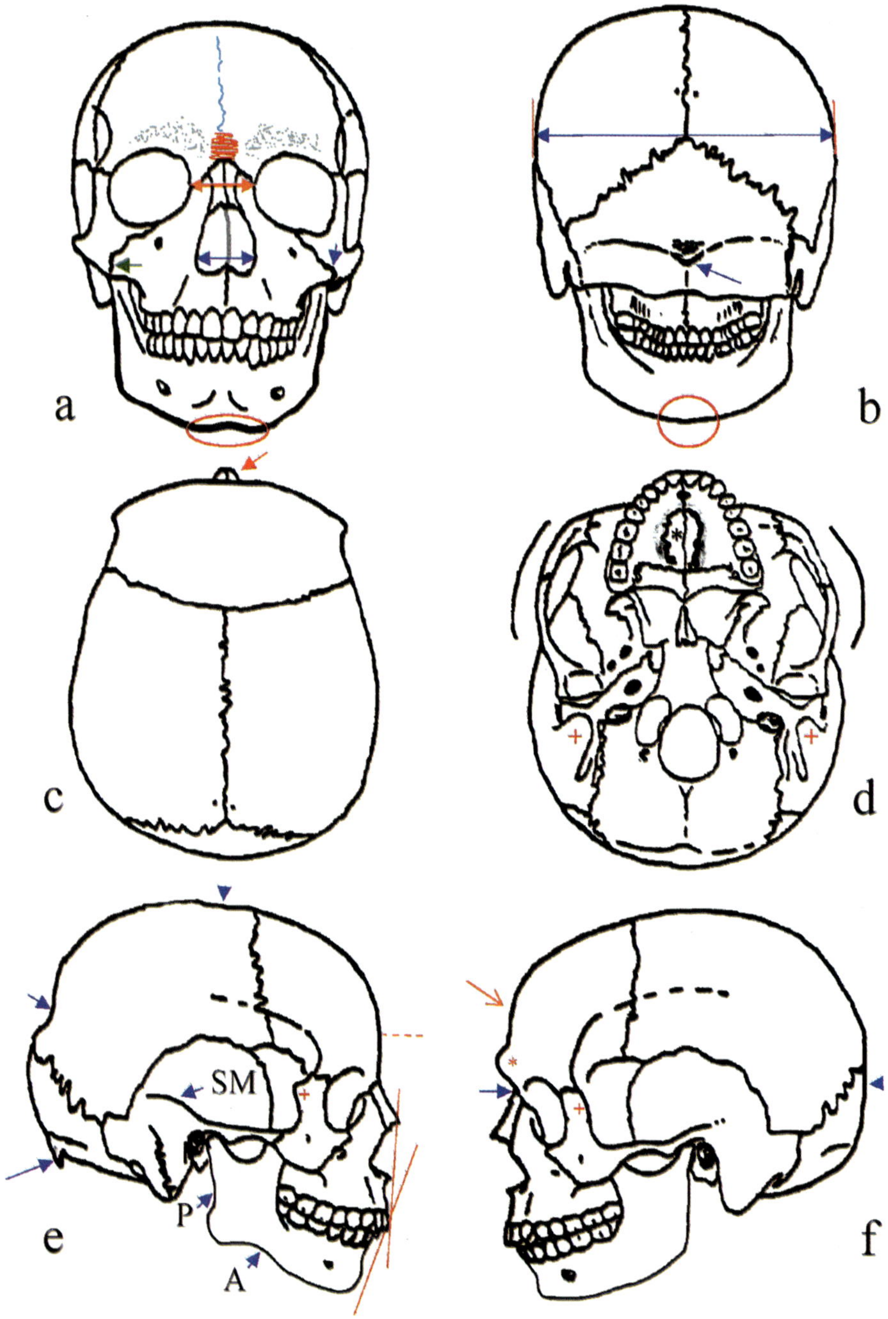

Figure 431. A=metopic suture (blue), supranasal suture (red horizontal lines), brow ridge (stippled), variation of zygomaxillary suture (green and blue arrows, interorbital breadth (orange double arrow), nasal width (blue double arrow), square chin (oval); B=cranial breadth (double arrow), inion spike (arrow), rounded chin (circle); C=projecting nasal bones viewed superiorly (arrow); D=torus palatinus (*), mastoid process (+), wide, curved zygomatic arches (curved lines); E=post-bregmatic depression (arrowhead), bathrocephaly (short arrow), inion spike (long arrow), supramastoid crest (SM), concave posterior border mandible (P), antegonial notch (A), posterior zygomatic tubercle (+), alveolar prognathism (red lines); F=curved frontal bone (red arrow), brow ridge (*), depression at nasion (blue arrow), no projection of posterior zygomatic tubercle (+), and flattened occipital region (arrowhead).

VIII. ORBIT SIZE AND SHAPE DIFFERENCES IN 12 KNOWN-IDENTITY ADULT CRANIA

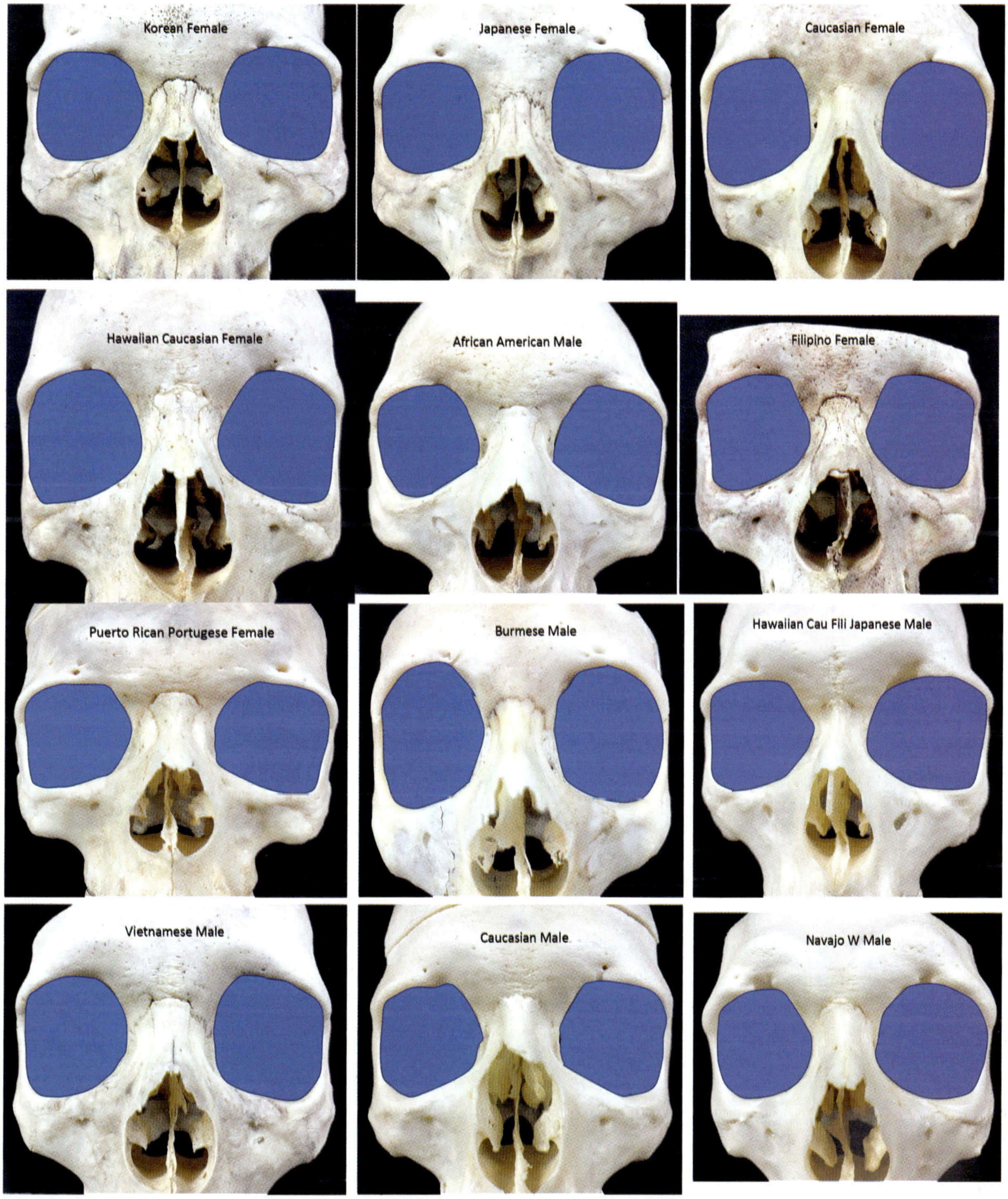

Figure 432. Differences in orbit size and shape in 12 known-identity adults.

IX. COMPARISON OF SOFT TISSUE AND CRANIAL FEATURES (CT) IN A EUROPEAN MALE

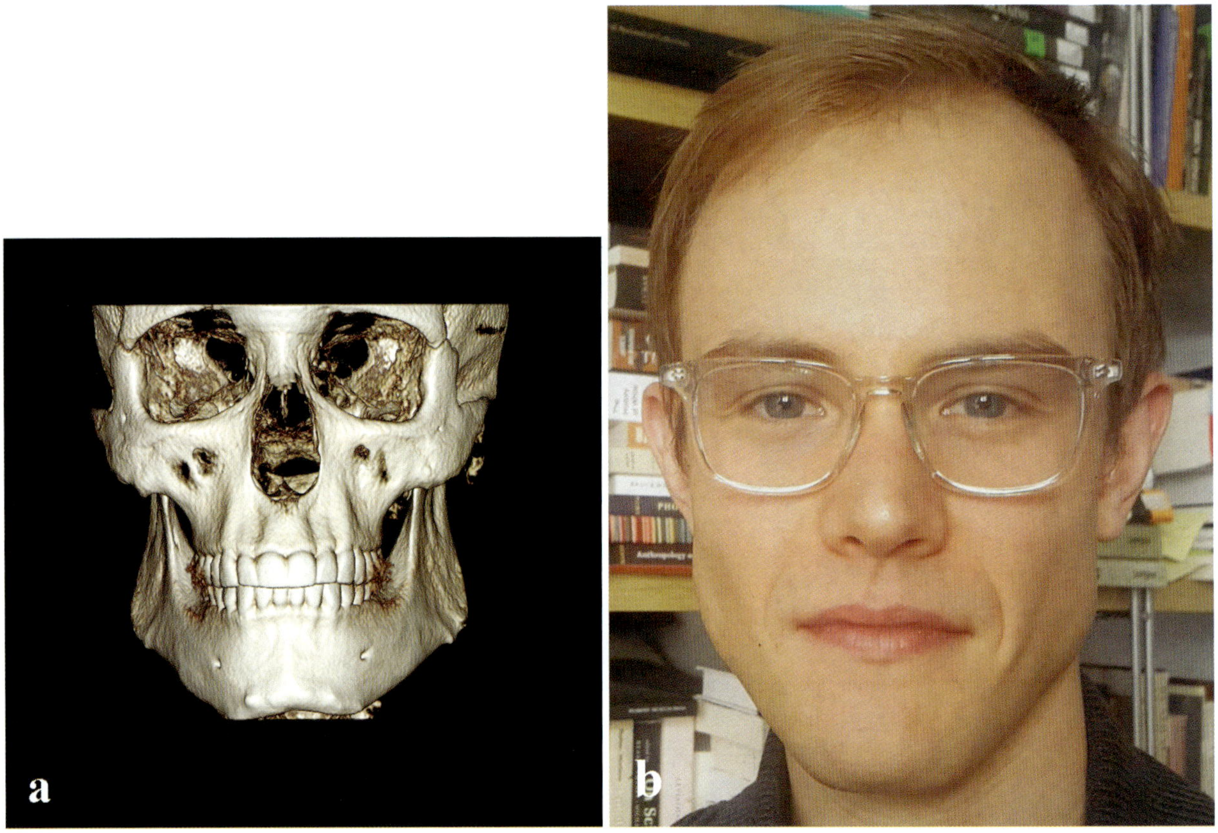

Figure 433a-b. Comparison of (a) Computed Tomography (CT) and (b) photograph showing marked gonial eversion and flaring, a bilobate chin, narrow nasal aperture and interorbital distance, and small malar tubercles in a young adult European male (photo and CT courtesy Paul Wolff Mitchell).

REFERENCES AND RECOMMENDED RESOURCES

Albert AM, Ricanek KR Jr., Patterson E. A Review of Literature on the Aging Adult Skull and Face: Implications for Forensic Science Research and Applications. *Forensic Science International 172:* 1–9, 2007.

Bass WM. *Human Osteology: A Laboratory and Field Manual* (5th ed.). Special Publication No. 2 of the Missouri Archaeological Society. Columbia, MO, 2005.

Brooks S, Brooks R, France D. Alveolar Prognathism Contour, An Aspect of Racial Identification. In: GW Gill and S Rhine (Editors). *Skeletal Attribution of Race.* Albuquerque NM: University of New Mexico: Maxwell Museum of Anthropological Papers No. 4, 1990.

(PDF) *Assessing Ancestry Through Nonmetric Traits of the Skull: A Test of Education and Experience.* Available from: https://www.researchgate.net/publication/36443034_Assessing_Ancestry_ThroughNonmetric_Traits_of_the_Skull_A_Test_of_Education_and_Experience [accessed January 01, 2020].

Brues AM. The once and future diagnosis of race. In *Skeletal Attribution of Race: Methods for Forensic Anthropology,* GW Gill and S Rhine (Eds.), pp. 1–8. Maxell Museum of Anthropology. Albuquerque, New Mexico, 1990.

Buikstra JE, Ubelaker DH. *Standards for Data Collection from Human Skeletal Remains.* Arkansas Archaeological Survey Research Series No. 44. Arkansas Archaeological Survey: Fayetteville, 1994.

Caufield PG. Tracing Technique and Identification of Landmarks. In A Jacobson and RL Jacobson (Editors). *Radiographic Cephalometry: From Basics to 3-D Imaging* (2nd ed., pp. 45–51). Hanover Park, IL: Quintessence, 2006.

Cericato OG, Franco A, Bittencourt MAV, Nunes MAP, Paranhos LR. Correlating Skeletal and Dental Developmental Stages using Radiographic Parameters. *Journal of Forensic and Legal Medicine 42:* 13–18, 2016.

Charalampakis A, Kourkoumelis G, Psari C, Antoniou V, Piagkou M, Demesticha T, Kotsiomitis E, Troupis T. The position of the mental foramen in dentate and edentulous mandibles: Clinical and surgical relevance. *Folia Morphologica 76*(4): 709–714, 2017. doi: 10.5603/FM.a2017.0042

Chae Sy, Sim HB, Kim MJ, Jang YH, Lee S-J, Kim DW, Lee WJ. Button Osteoma: A Review of Ten Cases. *Annals of Dermatology 27*(4): 394–397, 2015.

Charan Gowda BK, Sivapathasundharam B, Chatterji A, Chatterji BL. Histological appearance of postmortem pink teeth: Report of two cases. *Journal of Forensic Dental Sciences* 7(2): 168–170, 2015.

Dudzik B, Jantz RL. Misclassifications of Hispanics Using Fordisc3.1: Comparing Cranial Morphology in Asian and Hispanic Populations. *Journal of Forensic Sciences,* 2016. doi: 10.1111/1556-4029.13123

Duterloo HS, Planché P. (2011). *Handbook of Cephalometric Superimposition.* Hanover Park, IL: Quintessence, 2011.

Eshed V, Latimer B, Greenwald CM, Jellema LM, Rothschild BM, Wish-Baratz S, Hershkovitz I. Button osteoma: Its etiology and pathophysiology. *American Journal of Physical Anthropology 118*(3): 217–230, 2002.

Fiorellini JP, Kim DM, Ishikawa SO. The Tooth-Supporting Structures. In MG Newman, HH Takei, PR Klokkevold (Eds.), *Carranza's Clinical Periodontology* (10th ed., pp. 68–92). Elsevier, St. Louis, 2006.

France DL. Observational and metric analysis of sex in the skeleton. In: KJ Reichs (Editor), *Forensic Osteology: Advances in the Identification of Human Remains* (2nd ed.). Springfield, Charles C Thomas, Publisher, 1998.

Ghafari JG. The Moorrees Mesh Diagram: Proportionate Analysis of the Human Face. In A Jacobson and RL Jacobson (Editors). *Radiographic Cephalometry: From Basics to 3-D Imaging* (2nd ed., pp. 161–184). Hanover Park, IL: Quintessence, 2006.

Gill G. Challenge on the Frontier: Discerning American Indians from Whites osteologically. *Journal of Forensic Sciences 40:* 783–788, 1995.

Gill G. Continuous non-metric characteristics. In: *Skeletal Biology of the Ancient Rapanui* (Easter Islanders). VH Stefan and GW Gill (Editors). Cambridge Studies in Biological and Evolutionary Anthropology, Cambridge University Press, Cambridge, 2016.

Gill GW, Deeds CV. Temporal Changes in Nasal Sill Development Within the Northwestern Plains Region, pp. 263–270. In: *Skeletal Biology and Bioarchaeology of the Northwestern Plains.* GW Gill and RL Weathermon (Editors). Salt Lake City, University of Utah Press, 2008.

Hanihara T, Ishida H, Dodo Y. Os zygomaticum bipartitum: frequency distribution in major human populations. *Journal of Anatomy 192:* 539–555, 1998.

Hayashi K, Saitoi S, Mizoguchi I. Morphological Analysis of the Skeletal Remains of Japanese Females from the Ikenohata-Shichikencho Site. *European Journal of Orthodontics 34:* 575–581, 2012.

Heathcote GM, Bromage TG, Sava VJ, Hanson DB, Anderson BR. Enigmatic Cranial Superstructures Among Chamorro Ancestors from the Mariana Islands: Gross Anatomy and Microanatomy. *The Anatomical Record 297*(6): 1009–1021, 2014.

Heathcote, GM, Pietrusewsky M, Weiss E, Sava VJ, Anderson B, Ikehara-Quebral RM, Douglas TM, Ramírez-Aliaga JM, Matisoo-Smith EA, Stodder ALW, Walth CK, King CA, Hanson DB. Enigmatic Cranial Superstructures among Chamorro Ancestors from the Mariana Islands: Comparative Geographic Variation and a Proposal About Their Meaning. *The Journal of Island and Coastal Archaeology.* doi: 10.1080/15564894.2019.1638470

Hefner JT. *Assessing Nonmetric Cranial Traits Currently Used in Forensic Determination of Ancestry.* MA thesis, University of Florida, 2003.

Hefner JT. Cranial Nonmetric Variation and Estimating Ancestry. *Journal of Forensic Sciences 54*(5): 985–995, 2009.

Hefner JT. Cranial Morphoscopic Traits and the Assessment of American Black, American White, and Hispanic Ancestry, pp. 27–42. In GE Berg and SC Ta`ala (Eds.), *Biological Affinity in Forensic Identification of Human Skeletal Remains.* CRC Press, Boca Raton, 2015.

Hefner JT, Pilloud MA, Black CJ, Anderson BE. Morphoscopic Trait Expression in "Hispanic Populations. *Journal of Forensic Sciences,* 2015. doi: 10.1111/1556-4029.12826

Hefner JT. The Macromorphoscopic Databank. *American Journal of Physical Anthropology 166*(4): 994–1004, 2018. doi: 10.1002/ajpa.23492. Epub 2018 Apr 20.

Hefner JT, Linde KC. *Atlas of Human Cranial Macromorphoscopic Traits.* Academic Press, London, 2018.

Hefner JT, Spradley MK. Ancestry (forensic applications). *The International Encyclopedia of Biological Anthropology,* Wenda Trevathan (Editor). John Wiley & Sons, Inc., 2018.

Hooton EA. *The Indians of Pecos Pueblo.* Yale University Press, New Haven, 1930.

Houghton P. Rocker Jaws. *American Journal of Physical Anthropology 47*(3): 365–369, 1977.

Hrdlicka A. *Catalogue of human crania in the United States National Museum collections: Australians, Tasmanians, South African Bushmen, Hottentots, and Negro.* Government Printing Office, Washington (DC), 1928.

Jacobson A. Steiner Analysis. In A Jacobson and RL Jacobson (Editors). *Radiographic Cephalometry: From Basics to 3-D Imaging* (2nd ed., pp. 71–78). Hanover Park, IL: Quintessence, 2006.

Jantz RL. Cranial change in Americans: 1850–1975. *Journal of Forensic Sciences 46*(4): 784–787, 2001.

Jantz RL, Jantz LM. The remarkable change in Euro-American cranial shape and size. *Human Biology 88*(1): 56–64, 2016.

Jantz RL, Meadows Jantz L. Secular change in craniofacial morphology. *American Journal of Human Biology 12*(3): 327–338, 2000.

Jantz RL, Ousley SD. *Fordisc 3.0: Computerized forensic discriminant functions.* Version 3.1. Knoxville, TN: The University of Tennessee, 2005.

Jorde LB, Wooding SP. Genetic Variation, Classification, and "Race." *Nature Genetics Supplement 36*(11): s28–s33, 2004.

Kaifu Y. Tooth Wear and Compensatory Modification of the Anterior Dentoalveolar Complex in Humans. *American Journal of Physical Anthropology 111:* 369–392, 2000.

Kemkes-Grottenthaler A, Löbiga F, Stocka F. Mandibular ramus flexure and gonial eversion as morphologic indicators of sex. *Homo 53*(2): 97–11, 2002

Kim HJ, Choi Y, Yu MK, Lee KW, Min KS. Recognition and management of palatogingival groove for tooth survival: a literature review. *Restorative Dentistry and Endodontics 42*(2): 77–86, 2017.

Kogon SL. The prevalence, location, and conformation of palato-radicular grooves in maxillary incisors. *Journal of Periodontology 57:* 231–234, 1986.

Krogman WM. A guide to the identification of human skeletal material. *FBI Law Enforcement Bulletin 8*(8): 1–29, 1939.

Krogman WM. *The Skeleton in Forensic Medicine.* Graduate Medicine 17 (2 and 3), 1955.

Krogman WM. *The Human Skeleton in Forensic Medicine.* Springfield, Charles C Thomas, Publisher, 1962.

Langley NR, Jantz LM, Ousley SD, Jantz RL, Milner G. *Data Collection Procedures for Forensic Skeletal Material 2.0.* Forensic Anthropology Center, The University of Tennessee, Knoxville, 2016.

Langley NR, Dudzik B, Cloutier A. A Decision Tree for Nonmetric Sex Assessment from the Skull. *Journal of Forensic Sciences,* 2017. doi: 10.1111/1556-4029.13534

Luthur F. A Cephalometric Comparison of Medieval Skulls with a Modern Population. *European Journal of Orthodontics 15:* 315–325, 1993.

Machado CEP, Santiago BM, Lima LNC, Gonzales PS, Franco A, de Barros Vidal F, Aguilera IA., Guimaraes MA. *Forensic Science International 301:* e1–e7, 2019.

Maier CA, Zhang K, Manhein MH, Li X. Palate Shape and Depth: A Shape-Matching and Machine Learning Method for Estimating Ancestry from Human Skeletal Remains. *Journal of Forensic Sciences 60*(5): 1129–1134, 2015.

Mann RW. Stafne's Defects of the Human Mandible. Doctoral Dissertation, Department of Anthropology, University of Hawaii, Honolulu, 2001.

Mann RW, Hunt DR. Non-metric traits and anatomical variants that can mimic trauma in the human skeleton. *Forensic Science International 301:* 202–224, 2019.

Mann RW, Manabe J, Byrd JE. Relationship of the Parietal Foramen and Complexity of the Human Sagittal Suture. *International Journal of Morphology 27*(2): 553–564, 2009.

Mann RW, Jantz RL, Bass WM, Willey P. Maxillary Suture Obliteration: A Visual Method for Estimating Age in the Human Skeleton. *Journal of Forensic Sciences 36*(3): 781–791, 1991.

Mann RW, Hunt DR, Lozanoff S. *Photographic Regional Atlas of Non-metric Traits and Anatomical Variants in the Human Skeleton.* Charles C Thomas, Publisher, Springfield, 2016.

Margvelashvili A, Zollikofer CPE, Lordkipanidze D, Peltomaki T, Ponce de Leon MS. Tooth Wear and Dentoalveolar Modeling are Key Factors of Morphological Variation in Dmanisi Mandibles. *Proceedings of the National Academy of Sciences 110*(43): 17278–17283, 2013.

Martin R, Saller K. *Lehrbuch der anthropologie.* Gustav Fisher Verlag, Stuttgart, 1957.

McNamara JA, Brudon WL. *Orthodontics of Dentofacial Orthopedics.* Ann Arbor, MI: Needham, 2001.

Moore-Jansen PM, Ousley SD, Jantz RL. *Data Collection Procedures for Forensic Skeletal Material.* Report of Investigations no. 48, The University of Tennessee, Knoxville, Department of Anthropology, 1994.

Moorrees CFA. Twenty Centuries of Cephalometry. In A Jacobson and RL Jacobson (Editors). *Radiographic Cephalometry: From Basics to 3-D Imaging* (2nd ed., pp. 13–32). Hanover Park, IL: Quintessence, 2006.

Nguyen T, Profitt WR. The Decision-Making Process in Orthodontics. In LW Graber, Vanarsdall RL Jr., KWL Vig, GJ Huang (Editors). *Orthodontics Current Principles and Techniques* (6th ed., pp. 208–244). St. Louis, MO: Elsevier, 2017.

Ossenberg NS. Brief Communication: Cranial Nonmetric Trait Database on the Internet. *American Journal of Physical Anthropology 152:* 551–553, 2013.

OSTEOWARE, Standardized Skeletal Documentation Software. Smithsonian Institution, Washington, DC.

Pajevic T, Juloski J, Glisic B. Malocclusion from Prehistoric Times to the Medieval Times in Serbian Population: Dentoalveolar and Skeletal Relationship Comparisons in Samples. *Homo 70*(1): 31–43, 2019.

Plemons A, Hefner JT. Ancestry Estimation Using Macromorphoscopic Traits. *Academic Forensic Pathology 6*(3): 400–412, 2016.

Proffit WR, Fields HW, Sarver DM. *Contemporary Orthodontics* (5th ed., pp. 167–233). St. Louis, MO: Elsevier, 2012.

Rhine S. Nonmetric Skull Racing. In: GW Gill and S Rhine (Editors). *Skeletal Attribution of Race.* Albuquerque, NM: University of New Mexico: Maxwell Museum of Anthropological Papers No. 4, 1990.

Rogers M. An Analysis of Discrete Cranial Traits of Northwestern Plains Indians, pp. 177–199. In: *Skeletal Biology and Bioarchaeology of the Northwestern Plains.* GW Gill and RL Weathermon (Editors). Salt Lake City, University of Utah Press, 2008.

Saul JM, FP. Biker's Bones: An Avocational Syndrome. In Scott I. Fairgrieve (Editor), *Forensic Osteological Analysis: A Book of Case Studies,* pp. 23–250. Springfield, Charles C Thomas, 2010.

Schumacher FL. The Roentgenologic Study of the Mastoid Area. *Radiology 27*(6): 717–721, 1936.

Scientific Working Group for Forensic Anthropology (SWGANTH) *Ancestry Assessment* Issue Date: 06/12/2013.

Scott GR, Irish JD. *Human Crown and Root Morphology: The Arizona State University Dental Anthropology System.* Cambridge: Cambridge University Press, 2017.

Singh B, Kahlon SS, Narang RS, Gupta S, Khushbinder KK, Singh S, Kaur P. To Assess the Values of Gonial & Antegonial Angle on Panoramic Radiograph and their Role in the Gender Determination. *Journal of Dental and Oral Health 2*(4): 1–4, 2016. www.scientonline.org

Slice DE, Ross A. *3D-ID: Geometric Morphometric Classification of Crania for Forensic Scientists,* 2009. http://mitran-lab.amath.unc.edu/courses/MATH547/lessons/3d-id/3d_id_doc.pdf

Snow CE. *Early Hawaiians.* Lexington: University Press of Kentucky, 1974.

Stewart T. *Essentials of Forensic Anthropology.* Springfield, IL: Charles C Thomas, 1979.

Suchetha A, Heralgi R, Bharwani AG, Mundinamane D. Treatment of an intrabony osseous lesion associated with a palatoradicular groove. *Contemporary Clinical Dentistry 3*(S2): 260–263, 2012.

Sun L, Yuan L, Wang B, Zhang L, Shen G, Fang B. Changes of alveolar bone dehiscence and fenestration after augmented corticotomy-assisted orthodontic treatment: a CBCT evaluation. *Progress in Orthodontics 20*(1): 7, 2019. doi:10.1186/s40510-019-0259-z

The Chicago Manual of Style, 16th edition text. The University of Chicago, 2010.

Tos M, Stangerup SE. The Causes of Asymmetry of the Mastoid Air Cell System. *Acta Oto-Laryngologica 99*(5-6): 564–570, 2010.

Ubelaker DH. *Human Skeletal Remains: Excavation, Analysis, Interpretation* (2nd ed.). Manuals on Archeology, Washington, Taraxacum, 1989.

Utsuno H, Kageyama T, Uchida K, Ishii N, Minegishi S, Uemura K, Sakurada K. Establishment of a Predication Method for the Mid-facial Region of Unknown Human Mongoloid Remains. *Forensic Science International 288:* 297–303, 2018.

Walker P L Sexing skulls using discriminant function analysis of visually assessed traits. *American Journal of Physical Anthropology 136:* 39–50, 2008.

Weisler MI, Swindler D. Rocker jaws from the Marshall Islands: Evidence for interaction between eastern Micronesia and west Polynesia. *People and Culture in Oceania 18:* 23–33, 2002.

Weisensee KE, Jantz RL. Secular changes in craniofacial morphology of the Portuguese using geometric morphometrics. *American Journal of Physical Anthropology 145:* 548–559, 2011.

Wheat A. *Assessing Ancestry Through Nonmetric Traits of the Skull: A Test of Education and Experience.* Master's thesis, Texas State University-San Marcos, 2009.

White TD, Black MT, Folkens PA. *Human Osteology,* 3rd ed. Berkeley, Academic Press, 2012.

Withers JA, Brunsvold MA, Killjoy WJ, Rahe AJ. The relationship of palato-gingival grooves to localized periodontal disease. *Journal of Periodontology 52:* 41–44, 1981.

Wojciechowski T, Skadorwa T, Drozdz A, Ciszek B, Szopinski K. The radioanatomical assessment of the Körner's septum. *Anatomic Bases of Medical, Radiological and Surgical Techniques 41:* 669–673, 2019. https://doi.org/10.1007/s00276-018-2149-3

Recommended Resources

Algee-Hewitt, BFB, Hughes CE, Anderson BE. Temporal, geographic and identification trends in craniometric estimates of ancestry for persons of Latin American origin. *Forensic Anthropology 1*(1): 4–18, 2018.

Adams DC. Methods for shape analysis of landmark data from articulated structures. *Evolutionary Ecology Research 1:* 959–970, 1999.

Adams DC, Rohlf FJ, Slice DE. Geometric morphometrics: ten years of progress following the 'revolution'. *Italian Journal of Zoology 71*(12): 5–16, 2004.

Alvesalo L, Tigerstedt PMA. Heritabilities of human tooth dimensions. *Hereditas 77:* 311–318, 1974.

American Anthropological Association. 1998. AAA statement on race. *American Anthropology 100*(3): 712–713, 1998.

American Association of Physical Anthropologists. AAPA Statement on biological race. *American Journal of Physical Anthropology 101:* 569–570, 1996.

Berg GE, Kenyhercz MW. Introducing human mandible identification [(hu)MANid]: A free, web-based GUI to classify human mandibles. *Journal of Forensic Sciences.* doi:10.1111/1556-4029.13479, 2017.

Boas F. *Changes in the bodily form of descendants of immigrants.* New York: Columbia University Press, 1912.

Brace CL. Region does not mean 'race'—reality versus convention in forensic anthropology. *Journal of Forensic Sciences 40*(2): 171–175, 1995.

Brace CL, Ryan AS. Sexual dimorphism and human tooth size differences. *Journal of Human Evolution 9:* 417–435, 1980.

Berg GE. Sex Estimation of Unknown Human Skeletal Remains. In Natalie R. Langley and Maria Teresa A. Tersigni-Tarrant (Editors) *Forensic Anthropology: A Comprehensive Introduction* (2nd ed.), 143--161. Boca Raton, CRC Press, 2017.

Ditch LE, Rose JC. A multivariate dental sexing technique. *American Journal of Physical Anthropology 37:* 61–64, 1972.

Edgar HJH. Prediction of race using characteristics of dental morphology. *Journal of Forensic Sciences 50*(2): 1–5, 2005.

Edgar HJH. Microevolution of African American dental morphology. *American Journal of Physical Anthropology 132:* 535–544, 2007.

Edgar HJH. Biohistorical approaches to "race" in the United Stats: biological distances among African Americans, European Americans, and their ancestors. *American Journal of Physical Anthropology 139:* 58–67, 2009.

Edgar HJH. Estimation of ancestry using dental morphological characteristics. *Journal of Forensic Sciences 58*(S1): 3–8, 2013.

Falk D, Corruccini R. Efficacy of cranial versus dental measurements for separating human populations. *American Journal of Physical Anthropology 57:* 123–127, 1982.

Giran G, Paré A, Croisé B, Koudougou C, Mercier JM, Laure B, Corre P, Bertin H. Radiographic evaluation of percutaneous transfacial wiring versus open internal fixation for surgical treatment of unstable zygomatic bone fractures. *PLOS ONE 14*(8): e0220913, 2019. https://doi.org/10.1371/journal.pone. 0220913

Hanihara K. Racial characteristics in the dentition. *Journal of Dental Research 46*(5): 923–926, 1967.

Hanihara T. Morphological variation of major human populations based on nonmetric dental traits. *American Journal of Physical Anthropology 136*(2): 169–182, 2008.

Hanihara T, Ishida H. Metric dental variation of major human populations. *American Journal of Physical Anthropology 128:* 287–298, 2005.

Harris EF. A strategy for comparing odontometrics among groups. *Dental Anthropology 12*(1): 1–6, 1997.

Harris EF, Bailit HL. A principal components analysis of human odontometrics. *American Journal of Physical Anthropology 53:* 349–358, 1988.

Harvati K, Weaver TD. Human cranial anatomy and the differential preservation of population history and climate signatures. *Anatomical Record A. 288A:* 1225--1233, 2006.

Hauser G, DeStefano GF. *Epigenetic Variants of the Human Skull.* Verlag and Thieme, Stuttgart, 1989.

Hefner JT, Spradley MK, Anderson B. Ancestry assessment using random forest modeling. *Journal of Forensic Sciences Early View:* 1–7, 2014.

Hefner JT, Ousley SD. Statistical classification methods for estimating ancestry using morphoscopic traits. *Journal of Forensic Sciences Early View:* 1–8, 2014.

Howells WW. Cranial variation in man: a study by multivariate analysis of patterns of differences among recent human populations. *Papers of the Peabody Museum of Archaeology and Ethnology,* Vol. 67. Cambridge: Peabody Museum, Harvard University, 1973.

Howells WW. Skull shapes and the map: craniometric analyses in the dispersion of modern Homo. *Papers of the Peabody Museum of Archaeology and Ethnology,* Vol. 79. Cambridge, Mass.: Peabody Museum, 1989.

Hurst CV. Morphoscopic trait expressions used to identify Southwest Hispanics. *Journal of Forensic Sciences 57*(4): 859–865, 2012.

Irish JD. *Biological affinities of late Pleistocene through modern African aboriginal populations: the dental evidence.* PhD dissertation. Arizona State University, 1993.

Irish JD. Characteristic high- and low-frequency dental traits in sub-Saharan African populations. American *Journal of Physical Anthropology 102:* 455–467, 1997.

Irish JD. Ancestral dental traits in recent sub-Saharan Africans and the origins of modern humans. *Journal of Human Evolution 34:* 81–98, 1998.

Irish JD. Population continuity vs. discontinuity revisited: dental affinities among late Paleolithic through Christian-era Nubians. *American Journal of Physical Anthropology 128:* 520–535, 2005.

Irish JD. Who were the ancient Egyptians? Dental affinities among Neolithic through postdynastic peoples. *American Journal of Physical Anthropology 129:* 529–543, 2006.

Irish JD, Guatelli-Steinberg D. Ancient teeth and modern human origins: An expanded comparison of African Plio-Pleistocene and recent world dental samples. *Journal of Human Evolution 45:* 113–144, 2003.

Irish JD, Hemphill BE. An odontometrics investigation of Canary Islander origins. *Dental Anthropology 17*(1): 8–17, 2004.

Irish JD, Guatelli-Steinberg D, Legge SS, de Ruiter DJ, Berger LR. Dental morphology and the phylogenetic "place" of Australopithecus sediba. *Science 340:* 1233062-1-1233062-4, 2013.

Kenyhercz MW, Berg GE. Evaluating mixture discriminant analysis to classify human mandibles with (hu)MANid, a free, R-based GUI. In *New Perspectives in Forensic Human Skeletal Identification,* K.E Latham, E J Bartelink, and M Finnegan (Editors), pp. 35–43. Academic Press, London, 2018.

Kenyhercz MW, Klales AR, Rainwater CW, Fredette SM. The optimized summed scored attributes method for the classification of U.S. Blacks and Whites: A validation study. *Journal of Forensic Sciences 62:* 174–180, 2017.

Klales AR, Kenyhercz MW. Morphological assessment of ancestry using cranial macromorphoscopics. *Journal of Forensic Sciences 60:* 13–20, 2015.

L'Abbé EN, Van Rooyen C, Nawrocki SP, Becker PJ. An evaluation of non-metric cranial traits used to estimate ancestry in a South African sample. *Forensic Science International* 2011 Jun 15; 209(1-3): 195.e1–7. doi: 10.1016/j.forsciint.2011.04.002. Epub 2011 May 6.

Ousley SD, Jantz R, Freid D. Understanding race and human variation: why forensic anthropologists are good at identifying race. *American Journal of Physical Anthropology 139*(1): 68–76, 2009.

Ousley SD, Jones EB. Craniometric variation in the Aleutians: integrating morphological, molecular, spatial, and temporal data. *Human Biology 82*(5–6): 629–651, 2010.

Ousley SD, McKeown A. Three dimensional digitizing of human skulls as an archival procedure. In: Williams E, (Editor) *Human remains: Conservation, retrieval and analysis.* Oxford: Archaeopress, 2001.

Ousley S D, Jantz RL, Freid D. Understanding race and human variation: Why forensic anthropologists are good at identifying race. *American Journal of Physical Anthropology 139:* 68–76, 2009.

Ousley S D, Jantz RL. Fordisc 3 and statistical methods for estimating sex and ancestry. In *A Companion to Forensic Anthropology,* DC Dirkmaat (Editor), pp. 311–329. Wiley-Blackwell, Chichester, UK, 2012.

Perez SI, Monteiro LR. Nonrandom factors in modern human morphological diversification: a study of craniofacial variation in southern South American populations. *Evolution 63*(4): 978–993, 2009.

Pilloud MA, Hefner JT, Hanihara T, Hayashi A. The use of tooth crown measurements in the assessment of ancestry. *Journal of Forensic Sciences 59:* 1493–1500, 2014.

Relethford JH. Craniometric variation among modern human populations. *American Journal of Physical Anthropology 95:* 53–62, 1994.

Relethford JH. Boas and beyond: migration and craniometric variation. *American Journal of Human Biology 16:* 379–386, 2004.

Ross AH, Ubelaker DH, Falsetti AB. Craniometric variation in the Americas. *Human Biology 47*(6): 807–818, 2002.

Schmidt RW, Seguchi N, Thompson JL. Chinese immigrant population history in North America based on craniometric diversity. *Anthropological Science 119*(1): 9–19, 2011.

Scott, GR, Pilloud MA, Navega D, d'Oliveira Coelho J, Cunha E, Irish DJ rASUDAS: A new web-based application for estimating ancestry from tooth morphology. *Forensic Anthropology 1*(1): 18–31, 2018.

Spradley MK. *Biological anthropological aspects of the African diaspora: Geographic origins, secular trends, and plastic versus genetic influences utilizing craniometric data*. PhD dissertation. University of Tennessee Knoxville, 2006.

Spradley MK, Stull KE. Advancements in sex and ancestry estimation. In *New Perspectives in Forensic Human Skeletal Identification,* KE Latham, EJ Bartelink, and M Finnegan (Editors), pp. 13–22. Academic Press, London, 2018.

Spradley MK, Jantz RL. Ancestry estimation in forensic anthropology: Geometric morphometric versus standard and nonstandard interlandmark distance. *Journal of Forensic Sciences 61:* 892–897, 2016.

Spradley MK, Jantz RL, Robinson A, Peccerelli F. Demographic change and forensic identification: Problems in metric identification of Hispanic skeletons. *Journal of Forensic Sciences 53:* 21–28, 2008.

Stull KE, Kenyhercz MW, L'Abbé EN. Ancestry estimation in South Africa using craniometrics and geometric morphometrics. *Forensic Science International 245:* 206e1–206e7, 2014.

Spradley MK, Weisensee K. *Ancestry Estimation: The Importance, The History, and The Practice. In Forensic Anthropology: A Comprehensive Introduction* (2nd ed.), pp. 163–174. Natalie R. Langley and Maria Teresa A. Tersigni-Tarrant (Editors). Boca Raton, CRC Press, 2017.

Tappen NC. The development of the vermiculate pattern in the brow region of crania from Indian Knoll, Kentucky. *American Journal of Physical Anthropology 60*(4): 522–537, 1983.

Turner CG II. Three-rooted mandibular first permanent molars and the question of American Indian origins. *American Journal of Physical Anthropology 34:* 229–242, 1971.

Turner CG II. Dental evidence for the peopling of the Americas. In R Shutler, Jr, Editor. *Early Man in the New World.* Beverly Hills, CA: Sage Publications, pp. 147–157, 1983.

Turner CG II. The dental search for Native American origins. In: Kirk R, Szathmáry E (Editors). *Out of Asia. Canberra: Journal of Pacific History,* pp. 31– 78, 1985.

Turner CG II. Dentochronological separation estimates for Pacific Rim populations. *Science 232*(4754): 1140–1142, 1986.

Turner CG II. Major features of sundadonty and sinodonty, including suggestions about East Asian microevolution, population history, and Late Pleistocene relationships with Australian Aboriginals. *American Journal of Physical Anthropology 82:* 295–318, 1990.

Turner CG II, Nichol CR, Scott GR. Scoring procedures for key morphological traits of the permanent dentition: The Arizona State University dental anthropology system. In: Kelley MA, Larsen CS (Editors). *Advances in dental anthropology.* New York: Wiley-Liss, pp. 13–32, 1991.

Vitek CL. *A Critical Analysis of the Use of Non-Metric Traits for Ancestry Estimation among Two North American Population Samples.* Master's Thesis, University of Tennessee, 2012. https://trace.tennessee.edu/utk_gradthes/1218

INDEX

O

T

U

V

W

Z